# The Value of Natural Compounds as Therapeutic Agents

# The Value of Natural Compounds as Therapeutic Agents

Guest Editor

**Maria Beatrice Morelli**

MDPI

Basel • Beijing • Wuhan • Barcelona • Belgrade • Novi Sad • Cluj • Manchester

*Guest Editor*
Maria Beatrice Morelli
School of Pharmacy
University of Camerino
Camerino
Italy

*Editorial Office*
MDPI AG
Grosspeteranlage 5
4052 Basel, Switzerland

This is a reprint of the Special Issue, published open access by the journal *Biomolecules* (ISSN 2218-273X), freely accessible at: https://www.mdpi.com/journal/biomolecules/special_issues/Natural_Compound_Agent.

For citation purposes, cite each article independently as indicated on the article page online and as indicated below:

Lastname, A.A.; Lastname, B.B. Article Title. *Journal Name* **Year**, *Volume Number*, Page Range.

**ISBN 978-3-7258-5195-9 (Hbk)**
**ISBN 978-3-7258-5196-6 (PDF)**
**https://doi.org/10.3390/books978-3-7258-5196-6**

# Contents

# About the Editor

**Maria Beatrice Morelli**

Maria Beatrice Morelli is an Associate Professor in General Pathology at the School of Pharmacy, University of Camerino, where she also serves as Co-Director of the Postgraduate Master's program in Complementary and Integrated Therapies and the Advanced Course in Cannabis for Medical Use. Her research is focused on exploring the beneficial properties of natural compounds in human diseases, with a particular emphasis on cancer. She is also deeply interested in the role of ion channels and how their functions may be modulated by natural molecules. Throughout her career, she has gained extensive expertise in a wide range of cellular and molecular biology techniques, which she applies to investigate the therapeutic potential of natural agents in oncology and related fields.

# Preface

This Reprint is devoted to the study of naturally occurring substances and their relevance in human health and disease. Its scope strives to collect and highlight recent advances in the understanding of natural compounds and their biological effects at the cellular level. The aim of this Reprint is to provide a comprehensive overview of the mechanisms through which natural products, either as single defined molecules or complex extracts, exert their activity in different pathological contexts.

The motivation for this work arises from the increasing recognition that natural substances, which have long been used in traditional medicine, may represent a valuable source of therapeutic agents in modern pharmacology. Their potential roles in regulating key cellular processes, including proliferation, survival, apoptosis, autophagy, senescence, differentiation, migration, and invasion, make them of particular interest in conditions such as cancer and inflammation. Moreover, their involvement in the modulation of the cellular microenvironment further expands their therapeutic significance.

This Reprint is addressed to researchers, clinicians, and students who are engaged in the fields of pharmacology, oncology, immunology, and related biomedical sciences. It is our hope that this collection of articles will serve not only as a reference for current knowledge but also as an inspiration for future studies aiming to explore and validate the therapeutic potential of natural products.

**Maria Beatrice Morelli**
*Guest Editor*

biomolecules

MDPI

*Editorial*

# Exploring the Therapeutic Potential of Natural Compounds and Plant Extracts in Human Health

**Laura Zeppa [1,†], Cristina Aguzzi [2,†] and Maria Beatrice Morelli [2,*]**

1 School of Biosciences and Veterinary Medicine, University of Camerino, 62032 Camerino, Italy; laura.zeppa@unicam.it
2 School of Pharmacy, University of Camerino, 62032 Camerino, Italy; cristina.aguzzi@unicam.it
* Correspondence: mariabeatrice.morelli@unicam.it; Tel.: +39-0737403713
† These authors contributed equally to this work.

A growing body of evidence supports the various potential advantages of plant extracts and plant extract-based products in promoting human health and addressing a broad range of pathological conditions [1]. Over the past decades, scientific interest in natural compounds has surged, driven by the urgent need for safer, more sustainable, and cost-effective therapeutic alternatives. These bioactive compounds, often derived from plants used in traditional medicine, represent a vast and largely untapped reservoir of pharmacological agents with multifaceted biological activities [2].

Natural products are characterized by their chemical diversity and ability to modulate multiple molecular targets simultaneously, making them especially valuable in the treatment of complex diseases with multifactorial pathogenesis, such as cancer, cardiovascular disorders, metabolic syndromes, and neurodegenerative diseases [3–5]. Furthermore, many of these compounds exhibit favorable safety profiles, often accompanied by antioxidant, anti-inflammatory, immunomodulatory, and antimicrobial properties.

Advances in biotechnology, phytochemistry, and molecular pharmacology have significantly improved our ability to isolate, characterize, and understand the mechanisms of action of these natural compounds [6,7]. This has enabled not only the identification of novel drug leads but also the development of standardized extracts and nutraceutical formulations aimed at health promotion and disease prevention.

In this context, the integration of natural compounds into modern therapeutic strategies offers promising avenues for innovation in both preventive and personalized medicine. The present Special Issue aims to provide an overview of recent progress in this field, highlighting experimental, review, and clinical studies that explore the therapeutic potential of plant-derived molecules in diverse pathological contexts.

Three original studies explored the cardiovascular effects of different plant-derived compounds. Fernando et al. investigated phloridzin docosahexaenoate (PZ-DHA), an $\omega$-3 fatty acid ester derived from a flavonoid precursor, demonstrating its anti-angiogenic properties in both in vitro and in vivo models [8,9]. PZ-DHA significantly inhibited endothelial cell proliferation, migration, and tubule formation, while downregulating the Akt signaling pathway and VEGF165-induced RhoA activation (contribution 1). Emodin, a naturally occurring anthraquinone, was shown to mitigate pathological cardiac hypertrophy in hypertensive mice while also alleviating gut microbiota alterations associated with hypertension (contribution 2) [10] and contributing to the improvement of cardiac diseases in general [11,12]. Similarly, Obeidat et al. examined (+)-taxifolin (TAX), a flavonoid with antioxidant and anti-inflammatory properties, in a murine model of isoproterenol-induced

 https://doi.org/10.3390/biom15060774

acute myocardial injury. TAX treatment significantly reduced myocardial damage, oxidative stress, and inflammation, primarily through upregulation of the Nrf2/HO-1 pathway (contribution 3) [13].

Beyond cardiovascular implications, natural compounds also hold promise in addressing skeletal and joint disorders. Hecht et al. demonstrated that early postnatal administration of resveratrol preserved joint integrity and reduced pain in a mouse model of pseudoachondroplasia—a condition linked to mutations in the cartilage oligomeric matrix protein (contribution 4) [14].

The anti-inflammatory properties of natural compounds in the context of obesity were the focus of two complementary studies. Carpi et al. explored tanshinone IIA and cryptotanshinone, diterpenoids isolated from *Salvia miltiorrhiza* Bunge, in inflamed human adipocytes [15–17], where both compounds reduced proinflammatory cytokine and chemokine release, as well as monocyte recruitment (contribution 5). A review by Ngamsamer et al. addressed the beneficial roles of anthocyanins, water-soluble flavonoid pigments abundant in fruits and vegetables. These compounds have strong antioxidant effects, reducing reactive oxygen species and mitigating chronic inflammation [18,19]. Moreover, they were shown to regulate gut dysbiosis associated with obesity and to downregulate inflammatory cytokine production in both in vitro and in vivo models (contribution 6).

Several natural compounds also influence metabolic pathways and redox balance. In particular, plant-derived molecules are involved in the regulation of Nrf2, a transcription factor critical in oxidative stress responses and metabolic regulation. Moreover, natural products can stimulate heme oxygenase (HMOX) activity, an enzyme responsible for heme catabolism. This pathway leads to the production of bilirubin, a molecule with emerging roles in cytoprotection, antioxidative defense, and metabolic homeostasis. Flavonoids, curcumin, astragaloside, and vitamins are some of the natural products involved in the modulation of bilirubin and the metabolic Nrf-2 pathway. Other natural compounds can mimic the products of heme degradation or modulate this pathway directly, suggesting further therapeutic potential (contribution 7).

Cancer is another key area where natural compounds show promising effects. Ellagic acid, a polyphenol present in various fruits and nuts, has demonstrated anticancer activity by modulating numerous hallmarks of cancer—including cell proliferation, apoptosis, angiogenesis, and immune evasion [20]. A comprehensive review by Čižmáriková et al. details its mechanisms in cancer prevention and treatment (contribution 8). Another review focused on the synergistic use of polyphenols—both flavonoid and non-flavonoid types—with bortezomib (BTZ), a first-line proteasome inhibitor for multiple myeloma [21]. A total of nine flavonoids (e.g., icariin, EGCG, morin) and four non-flavonoids (e.g., resveratrol, curcumin) were identified as enhancing BTZ's anticancer effects, underscoring the role of dietary components in augmenting conventional chemotherapy (contribution 9).

Some natural products exhibit a broad spectrum of biological activities. A review by Tematio Fouedjou et al. presented the therapeutic properties of various Cordyline species, traditionally used in folk medicine. These plants demonstrated antimicrobial, antiparasitic, antioxidant, antiproliferative, and organ-protective effects (contribution 10). Similarly, date seeds (*Phoenix dactylifera* L.) were highlighted for their high phenolic content and diverse pharmacological actions, including anti-inflammatory, antioxidant, antibacterial, antiviral, antidiabetic, and anticancer effects [22,23] (contribution 11).

Lastly, the clinical potential of natural compounds has been investigated in human studies. Al-Jubori et al. compiled clinical trials assessing different species of gum Arabic in the management of diseases such as sickle cell anemia, rheumatoid arthritis, periodontitis, metabolic and gastrointestinal disorders, and kidney diseases. These results highlight the transition of natural compounds from traditional uses to evidence-based applications in clinical settings (contribution 12).

In conclusion, the studies presented in this Special Issue collectively reinforce the therapeutic promise of natural compounds and plant-based products across a wide range of diseases. Continued research into their mechanisms of action, synergistic potential with standard therapies, and clinical translation will further validate their role in modern medicine.

**Author Contributions:** Conceptualization, M.B.M.; writing—original draft preparation, L.Z. and C.A.; writing—review and editing, C.A. and M.B.M.; supervision, M.B.M. All authors have read and agreed to the published version of the manuscript.

**Conflicts of Interest:** The authors declare no conflicts of interest.

## Abbreviations

The following abbreviations are used in this manuscript:

| | |
|---|---|
| PZ-DHA | phloridzin docosahexaenoate |
| TAX | taxifolin |
| HMOX | heme oxygenase |
| BTZ | bortezomib |

**List of Contributions**


1. Fernando, W.; MacLean, E.; Monro, S.; Power Coombs, M.R.; Marcato, P.; Rupasinghe, H.P.V.; Hoskin, D.W. Phloridzin Docosahexaenoate, an Omega-3 Fatty Acid Ester of a Flavonoid Precursor, Inhibits Angiogenesis by Suppressing Endothelial Cell Proliferation, Migration, and Differentiation. *Biomolecules* **2024**, *14*, 769. https://doi.org/10.3390/biom14070769.
2. Evans, L.; Price, T.; Hubert, N.; Moore, J.; Shen, Y.; Athukorala, M.; Frese, S.; Martinez-Guryn, K.; Ferguson, B.S. Emodin Inhibited Pathological Cardiac Hypertrophy in Response to Angiotensin-Induced Hypertension and Altered the Gut Microbiome. *Biomolecules* **2023**, *13*, 1274. https://doi.org/10.3390/biom13091274.
3. Obeidat, H.M.; Althunibat, O.Y.; Alfwuaires, M.A.; Aladaileh, S.H.; Algefare, A.I.; Almuqati, A.F.; Alasmari, F.; Aldal'in, H.K.; Alanezi, A.A.; Alsuwayt, B.; et al. Cardioprotective Effect of Taxifolin against Isoproterenol-Induced Cardiac Injury through Decreasing Oxidative Stress, Inflammation, and Cell Death, and Activating Nrf2/HO-1 in Mice. *Biomolecules* **2022**, *12*, 1546. https://doi.org/10.3390/biom12111546.
4. Hecht, J.T.; Veerisetty, A.C.; Patra, D.; Hossain, M.G.; Chiu, F.; Mobed, C.; Gannon, F.H.; Posey, K.L. Early Resveratrol Treatment Mitigates Joint Degeneration and Dampens Pain in a Mouse Model of Pseudoachondroplasia (PSACH). *Biomolecules* **2023**, *13*, 1553. https://doi.org/10.3390/biom13101553.
5. Carpi, S.; Quarta, S.; Doccini, S.; Saviano, A.; Marigliano, N.; Polini, B.; Massaro, M.; Carluccio, M.A.; Calabriso, N.; Wabitsch, M.; et al. Tanshinone IIA and Cryptotanshinone Counteract Inflammation by Regulating Gene and miRNA Expression in Human SGBS Adipocytes. *Biomolecules* **2023**, *13*, 1029. https://doi.org/10.3390/biom13071029.
6. Ngamsamer, C.; Sirivarasai, J.; Sutjarit, N. The Benefits of Anthocyanins against Obesity-Induced Inflammation. *Biomolecules* **2022**, *12*, 852. https://doi.org/10.3390/biom12060852.
7. Jayanti, S.; Vitek, L.; Verde, C.D.; Llido, J.P.; Sukowati, C.; Tiribelli, C.; Gazzin, S. Role of Natural Compounds Modulating Heme Catabolic Pathway in Gut, Liver, Cardiovascular, and Brain Diseases. *Biomolecules* **2024**, *14*, 63. https://doi.org/10.3390/biom14010063.
8. Čižmáriková, M.; Michalková, R.; Mirossay, L.; Mojžišová, G.; Zigová, M.; Bardelčíková, A.; Mojžiš, J. Ellagic Acid and Cancer Hallmarks: Insights from Experimental Evidence. *Biomolecules* **2023**, *13*, 1653. https://doi.org/10.3390/biom13111653.
9. Ding, K.; Jiang, W.; Jia, H.; Lei, M. Synergistically Anti-Multiple Myeloma Effects: Flavonoid, Non-Flavonoid Polyphenols, and Bortezomib. *Biomolecules* **2022**, *12*, 1647. https://doi.org/10.3390/biom12111647.

10. Tematio Fouedjou, R.; Tsakem, B.; Siwe-Noundou, X.; Dongmo Fogang, H.P.; Tiombou Donkia, A.; Kemvoufo Ponou, B.; Poka, M.; Demana, P.H.; Teponno, R.B.; Azefack Tapondjou, L. Ethnobotany, Phytochemistry, and Biological Activities of the Genus Cordyline. *Biomolecules* **2023**, *13*, 1783. https://doi.org/10.3390/biom13121783.
11. Alkhoori, M.A.; Kong, A.S.-Y.; Aljaafari, M.N.; Abushelaibi, A.; Erin Lim, S.-H.; Cheng, W.-H.; Chong, C.-M.; Lai, K.-S. Biochemical Composition and Biological Activities of Date Palm (*Phoenix dactylifera* L.) Seeds: A Review. *Biomolecules* **2022**, *12*, 1626. https://doi.org/10.3390/biom12111626.
12. Al-Jubori, Y.; Ahmed, N.T.B.; Albusaidi, R.; Madden, J.; Das, S.; Sirasanagandla, S.R. The Efficacy of Gum Arabic in Managing Diseases: A Systematic Review of Evidence-Based Clinical Trials. *Biomolecules* **2023**, *13*, 138. https://doi.org/10.3390/biom13010138.

## References


1. Veiga, M.; Costa, E.M.; Silva, S.; Pintado, M. Impact of plant extracts upon human health: A review. *Crit. Rev. Food Sci. Nutr.* **2020**, *60*, 873–886. [CrossRef] [PubMed]
2. Atanasov, A.G.; Zotchev, S.B.; Dirsch, V.M.; International Natural Product Sciences Taskforce; Supuran, C.T. Natural products in drug discovery: Advances and opportunities. *Nat. Rev. Drug Discov.* **2021**, *20*, 200–216. [CrossRef] [PubMed]
3. Nan, Y.; Su, H.; Zhou, B.; Liu, S. The function of natural compounds in important anticancer mechanisms. *Front. Oncol.* **2023**, *12*, 1049888. [CrossRef] [PubMed]
4. Ansari, P.; Reberio, A.D.; Ansari, N.J.; Kumar, S.; Khan, J.T.; Chowdhury, S.; Abd El-Mordy, F.M.; Hannan, J.M.A.; Flatt, P.R.; Abdel-Wahab, Y.H.A.; et al. Therapeutic Potential of Medicinal Plants and Their Phytoconstituents in Diabetes, Cancer, Infections, Cardiovascular Diseases, Inflammation and Gastrointestinal Disorders. *Biomedicines* **2025**, *13*, 454. [CrossRef]
5. Pirintsos, S.; Panagiotopoulos, A.; Bariotakis, M.; Daskalakis, V.; Lionis, C.; Sourvinos, G.; Karakasiliotis, I.; Kampa, M.; Castanas, E. From Traditional Ethnopharmacology to Modern Natural Drug Discovery: A Methodology Discussion and Specific Examples. *Molecules* **2022**, *27*, 4060. [CrossRef]
6. de la Peña, I.; Afable, T.; Dahilig-Talan, V.R.; Cruz, P. Review of Plant Extracts and Active Components: Mechanisms of Action for the Treatment of Obesity-Induced Cognitive Impairment. *Brain Sci.* **2023**, *13*, 929. [CrossRef]
7. Kumar, A.P.N.; Kumar, M.; Jose, A.; Tomer, V.; Oz, E.; Proestos, C.; Zeng, M.; Elobeid, T.; Sneha, K.; Oz, F. Major Phytochemicals: Recent Advances in Health Benefits and Extraction Method. *Molecules* **2023**, *28*, 887. [CrossRef]
8. Matesanz, N.; Park, G.; McAllister, H.; Leahey, B.; Devine, A.; McVeigh, G.; Gardiner, T.A.; McDonald, D.M. Docosahexaenoic acid improves the nitroso-redox balance and reduces VEGF-mediated angiogenic signaling in microvascular endothelial cells. *Investig. Ophthalmol. Vis. Sci.* **2010**, *51*, 6815–6825. [CrossRef]
9. Zhang, G.; Panigrahy, D.; Mahakian, L.M.; Yang, J.; Liu, J.-Y.; Stephen Lee, K.S.; Wettersten, H.I.; Ulu, A.; Hu, X.; Tam, S.; et al. Epoxy metabolites of docosahexaenoic acid (DHA) inhibit angiogenesis, tumor growth, and metastasis. *Proc. Natl. Acad. Sci. USA* **2013**, *110*, 6530–6535. [CrossRef]
10. Evans, L.W.; Bender, A.; Burnett, L.; Godoy, L.; Shen, Y.; Staten, D.; Zhou, T.; Angermann, J.E.; Ferguson, B.S. Emodin and emodin-rich rhubarb inhibits histone deacetylase (HDAC) activity and cardiac myocyte hypertrophy. *J. Nutr. Biochem* **2020**, *79*, 108339. [CrossRef]
11. Xiao, D.; Zhang, Y.; Wang, R.; Fu, Y.; Zhou, T.; Diao, H.; Wang, Z.; Lin, Y.; Li, Z.; Wen, L.; et al. Emodin alleviates cardiac fibrosis by suppressing activation of cardiac fibroblasts via upregulating metastasis associated protein 3. *Acta Pharm. Sin. B* **2019**, *9*, 724–733. [CrossRef]
12. Gu, Z.; Zhang, S.; Zhao, S.; Cui, Y.; Sun, L. Emodin improves the cardiac function in the rats with chronic heart failure through regulation of the miR-26b-5p/PTEN pathway. *Arch. Med. Sci.* **2020**, *20*, 655–663. [CrossRef] [PubMed]
13. Abukhalil, M.H.; Al-Alami, Z.; Alfwuaires, M.A.; Imran, M.R.; Aladaileh, S.H.; Althunibat, O.Y. Taxifolin Protects Against 5-Fluorouracil-Induced Cardiotoxicity in Mice Through Mitigating Oxidative Stress, Inflammation, and Apoptosis: Possible Involvement of Sirt1/Nrf2/HO-1 Signaling. *Cardiovasc. Toxicol.* **2025**, *25*, 455–470. [CrossRef] [PubMed]
14. Hecht, J.T.; Veerisetty, A.C.; Wu, J.; Coustry, F.; Hossain, M.G.; Chiu, F.; Gannon, F.H.; Posey, K.L. Primary Osteoarthritis Early Joint Degeneration Induced by Endoplasmic Reticulum Stress Is Mitigated by Resveratrol. *Am. J. Pathol.* **2021**, *191*, 1624–1637. [CrossRef]
15. Jung, D.Y.; Suh, N.; Jung, M.H. Tanshinone 1 prevents high fat diet-induced obesity through activation of brown adipocytes and induction of browning in white adipocytes. *Life Sci.* **2022**, *298*, 120488. [CrossRef]
16. Jung, D.Y.; Kim, J.H.; Jung, M.H. Anti-Obesity Effects of Tanshinone I from *Salvia miltiorrhiza* Bunge in Mice Fed a High-Fat Diet through Inhibition of Early Adipogenesis. *Nutrients* **2020**, *12*, 1242. [CrossRef] [PubMed]

17. Ding, B.; Lin, C.; Liu, Q.; He, Y.; Ruganzu, J.B.; Jin, H.; Peng, X.; Ji, S.; Ma, Y.; Yang, W. Tanshinone IIA attenuates neuroinflammation via inhibiting RAGE/NF-κB signaling pathway in vivo and in vitro. *J. Neuroinflamm.* **2020**, *17*, 302. [CrossRef]
18. Santamarina, A.B.; Calder, P.C.; Estadella, D.; Pisani, L.P. Anthocyanins ameliorate obesity-associated metainflammation: Preclinical and clinical evidence. *Nutr. Res.* **2023**, *114*, 50–70. [CrossRef]
19. Fragoso-Medina, J.A.; López Vaquera, S.R.; Domínguez-Uscanga, A.; Luna-Vital, D.; García, N. Single anthocyanins effectiveness modulating inflammation markers in obesity: Dosage and matrix composition analysis. *Front. Nutr.* **2023**, *10*, 1255518. [CrossRef]
20. Mohammadinejad, A.; Mohajeri, T.; Aleyaghoob, G.; Heidarian, F.; Kazemi Oskuee, R. Ellagic acid as a potent anticancer drug: A comprehensive review on in vitro, in vivo, in silico, and drug delivery studies. *Biotechnol. Appl. Biochem.* **2022**, *69*, 2323–2356. [CrossRef]
21. Rani, S.; Gupta, S.; Tejavath, K.K.; Gupta, U. Effect of combination of polyphenols, polysaccharide, and sodium selenite on bortezomib anti-cancer action. *Int. J. Biol. Macromol.* **2025**, *289*, 138809. [CrossRef] [PubMed]
22. Al-Karmadi, A.; Okoh, A.I. An Overview of Date (*Phoenix dactylifera*) Fruits as an Important Global Food Resource. *Foods* **2024**, *13*, 1024. [CrossRef] [PubMed]
23. Qadir, A.; Shakeel, F.; Ali, A.; Faiyazuddin, M. Phytotherapeutic potential and pharmaceutical impact of *Phoenix dactylifera* (date palm): Current research and future prospects. *J. Food Sci. Technol.* **2020**, *57*, 1191–1204. [CrossRef] [PubMed]

biomolecules

MDPI

Article

# Phloridzin Docosahexaenoate, an Omega-3 Fatty Acid Ester of a Flavonoid Precursor, Inhibits Angiogenesis by Suppressing Endothelial Cell Proliferation, Migration, and Differentiation

**Wasundara Fernando [1], Emma MacLean [2], Susan Monro [3], Melanie R. Power Coombs [3], Paola Marcato [1], H. P. Vasantha Rupasinghe [1,4] and David W. Hoskin [1,*]**

1 Department of Pathology, Faculty of Medicine, Dalhousie University, Halifax, NS B3H 4R2, Canada; wasufer@dal.ca (W.F.); paola.marcato@dal.ca (P.M.); vrupasinghe@dal.ca (H.P.V.R.)
2 Department of Medical Sciences, Faculty of Medicine, Dalhousie University, Halifax, NS B3H 4R2, Canada; emma.maclean@dal.ca
3 Department of Biology, Faculty of Science, Acadia University, Wolfville, NS B4P 2R6, Canada; susan.monro@acadiau.ca (S.M.); melanie.coombs@acadiau.ca (M.R.P.C.)
4 Department of Plant, Food, and Environmental Sciences, Faculty of Agriculture, Dalhousie University, Truro, NS B3H 4R2, Canada
* Correspondence: d.w.hoskin@dal.ca; Tel.: +1-902-494-4538

**Abstract:** Angiogenesis is a normal physiological process that also contributes to diabetic retinopathy-related complications and facilitates tumor metastasis by promoting the hematogenic dissemination of malignant cells from solid tumors. Here, we investigated the in vitro, ex vivo, and in vivo anti-angiogenic activity of phloridzin docosahexaenoate (PZ-DHA), a novel ω-3 fatty acid ester of a flavonoid precursor. Human umbilical vein endothelial cells (HUVEC) and human dermal microvascular endothelial cells (HMVEC) treated with a sub-cytotoxic concentration of PZ-DHA to assess in vitro anti-angiogenic activity showed impaired tubule formation on a Matrigel matrix. Ex vivo angiogenesis was measured using rat thoracic aortas, which exhibited reduced vessel sprouting and tubule formation in the presence of PZ-DHA. Female BALB/c mice bearing $VEGF_{165}$- and basic fibroblast growth factor-containing Matrigel plugs showed a significant reduction in blood vessel development following PZ-DHA treatment. PZ-DHA inhibited HUVEC and HMVEC proliferation, as well as the migration of HUVECs in gap closure and trans-well cell migration assays. PZ-DHA inhibited upstream and downstream components of the Akt pathway and vascular endothelial growth factor ($VEGF_{165}$)-induced overexpression of small molecular Rho GTPases in HUVECs, suggesting a decrease in actin cytoskeletal-mediated stress fiber formation and migration. Taken together, these findings reveal the potential of combined food biomolecules in PZ-DHA to inhibit angiogenesis.

**Keywords:** angiogenesis; flavonoid; fatty acid; esterification; signal transduction

## 1. Introduction

Angiogenesis, which is the formation of new blood vessels and capillaries from existing vasculature, depends on a complex and well-maintained balance between pro-angiogenic and anti-angiogenic factors. Angiogenesis occurs during wound healing, embryogenesis, and cyclic changes in the female reproductive system [1–3]. However, angiogenesis is also involved in disease states such as diabetic retinopathy, and the progression and metastasis of solid tumors [4,5].

Diabetic retinopathy is a secondary microvascular complication of diabetes mellitus caused by abnormal growth of retinal blood vessels, which is associated with multiple intraocular complications such as macular edema and choroidal neovascularization [6]. Cancer is another major disease that depends largely on angiogenesis for its progression. The supply of oxygen and nutrients is vital for the continuous growth and metastatic spread of solid tumors. In fact, without supporting vasculature, the size of a solid tumor is restricted to 1–2 $mm^3$ [7]. The interior of a solid tumor is hypoxic as a result of the tumor vasculature being located, for the most part, around the tumor. This leads to the stabilization of the transcription factor, hypoxia-inducible factor-1 alpha (HIF-1$\alpha$), leading to the activation of target genes such as vascular endothelial growth factor (VEGF) [8,9]. Although tumor hypoxia does not contribute to the unusual elevation of interstitial pressure in the tumor microenvironment [10], hypoxia-induced VEGF activation induces the growth of new blood vessels in the tumor interior [8,9]. Although the exact mechanisms and pathways that regulate the increased interstitial pressure in the tumor interior are not clearly delineated, current knowledge suggests that VEGF-induced formation of large numbers of leaky blood vessels with an irregular shape, as well as fibroblast-mediated tumor contractility, play a significant role [11,12]. In addition, many studies suggest that increased interstitial pressure within solid tumors restricts the delivery of chemotherapeutic drugs [11,13]. The formation of new blood vessels also serves as the main mechanism by which tumor cells enter the systemic circulation and travel to other sites in the body. Therefore, inhibition of angiogenesis is predicted to suppress cancer progression by limiting the metastatic spread of cancer cells [7,14].

Blocking receptors such as VEGFR2 prevents the binding of principal pro-angiogenic signaling molecules and is a common approach to inhibit angiogenesis [15]. Endothelial cell survival pathways such as Akt signaling have also been researched as therapeutic targets in the development of anti-angiogenic drugs [16]. Flavonoids demonstrate anti-angiogenic activity through the inhibition of Akt signaling and VEGFR2-mediated angiogenic signaling in endothelial cells [17,18]. Dietary $\omega$-3 fatty acids such as docosahexaenoic acid (DHA) and its metabolites have proven to have anti-angiogenic activity [19,20]. Moreover, flavonoid-fatty acid conjugates show at least comparable, if not improved, anti-angiogenic activity when compared to parent flavonoids [21].

PZ-DHA combines a flavonoid precursor found in apple peels, known as phloridzin (PZ), with DHA through an enzyme-catalyzed acylation reaction. Supplementary Figure S1 shows the chemical structure of PZ-DHA. In our previous studies, we have shown that PZ-DHA possesses selective cytotoxic activity toward breast cancer cells, while sparing normal epithelial cells [22], and inhibits breast cancer cell metastasis in mice [23]. PZ-DHA also inhibits the growth of liver cancer cells [24] and T-cell acute lymphoblastic leukemia cells [25]. In this study, to investigate a potential mechanism for PZ-DHA-induced anti-metastatic activity, the impact of PZ-DHA on in vitro proliferation, migration, and tubule formation by human umbilical vein endothelial cells (HUVECs) and human microvascular vein endothelial cells (HMVECs) was tested. In addition, ex vivo anti-angiogenic activity was determined using thoracic aortic sections harvested from male Wistar rats. Furthermore, BALB/c female mice implanted with VEGF- and basic fibroblast growth factor (bFGF)-containing Matrigel plugs were used to evaluate the in vivo anti-angiogenic activity of systemically administered PZ-DHA. Finally, western blot analysis was employed to determine the effect of PZ-DHA on small molecular Rho GTPase signaling and phosphoinositide-dependent protein kinase 1 (PDK1), cyclin D3, and mammalian target of rapamycin (mTOR) activation.

## 2. Materials and Methods

### 2.1. Reagents and Chemicals

The horse radish peroxidase (HRP)/3,3′-Diaminobenzidine (DAB) detection system was purchased from Agilent Technologies (Mississauga, ON, Canada). Phenol red-free Matrigel was obtained from Corning Life Sciences (Tewkbury, MA, USA). The ECMatrix™ assay kit (EMD Millipore, Temecula, CA, USA) and endothelial basal medium (EBM)/supplements were purchased from Lonza Inc. (Walkersville, MD, USA). DHA was purchased from Nu-Chek Prep Inc. (Elysian, MN, USA). Human bFGF and human VEGF-165 were purchased from PeproTech (Rocky Hill, NJ, USA). RNase was obtained from Qiagen Inc. (Mississauga, ON, Canada). The Diff-Quik staining kit was purchased from Siemens Healthcare Diagnostics (Los Angeles, CA, USA). Aprotinin, 30% Brij 23 solution, Dulbecco's Modified Eagle Medium (DMEM), Drabkin's reagent, human hemoglobin, leupeptin, mitomycin C from *Streptomyces caespitosus*, Nonidet P-40 (NP-40), pepstatin A, phenazine methosulfate (PMS), phenylmethylsulfonyl fluoride (PMSF), PZ, porcine gelatin, sodium deoxycholate, and sodium fluoride (NaF) were purchased from Sigma-Aldrich (Oakville, ON, Canada).

### 2.2. Antibodies

Anti-CDK4 rabbit monoclonal Ab, anti-cyclinD3 rabbit monoclonal Ab, anti-total-PDK1 rabbit monoclonal antibody (Ab), anti-phospho-PDK1 (Ser241) rabbit monoclonal Ab, anti-total-mTOR rabbit monoclonal Ab, anti-phospho-mTOR (Ser2448) rabbit monoclonal Ab, anti-RhoA rabbit monoclonal Ab, anti-Rac1/2/3 rabbit monoclonal Ab, anti-Cdc42 rabbit monoclonal Ab, HRP-conjugated rabbit anti-β-actin, and HRP-conjugated donkey anti-rabbit Ab were purchased from Cell Signaling Technology Inc. (Danvers, MA, USA).

### 2.3. Cells and Cell Culture Conditions

HUVECs and HMVECs purchased from Lonza Inc. (Walkersville, MD, USA) were maintained in EBM, supplemented with fetal bovine serum, recombinant human bFGF, ascorbic acid, recombinant long $R^3$-insulin-like growth factor 1, recombinant human epidermal growth factor, recombinant human VEGF, heparin, hydrocortisone, gentamicin sulfate, and amphotericin B, according to the supplier's instructions. The cells were maintained at 37 °C in a humidified incubator containing 5% carbon dioxide. Cells were sub-cultured every 4–5 days.

### 2.4. Animals

Ethics approval for animal use was obtained from the Dalhousie University Committee on Laboratory Animals in accordance with Canadian Council for Animal Care guidelines. Six-to-eight-week-old female BALB/c female mice were purchased from Charles River Canada (Lasalle, QC, Canada). Mice were fed a regular rodent diet and water was supplied ad libitum.

### 2.5. Oregon Green 488 Staining

HUVECs and HMVECs were seeded into culture and synchronized. Adherent cells were stained with 1.25 µM Oregon Green 488 in serum-free DMEM for 45 min. The incubation was continued in a complete growth medium for 2 h to promote cell recovery. Cells were treated with sub-cytotoxic concentrations (10 or 20 µM) of PZ, DHA, PZ-DHA, vehicle, or medium alone and cultured in the dark for 72 h. Oregon Green 488 fluorescence of treated cells was measured using a FACSCalibur instrument (BD Biosciences, Mississauga, ON, Canada) in comparison to vehicle control and a non-proliferating control. The number of cell divisions (n) that took place was calculated using the formula: $MCF_{control} = 2^n \times MCF_{treatment}$ where MCF is the mean channel fluorescence.

### 2.6. Cell Cycle Analysis

Synchronized cells were seeded, and adherent cells were treated with a sub-cytotoxic concentration (10 µM) of PZ, DHA, PZ-DHA, vehicle, or medium alone, and cultured for 72 h. Cells were harvested, rinsed, and resuspended in ice-cold PBS. While vortexing, ice-cold 70% ethanol was added drop-by-drop and tube contents were incubated for at least 24 h at −20 °C to allow fixing. Fixed cells were washed with PBS and resuspended in cell cycle staining solution (0.1% $v/v$ Triton-X-100 and 2 µL/mL DNA-free Rnase A in 1 × PBS) containing 20 µL/mL propidium iodide and incubated for 30 min at room temperature. Flow cytometric analysis was performed using a FACSCalibur instrument.

### 2.7. Gap Closure Assays

HUVECs and HMVECs (10,000 cells/100 µL) were seeded in 2-well culture inserts (Ibidi GmbH, Martinsried, Germany) placed in 6-well plates and adherent cells were treated with 10 μg/mL mitomycin C in serum-free DMEM medium for 2 h at 37 °C to inhibit cell proliferation. Cells were allowed to recover for 12 h in EBM-containing serum and treated with PZ, DHA, PZ-DHA (10 μM), vehicle, or medium alone for 24 h. Culture inserts were removed, and the gaps were periodically photographed, starting at t = 0 h, until completely closed by medium-treated cells (t = 20 h).

### 2.8. Trans-Well Cell Migration Assay

HUVECs were seeded and treated with 10 μM of PZ, DHA, PZ-DHA, vehicle, or medium alone for 24 h, and treatments were continued in serum-free medium for another 6 h. Cells (50,000) were resuspended in 50 μL warm serum-free medium and loaded into the wells of the top chamber of co-culture inserts (Thermo Fisher Scientific, Mississauga, ON, Canada). Cells were allowed to migrate through an 8 μm porous membrane for 22 h. Migrated cells were stained using a Diff-Quik™ staining set.

### 2.9. Western Blot Analysis

Adherent HUVECs were treated with 10 μM of PZ, DHA, PZ-DHA, vehicle, or medium alone for 72 h. Cells were harvested and incubated in ice-cold lysis buffer [50 mM Tris (pH 7.5), 150 mM sodium chloride, 50 mM disodium hydrogen phosphate, 0.25% sodium deoxycholate (*w*/*v*), 0.1% NP-40 (*v*/*v*), 100 μM sodium orthovanadate, 10 mM NaF, 5 mM ethylenediaminetetraacetic acid, and 5 mM ethylene glycol tetraacetic acid containing freshly added protease inhibitors (1 mM PMSF, 10 μg/mL aprotinin, 5 μg/mL leupeptin, 10 μM phenylarsine oxide, 1 mM dithiothreitol, and 5 μg/mL pepstatin) for 15 min. Cell lysates were clarified by centrifugation and the protein concentration was determined by Bradford assay. After electrophoresis, proteins were transferred to nitrocellulose membranes and blots were incubated in 5% non-fat milk or 5% BSA for 1 h at room temperature to block nonspecific binding. Blots were probed overnight at 4 °C with primary Ab against the protein of interest. Then, the blots were washed thoroughly with Tween-TBS and probed with HRP-conjugated donkey anti-rabbit or goat anti-mouse IgG Ab for 1 h at RT. Even protein loading was confirmed by probing the blots with HRP-conjugated rabbit anti-β actin Ab or HRP-conjugated rabbit anti-α tubulin Ab. The proteins of interest were visualized by a ChemidocTouch™ imaging system (Bio-Rad Laboratories, Mississauga, ON, Canada).

### 2.10. In Vitro Angiogenesis Assay

In vitro angiogenesis was studied using a commercially available in vitro angiogenesis ECMatrix™ assay kit (EMD Millipore, Temecula, CA, USA), according to the manufacturer's instructions. Briefly, 9 parts of ECMatrix were mixed with one part of the diluent buffer and 10 μL of the mixture was added to the inner well of the μ-slide angiogenesis plate (Ibidi GmbH, Martinsried, Germany). The plate was incubated at 37 °C for 1 h. HUVECs or HMVECs (7500 cells) treated with 10 μM (HMVECs) or 20 μM (HUVECs) PZ, DHA, PZ-DHA, vehicle, or medium alone for 72 h were resuspended in 50 μL of EGM and seeded onto polymerized ECMatrix. Tubule formation by HMVECs and HUVECs was monitored and photographed after 4 h and 6 h, respectively. Supplementary Table S1 shows the scoring scheme.

### 2.11. Ex Vivo Angiogenesis Assay

Aortas from adult male Wistar rats were cleaned using sterile saline and cut into 1 mm × 3 mm sections. Aorta sections were then embedded in 200 μL Matrigel and incubated at 37 °C for 1 h. Matrigel was covered with 200 μL EBM and incubated overnight at 37 °C (day 0). On Day 1, aorta sections embedded in Matrigel were treated with PZ, DHA, PZ-DHA (20 μM), vehicle, or medium alone for 8 days. Medium/treatment was changed on Day 4. The development of tubules from aortic endothelium was monitored and photographed on Days 5 and 8.

### 2.12. Matrigel Plug Assay

Phenol red-free Matrigel (300 µL) containing human $VEGF_{165}$ (2 µg/mL) and bFGF (2 µg/mL) was implanted by subcutaneous injection on both the left and right sides along the mid-dorsal line of the lower posterior area of BALB/c female mice (Day 0). On Day 1, mice were randomly assigned into two groups (11 mice/group) and 5 doses of saline or PZ-DHA (100 mg/kg) were administered by intraperitoneal injection every second day for 9 days. After 9 days, Matrigel plugs were harvested and photographed. Hemoglobin concentration in Matrigel plugs was determined using the cyanmethemoglobin method. Briefly, a human hemoglobin (0.717 mg/mL) solution made in Drabkin's reagent (Sigma-Aldrich, Oakville, ON, Canada) containing 0.0005% $v/v$ 30% Brij 23 solution was used to generate the standard curve of cyanmethemoglobin ($R^2 = 1.00$). Matrigel plugs were homogenized in 500 µL of Drabkin's reagent and homogenates were centrifuged at $9600\times g$ at 4 °C for 6 min. The supernatant was collected and transferred into 96-well plates in triplicate. Absorbance was measured at 540 nm and cyanmethemoglobin concentration in Matrigel plugs was calculated.

### 2.13. Statistical Analysis

Three independent experiments were performed for each assay and the mean of the three experiments was calculated. A one-way ANOVA multiple means comparison method was performed and the differences between means were compared using Tukey's post-mean comparison method. The analysis was considered significant at the following levels: * $p < 0.05$, ** $p < 0.01$, *** $p < 0.001$.

## 3. Results

### 3.1. PZ-DHA Inhibits Angiogenesis In Vitro and Ex Vivo

The concentrations of PZ, DHA, and PZ-DHA for HUVECs and HMVECs used in this study were confirmed to be sub-cytotoxic using 7AAD cell viability assays (Supplementary Figure S2). This was to ensure that there would be no inadvertent activation of cell death pathways while determining the impact of PZ-DHA on angiogenesis. In vitro angiogenesis was scored according to the stage/complexity of the tubule formation by HUVECs and HMVECs treated with vehicle, PZ, DHA, or PZ-DHA. Treatment with PZ-DHA (20 µM) decreased the in vitro HUVEC angiogenesis by 6.5-fold; however, PZ alone did not inhibit tubule formation by HUVECs (Figure 1A). The inhibitory effect of PZ-DHA was significantly greater than either PZ or DHA alone. PZ-DHA (10 µM), as well as DHA, significantly attenuated the in vitro tubule formation by HMVECs by 2.8-fold and 1.8-fold, respectively (Figure 1B). Again, PZ-DHA had a greater impact on tubule formation than PZ or DHA alone. In preliminary experiments, PZ-DHA also suppressed ex vivo angiogenesis. As shown in Supplementary Figure S3, the sprouting of microvessels from rat aorta endothelium embedded in a Matrigel matrix was reduced in the presence of 20 µM PZ-DHA. Upon extended incubation, PZ-DHA-treated cells aligned on the Matrigel matrix but did not differentiate to form tubules.

### 3.2. PZ-DHA Inhibits In Vivo Angiogenesis in BALB/c Female Mice

The impact of PZ-DHA on in vivo angiogenesis was investigated using a Matrigel plug assay performed in BALB/c female mice. VEGF- and bFGF-induced angiogenesis in Matrigel plugs was inhibited by intraperitoneal administration of PZ-DHA (Figure 2A). Hemoglobin in Matrigel plugs was converted into a cyanmethemoglobin complex by a reaction with cyanide ions in Drabkin's reagent. A 2.3-fold reduction in the formation of cyanmethemoglobin complex was noted in Matrigel plugs excised from PZ-DHA-treated mice (Figure 2B). PZ-DHA-induced reduction in the growth of blood vessels in the body wall (around the Matrigel plugs) was also observed (Figure 2C).

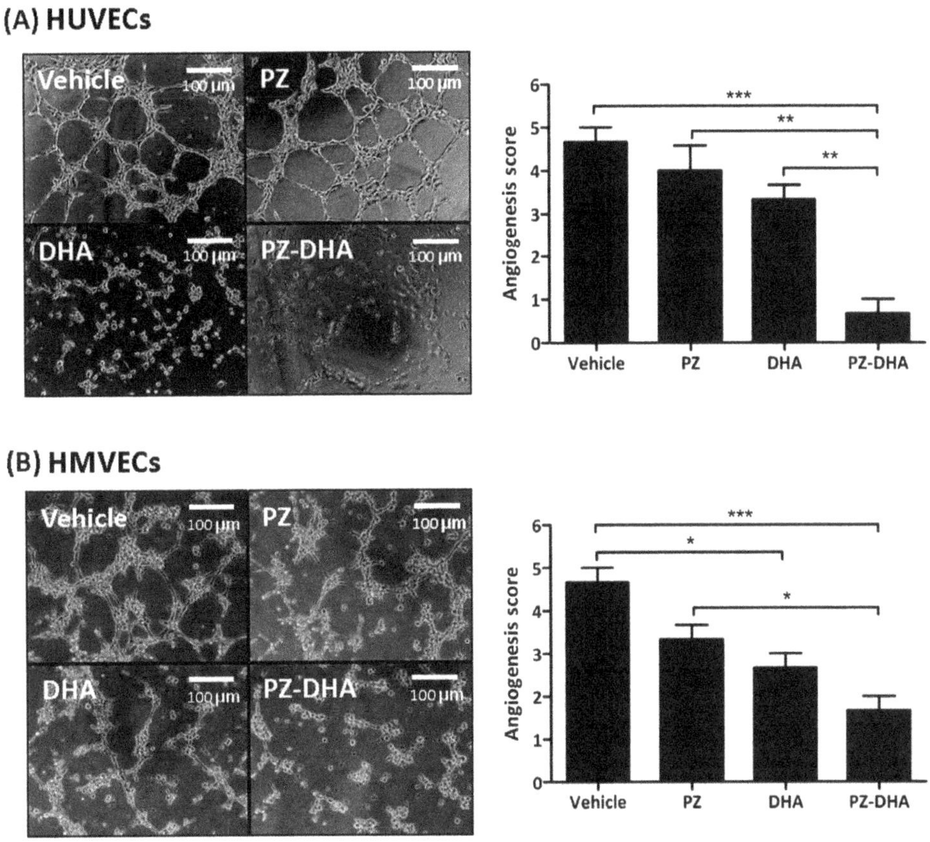


**Figure 1.** PZ-DHA inhibits angiogenesis in vitro. PZ, DHA, PZ-DHA (HUVECs: 20 μM, HMVECs: 10 μM), vehicle-, or medium-treated HMVECs or HUVECs were harvested, and 7500 cells were resuspended in 50 μL of EGM and seeded onto a polymerized ECMatrix. (**A**) HMVEC and (**B**) HUVEC tubule formation were monitored and photographed after 4 h and 6 h, respectively. Images were analyzed and tube formation was quantified according to the complexity of the tube network. Mean angiogenesis scores ± SEM were calculated from 3 independent experiments. Statistical analysis was performed using the ANOVA multiple means comparison method and differences among means were compared using Tukey's post-mean comparison method; * $p < 0.05$, ** $p < 0.01$, and *** $p < 0.001$.

### 3.3. *PZ-DHA Inhibits the Proliferation of HUVECs and HMVECs*

We next determined the mechanism(s) that might account for the inhibitory effect of PZ-DHA on in vitro, ex vivo, and in vivo angiogenesis. As shown in Figure 3A, 10 μM PZ-DHA decreased HUVEC proliferation by 1.7-fold, and by 2.8-fold when HUVECS were treated with 20 μM PZ-DHA. DHA alone also had an antiproliferative effect on HUVECs. PZ-DHA at 10 μM suppressed HMVEC proliferation by 3.4-fold (Figure 3B). Neither DHA nor PZ alone affected HMVEC proliferation. Figure 3C shows that 10 μM PZ-DHA arrested the HUVEC cell cycle at $G_0/G_1$ while significantly decreasing the number of cells progressing to the $G_2/M$ phase. The progression of a cell through G1 phase of the cell cycle is regulated by several cyclins and CDKs, including cyclin D and CDK4. PZ-DHA (10 μM) significantly decreased cyclin D3 levels in HUVECs by 28% (Figure 3D), and the expression of CDK4 (Figure 3E) was suppressed by both DHA and PZ-DHA by 47% and 72%, respectively; however, PZ by itself had no effect on HUVEC and HMVEC proliferation or expression of any of the cell cycle regulatory proteins tested. These findings suggest that the antiproliferative effect of PZ-DHA on endothelial cells contributes to its anti-angiogenic activity.

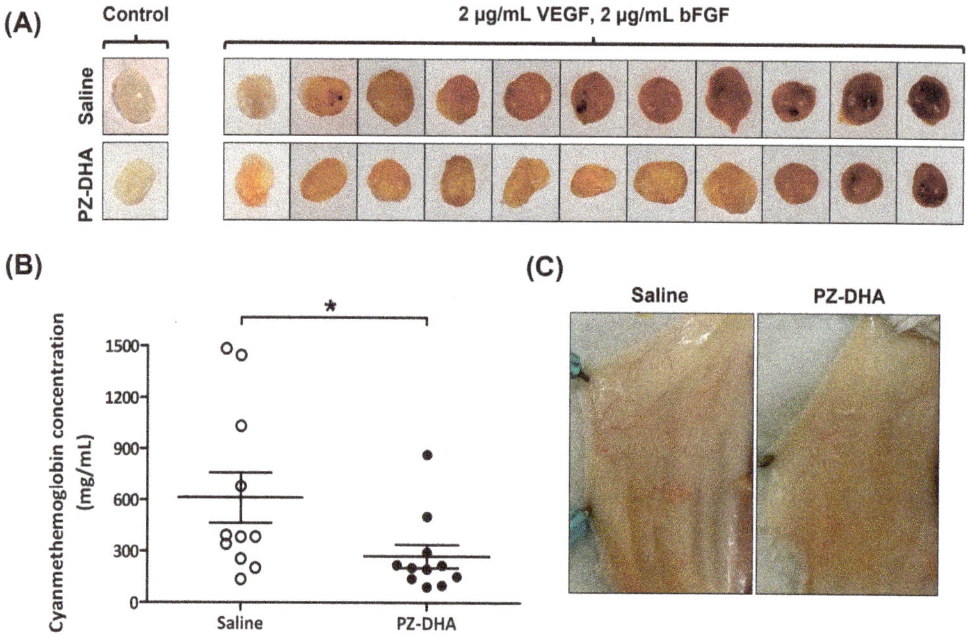


**Figure 2.** PZ-DHA inhibits in vivo angiogenesis in female BALB/c mice. Phenol red-free Matrigel (300 μL) mixed with/without human $VEGF_{165}$ (2 μg/mL) and bFGF (2 μg/mL) were subcutaneously implanted on both the left and right sides along the mid-dorsal line of the lower posterior area of BALB/c female mice (Day 0). On Day 1, mice were randomly assigned into 2 groups (11 mice/group) and saline or PZ-DHA (100 mg/kg) was administered by intraperitoneal injection. Altogether, saline or PZ-DHA was administered every second day (Days 1, 3, 5, 7, and 9) for 9 days. Mice were euthanized. (**A**) Matrigel plugs were harvested and photographed. Control Matrigel plugs are shown to confirm that angiogenesis did not occur in the absence of $VEGF_{165}$ and bFGF. (**B**) Mean hemoglobin concentration ± SEM of $VEGF_{165}$- and bFGF-containing Matrigel plugs from saline- and PZ-DHA-treated mice was determined using the cyanmethemoglobin method. Statistical differences between means were compared using Student's $t$-test; * $p < 0.05$. (**C**) Growth of blood vessels on the body wall, toward the Matrigel plugs, was also photographed.

### *3.4. PZ-DHA Inhibits the Migration of HUVECs and HMVECs*

Endothelial cell migration is important during early angiogenesis and subsequent tubule formation [26,27]. PZ-DHA (10 μM) reduced the migration of HUVECs and HMVECs by 59% (Figure 4A) and 56% (Figure 4B), respectively, in a gap closure assay. PZ-DHA-induced inhibition of HUVEC migration was further investigated using a transwell cell migration assay. The migration of serum-starved-HUVECs showed a significant 40% inhibition in the presence of 10 μM PZ-DHA. This inhibitory effect of PZ-DHA was significantly greater than that of either parent compound; however, the polygonal endothelial cellular shape was restored by vehicle-, PZ-, and DHA-treated HUVEC following migration. In contrast, the shape of PZ-DHA-treated HUVECs was still distorted after migration (Figure 4C). These data suggest the inhibitory effect of PZ-DHA on endothelial cell migration is a factor in its anti-angiogenic activity.

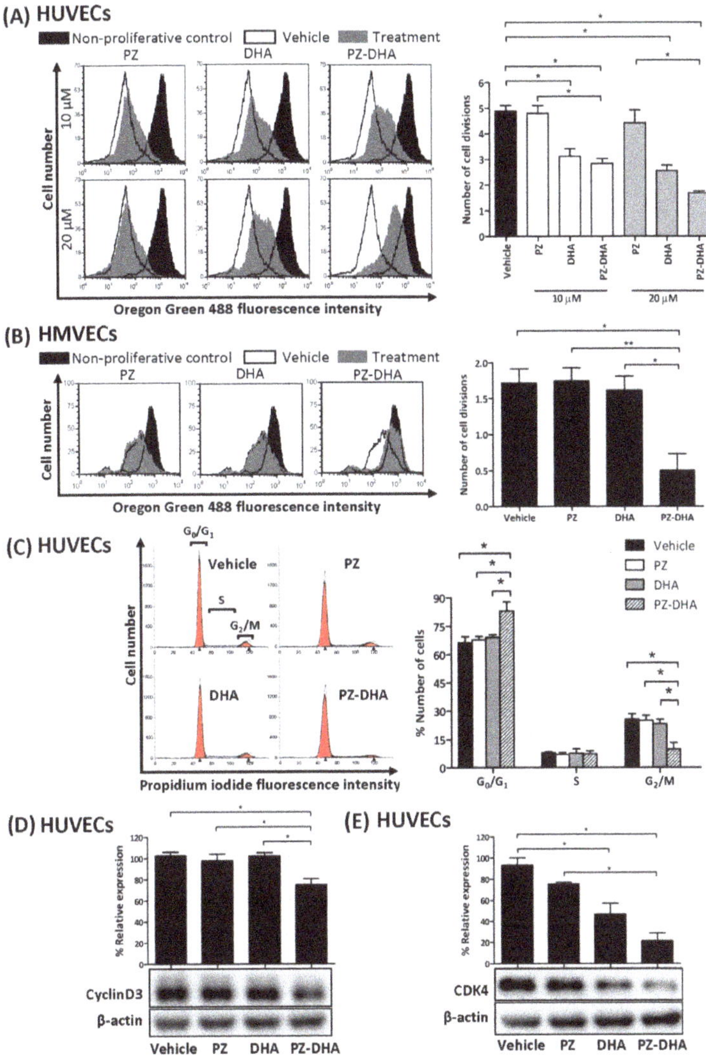


**Figure 3.** PZ-DHA inhibits the in vitro proliferation of HUVECs and HMVECs. Oregon Green 488-stained (**A**) HUVECs and (**B**) HMVECs were treated with PZ, DHA, PZ-DHA (HUVECs: 10 or 20 µM, HMVECs: 10 µM), vehicle, or medium and cultured for 72 h at 37 °C. At the end of incubation, cells were harvested and analyzed by flow cytometry. Data shown are representative histograms and mean number of cell divisions ± SEM. (**C**) HUVECs were treated with PZ, DHA, PZ-DHA (10 µM), vehicle, or medium alone and cultured for 72 h at 37 °C. Cells were fixed and stained with PI in the presence of RNase for analysis by flow cytometry. Representative histograms were generated and the mean % number ± SEM of cells in each phase of the cell cycle was calculated. (**D**) Cyclin D3 and (**E**) CDK4 expression were determined using western blot analysis of protein-rich cell lysates HUVECs that were treated with PZ, DHA, PZ-DHA (10 µM), vehicle, or medium alone for 72 h. Statistical analysis of data from 3 independent experiments was performed using the one-way ANOVA multiple means comparison method, and differences among means were compared using Tukey's post-mean comparison method. * $p < 0.05$, ** $p < 0.01$. (Original Western Blot Images see supplementary Figure S4).

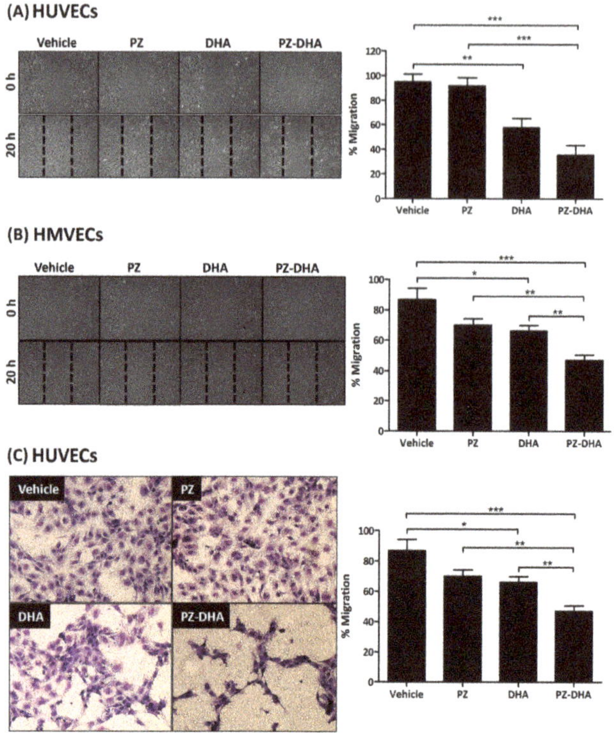


**Figure 4.** PZ-DHA inhibits the in vitro migration of HUVECs and HMVECs. (**A**) HMVECs and (**B**) HUVECs in cell culture inserts were treated with PZ, DHA, PZ-DHA (10 μM), vehicle, or medium alone and cultured for 24 h. Inserts were removed and the number of cells that migrated into the gap was quantified. Representative pictures of cells in the gap and mean % migrated cells ± SEM are shown. (**C**) HUVECs were seeded and treated with PZ, DHA, PZ-DHA (10 μM), vehicle, or medium alone and cultured for 24 h. Treated cells were serum-starved and migration toward serum through a porous membrane was determined. Representative pictures of migrated cells and mean % migration ± SEM are shown. Statistical analysis of data from 3 independent experiments was carried out using the ANOVA multiple means comparison statistical method and differences among means were compared using Tukey's test; * $p < 0.05$, ** $p < 0.01$, *** $p < 0.001$.

### *3.5. PZ-DHA Inhibits PDK1 and mTOR Activation and VEGF$_{165}$-Induced Small Molecular Rho GTPase Signaling in HUVECs*

The PDK1-activated phosphatidylinositol 3-kinase (PI3K)/Akt/mTOR pathway is involved in blood vessel growth and maintenance, which also involves Rho GTPase signaling, which is often activated by VEGF$_{165}$ [28–30]. PZ-DHA inhibited upstream (PDK1) and downstream (mTOR) components of the Akt signaling pathway in HUVECs. As shown in Figure 5A, 10 μM PZ-DHA inhibited the phosphorylation of PDK1 at Ser241. The downstream phosphorylation of mTOR at Ser2448 was also decreased in the presence of 10 μM PZ-DHA (Figure 5B).

Figure 6A shows that 10 μM PZ-DHA, as well as the same concentration of DHA, significantly inhibited RhoA expression in HUVECs (Figure 6A). Cdc42 expression was also decreased following 10 μM PZ-DHA or DHA treatment (Figure 6B). In contrast, Rac1/2/3 expression remained unchanged in the presence of 10 μM PZ-DHA (Figure 6C). VEGF$_{165}$-induced RhoA activation was not reduced in the presence of 10 μM PZ, DHA, or PZ-DHA (Figure 6D); however, VEGF$_{165}$-induced Cdc42 overexpression was reduced by 10 μM PZ-DHA (Figure 6E). VEGF$_{165}$-induced Rac1/2/3 expression was also significantly inhibited in the presence of 10 µM PZ-DHA, although there was no effect by either parent compound alone (Figure 6F). These findings suggest that PZ-DHA-mediated inhibition of key angiogenesis-associated signal transduction pathways is associated with its anti-angiogenic activity.

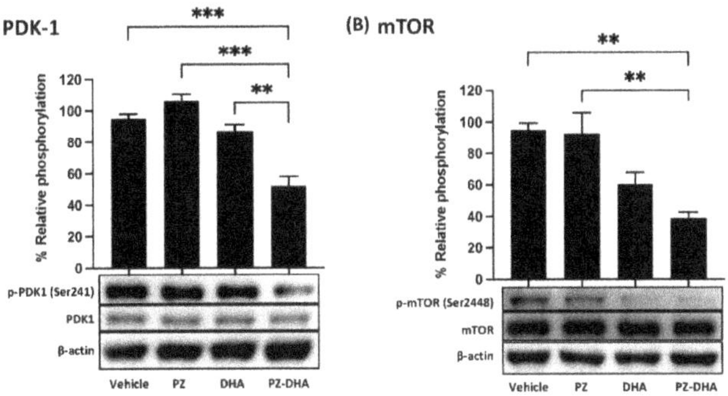


**Figure 5.** PZ-DHA inhibits PDK1 and mTOR phosphorylation in HUVECs. HUVECs were treated with PZ, DHA, PZ-DHA (10 µM), or vehicle alone and cultured for 72 h. Cells were harvested and protein-rich cell lysates were prepared for western blot analysis of (**A**) phospho-PDK1 (Ser241)/total-PDK1 expression or (**B**) phospho-mTOR (Ser2448)/total-mTOR expression. Equal protein loading was confirmed by β-actin expression. Data shown are mean % relative expression ± SEM from 3 independent experiments. The ANOVA multiple means comparison statistical method was performed and differences among means were compared using Tukey's test; ** $p < 0.01$, *** $p < 0.001$. (Original Western Blot Images see supplementary Figure S5).

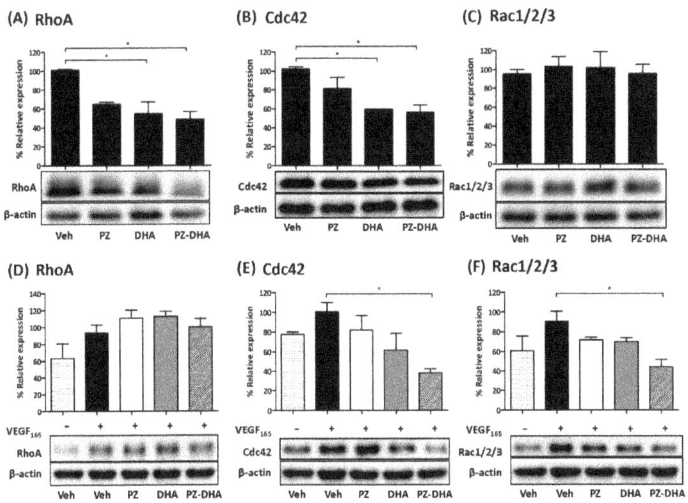


**Figure 6.** PZ-DHA inhibits endogenous and VEGF-induced small GTPase signaling in HUVECs. HUVECs were treated with PZ, DHA, PZ-DHA (10 µM), or vehicle alone and cultured for 72 h in the presence or absence of $VEGF_{165}$. Cells were harvested and protein-rich cell lysates were prepared for western blot analysis of (**A**,**D**) RhoA expression, (**B**,**E**) Cdc42 expression, and (**C**,**F**) Rac1/2/3 expression. Equal protein loading was confirmed by β-actin expression. Data shown are mean % relative expression ± SEM from 3 independent experiments. The ANOVA multiple means comparison statistical method was performed and differences among means were compared using Tukey's test; * $p < 0.05$. (Original Western Blot Images see supplementary Figure S6).

## 4. Discussion

Angiogenesis plays an important role in several pathological conditions including chronic inflammation and cancer [4,5]. In the current study, the effects of PZ-DHA on different aspects of angiogenesis, including endothelial cell proliferation, migration, and differentiation, were tested in vitro, ex vivo, and in vivo. PZ-DHA attenuated in vitro tubule formation by HUVECs and HMVECs, microvessel sprouting from rat aortic endothelium, and VEGF- and bFGF-induced angiogenesis in Matrigel plugs implanted in BALB/c mice. These findings are consistent with anti-angiogenic effects of other flavonoids such as quercetin [31], apigenin [32], and epigallocatechin gallate [33], which suppress angiogenesis via their inhibition of Akt phosphorylation, HIF-1$\alpha$ signaling, and expression of cell adhesion molecules.

The proliferation of HUVECs and HMVECs was attenuated by sub-cytotoxic concentrations of PZ-DHA, causing inhibition of S phase entry of HUVECs because of $G_0/G_1$ cell cycle arrest. At a molecular level, the expression of cyclin D3 and CDK4 was inhibited by PZ-DHA, suggesting a decreased formation of cyclin D/CDK4 complex. This observation was in line with the increased accumulation of HUVECs in the $G_0/G_1$ phase. DHA, one of the parent compounds, did not alter the expression of cyclin D3 or the number of HUVECs passing the G1 checkpoint to enter the S phase; however, the expression of CDK4 was downregulated by DHA. In contrast, Kim and colleagues showed that DHA induces $G_0/G_1$ cell cycle arrest, causes an increase in the sub-$G_1$ peak, and increases staining with Annexin-V-FITC/PI, all of which are consistent with HUVEC apoptosis [34]. Our findings differ, possibly because an apoptosis-inducing concentration (40 μM) of DHA was used by Kim et al., whereas our study was conducted using sub-cytotoxic concentrations of DHA (10–20 μM). The fate of a cell during stress conditions is determined by an interplay of cell cycle regulators and pro-apoptotic factors [35,36]. Induction of apoptotic signals vis p53 activation may have played a partial role through the p53-p21 axis in the $G_0/G_1$ cell cycle arrest [34]. $G_1/S$ transition is also regulated by a type of GTPase family protein known as Rho [37], which was inhibited by PZ-DHA. Furthermore, the accumulation of $p21^{WAF1/CIP1}$ inhibits CDK activity, which negatively regulates the $G_1/S$ transition [38]. In summary, both the inhibition of cell cycle regulatory proteins and small molecular Rho GTPase molecules have been attributed to PZ-DHA-induced anti-proliferative activity in endothelial cells.

Rho GTPase signaling regulates endothelial cell rearrangement and organization during angiogenesis [39,40]. $VEGF_{165}$-dependent Rac activation via Rac1 guanine nucleotide exchange factor DOCK4 is necessary for Cdc42 activation, filopodia, and lumen formation [41]. PZ-DHA down-regulated $VEGF_{165}$-induced Cdc42 and Rac1/2/3 expression by HUVECs but did not affect the expression of RhoA in response to $VEGF_{165}$. However, PZ-DHA significantly decreased tubule formation by HUVECs and HMVECs in vitro. Preliminary experiments suggest that PZ-DHA also inhibited $VEGF_{165}$-induced sprouting of tubules from rat aortic sections ex vivo. These findings indicate that PZ-DHA-induced anti-angiogenic activity was not necessarily mediated solely through RhoA-dependent mechanisms, but possibly in combination with inhibition of $VEGF_{165}$-stimulated Cdc42 and Rac1/2/3.

PZ-DHA inhibited the migration and expression of RhoA and Cdc42, but not Rac1/2/3, in HUVECs, suggesting inhibition of filopodia-driven HUVEC migration. PZ-DHA significantly and more effectively suppressed the migration of HUVECs in trans-well cell migration assays, compared to its parent compounds. The polygonal cellular shape was restored following the migration of vehicle-, PZ-, and DHA-treated HUVEC; but not in PZ-DHA-treated cells. The process of angiogenesis is composed of multiple steps that include changes in the integrity of the endothelial cell barrier, basement membrane degradation, endothelial cell proliferation and migration, morphogenesis, and capillary formation, all of which involve distinct roles by RhoA, Cdc42, and Rac1/2/3 [42]. Well-known cell survival and proliferation pathways such as the PDK1-activated PI3K/Akt/mTOR pathway and mitogen-activated protein kinase signaling pathways regulate endothelial cell proliferation and assembly [43,44]. Crosstalk between cell survival pathways and other signaling cascades, such as Rho GTPase and matrix metalloproteinase signaling, regulate angiogenesis [45,46].

Induction of the PDK1-activated PI3K/Akt/mTOR pathway impacts multiple downstream signal transduction pathways involved in endothelial cell proliferation (cell cycle regulators), migration through actin reorganization (small molecular RhoGTPases), and eventual blood vessel development [47–49]. At a sub-cytotoxic concentration, PZ-DHA suppressed PDK1 phosphorylation upstream of Akt and cyclin D3 and mTOR phosphorylation downstream from Akt. Currently, it is not known whether the phosphorylation of endothelial cell Akt is directly inhibited by PZ-DHA. However, the DHA component of PZ-DHA inhibits the phosphorylation of Akt at ser473 and thr308, as well as mTOR phosphorylation, in human prostate cancer cells [50].

Figure 7 depicts a model for the impact of PZ-DHA on angiogenesis-associated signal transduction that is consistent with our findings. The effect of PZ-DHA at the receptor binding level is yet to be understood.

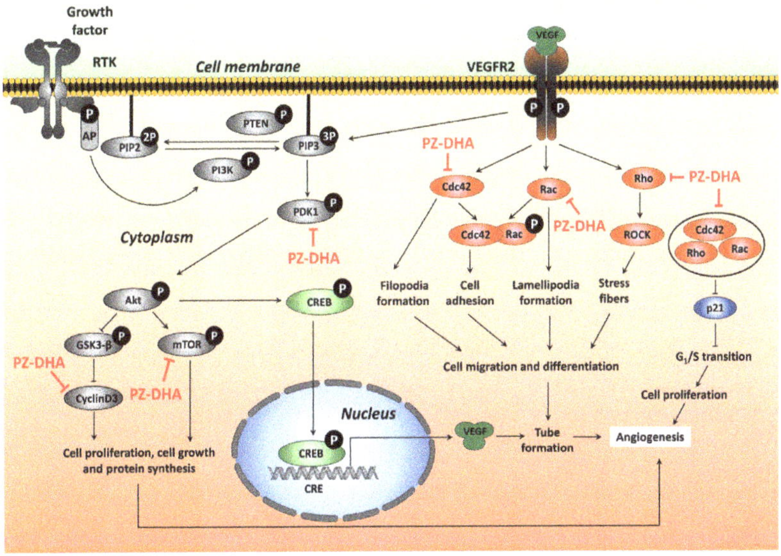


**Figure 7.** Proposed scheme for the anti-angiogenic activities of PZ-DHA. PZ-DHA inhibits angiogenesis by inhibiting endothelial cell proliferation, migration, and tubule formation. The antiproliferative effects of PZ-DHA on endothelial cells are mediated through Rho GTPase-driven $G_1$ arrest through the p53-p21 axis, as well as inhibition of upstream and downstream components of the PI3K/Akt/mTOR signaling pathway. PZ-DHA inhibits endothelial cell migration and differentiation by inhibiting VEGF-induced activation of small molecular Rho GTPase and activation of PDK1, cyclin D3, and mTOR. Akt: protein kinase B; AP: activator protein; Cdc42, Rac1/2/3, and RhoA: Rho family small GTPases; CRE: cAMP (cyclic adenosine monophosphate) response element; CREB: cAMP response element binding protein; GSK3-β: glycogen synthase kinase 3-β; mTOR: mammalian target of rapamycin; p21: cyclin-dependent kinase inhibitor 1; PDK1: phosphoinositide-dependent kinase-1; PI3K: phosphoinositide 3-kinase; PIP2: phosphatidylinositol-4,5-bisphosphate; PIP3: phosphatidylinositol-3,4,5-trisphosphate; PTEN: phosphatase and tensin homolog; PZ-DHA: phloridzin docosahexaenoate; ROCK: Rho-associated protein kinase; RTK: receptor tyrosine kinase; VEGF: vascular endothelial growth factor; VEGFR2: vascular endothelial growth factor receptor.

Irregularity in shape, high permeability, and poorly supported pericytes are key features of tumor-associated blood vessels [51,52]. This leads to the formation of leaky and contorted vasculature, which is associated with the increased interstitial fluid pressure within the tumor that facilitates passive migration of cancer cells and poor penetration of the tumor by chemotherapeutic drugs [11,13]. Although it was initially thought that angiogenesis blockade eliminates cancer cells by oxygen and nutrient starvation, recent experience suggests that "normalization" of tumor-associated angiogenesis is more effective than its inhibition since this approach corrects elevated interstitial fluid pressure, thereby allowing chemotherapeutic drug penetration and reduced cancer cell metastasis [53]. PZ-DHA-mediated inhibition of upstream (PDK1) and downstream (mTOR) components of the Akt signaling pathway may correct the abnormal vasculature in tumors since inhibition of PI3K/Akt/mTOR signaling promotes endothelial cell elongation associated with proper regulation of angiogenesis [54]. However, confirmation awaits further investigation of the effect of PZ-DHA on abnormal angiogenesis in tumors. We conclude that in addition to targeting breast cancer cells [23], PZ-DHA may also interfere with the angiogenesis-related hematogenic spread of breast cancer cells, thereby contributing to reduced metastasis of PZ-DHA-treated mammary carcinoma cells in a mouse model of metastatic breast cancer [23].

**Supplementary Materials:** The following supporting information can be downloaded at: https://www.mdpi.com/article/10.3390/biom14070769/s1, Figure S1: Chemical structure of PZ-DHA; Figure S2: Determination of subcytotoxic concentrations of PZ-DHA for HMVECs and HUVECs in vitro; Figure S3: PZ-DHA inhibits angiogenesis ex vivo; Table S1: Angiogenesis pattern and scoring. Figures S4–S6: Original western blot images for Figures 3, 5 and 6.

**Author Contributions:** Conception and design: W.F., H.P.V.R. and D.W.H. Development of methodology: W.F., H.P.V.R. and D.W.H. Acquisition of data: W.F., E.M., S.M. and M.R.P.C. Analysis and interpretation of data: W.F., E.M., S.M. and M.R.P.C. Writing, review and revision of the manuscript: W.F., E.M., P.M., H.P.V.R. and D.W.H. Study supervision: P.M., M.R.P.C., H.P.V.R. and D.W.H. All authors have read and agreed to the published version of the manuscript.

**Funding:** W.F. was supported by a Cancer Research Training Program Award from the Beatrice Hunter Cancer Research Institute through funds provided by the Dalhousie Medical Research Foundation Edward F. Crease Memorial Graduate Studentship in Cancer Research, CIBC Graduate Scholarship in Medical Research, and the Canadian Cancer Society/Canadian Breast Cancer Foundation-Queen Elizabeth II Foundation Endowed Chair in Breast Cancer Research held by D.H. This project was funded by the Canadian Cancer Society/Canadian Breast Cancer Foundation-Queen Elizabeth II Foundation Endowed Chair in Breast Cancer Research held by DH and a Discovery Grant from the Natural Sciences and Engineering Research Council of Canada (RGPIN2016 05369) to V.R.

**Institutional Review Board Statement:** The animal study protocol was approved by the Ethics Committee of Dalhousie University (protocol code 17-069 approved 1 July 2017).

**Informed Consent Statement:** Not applicable.

**Data Availability Statement:** Data available on request.

**Acknowledgments:** We would like to recognize technical assistance from Javad Ghassemi-Rad.

**Conflicts of Interest:** The authors have no conflicts of interest to declare.

## References

1. Góth, M.I.; Hubina, E.; Raptis, S.; Nagy, G.M.; Tóth, B.E. Physiological and pathological angiogenesis in the endocrine system. *Microsc. Res. Tech.* **2003**, *60*, 98–106. [CrossRef] [PubMed]
2. Kurz, H. Physiology of angiogenesis. *J. Neurooncol.* **2000**, *50*, 17–35. [CrossRef]
3. Seaman, S.; Stevens, J.; Yang, M.Y.; Logsdon, D.; Graff-Cherry, C.; St Croix, B. Genes that distinguish physiological and pathological angiogenesis. *Cancer Cell* **2007**, *11*, 539–554. [CrossRef]
4. Chung, A.S.; Ferrara, N. Developmental and Pathological Angiogenesis. *Annu. Rev. Cell Dev. Biol.* **2011**, *27*, 563–584. [CrossRef]
5. Papetti, M.; Herman, I.M. Mechanisms of normal and tumor-derived angiogenesis. *Am. J. Physiol.-Cell Physiol.* **2002**, *282*, C947–C970. [CrossRef]

6. Crawford, T.; Alfaro, D., III; Kerrison, J.; Jablon, E. Diabetic Retinopathy and Angiogenesis. *Curr. Diabetes Rev.* **2009**, *5*, 8–13. [CrossRef]
7. Folkman, J. Role of angiogenesis in tumor growth and metastasis. *Semin. Oncol.* **2002**, *29*, 15–18. [CrossRef]
8. Chua, Y.L.; Dufour, E.; Dassa, E.P.; Rustin, P.; Jacobs, H.T.; Taylor, C.T.; Hagen, T. Stabilization of Hypoxia-inducible Factor-1α Protein in Hypoxia Occurs Independently of Mitochondrial Reactive Oxygen Species Production. *J. Biol. Chem.* **2010**, *285*, 31277–31284. [CrossRef]
9. Liu, W.; Shen, S.-M.; Zhao, X.-Y.; Chen, G.-Q. Targeted genes and interacting proteins of hypoxia inducible factor-1. *Int. J. Biochem. Mol. Biol.* **2012**, *3*, 165–178.
10. Boucher, Y.; Lee, I.; Jain, R.K. Lack of General Correlation between Interstitial Fluid Pressure and Oxygen Partial Pressure in Solid Tumors. *Microvasc. Res.* **1995**, *50*, 175–182. [CrossRef] [PubMed]
11. Heldin, C.-H.; Rubin, K.; Pietras, K.; Östman, A. High interstitial fluid pressure—An obstacle in cancer therapy. *Nat. Rev. Cancer* **2004**, *4*, 806–813. [CrossRef] [PubMed]
12. Nagy, J.A.; Chang, S.-H.; Dvorak, A.M.; Dvorak, H.F. Why are tumour blood vessels abnormal and why is it important to know? *Br. J. Cancer* **2009**, *100*, 865–869. [CrossRef] [PubMed]
13. Lunt, S.J.; Fyles, A.; Hill, R.P.; Milosevic, M. Interstitial fluid pressure in tumors: Therapeutic barrier and biomarker of angiogenesis. *Futur. Oncol.* **2008**, *4*, 793–802. [CrossRef] [PubMed]
14. Bielenberg, D.R.; Zetter, B.R. The Contribution of Angiogenesis to the Process of Metastasis. *Cancer J.* **2015**, *21*, 267–273. [CrossRef] [PubMed]
15. Falcon, B.L.; Chintharlapalli, S.; Uhlik, M.T.; Pytowski, B. Antagonist antibodies to vascular endothelial growth factor receptor 2 (VEGFR-2) as anti-angiogenic agents. *Pharmacol. Ther.* **2016**, *164*, 204–225. [CrossRef]
16. Jiang, B.-H.; Liu, L.-Z. AKT signaling in regulating angiogenesis. *Curr. Cancer Drug Targets* **2008**, *8*, 19–26. [CrossRef] [PubMed]
17. Fernando, W.; Rupasinghe, H.P.V.; Hoskin, D.W. Regulation of Hypoxia-inducible Factor-1α and Vascular Endothelial Growth Factor Signaling by Plant Flavonoids. Mini Rev. *Med. Chem.* **2015**, *15*, 479–489.
18. Mirossay, L.; Varinská, L.; Mojžiš, J. Antiangiogenic effect of flavonoids and chalcones: An update. *Int. J. Mol. Sci.* **2018**, *19*, 27. [CrossRef]
19. Matesanz, N.; Park, G.; McAllister, H.; Leahey, B.; Devine, A.; McVeigh, G.; Gardiner, T.A.; McDonald, D.M. Docosahexaenoic acid improves the nitroso-redox balance and reduces VEGF-mediated angiogenic signaling in microvascular endothelial cells. *Invest. Ophthalmol. Vis. Sci.* **2010**, *51*, 6815–6825. [CrossRef]
20. Zhang, G.; Panigrahy, D.; Mahakian, L.M.; Yang, J.; Liu, J.-Y.; Stephen Lee, K.S.; Wettersten, H.I.; Ulu, A.; Hu, X.; Tam, S.; et al. Epoxy metabolites of docosahexaenoic acid (DHA) inhibit angiogenesis, tumor growth, and metastasis. *Proc. Natl. Acad. Sci. USA* **2013**, *110*, 6530–6535. [CrossRef]
21. Sun, C.Q.; Johnson, K.D.; Wong, H.; Foo, L.Y. Biotransformation of Flavonoid Conjugates with Fatty Acids and Evaluations of Their Functionalities. *Front. Pharmacol.* **2017**, *8*, 759. [CrossRef] [PubMed]
22. Fernando, W.; Coombs, M.R.P.; Hoskin, D.W.; Rupasinghe, H.P.V. Docosahexaenoic acid-acylated phloridzin, a novel polyphenol fatty acid ester derivative, is cytotoxic to breast cancer cells. *Carcinogenesis* **2016**, *37*, 1004–1013. [CrossRef] [PubMed]
23. Fernando, W.; Coyle, K.; Marcato, P.; Rupasinghe, H.P.V.; Hoskin, D.W. Phloridzin docosahexaenoate, a novel fatty acid ester of a plant polyphenol, inhibits mammary carcinoma cell metastasis. *Cancer Lett.* **2019**, *465*, 68–81. [CrossRef] [PubMed]
24. Nair, S.V.G.; Ziaullah Rupasinghe, H.P.V. Fatty acid esters of phloridzin induce apoptosis of human liver cancer cells through altered gene expression. *PLoS ONE* **2014**, *9*, e107149. [CrossRef] [PubMed]
25. Arumuggam, N.; Melong, N.; Too, C.K.; Berman, J.N.; Rupasinghe, H.P.V. Phloridzin docosahexaenoate, a novel flavonoid derivative, suppresses growth and induces apoptosis in T-cell acute lymphoblastic leukemia cells. *Am. J. Cancer Res.* **2017**, *7*, 2452–2464. [PubMed]
26. Cheng, H.W.; Chen, Y.F.; Wong, J.M.; Weng, C.W.; Chen, H.Y.; Yu, S.L.; Chen, H.W.; Yuan, A.; Chen, J.J.W. Cancer cells increase endothelial cell tube formation and survival by activating the PI3K/Akt signalling pathway. *J. Exp. Clin. Cancer Res.* **2017**, *36*, 27. [CrossRef] [PubMed]
27. Lamalice, L.; Le Boeuf, F.; Huot, J. Endothelial Cell Migration During Angiogenesis. *Circ. Res.* **2007**, *100*, 782–794. [CrossRef] [PubMed]
28. El Baba, N.; Farran, M.; Khalil, E.A.; Jaafar, L.; Fakhoury, I.; El-Sibai, M. The Role of Rho GTPases in VEGF Signaling in Cancer Cells. *Anal. Cell. Pathol.* **2020**, *2020*, 2097214. [CrossRef] [PubMed]
29. Shibuya, M. Vascular Endothelial Growth Factor (VEGF) and Its Receptor (VEGFR) Signaling in Angiogenesis: A Crucial Target for Anti- and Pro-Angiogenic Therapies. *Genes Cancer* **2011**, *2*, 1097–1105. [CrossRef]
30. Shiojima, I.; Walsh, K. Role of Akt signaling in vascular homeostasis and angiogenesis. *Circ. Res.* **2002**, *90*, 1243–1250. [CrossRef]
31. Pratheeshkumar, P.; Budhraja, A.; Son, Y.-O.; Wang, X.; Zhang, Z.; Ding, S.; Wang, L.; Hitron, A.; Lee, J.-C.; Xu, M.; et al. Quercetin inhibits angiogenesis mediated human prostate tumor growth by targeting VEGFR-2 regulated AKT/mTOR/P70S6K signaling pathways. *PLoS ONE* **2012**, *7*, e47516. [CrossRef]
32. Bassino, E.; Antoniotti, S.; Gasparri, F.; Munaron, L. Effects of flavonoid derivatives on human microvascular endothelial cells. *Nat. Prod. Res.* **2016**, *30*, 2831–2834. [CrossRef] [PubMed]

33. Moyle, C.W.A.; Cerezo, A.B.; Winterbone, M.S.; Hollands, W.J.; Alexeev, Y.; Needs, P.W.; Kroon, P.A. Potent inhibition of VEGFR-2 activation by tight binding of green tea epigallocatechin gallate and apple procyanidins to VEGF: Relevance to angiogenesis. *Mol. Nutr. Food Res.* **2015**, *59*, 401–412. [CrossRef] [PubMed]
34. Kim, H.J.; Vosseler, C.A.; Weber, P.C.; Erl, W. Docosahexaenoic acid induces apoptosis in proliferating human endothelial cells. *J. Cell. Physiol.* **2005**, *204*, 881–888. [CrossRef]
35. Pucci, B.; Kasten, M.; Giordano, A. Cell cycle and apoptosis. *Neoplasia* **2000**, *2*, 291–299. [CrossRef] [PubMed]
36. Pietenpol, J.A.; Stewart, Z.A. Cell cycle checkpoint signaling: Cell cycle arrest versus apoptosis. *Toxicology* **2002**, *181–182*, 475–481. [CrossRef]
37. David, M.; Petit, D.; Bertoglio, J. Cell cycle regulation of Rho signaling pathways. *Cell Cycle* **2012**, *11*, 3003–3010. [CrossRef] [PubMed]
38. Niculescu, A.B.; Chen, X.; Smeets, M.; Hengst, L.; Prives, C.; Reed, S.I. Effects of p21(Cip1/Waf1) at both the G1/S and the G2/M cell cycle transitions: pRb is a critical determinant in blocking DNA replication and in preventing endoreduplication. *Mol. Cell. Biol.* **1998**, *18*, 629–643. [CrossRef] [PubMed]
39. Van Nieuw Amerongen, G.P.; Koolwijk, P.; Versteilen, A.; Van Hinsbergh, V.W.M. Involvement of RhoA/Rho Kinase Signaling in VEGF-Induced Endothelial Cell Migration and Angiogenesis In Vitro. *Arter. Thromb. Vasc. Biol.* **2003**, *23*, 211–217. [CrossRef]
40. Fryer, B.H.; Field, J. Rho, Rac, Pak and angiogenesis: Old roles and newly identified responsibilities in endothelial cells. *Cancer Lett.* **2005**, *229*, 13–23. [CrossRef]
41. Abraham, S.; Scarcia, M.; Bagshaw, R.D.; McMahon, K.; Grant, G.; Harvey, T.; Yeo, M.; Esteves, F.O.G.; Thygesen, H.H.; Jones, P.F.; et al. A Rac/Cdc42 exchange factor complex promotes formation of lateral filopodia and blood vessel lumen morphogenesis. *Nat. Commun.* **2015**, *6*, 7286. [CrossRef] [PubMed]
42. van der Meel, R.; Symons, M.H.; Kudernatsch, R.; Kok, R.J.; Schiffelers, R.M.; Storm, G.; Gallagher, W.M.; Byrne, A.T. The VEGF/Rho GTPase signalling pathway: A promising target for anti-angiogenic/anti-invasion therapy. Drug Discov. *Today* **2011**, *16*, 219–228. [CrossRef] [PubMed]
43. Liu, Z.-J.; Xiao, M.; Balint, K.; Soma, A.; Pinnix, C.C.; Capobianco, A.J.; Velazquez, O.C.; Herlyn, M. Inhibition of endothelial cell proliferation by Notch1 signaling is mediated by repressing MAPK and PI3K/Akt pathways and requires MAML1. *FASEB J.* **2006**, *20*, 1009–1011. [CrossRef]
44. Wang, S.; Amato, K.R.; Song, W.; Youngblood, V.; Lee, K.; Boothby, M.; Brantley-Sieders, D.M.; Chen, J. Regulation of endothelial cell proliferation and vascular assembly through distinct mTORC2 signaling pathways. *Mol. Cell. Biol.* **2015**, *35*, 1299–1313. [CrossRef] [PubMed]
45. Boyd, P.J.; Doyle, J.; Gee, E.; Pallan, S.; Haas, T.L. MAPK signaling regulates endothelial cell assembly into networks and expression of MT1-MMP and MMP-2. *Am. J. Physiol. Physiol.* **2005**, *288*, C659–C668. [CrossRef] [PubMed]
46. Mavria, G.; Vercoulen, Y.; Yeo, M.; Paterson, H.; Karasarides, M.; Marais, R.; Bird, D.; Marshall, C.J. ERK-MAPK signaling opposes Rho-kinase to promote endothelial cell survival and sprouting during angiogenesis. *Cancer Cell* **2005**, *9*, 33–44. [CrossRef] [PubMed]
47. Somanath, P.R.; Razorenova, O.V.; Chen, J.; Byzova, T.V. Akt1 in endothelial cell and angiogenesis. *Cell Cycle* **2006**, *5*, 512–518. [CrossRef]
48. Ma, J.; Sawai, H.; Ochi, N.; Matsuo, Y.; Xu, D.; Yasuda, A.; Takahashi, H.; Wakasugi, T.; Takeyama, H. PTEN regulate angiogenesis through PI3K/Akt/VEGF signaling pathway in human pancreatic cancer cells. *Mol. Cell. Biochem.* **2009**, *331*, 161–171. [CrossRef]
49. Yao, L.; Romero, M.J.; Toque, H.A.; Yang, G.; Caldwell, R.B.; Caldwell, R.W. The role of RhoA/Rho kinase pathway in endothelial dysfunction. *J. Cardiovasc. Dis. Res.* **2010**, *1*, 165–170. [CrossRef]
50. Shin, S.; Jing, K.; Jeong, S.; Kim, N.; Song, K.-S.; Heo, J.-H.; Park, J.-H.; Seo, K.-S.; Han, J.; Park, J.-I.; et al. The omega-3 polyunsaturated fatty acid DHA induces simultaneous apoptosis and authphagy via mitochondrial ROS-mediated Akt-mTOR signaling in prostate cancer cells expressing mutant p53. *Biomed Res. Int.* **2013**, *2013*, 568671. [CrossRef]
51. Raza, A.; Franklin, M.J.; Dudek, A.Z. Pericytes and vessel maturation during tumor angiogenesis and metastasis. *Am. J. Hematol.* **2010**, *85*, 593–598. [CrossRef] [PubMed]
52. Fakhrejahani, E.; Toi, M. Tumor angiogenesis: Pericytes and maturation are not to be ignored. *J. Oncol.* **2012**, *2012*, 261750. [CrossRef] [PubMed]
53. Goel, S.; Duda, D.G.; Xu, L.; Munn, L.L.; Boucher, Y.; Fukumura, D.; Jain, R.K. Normalization of the vasculature for treatment of cancer and other diseases. *Physiol. Rev.* **2011**, *91*, 1071–1121. [CrossRef]
54. Tsuji-Tamura, K.; Ogawa, M. Inhibition of the PI3K-Akt and mTORC1 signaling pathways promotes the elongation of vascular endothelial cells. *J. Cell Sci.* **2016**, *129*, 1165–1178. [CrossRef]

*biomolecules*

Communication

# Emodin Inhibited Pathological Cardiac Hypertrophy in Response to Angiotensin-Induced Hypertension and Altered the Gut Microbiome

**Levi Evans [1,2], Tori Price [3], Nathaniel Hubert [3], Julia Moore [3], Yiqui Shen [1], Maheshi Athukorala [1], Steven Frese [1], Kristina Martinez-Guryn [3] and Bradley S. Ferguson [1,2,4,*]**

1 Department of Nutrition, University of Nevada, Reno, Reno, NV 89557, USA; sfrese@unr.edu (S.F.)
2 Environmental Sciences Program, University of Nevada, Reno, Reno, NV 89557, USA
3 Biomedical Sciences Program, Midwestern University, Downers Grove, IL 60515, USA
4 Center of Biomedical Research Excellence for Molecular and Cellular Signal Transduction in the Cardiovascular System, University of Nevada, Reno, Reno, NV 89557, USA
* Correspondence: bferguson@unr.edu; Tel.: +1-775-784-6278

**Abstract:** Objective: Evidence suggests that food bioactives affect the epigenome to prevent pathological cardiac hypertrophy. Recently, we showed that emodin, an anthraquinone, attenuated pathological cardiac hypertrophy and histone deacetylase (HDAC) activity. However, we only examined the cardioprotective effects of emodin's parent compound and not those of emodin metabolites or of emodin–gut microbiome interactions. The microbiome has emerged as a key player in chronic diseases such as metabolic and cardiac disease. Thus, we hypothesized that emodin could reverse hypertension-induced changes in microbial communities. Methods: Normo- and hypertensive (angiotensin II) C57/BL6 female mice were randomly assigned to receive a vehicle (Veh; DMSO:PEG 1:1) or emodin (Emod; 30 mg/kg) for 14 days. Body weights were collected pre- and post-treatment, and blood pressure was assessed via tail cuff. At the study's end, the mice were euthanized and assessed for their heart weights. In addition, stool samples and cecal contents were collected to elucidate changes in the microbial populations using 16S rRNA sequencing. Lastly, the tissue was lysed, and RNA was isolated for qPCR. One-way ANOVA with Tukey's post hoc test was performed unless otherwise specified, and $p < 0.05$ was considered significant. Results: Emodin significantly attenuated cardiac hypertrophy in the female mice. No significant changes were observed in body weight or systolic blood pressure in response to hypertension or emodin. Lastly, analysis suggests that hypertension altered the microbiome in the cecum and cecal content, with additional evidence to support that emodin affects gut microbiota in the feces and colon. Conclusions: Our data demonstrate that emodin attenuates pathological hypertrophy in female mice. Future research is needed to dissect if changes in the microbiome contributes to emodin-mediated attenuation in cardiac remodeling.

**Keywords:** emodin; microbiome; hypertension

## 1. Introduction

Hypertension is a major public health concern that contributes to the development of heart failure [1]. Hypertension impacts vascular resistance and loads the heart, which contributes to scar tissue formation (fibrosis) and heart enlargement (hypertrophy). This ultimately results in cardiac dysfunction and heart failure [1]. Technological advancements have enhanced our understanding of diet in (cardiovascular disease) CVD and health. This is evident in recent studies of the microbiome, which has emerged as a major player in CVD pathogenesis [2–4]. In fact, distinct microbial communities can be found in humans at risk of CVD [5] with dyslipidemia [2], atrial fibrillation [6], hypertension [4], and heart failure [7]. More importantly, angiotensin II-induced hypertension only increased pathological cardiac hypertrophy, fibrosis, and dysfunction in conventionally raised mice but not germ-free mice, suggesting that the microbiome is essential for hypertension-induced changes in the myocardium [8]. Consistent with this, fecal microbiota transplant (FMT) from hypertensive rats led to hypertension and dysbiosis in non-hypertensive rats [9]. Conversely, therapies that target and normalize the microbiome, similar to healthy controls, improve hypertension in rodents and humans [10–12]. As the diet plays a major role in regulating the microbiome, dietary components have the potential to regulate and maintain microbial homeostasis to prevent or ameliorate CVD and myocardial dysfunction.

According to the Global Burden of Disease study, poor-quality diets have contributed to the rise in hypertension, heart failure, and other cardiovascular disease components [13]. Indeed, diet is now considered the number one risk factor for preventable diseases in the United States and other developed nations [13]. Understanding how nutrients and other dietary components regulate cellular fate to promote health and longevity affords clinicians an opportunity to identify dietary strategies that can reduce and potentially reverse CVDs including hypertension and heart failure. This can be seen in the literature where the increased intake of fruits and vegetables is linked to cardioprotection [14,15], and in which studies identified fiber, vitamins, minerals, and phytochemicals within these foods as capable of preventing and potentially treating pathological cardiac hypertrophy, fibrosis, and dysfunction [16–18].

Phytochemicals have emerged over the last decade for the prevention and treatment of CVD [19,20]. Indeed, our group has published that emodin, found in rhubarb, buckthorn, knotweed, beans, and cabbage, can inhibit pathological cardiac remodeling [21]. Other groups have also highlighted the cardioprotective actions of emodin in the heart. For instance, Wu et al. [22] reported that emodin improved left ventricular function via increased phosphorylation of protein kinase B/glycogen synthase kinase-3 beta (pAkt/pGSK-3β) in diabetic rats with cardiomyopathy [22]. Emodin was also shown to attenuate activation of the NLR family pyrin domain containing 3 (NLRP3) inflammasome that led to decreased cardiomyocyte death in response to ischemia-reperfusion (I/R) injury [23]. As mentioned above, we reported emodin inhibited pathological cardiomyocyte hypertrophy by attenuating Zn-dependent HDACs, while others have shown improvements via activation of $NAD^+$-dependent HDACs [21,24], thus indicating that emodin can regulate the genome via epigenetic interactions.

Historically, many studies focused on the actions of the parent compound, e.g. emodin, and examined direct actions within cell culture or through intraperitoneal injections. Few studies have examined the impact of emodin on the microbiome [25], nor have these studies elucidated emodin-microbiome interactions in a model of CVD. This is an important consideration given the recent evidence highlighting the role for phytochemicals in the regulation of the gut microbiome [26]. Indeed, several recent reports have highlighted the potential for phytochemical-gut microbiome interactions that change microbial diversity, and attenuate gut inflammation and dysbiosis, which can impact cardiovascular phenotypes that include improving arterial stiffness commonly linked with aging, obesity, and hypertension [26].

As the microbiome has been implicated in modulating phytochemical efficacy [27], it is imperative to elucidate the effects of emodin in the gut. Thus, the goal of this study was to examine the effects of emodin on the microbiome of hypertensive and healthy mice. Again, we observed that emodin attenuated pathological hypertrophy in hypertensive female mice with no changes in blood pressure. We also report several novel and interesting observations in this study. First, hypertension decreased microbial diversity and altered the microbiome, in which the abundance of bacteria like *Lachnospiraceae x4554* was negatively correlated with heart size. Second, emodin decreased bacterial diversity in healthy mice three days post-ingestion yet increased the abundance of beneficial bacteria (e.g., *Bacteroides thetaiotamicron*). Third, emodin normalized the microbiome in hypertensive animals similar to the control mice and improved bacterial composition, which favored leanness (e.g., *Roseburia* or *Akkermansia*) and cardioprotection (e.g., *Akkermansia*). Lastly, the microbial composition was dependent on the route of emodin administration, in which the gavage-dosed animals displayed an increase in selective protective bacteria (e.g., *Roseburia*), while the intraperitoneally dosed mice exhibited an increase in other protective bacterial species (e.g., *Bacteroides thetaiotamicron*), suggesting that liver and bacterial metabolism of emodin affect microbial communities. While correlative, these findings suggest that the cardioprotective actions of emodin are driven, in part, through the regulation of the microbiome.

## 2. Materials and Methods

Animal experiments: All animal studies were performed in accordance with the University of Nevada Reno IACUC. Female C57BL/6 mice were purchased from Jackson Labs and arrived at nine weeks of age. The animals were housed in the Office of Animal Resources at the University of Nevada in Reno. One week post acclimation, the mice (10 weeks of age) were randomly assigned to receive a sham operation (Sham) or a subcutaneous 14-day micro-osmotic pump (Alzet, Cupertino, CA, model 1002) that contained angiotensin II (AngII, Bachem, Bubendorf, Switzerland), which was administered at 1.5 μg/kg/min. The mice that were assigned to receive AngII were further randomly assigned to receive an intraperitoneal (i.p.) injection of a vehicle control (1:1 dimethyl sulfoxide:polyethylene glycol-300 (DMSO, Pharmco-AAPER; PEG-300, Acros Organics, Waltham, MA) or emodin (SelleckChem, Houston, TX, S2295) at 30 mg/kg/day (dissolved in a 1:1 solution of DMSO:PEG-300)). The Sham group further received a vehicle control via IP injection. The 14-day experiment contained three different groups: non-hypertensive sham control (Sham); hypertension plus vehicle control (Hypert); and hypertension plus emodin at 30 mg/kg/day (Hypert + Emodin). In all, n = 8 mice/group were used in these studies. The mice were weighed and dosed with a vehicle or emodin via IP injection each day for 14 days. The mice were acclimated to a tail-cuff blood pressure system (Coda High Throughput System, Kent Scientific, Torrington, CT) for 3 days prior to systolic blood pressure measurements at the study's end. The ceca and hearts were dissected at the end of the study. The whole heart and left ventricle weights were assessed. The tibia was dissected and measured. The heart weight (HW) and left ventricle (LV) weight were normalized to the tibia length as well as the body weight. The ceca and ceca content were used for 16S rRNA sequencing (n = 3/group).

Nine-week-old C57BL/6 mice were ordered from Jackson Labs and allowed to acclimate. Ten-week old healthy mice were randomly assigned to receive the vehicle (1:1, DMSO:PEG-300) or emodin at 30 mg/kg/day via oral gavage for seven days. The mice were weighed and dosed every day for seven days. Fecal samples were collected every day for seven days. In total, n = 4/treatment group were used for the gavage studies. At the study's end, the gastrointestinal tract was dissected into individual components of the small and large intestine as well as the cecum. Tissues were immediately flash-frozen and sent to Midwestern University for DNA extraction and 16S sequencing. Daily feces were also analyzed via 16S rRNA amplicon sequencing.

Real-time quantitative polymerase chain reaction (rt-qPCR): RNA was isolated from the left ventricles of the mice from the hypertension experiments via QIAzol (Qiagen, Hilden, Germany) as previously described [21]. The RNA concentration was determined via a NanoDrop Spectrometry ND1000, and 500 ng RNA was reverse transcribed to cDNA using a Verso cDNA Synthesis Kit (ThermoFisher Scientific, Waltham, MA). The RT-qPCR was examined with Apex qPCR GREEN Master Mix (Genesee Scientific, 42–120) using primers for atrial natriuretic peptide (ANP) (GCC GGT AGA AGA TGA GGT CAT, GCT TCC TCA GTC TGC TCA CTC) and B-type natriuretic peptide (BNP) (CGC TGG GAG GTC ACT CCT AT, GCT CTG GAG ACT GGC TAG GAC TT). Gene expression was examined using the BioRad CF96X real-time instrument.

DNA extraction and 16S rRNA sequencing: The small intestine was separated into the duodenum (first 4 cm distal of the stomach), jejunum (beginning 6 cm distal of the stomach, with the next 6 cm collected as well), and ileum (8 cm proximal of the cecum, with the next 6 cm collected as well). These tissues as well as the colon and cecum were placed in o-ring tubes with silica beads along with a DNA lysis buffer and bead beaten for 2 min. Subsequently, DNA extraction was performed as previously reported [28]. The DNA was submitted to Argonne National Laboratory for 16S rRNA amplicon sequencing conducted via Illumina Mi-Seq. Quality filtering and downstream analyses were performed using previously published protocols [29]. Any sequences that were found to be unknown were blasted to determine their origins. If sequences were determined to be eukaryotic, then they were removed from the analysis. Illumina-utils software was used to demultiplex, merge paired reads, and quality filter sequence data. Minimum entropy decomposition was used to assess the quality-filtered reads. High-resolution oligotypes were generated using Shannon entropy [30]. GAST was used to assign taxonomies to oligotypes as previously described [29]. Shannon diversity analysis was performed using QIIME software. Anvi'o was used to generate heat maps displaying taxonomic differences between the diet and treatment groups. The normalized relative taxonomic abundance at the family level was compared via Kruskal–Wallis testing based on (1) diet and (2) treatment, encompassing all gut regions and also within each gut region (FDR correction $p < 0.05$) in QIIME software. In some cases, groups contained n = 2 due to samples failing the sequencing run or having too few sequences to be incorporated into the analysis. Uploaded data: https://www.ncbi.nlm.nih.gov/bioproject/PRJNA1003954 (accessed on 9 August 2023).

Statistical analysis: Statistical differences in gene expression and gross anatomy was determined via one-way ANOVA with Tukey's post-hoc analysis to compare between groups. Significance was set at $p < 0.05$. Statistical analyses for 16S sequencing are described in the previous section.

## 3. Results

### 3.1. Emodin Inhibits Pathological Cardiac Hypertrophy in Response to Hypertension

Angiotensin II (AngII) participates in the renin angiotensin system (RAS) and targets angiotensin receptors, which subsequently activates vasoconstriction and the signal transduction pathways in the heart [31]. As a result, AngII induces systemic hypertension and causes hallmarks of heart failure, including cardiac hypertrophy and fibrosis [31]. To begin our studies, we tested the postulate that emodin would attenuate hypertension-induced pathological cardiac hypertrophy. Our data demonstrate that the mice treated with AngII experienced an increase in systolic blood pressure, heart weight (HW), and left ventricular weight (LVW) compared with the sham vehicle control group (Sham) (Table 1). Moreover, the HW and LVW remained significantly increased when normalized to either the tibia length or body weight (BW) (Table 1). Consistent with this increase in cardiac hypertrophy, we further demonstrate that the hypertensive mice had increased gene expression of hypertrophy markers, natriuretic peptides ANP and BNP (Table 1). Notably, i.p. delivery of emodin (Hypert + Emodin) attenuated the HW and LVW with and without normalization to the tibia length and BW when compared with the hypertensive mice (Table 1). No changes were observed in BW or systolic blood pressure between the emodin-treated and hypertensive mice (Table 1). Emodin further attenuated the upregulation of fibrotic genes collagen 1 (Col1) and connective tissue growth factor (CTGF) (Table 1). These data suggest that emodin was cardioprotective in this hypertension model.

**Table 1.** Anthropomorphic analysis of hypertensive (hypert) mice with and without emodin. One-way ANOVA with Tukey's post hoc test was used to detect significance. [a] $p < 0.05$.

| | Hypertension Experiment | | |
|---|---|---|---|
| | **Sham** | **Hypert** | **Hypert ± Emodin** |
| **Study End BW (g)** | 21.16 ± 0.87 | 21.25 ± 0.56 | 20.09 ± 1.45 |
| **Change in BW (g)** | 2.09 ± 0.14 | 2.43 ± 0.33 | 1.19 ± 0.25 |
| **Systolic Blood Pressure (mmHg)** | 123.98 ± 9.36 | 149.32 ± 14.77 [a] | 137.59 ± 8.99 [a] |
| **Heart Weight (mg)** | 83.00 ± 1.82 | 104.25 ± 3.20 [a] | 79.40 ± 4.39 |
| **Left Ventricle Weight (mg)** | 68.00 ± 2.32 | 86.50 ± 2.33 [a] | 67.00 ± 3.69 |
| **HW (mg)/TL (mm)** | 4.76 ± 0.06 | 6.06 ± 0.14 [a] | 4.55 ± 0.23 |
| **LVW (mg)/TL (mm)** | 3.90 ± 0.13 | 5.03 ± 0.10 [a] | 3.84 ± 0.20 |
| **HW (mg)/BW (g)** | 3.92 | 4.91 [a] | 3.95 |
| **LVW (mg)/BW (g)** | 3.21 | 4.1 [a] | 3.33 |
| **ANP/18S $2^{-\Delta\Delta Ct}$** | 1.06 ± 0.25 | 4.80 ± 2.29 [a] | 2.63 ± 1.82 |
| **BNP/18S $2^{-\Delta\Delta Ct}$** | 1.06 ± 0.30 | 4.11 ± 1.05 [a] | 2.59 ± 1.16 |
| **Col1/18S $2^{-\Delta\Delta Ct}$** | 1.00 ± 0.08 | 1.60 ± 0.18 [a] | 1.11 ± 0.26 |
| **CTFG/18S $2^{-\Delta\Delta Ct}$** | 1.03 ± 0.21 | 3.87 ± 2.06 [a] | 2.12 ± 0.73 |
| **18S $2^{-\Delta\Delta Ct}$** | 1.03 ± 0.22 | 0.84 ± 0.17 | 0.85 ± 0.16 |

### *3.2. Hypertension and Pathological Cardiac Hypertrophy Are Linked to Changes in the Gut Microbiome and Normalized with Emodin*

The microbiome has been implicated in CVD development [8]. Moreover, it has been suggested that phytochemicals interact with the microbiome to protect the heart [27]. As we observed that emodin attenuated hypertension-induced cardiac hypertrophy, we next sought to examine how emodin regulated microbes within the gut of hypertensive mice. Using 16S sequencing, we report altered microbial composition in the ceca of hypertensive animals and hypertensive mice treated with emodin (Figure 1A). Moreover, we report that *Lachnospiraceae x4554* was negatively correlated with the heart weight, left ventricle (LV) weight, and LV weight normalized to the tibia length (Figure 1B). Interestingly, emodin normalized *Lachnospiraceae x4554* back to levels similar to the healthy (Sham) mice (Figure 1A). We further observed that *Akkermansia*, which has been previously associated with anti-obesity and anti-diabetes properties [32,33], was significantly increased in the emodin-treated hypertensive mice compared with the hypertensive mice in the cecum (Figure 1A) and cecal content (Figure 2A). Of note, we further observed that hypertension and emodin altered the microbial abundance of *Ruminococcaceae* in the cecal content and that *Ruminococcaceae* was negatively correlated with pathological cardiac hypertrophy (Figure 2B). Combined, these data suggest that hypertension or changes in heart size may negatively affect the microbiome and that emodin can normalize the expression of protective bacteria (e.g., *Akkermansia*).

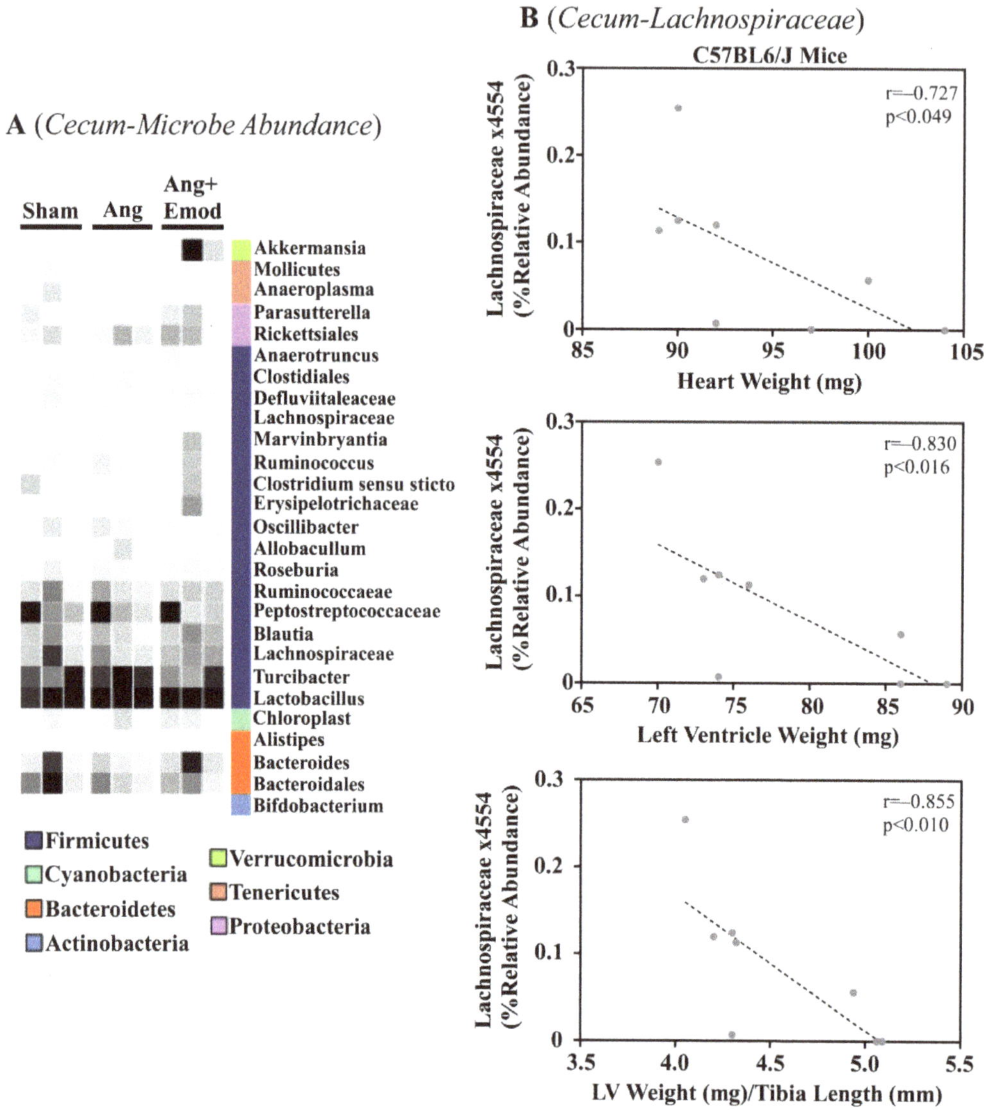


**Figure 1.** Angiotensin II-induced changes in the cecum microbiome were altered with emodin treatment. Ten-week-old C57BL6/J male mice were treated with vehicle (DMSO:PEG 200), angiotensin II (Ang) (1.5 g/kg/min), or Ang + emodin (Ang + Emod; 30 mpk). Cecal content was removed and cecum collected for 16S rRNA sequencing. (**A**) A heatmap showing changes in the microbiome and (**B**) Spearman correlation analysis performed to identify bacteria that negatively correlated with heart weight, left ventricle (LV) weight, and LV weight normalized to tibia length (LV/TL). Color intensity indicates abundance in the heatmap, with darker colors representing more abundance.

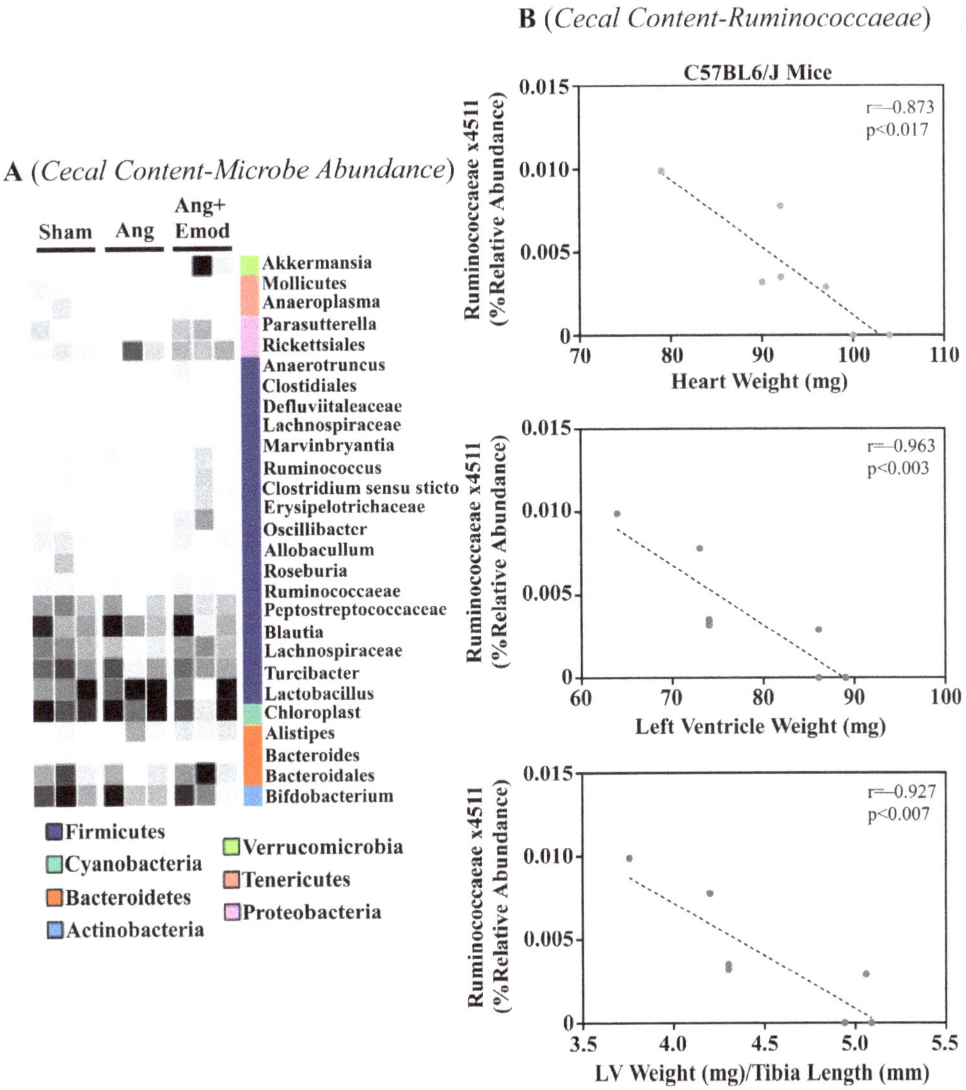


**Figure 2.** Angiotensin II-induced changes in the cecal microbiome were altered with emodin treatment. Ten-week old C57BL6/J male mice were treated with vehicle (DMSO:PEG 200), Angiotensin ii (Ang) (1.5 g/kg/min) or Ang + emodin (Ang + Emod; 30 mpk). Cecal content was scraped from the cecum and cecal content sent for 16S rRNA seqeuncing. (**A**) A heatmap showing changes in the microbiome and (**B**) Spearman correlation analysis was performed to identify bacteria that negatively correlated with heart weight, left ventricle (LV) weight and LV weight normalized to tibia length (LV/TL). Color intensity indicates abundance in the heatmap, with darker colors representing more abundance.

### *3.3. Emodin Rapidly Changes Microbial Diversity and Bacterial Composition throughoutthe Gut*

The data above demonstrate that hypertension affects the microbiome, and emodin can normalize these pathogenic actions. As a next step, we examined how quickly emodin could alter the microbiome in healthy mice by collecting fecal samples daily from mice gavaged with emodin (30 mg/kg/day) or placebo control for up to seven days (Figure 3A). The 16S sequencing showed that emodin significantly reduced microbial diversity through three days of emodin treatment (Figure 3B). Interestingly, these changes appeared to diminish by day 7, although they were still reduced (Figure 3B). This is in keeping with other findings that changes in the microbiome are dynamic and can return to homeostasis of the host [34,35].

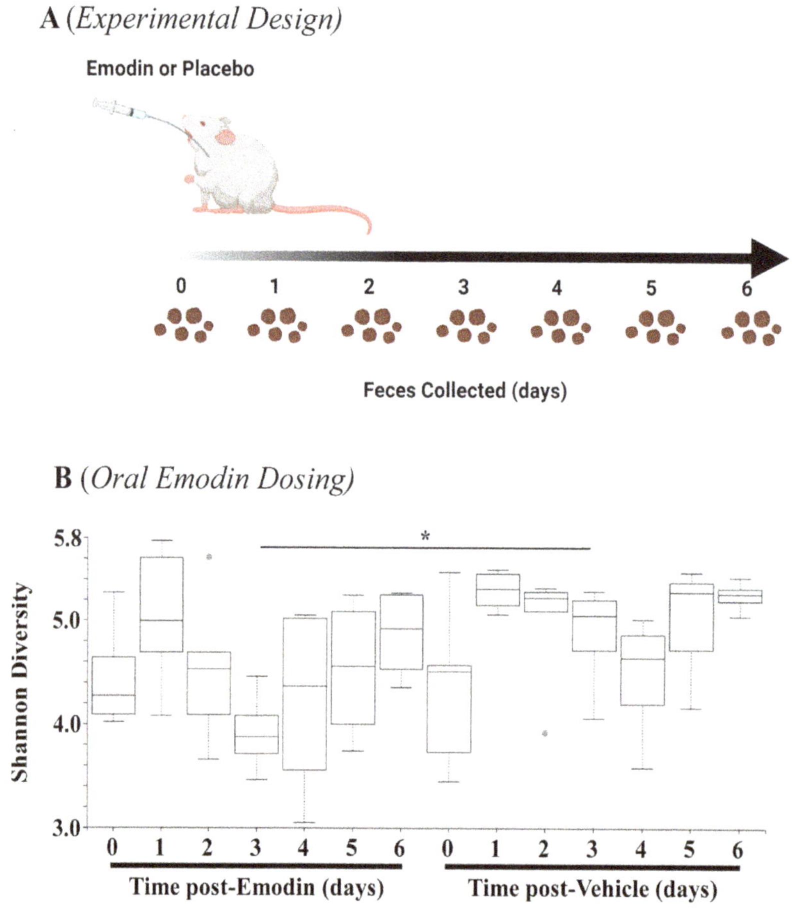


**Figure 3.** Fecal microbial diversity was attenuated with emodin. (**A**) Ten-week-old C57BL6/J mice were gavaged with vehicle control (DMSO:PEG) or emodin (30 mpk) daily with feces collected daily. (**B**) Shannon diversity analysis was performed using QIIME software for days 0–6 of the experiment comparing emodin (n = 4) vs. vehicle control (n = 5) (day 3, H = 4.9, * $p$ = 0.0275).

While the fecal samples (Figure 3B), ceca (Figure 1), and cecal content (Figure 2) suggest that emodin alters the microbiome, it does not show if these changes are localized or take place throughout the digestive tract. As such, we investigated the effects of emodin on microbial communities throughout the entire GI tract. We report that emodin reduced the abundances of select taxa in the duodenum, jejunum, ileum, cecum, cecum content, colon, and feces after seven days of dosing (Figure 4). For example, *Lactobacillus* was enriched throughout the entire GI tract and in the feces of the vehicle-treated mice but not in the emodin-treated mice (Figure 4). Furthermore, emodin reduced *Peptostreptococcaceae* in the duodenum, ileum, jejunum, and in the colon (Figure 4). Notably, an increased abundance of

*Peptostreptococcaceae* has been observed in diet-induced obese mice in the jejunum [29]. Not all of these changes were generalizable across the gut where, for instance, *Ruminococcaceae* was enriched in the colons of the vehicle-treated mice but not the emodin-treated mice (Figure 4). These data demonstrate that emodin can rapidly change the microbiome and that these anti-microbial actions can occur across the GI tract as well as within selective regions of the gut. While emodin decreased the abundance of some microbes, enrichment was also observed. For example, *Allobaculum* was enriched in the jejunum and feces through day 3 for the mice supplemented with emodin. Most notably, emodin increased the taxonomical abundance of *Akkermansia* in the colon and feces (Figures 3B and 4). This also occurred in the ceca of the hypertensive emodin-treated mice above (Figure 1). Additionally, *Roseburia* was enriched by emodin in the jejunum, ileum, and colon (Figure 3). Similar to *Akkermansia*, *Roseburia* has also been implicated in obesity prevention and heart health, and this has been attributed to increasing short-chain fatty acid production [36].

(*Microbe Abundance*)

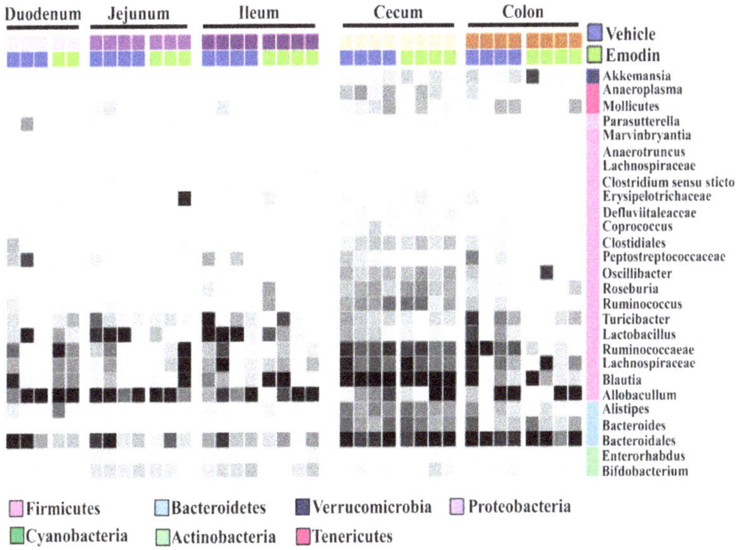


**Figure 4.** Emodin affected the microbiome of the ceca and colons in mice. 10-week old C57BL6/J mice were gavaged with vehicle control (DMSO:PEG) or emodin (30 mpk) daily and fecal content collected daily. At the study end (day 7) the duodenum, jejunum, ileum, cecum and colon were collected and 16S rRNA sequencing used to examine microbial changes. A heatmap was used to visualize changes across the gut and in response to emodin treatment. Color intensity indicates abundance in the heatmap, with darker colors representing more abundance.

## 4. Discussion

In this report, we showed that emodin attenuated hypertension-induced cardiac hypertrophy without changing systolic blood pressure or body weight in female mice. This is consistent with a reduction in ANP and BNP, which are two biomarkers of heart failure [37]. In addition, we report novel findings that (1) emodin increased the abundance of cardioprotective bacteria (e.g., *Akkermansia*) in hypertensive mice, and (2) emodin rapidly (within 3 days) decreased microbial diversity while increasing the abundance of cardioprotective bacteria, specifically in the colon (e.g., *Akkermansia*), as well as health-promoting bacteria (e.g., *Roseburia*) throughout the gut. These data show that microbial changes occur rapidly with dietary intervention and that emodin can improve the bacterial abundance of known cardioprotective species in both healthy and hypertensive mice.

Cross-sectional human studies show a relationship between hypertension and the gut microbiome in humans [38–40]. Indeed, patients with higher blood pressure or hypertension have a higher abundance of gram-negative bacteria such as Klebsiella, Parabacteroides, and Prevotella [38–40]. Moreover, Klebsiella and Streptococcaceae were observed to be positively correlated in patients with high blood pressure [40]. Consistent with this, germ-free (GF) mice dosed with fecal matter from hypertensive donors developed high systolic and diastolic blood pressure compared to GF counterparts with normotensive FMT [3,41]. Lastly, dysbiosis has been described in several hypertensive rodent models, including the angiotensin ii-infused model similar to what we employed in our study design [9,10,42]. Given this, it isn't surprising that we observed changes in gut bacteria in hypertensive mice, which were either reversed or partially restored with emodin treatment.

*Akkermansia* is a commensal microorganism that has gained much attention; this microbe is negatively correlated with obesity and type 2 diabetes [43]. In fact, probiotic supplementation of *Akkermansia* attenuated obesity and diabetes in rodents and humans [32,33]. As obesity and diabetes both increase the risk of developing CVDs [1], *Akkermansia* is likely to protect the heart by ameliorating metabolic disease. Additionally, *Akkermansia* also acts as an anti-inflammatory microbe; inflammation is a common driver of CVD and heart dysfunction [44,45]. Consistent with this, *Akkermansia* supplementation reduced inflammation-induced damage and atherosclerosis in the hearts of Apolipoprotein E deficient mice fed a Western Diet [46]. It should also be noted that rhubarb, a plant rich in emodin, also increased *Akkermansia* abundance and this was linked to inhibition of metabolic disease [47]. Here, we report that emodin enriched *Akkermansia*, independent of the route of administration (i.e., gavage vs. i.p.), and this increase was linked to attenuation of hypertension-induced cardiac hypertrophy. Combined, our data suggest that emodin acts as a prebiotic of *Akkermansia*. Future studies that look at the cardioprotective actions of *Akkermansia* as a probiotic in our hypertension model would be of interest, as would in vitro analysis to determine if emodin directly improves *Akkermansia* growth.

Similar to *Akkermansia*, *Roseburia* is a microbe that has been linked to health promotion, and a diet rich in complex carbohydrates drives the increase in *Roseburia* abundance [36]. Of interest, *Roseburia* is a short chain fatty acid producing bacteria located in the intestine [48]. Short-chain fatty acids like butyrate have been reported to inhibit histone deacetylase (HDAC) enzymes; butyrate also attenuates pathological cardiac remodeling and improves cardiac function via HDAC inhibition [49,50]. Specifically, orally supplemented sodium butyrate attenuated high-fat diet-induced dysmetabolic symptoms including cardiac dysfunction, hypertrophy, fibrosis and apoptosis in mice [50]. Similar results were reported with streptozotocin-induced diabetic cardiomyopathy [49]. Both studies showed that butyrate inhibited HDAC activity in the heart [49,50]. *Roseburia* supplementation was also shown to decrease atherosclerotic lesions in germ-free mice, improve gut permeability and reduce inflammation compared to control counterparts [51]. Just as interesting, *Roseburia* inoculation increased histone H3 acetylation and gene expression in the gut [51]. However, the authors reported no change in epigenetic marks in the aortas [51]. These data suggest that the butyrate-producing microbe *Roseburia* was able to regulate local but not systemic epigenetic marks. Consistent with these HDAC inhibitory actions, our previous work showed that emodin reduced HDAC activity in the hearts of mice [21]. Moreover, in this report we showed that emodin increased *Roseburia* abundance, suggesting that emodin-mediated HDAC inhibition may, in part, be due to *Roseburia*-induced short chain fatty acids in addition to emodin inhibitory actions. It would be interesting to assess HDAC activity and pathological cardiac remodeling in germ-free mice hypertensive mice treated with emodin and with and without *Roseburia* inoculation.

Unlike the other bacteria discussed above, *Peptostreptococcaceae* was more abundant in mice treated with vehicle control than those gavaged with emodin. The role for *Peptostreptococcaceae* in human health remains unclear. For example, individuals with the autoimmune condition, primary immune thrombocytopenia, had lower levels of *Peptostreptococcaceae* compared to healthy controls [52]. Further, *Peptostreptococcaceae* correlated with circulating

N-acyl-ethanolamines and 2-monoacyl-glycerols after omega fatty acid consumption [53]. These two reports would make it seem as if *Peptostreptococcaceae* may confer beneficial outcomes. However, others have shown that *Peptostreptococcaceae* is enriched with high fat diet [54,55], specifically in the jejunum [29]. In fact, *Peptostreptococcaceae* was reduced in ApoE knockout mice that consumed a high fat diet treated with the cholesterol-lowering agent, α-Cyclodextrin [55], suggesting that lowering *Peptostreptococcaceae* is beneficial in this model. Here, we report that emodin reduced *Peptostreptococcaceae* abundance and is cardioprotective, consistent with findings from high fat diet studies. These mixed reports demonstrate that further studies on *Peptostreptococcaceae* in heart health are needed.

Our data also show a strong negative correlation between cardiac remodeling and the relative abundance of *Lachnospiraceae x4554*. The Lachnospiraceae family of microbes, which includes the *Roseburia* and *Ruminococcaceae* genera, is anaerobic and belongs to the Fermicutes phylum [56]. The role of *Lachnospiraceae* in health and disease, including metabolic dysfunction and diseases of the liver, kidney, and gut, was recently reviewed [56]. Discussed therein were several reports that found Lachnospiraceae genera, particularly *Blautia*, were associated with the aforementioned diseases [57–60]. For example, *Blautia* was enriched in rats with chronic kidney disease induced via a 5/6 nephrectomy [57] which, similar to the angiotensin II model we used, increased cardiac remodeling through hypertension. Indeed, Feng et al. associated the Lachnospiraceae genus *Blautia* with plasma metabolites of disease, including those of cardiovascular disease (e.g., trimethylamine N-oxide (TMAO)), and nephritis [57]. However, at least 58 genera and unclassified strains make up the Lachnospiraceae family of microbes [56], including the cardioprotective bug discussed above, *Roseburia*, and a bug we found to negatively correlate with pathological heart enlargement: *Lachnospiraceae x4554*. Finally, the Lachnospiraceae family of microbes contributes to the production of short-chain fatty acids such as the above characterized HDAC inhibitor butyrate [56]. Therefore, these data and ours suggest that Lachnospiraceae can be both protective and deleterious in the heart, depending on the strain.

## 5. Conclusions

In conclusion, emodin conferred cardioprotection in a model of hypertension and these cardioprotective benefits were positively linked with Akkermansia, Rosuburia, and Allobaculum abundance; and negatively associated with Peptostreptococcaceae enrichment. Moreover, emodin reduced microbiota diversity, which is consistent with reports of other phytochemicals that have been shown to have anti-microbial properties [61]. Lastly, consistent with previous reports, we show that hypertension markedly changed the microbiome, which was partially restored with emodin treatment. While our data suggest that the microbiome plays a role in the anti-hypertrophic actions of emodin, this report is correlative and further research involving work with germ-free animals, FMT and in vitro assays are needed to define the bacteria important for emodin-mediated actions in the heart.

**Author Contributions:** Conceptualization, B.S.F. and L.E.; methodology, L.E., T.P., N.H., J.M., Y.S. and M.A.; software, N.H.; formal analysis, L.E., K.M.-G., S.F., N.H. and B.S.F.; data curation, N.H., K.M.-G., S.F. and L.E.; writing—original draft preparation, L.E., K.M.-G., T.P. and B.S.F.; writing—review and editing, B.S.F., S.F. and K.M.-G.; funding acquisition, B.S.F. and K.M.-G. All authors have read and agreed to the published version of the manuscript.

**Funding:** This research was funded by the USDA NIFA (Hatch-NEV00767), the Dennis Meiss & Janet Ralston Fund for Nutri-epigenetic Research, the National Institute for General Medical Sciences (NIGMS) of the NIH (P20 GM130459), the National Heart, Lung, and Blood Institute of the NIH (R15 HL143496), National Institute of Aging of the NIH (R21 AG077248) and NSF EPSCOR Track II (OIA-1826801) to B.S.F and National Institute of Diabetes and Digestive and Kidney Diseases of the NIH (R21 DK129890-02) to K.M.G.

**Institutional Review Board Statement:** Approval protocol #20-06-1019-1; University of Nevada, Reno.

**Informed Consent Statement:** Not applicable.

**Data Availability Statement:** Data will be made freely available. See link in methods for Microbiome data.

**Conflicts of Interest:** The authors declare no conflict of interest.

## Abbreviations

HDAC (histone deacetylase); Veh (Vehicle); Emod (Emodin); CVD (Cardiovascular Disease); Akt/GSK-3β (Protein kinase B/glycogen synthase kinase 3 beta); NLRP3 (NLR family pyrin domain containing 3); I/R (Ischemia/Reperfusion); AngII (angiotensin ii); HW (heart weight); LV (left ventricle); BW (body weight); LVW (left ventricular weight); LV/TL (left ventricle/tibia length); ANP (atrial natriuretic peptide); BNP (brain natriuretic peptide); RAS (renin angiotensin system); Col1 (collagen 1); CTGF (connective tissue growth factor); TMAO (trimethylamine N-oxide); GF (germ-free); FMT (fecal matter transplant)

## References


1. Virani, S.S.; Alonso, A.; Benjamin, E.J.; Bittencourt, M.S.; Callaway, C.W.; Carson, A.P.; Chamberlain, A.M.; Chang, A.R.; Cheng, S.; Delling, F.N.; et al. Heart disease and stroke statistics—2020 update: A report from the American Heart Association. *Circulation* **2020**, *141*, e139–e596. [CrossRef] [PubMed]
2. Fu, J.; Bonder, M.J.; Cenit, M.C.; Tigchelaar, E.F.; Maatman, A.; Dekens, J.A.M.; Brandsma, E.; Marczynska, J.; Imhann, F.; Weersma, R.K.; et al. The gut microbiome contributes to a substantial proportion of the variation in blood lipids. *Circ. Res.* **2015**, *117*, 817–824. [CrossRef] [PubMed]
3. Jie, Z.; Xia, H.; Zhong, S.L.; Feng, Q.; Li, S.; Liang, S.; Zhong, H.; Liu, Z.; Gao, Y.; Zhao, H.; et al. The gut microbiome in atherosclerotic cardiovascular disease. *Nat. Commun.* **2017**, *8*, 845. [CrossRef]
4. Yan, Q.; Gu, Y.; Li, X.; Yang, W.; Jia, L.; Chen, C.; Han, X.; Huang, Y.; Zhao, L.; Li, P.; et al. Alterations of the gut microbiome in hypertension. *Front. Cell. Infect. Microbiol.* **2017**, *7*, 381. [CrossRef]
5. Kelly, T.N.; Bazzano, L.A.; Ajami, N.J.; He, H.; Zhao, J.; Petrosino, J.F.; Correa, A.; He, J. Gut Microbiome Associates with Lifetime Cardiovascular Disease Risk Profile among Bogalusa Heart Study Participants. *Circ. Res.* **2016**, *119*, 956–964. [CrossRef] [PubMed]
6. Zuo, K.; Li, J.; Wang, P.; Liu, Y.; Liu, Z.; Yin, X.; Liu, X.; Yang, X. Duration of Persistent Atrial Fibrillation Is Associated with Alterations in Human Gut Microbiota and Metabolic Phenotypes. *mSystems* **2019**, *4*, e00422-19. [CrossRef] [PubMed]
7. Hayashi, T.; Yamashita, T.; Watanabe, H.; Kami, K.; Yoshida, N.; Tabata, T.; Emoto, T.; Sasaki, N.; Mizoguchi, T.; Irino, Y.; et al. Gut microbiome and plasma microbiome-related metabolites in patients with decompensated and compensated heart failure. *Circ. J.* **2019**, *83*, 182–192. [CrossRef]
8. Karbach, S.H.; Schönfelder, T.; Brandão, I.; Wilms, E.; Hörmann, N.; Jäckel, S.; Schüler, R.; Finger, S.; Knorr, M.; Lagrange, J.; et al. Gut Microbiota Promote Angiotensin II-Induced Arterial Hypertension and Vascular Dysfunction. *J. Am. Heart Assoc.* **2016**, *5*, e003698. [CrossRef]
9. Adnan, S.; Nelson, J.W.; Ajami, N.J.; Venna, V.R.; Petrosino, J.F.; Bryan, R.M.; Durgan, D.J. Alterations in the gut microbiota can elicit hypertension in rats. *Physiol. Genomics* **2017**, *49*, 96–104. [CrossRef]
10. Yang, T.; Santisteban, M.M.; Rodriguez, V.; Li, E.; Ahmari, N.; Carvajal, J.M.; Zadeh, M.; Gong, M.; Qi, Y.; Zubcevic, J.; et al. Gut Dysbiosis is Linked to Hypertension. *Hypertension* **2015**, *65*, 1331–1340. [CrossRef]
11. Galla, S.; Chakraborty, S.; Cheng, X.; Yeo, J.; Mell, B.; Zhang, H.; Mathew, A.V.; Vijay-Kumar, M.; Joe, B. Disparate effects of antibiotics on hypertension. *Physiol. Genomics* **2018**, *50*, 837–845. [CrossRef]
12. Huang, R.; Wang, K.; Hu, J. Effect of probiotics on depression: A systematic review and meta-analysis of randomized controlled trials. *Nutrients* **2016**, *8*, 483. [CrossRef] [PubMed]
13. Afshin, A.; Sur, P.J.; Fay, K.A.; Cornaby, L.; Ferrara, G.; Salama, J.S.; Mullany, E.C.; Abate, K.H.; Abbafati, C.; Abebe, Z.; et al. Health effects of dietary risks in 195 countries, 1990–2017: A systematic analysis for the Global Burden of Disease Study 2017. *Lancet* **2019**, *393*, 1958–1972. [CrossRef]
14. Wang, X.; Ouyang, Y.; Liu, J.; Zhu, M.; Zhao, G.; Bao, W.; Hu, F.B. Fruit and vegetable consumption and mortality from all causes, cardiovascular disease, and cancer: Systematic review and dose-response meta-analysis of prospective cohort studies. *BMJ* **2014**, *349*, g4490. [CrossRef] [PubMed]
15. Aune, D.; Giovannucci, E.; Boffetta, P.; Fadnes, L.T.; Keum, N.N.; Norat, T.; Greenwood, D.C.; Riboli, E.; Vatten, L.J.; Tonstad, S. Fruit and vegetable intake and the risk of cardiovascular disease, total cancer and all-cause mortality-A systematic review and dose-response meta-analysis of prospective studies. *Int. J. Epidemiol.* **2017**, *46*, 1029–1056. [CrossRef] [PubMed]
16. Levitan, E.B.; Lewis, C.E.; Tinker, L.F.; Eaton, C.B.; Ahmed, A.; Manson, J.E.; Snetselaar, L.G.; Martin, L.W.; Trevisan, M.; Howard, B.V.; et al. Mediterranean and DASH diet scores and mortality in women with heart failure: The Women's Health Initiative. *Circ. Heart Fail.* **2013**, *6*, 1116–1123. [CrossRef]

17. Pounis, G.; Costanzo, S.; Di Giuseppe, R.; De Lucia, F.; Santimone, I.; Sciarretta, A.; Barisciano, P.; Persichillo, M.; De Curtis, A.; Zito, F.; et al. Consumption of healthy foods at different content of antioxidant vitamins and phytochemicals and metabolic risk factors for cardiovascular disease in men and women of the Moli-sani study. *Eur. J. Clin. Nutr.* **2013**, *67*, 207–213. [CrossRef]
18. Tripp, M.L.; Dahlberg, C.J.; Eliason, S.; Lamb, J.J.; Ou, J.J.; Gao, W.; Bhandari, J.; Graham, D.; Dudleenamjil, E.; Babish, J.G. A Low-Glycemic, Mediterranean Diet and Lifestyle Modification Program with Targeted Nutraceuticals Reduces Body Weight, Improves Cardiometabolic Variables and Longevity Biomarkers in Overweight Subjects: A 13-Week Observational Trial. *J. Med. Food* **2019**, *22*, 479–489. [CrossRef]
19. Islam, M.A.; Alam, F.; Solayman, M.; Khalil, M.I.; Kamal, M.A.; Gan, S.H. Dietary Phytochemicals: Natural Swords Combating Inflammation and Oxidation-Mediated Degenerative Diseases. *Oxid. Med. Cell. Longev.* **2016**, *2016*, 5137431. [CrossRef]
20. Komici, K.; Conti, V.; Davinelli, S.; Bencivenga, L.; Rengo, G.; Filippelli, A.; Ferrara, N.; Corbi, G. Cardioprotective Effects of Dietary Phytochemicals on Oxidative Stress in Heart Failure by a Sex-Gender-Oriented Point of View. *Oxid. Med. Cell. Longev.* **2020**, *2020*, 2176728. [CrossRef]
21. Evans, L.W.; Bender, A.; Burnett, L.; Godoy, L.; Shen, Y.; Staten, D.; Zhou, T.; Angermann, J.E.; Ferguson, B.S. Emodin and emodin-rich rhubarb inhibits histone deacetylase (HDAC) activity and cardiac myocyte hypertrophy. *J. Nutr. Biochem.* **2020**, *79*, 108339. [CrossRef] [PubMed]
22. Wu, Z.; Chen, Q.; Ke, D.; Li, G.; Deng, W. Emodin protects against diabetic cardiomyopathy by regulating the AKT/GSK-3β Signaling pathway in the Rat Model. *Molecules* **2014**, *19*, 14782–14793. [CrossRef] [PubMed]
23. Ye, B.; Chen, X.; Dai, S.; Han, J.; Liang, X.; Lin, S.; Cai, X.; Huang, Z.; Huang, W. Emodin alleviates myocardial ischemia/reperfusion injury by inhibiting gasdermin D-mediated pyroptosis in cardiomyocytes. *Drug Des. Devel. Ther.* **2019**, *13*, 975–990. [CrossRef] [PubMed]
24. Gao, J.; Zhang, K.; Wang, Y.; Guo, R.; Liu, H.; Jia, C.; Sun, X.; Wu, C.; Wang, W.; Du, J.; et al. A machine learning-driven study indicates emodin improves cardiac hypertrophy by modulation of mitochondrial SIRT3 signaling. *Pharmacol. Res.* **2020**, *155*, 104739. [CrossRef] [PubMed]
25. Zeng, Y.Q.; Dai, Z.; Lu, F.; Lu, Z.; Liu, X.; Chen, C.; Qu, P.; Li, D.; Hua, Z.; Qu, Y.; et al. Emodin via colonic irrigation modulates gut microbiota and reduces uremic toxins in rats with chronic kidney disease. *Oncotarget* **2016**, *7*, 17468. [CrossRef]
26. De Bruyne, T.; Steenput, B.; Roth, L.; De Meyer, G.R.Y.; Dos Santos, C.N.; Valentová, K.; Dambrova, M.; Hermans, N. Dietary polyphenols targeting arterial stiffness: Interplay of contributing mechanisms and gut microbiome-related Metabolism. *Nutrients* **2019**, *11*, 578. [CrossRef]
27. Evans, L.W.; Athukorala, M.; Martinez-Guryn, K.; Ferguson, B.S. The role of histone acetylation and the microbiome in phytochemical efficacy for cardiovascular diseases. *Int. J. Mol. Sci.* **2020**, *21*, 4006. [CrossRef]
28. Wang, Y.; Hoenig, J.D.; Malin, K.J.; Qamar, S.; Petrof, E.O.; Sun, J.; Antonopoulos, D.A.; Chang, E.B.; Claud, E.C. 16S rRNA gene-based analysis of fecal microbiota from preterm infants with and without necrotizing enterocolitis. *ISME J.* **2009**, *3*, 944–954. [CrossRef]
29. Martinez-Guryn, K.; Hubert, N.; Frazier, K.; Urlass, S.; Musch, M.W.; Ojeda, P.; Pierre, J.F.; Miyoshi, J.; Sontag, T.J.; Cham, C.M.; et al. Small Intestine Microbiota Regulate Host Digestive and Absorptive Adaptive Responses to Dietary Lipids. *Cell Host Microbe* **2018**, *23*, 458–469.e5. [CrossRef]
30. Eren, A.M.; Maignien, L.; Sul, W.J.; Murphy, L.G.; Grim, S.L.; Morrison, H.G.; Sogin, M.L. Oligotyping: Differentiating between closely related microbial taxa using 16S rRNA gene data. *Methods Ecol. Evol.* **2013**, *4*, 1111–1119. [CrossRef]
31. Tham, Y.K.; Bernardo, B.C.; Ooi, J.Y.Y.; Weeks, K.L.; McMullen, J.R. Pathophysiology of cardiac hypertrophy and heart failure: Signaling pathways and novel therapeutic targets. *Arch. Toxicol.* **2015**, *89*, 1401–1438. [CrossRef]
32. Plovier, H.; Everard, A.; Druart, C.; Depommier, C.; Van Hul, M.; Geurts, L.; Chilloux, J.; Ottman, N.; Duparc, T.; Lichtenstein, L.; et al. A purified membrane protein from Akkermansia muciniphila or the pasteurized basscterium improves metabolism in obese and diabetic mice. *Nat. Med.* **2017**, *23*, 107–113. [CrossRef] [PubMed]
33. Depommier, C.; Everard, A.; Druart, C.; Plovier, H.; Van Hul, M.; Vieira-Silva, S.; Falony, G.; Raes, J.; Maiter, D.; Delzenne, N.M.; et al. Supplementation with Akkermansia muciniphila in overweight and obese human volunteers: A proof-of-concept exploratory study. *Nat. Med.* **2019**, *25*, 1096–1103. [CrossRef] [PubMed]
34. David, L.A.; Maurice, C.F.; Carmody, R.N.; Gootenberg, D.B.; Button, J.E.; Wolfe, B.E.; Ling, A.V.; Devlin, A.S.; Varma, Y.; Fischbach, M.A.; et al. Diet rapidly and reproducibly alters the human gut microbiome. *Nature* **2014**, *505*, 559–563. [CrossRef]
35. De Meij, T.G.J.; Budding, A.E.; De Groot, E.F.J.; Jansen, F.M.; Kneepkens, C.M.F.; Benninga, M.A.; Penders, J.; Van Bodegraven, A.A.; Savelkoul, P.H.M. Composition and stability of intestinal microbiota of healthy children within a Dutch population. *FASEB J.* **2016**, *30*, 1512–1522. [CrossRef] [PubMed]
36. Tamanai-Shacoori, Z.; Smida, I.; Bousarghin, L.; Loreal, O.; Meuric, V.; Fong, S.B.; Bonnaure-Mallet, M.; Jolivet-Gougeon, A. Roseburia spp.: A marker of health? *Future Microbiol.* **2017**, *12*, 157–170. [CrossRef]
37. Koshman, Y.E.; Patel, N.; Chu, M.; Iyengar, R.; Kim, T.; Ersahin, C.; Lewis, W.; Heroux, A.; Samarel, A.M. Regulation of connective tissue growth factor gene expression and fibrosis in human heart failure. *J. Card. Fail.* **2013**, *19*, 283–294. [CrossRef] [PubMed]
38. Dan, X.; Mushi, Z.; Baili, W.; Han, L.; Enqi, W.; Huanhu, Z.; Shuchun, L. Differential analysis of hypertension-associated intestinal microbiota. *Int. J. Med. Sci.* **2021**, *18*, 3748. [CrossRef]
39. Sun, S.; Lulla, A.; Sioda, M.; Winglee, K.; Wu, M.C.; Jacobs, D.R.; Shikany, J.M.; Lloyd-Jones, D.M.; Launer, L.J.; Fodor, A.A.; et al. Gut microbiota composition and blood pressure: The CARDIA study. *Hypertension* **2019**, *73*, 998–1006. [CrossRef]

40. Verhaar, B.; Collard, D.; Prodan, A.; Levels, J.H.; Zwinderman, A.; Snijder, M.; Vogt, L.; Peters, M.J.; Muller, M.; Nieuwdorp, M.; et al. Associations between gut microbiome, short chain fatty acids and blood pressure across ethnic groups: The HELIUS study. *Eur. Heart J.* **2020**, *41*, ehaa946.2701. [CrossRef]
41. Li, J.; Zhao, F.; Wang, Y.; Chen, J.; Tao, J.; Tian, G.; Wu, S.; Liu, W.; Cui, Q.; Geng, B.; et al. Gut microbiota dysbiosis contributes to the development of hypertension. *Microbiome* **2017**, *5*, 14. [CrossRef] [PubMed]
42. Mell, B.; Jala, V.R.; Mathew, A.V.; Byun, J.; Waghulde, H.; Zhang, Y.; Haribabu, B.; Vijay-Kumar, M.; Pennathur, S.; Joe, B. Evidence for a link between gut microbiota and hypertension in the Dahl rat. *Physiol. Genomics* **2015**, *47*, 187–197. [CrossRef] [PubMed]
43. Dao, M.C.; Everard, A.; Aron-Wisnewsky, J.; Sokolovska, N.; Prifti, E.; Verger, E.O.; Kayser, B.D.; Levenez, F.; Chilloux, J.; Hoyles, L.; et al. Akkermansia muciniphila and improved metabolic health during a dietary intervention in obesity: Relationship with gut microbiome richness and ecology. *Gut* **2016**, *65*, 426–436. [CrossRef] [PubMed]
44. Everard, A.; Belzer, C.; Geurts, L.; Ouwerkerk, J.P.; Druart, C.; Bindels, L.B.; Guiot, Y.; Derrien, M.; Muccioli, G.G.; Delzenne, N.M.; et al. Cross-talk between Akkermansia muciniphila and intestinal epithelium controls diet-induced obesity. *Proc. Natl. Acad. Sci. USA* **2013**, *110*, 9066–9071. [CrossRef]
45. Willerson, J.T.; Ridker, P.M. Inflammation as a cardiovascular risk factor. *Circulation* **2004**, *109*, II-2–II-10. [CrossRef]
46. Li, J.; Lin, S.; Vanhoutte, P.M.; Woo, C.W.; Xu, A. Akkermansia muciniphila protects against atherosclerosis by preventing metabolic endotoxemia-induced inflammation in Apoe-/- Mice. *Circulation* **2016**, *133*, 2434–2446. [CrossRef]
47. Régnier, M.; Rastelli, M.; Morissette, A.; Suriano, F.; Le Roy, T.; Pilon, G.; Delzenne, N.M.; Marette, A.; Van Hul, M.; Cani, P.D. Rhubarb Supplementation Prevents Diet-Induced Obesity and Diabetes in Association with Increased Akkermansia muciniphila in Mice. *Nutrients* **2020**, *12*, 2932. [CrossRef]
48. Louis, P.; Flint, H.J. Diversity, metabolism and microbial ecology of butyrate-producing bacteria from the human large intestine. *FEMS Microbiol. Lett.* **2009**, *294*, 1–8. [CrossRef]
49. Chen, Y.; Du, J.; Zhao, Y.T.; Zhang, L.; Lv, G.; Zhuang, S.; Qin, G.; Zhao, T.C. Histone deacetylase (HDAC) inhibition improves myocardial function and prevents cardiac remodeling in diabetic mice. *Cardiovasc. Diabetol.* **2015**, *14*, 99. [CrossRef]
50. Zhang, L.; Du, J.; Yano, N.; Wang, H.; Zhao, Y.T.; Dubielecka, P.M.; Zhuang, S.; Chin, Y.E.; Qin, G.; Zhao, T.C. Sodium Butyrate Protects Against High Fat Diet-Induced Cardiac Dysfunction and Metabolic Disorders in Type II Diabetic Mice. *J. Cell. Biochem.* **2017**, *118*, 2395–2408. [CrossRef]
51. Kasahara, K.; Krautkramer, K.A.; Org, E.; Romano, K.A.; Kerby, R.L.; Vivas, E.I.; Mehrabian, M.; Denu, J.M.; Bäckhed, F.; Lusis, A.J.; et al. Interactions between Roseburia intestinalis and diet modulate atherogenesis in a murine model. *Nat. Microbiol.* **2018**, *3*, 1461–1471. [CrossRef]
52. Liu, C.; Cheng, L.; Ji, L.; Li, F.; Zhan, Y.; Wu, B.; Ke, Y.; Chen, P.; Hua, F.; Yuan, L.; et al. Intestinal microbiota dysbiosis play a role in pathogenesis of patients with primary immune thrombocytopenia. *Thromb. Res.* **2020**, *190*, 11–19. [CrossRef] [PubMed]
53. Castonguay-Paradis, S.; Lacroix, S.; Rochefort, G.; Parent, L.; Perron, J.; Martin, C.; Lamarche, B.; Raymond, F.; Flamand, N.; Di Marzo, V.; et al. Dietary fatty acid intake and gut microbiota determine circulating endocannabinoidome signaling beyond the effect of body fat. *Sci. Rep.* **2020**, *10*, 15975. [CrossRef] [PubMed]
54. Liu, D.; Huang, J.; Luo, Y.; Wen, B.; Wu, W.; Zeng, H.; Zhonghua, L. Fuzhuan Brick Tea Attenuates High-Fat Diet-Induced Obesity and Associated Metabolic Disorders by Shaping Gut Microbiota. *J. Agric. Food Chem.* **2019**, *67*, 13589–13604. [CrossRef]
55. Sakurai, T.; Sakurai, A.; Chen, Y.; Vaisman, B.L.; Amar, M.J.; Pryor, M.; Thacker, S.G.; Zhang, X.; Wang, X.; Zhang, Y.; et al. Dietary α-cyclodextrin reduces atherosclerosis and modifies gut flora in apolipoprotein E-deficient mice. *Mol. Nutr. Food Res.* **2017**, *61*, 1600804. [CrossRef] [PubMed]
56. Vacca, M.; Celano, G.; Calabrese, F.M.; Portincasa, P.; Gobbetti, M.; De Angelis, M. The controversial role of human gut lachnospiraceae. *Microorganisms* **2020**, *8*, 573. [CrossRef] [PubMed]
57. Feng, Y.L.; Cao, G.; Chen, D.Q.; Vaziri, N.D.; Chen, L.; Zhang, J.; Wang, M.; Guo, Y.; Zhao, Y.Y. Microbiome–metabolomics reveals gut microbiota associated with glycine-conjugated metabolites and polyamine metabolism in chronic kidney disease. *Cell. Mol. Life Sci.* **2019**, *76*, 4961–4978. [CrossRef]
58. Lippert, K.; Kedenko, L.; Antonielli, L.; Kedenko, I.; Gemeier, C.; Leitner, M.; Kautzky-Willer, A.; Paulweber, B.; Hackl, E. Gut microbiota dysbiosis associated with glucose metabolism disorders and the metabolic syndrome in older adults. *Benef. Microbes* **2017**, *8*, 545–556. [CrossRef]
59. Shen, F.; Zheng, R.D.; Sun, X.Q.; Ding, W.J.; Wang, X.Y.; Fan, J.G. Gut microbiota dysbiosis in patients with non-alcoholic fatty liver disease. *Hepatobiliary Pancreat. Dis. Int.* **2017**, *16*, 375–381. [CrossRef]
60. Djekic, D.; Shi, L.; Brolin, H.; Carlsson, F.; Särnqvist, C.; Savolainen, O.; Cao, Y.; Bäckhed, F.; Tremaroli, V.; Landberg, R.; et al. Effects of a Vegetarian Diet on Cardiometabolic Risk Factors, Gut Microbiota, and Plasma Metabolome in Subjects with Ischemic Heart Disease: A Randomized, Crossover Study. *J. Am. Heart Assoc.* **2020**, *9*, e016518. [CrossRef]
61. Dong, X.; Fu, J.; Yin, X.; Cao, S.; Li, X.; Lin, L.; Ni, J. Emodin: A Review of its Pharmacology, Toxicity and Pharmacokinetics. *Phyther. Res.* **2016**, *30*, 1207–1218. [CrossRef] [PubMed]

biomolecules

MDPI

Article

# Cardioprotective Effect of Taxifolin against Isoproterenol-Induced Cardiac Injury through Decreasing Oxidative Stress, Inflammation, and Cell Death, and Activating Nrf2/HO-1 in Mice

**Heba M. Obeidat [1], Osama Y. Althunibat [1], Manal A. Alfwuaires [2], Saleem H. Aladaileh [1,3], Abdulmohsen I. Algefare [2], Afaf F. Almuqati [4], Fawaz Alasmari [5], Hammad Khalifeh Aldal'in [6], Abdulkareem A. Alanezi [7], Bader Alsuwayt [3] and Mohammad H. Abukhalil [1,8,*]**

1 Department of Medical Analysis, Princess Aisha Bint Al-Hussein College of Nursing and Health Sciences, Al-Hussein Bin Talal University, Ma'an 71111, Jordan
2 Department of Biological Sciences, Faculty of Science, King Faisal University, Al-Ahsa 31982, Saudi Arabia
3 Department of Pharmacy Practice, College of Pharmacy, University of Hafr Al-Batin, Hafr Al-Batin 31991, Saudi Arabia
4 Department of Pharmaceutical Chemistry, College of Pharmacy, University of Hafr Al-Batin, Hafr Al-Batin 31991, Saudi Arabia
5 Department of Pharmacology and Toxicology, College of Pharmacy, King Saud University, Riyadh 11451, Saudi Arabia
6 Department of Medical Support, Al-Karak University College, Al-Balqa Applied University, Al-Karak 19117, Jordan
7 Department of Pharmaceutics, College of Pharmacy, University of Hafr Al-Batin, Hafr Al-Batin 31991, Saudi Arabia
8 Department of Biology, College of Science, Al-Hussein Bin Talal University, Ma'an 71111, Jordan
* Correspondence: mabukhalil@ahu.edu.jo

**Abstract:** Oxidative stress and inflammation are key components in cardiovascular diseases and heart dysfunction. Herein, we evaluated the protective effects of (+)-taxifolin (TAX), a potent flavonoid with significant antioxidant and anti-inflammatory actions, on myocardial oxidative tissue injury, inflammation, and cell death, using a mouse model of isoproterenol (ISO)-induced acute myocardial injury. Mice were given TAX (25 and 50 mg/kg, orally) for 14 days before receiving two subsequent injections of ISO (100 mg/kg, s.c.) at an interval of 24 h on the 15th and 16th days. The ISO-induced cardiac tissue injury was evidenced by increased serum creatine kinase-MB (CK-MB), cardiac troponin I (cTnI), and lactate dehydrogenase (LDH), along with several histopathological changes. The ISO also induced increased malondialdehyde (MDA) with concomitant declined myocardial glutathione level and antioxidant enzymes activities. Moreover, ISO-induced heart injury was accompained with elevated cardiac NF-κB p65, TNF-α, IL-1β, Bax, and caspase-3, as well as decreased Bcl-2, Nrf2, and HO-1. Remarkably, TAX reduced the severity of cardiac injury, oxidative stress, inflammation, and cell death, while enhancing antioxidants, Bcl-2, and Nrf2/HO-1 signaling in ISO-injected mice. In conclusion, TAX protects against ISO-induced acute myocardial injury via activating the Nrf2/HO-1 signaling pathway and attenuating the oxidative tissue injury and key regulators of inflammatory response and apoptosis. Thus, our findings imply that TAX may constitute a new cardioprotective therapy against acute MI, which undoubtedly deserves further exploration in upcoming human trials.

**Keywords:** taxifolin; isoproterenol; inflammation; myocardial injury; Nrf2; oxidative stress

## 1. Introduction

Cardiovascular diseases (CVDs) remain one of the most significant causes of death globally, with an estimated 32% of all deaths worldwide [1,2]. Myocardial infarction (MI) is triggered by a reduction in myocardial blood supply caused by an abrupt blockage of the coronary arteries [3]. Ischemic heart damage is caused primarily by a sudden lack of oxygen and nutrients, as well as the accumulation of metabolic byproducts, eventually culminating in cell injury and death [4]. Despite the broad availability of therapeutic options, MI is still a leading cause of morbidity and death globally; hence, new therapeutic tools for the prevention/treatment of MI are needed. Isoproterenol (ISO) is a common treatment of bradycardia and heart block by acting as a non-selective beta agonist; however, it can lead to functional and structural changes in the myocardium resembling pathological changes observed in acute MI when given at high doses [5,6]. The molecular mechanism by which ISO-induced myocardial injury is complex and multifactorial, with oxidative stress assumed to be the main fundamental mechanism producing pathologic complications involved in ISO cardiotoxicity [6,7].

Oxidative stress can induce tissue injury and plays a key role in the initiation of pathologic alterations in the myocardium, including inflammation and cell death, which eventually leads to the development of cardiac dysfunction and injury [6,7]. The persistently reactive oxygen species (ROS) generation coupled with the dysfunction of important antioxidant defense mechanisms can provoke myocardial inflammation through activating NF-κB and may trigger the stimulation of pro-apoptotic factors in cardiomyocytes, which mediate caspase-dependent and -independent apoptotic cell death [8]. Therefore, mitigation of oxidative stress and inflammation, and activation of cytoprotective pathways, such as the nuclear factor erythroid 2-related factor 2 (Nrf2) signaling pathway, could be a valuable strategy for the prevention and/or treatment of MI and perhaps other CVDs. Indeed, Nrf2 is an important nuclear transcriptional factor that regulates redox homeostasis by activating a variety of antioxidant and detoxifying enzymes [9]. Several studies have indicated that Nrf2 activation prevented cardiac dysfunction and attenuated oxidative tissue injury, inflammatory responses, hypertrophy, and cell death in ISO-intoxicated animals [10–15].

Multiple lines of evidence demonstrate that several natural compounds have shown preventive and therapeutic properties against ISO-induced cardiac injury via modulating ROS overproduction, inflammation, and Nrf2/heme oxygenase-1 (HO-1) signaling [10,11,16–18]. (+)-Taxifolin (TAX; $C_{15}H_{12}O_7$, Figure 1) is a natural flavonoid present mainly in olive oil, grapes, and onions, and one which possesses several pharmacological actions, including radical scavenging and anti-inflammatory properties [19,20]. Indeed, TAX was shown to prevent against acrylamide-induced heart damage in rats by decreasing oxidative stress and enhancing antioxidants [21]. Furthermore, TAX protected against renal damage induced by unilateral ureteral obstruction (UUO) in rats via suppressing oxidative tissue injury, inflammation, and fibrosis [22]. Additionally, TAX alleviated myocardial damage triggered by ischemia/reperfusion condition by attenuating oxidative stress and modulating the apoptotic pathway [23]. A recent study showed that TAX prevented cadmium-induced renal damage by modulating redox status, inflammatory response, and apoptotic cell death, as well as by upregulating the Nrf2/HO-1 signaling pathway. An in vitro study showed that TAX protected against oxidative stress-induced retinal pigment epithelium (RPE) cell death through the activation of Nrf2 and its regulated phase II antioxidant enzymes [24]. Moreover, TAX prevented alcohol-induced hepatic apoptosis in mice through modulating oxidative stress, NF-κB-mediated inflammation, and PI3K/Akt signaling pathway [25].

Although multiple therapeutic benefits have been explored, the mechanism behind TAX's therapeutic impact on ISO-triggered heart damage is yet to be identified. We hypothesized that TAX would be a novel strategy for protecting against the cardiac injury induced by ISO via suppressing myocardial oxidative damage, inflammation, and cell death, and activating Nrf2/HO-1 signaling. Therefore, we looked into the influence of TAX on the markers of oxidative stress, inflammatory response, and apoptosis, as well as the Nrf2/HO-1 signaling pathway in the cardiac tissue of ISO-intoxicated mice. The results of this study may provide a new mechanistic insight into the protective actions of TAX against ISO cardiotoxicity, as well as a potential strategy for CVDs and heart failure prevention.

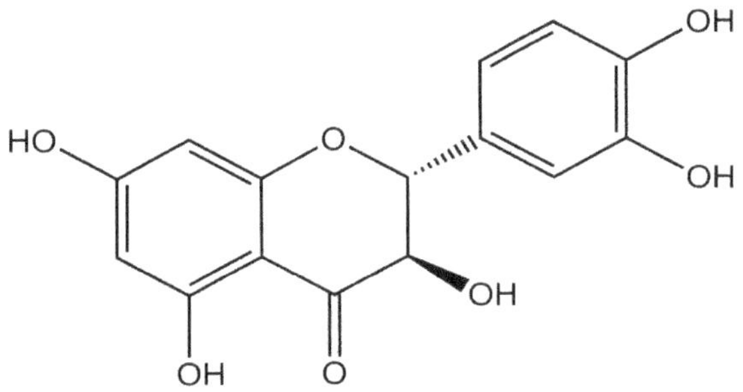

Figure 1. Chemical structure of TAX.

## 2. Materials and Methods

### 2.1. Reagents and Chemicals

The ISO was procured from Sigma (St. Louis, MO, USA). The TAX was purchased from Biosynth Carbosynth (Berkshire, UK). A Cardiac troponin I (cTnI) ELISA kit was obtained from Kamiya (Tukwila, WA, USA). Kits for creatine kinase-MB (CK-MB) and lactate dehydrogenase (LDH) activities were procured from Spinreact (Girona, Spain). A HO-1 ELISA kit was obtained from MyBioSource (San Diego, CA, USA). The ELISA kits for tumor necrosis factor-α (TNF-α) and interleukin-1β (IL-1β) were supplied by R&D Systems (Minneapolis, MN, USA). Anti-NF-κB p65, anti-caspase-3, and anti-Nrf2 antibodies were obtained from ThermoFisher (Waltham, MA, USA; 1:100 dilution), while anti-Bax and anti-Bcl-2 antibodies were supplied by Abcam (Cambridge, MA, USA; 1:100 dilution).

### 2.2. Animals and Treatment

A mouse model of ISO-triggered acute myocardial injury was used here to examine the cardioprotective action of TAX. Animals were obtained from Jordan University of Science and Technology (JUST), Irbid, Jordan. All animal-related protocols including care, handling, and treatment were verified by an Al-Hussein Bin Talal University panel of animal research and ethics and were in conformity with the National Institute of Health standards (NIH publication No. 85-23, revised 2011). The experiment included 30 Swiss albino mice (25–30 g), which were housed under standard conditions in an environment-controlled room and given free access to food and water. The vehicle solutions of physiological saline and 0.5% DMSO were used to dissolve ISO [12] and TAX [26], respectively.

Following a week of acclimatization, the selected mice were separated into five groups ($n$ = 6), as follows. The first group included the control mice which were given 0.5% DMSO for 14 days, using oral gavage, followed by two subcutaneous (s.c) injections of physiological saline on days 15 and 16; the second group (TAX) were given TAX (50 mg/kg) for 14 days followed by two 24-h interval sequential injections (s.c) of physiological saline; groups III (ISO), IV (TAX 25 mg/kg and ISO), and V (TAX 50 mg/kg and ISO) received oral 0.5% DMSO, 25 mg/kg TAX, or 50 mg/kg TAX, respectively, for 14 days before they were injected twice with ISO (100 mg/kg, s.c.) on the 15th and 16th days. The TAX and ISO dosages were chosen based on relevant articles by Algefare [26] and Abukhalil et al. [11], respectively.

At the end of the experiment, anesthesia was induced by injecting mice with ketamine/xylazine (100 mg/kg and 10 mg/kg, respectively, i.p), followed by blood collection via cardiac puncture and immediate heart extraction and processing. The blood samples were left standing until full coagulation occurred, then centrifuged for serum preparation and aliquoted at −20 °C for cardiac marker analysis. The excised hearts were dissected and rinsed in 50 mM (pH 7.0) of cold phosphate buffer saline (PBS). Some heart pieces were preserved in 10% neutral buffered formalin (NBF) for histological evaluation, while others were homogenized in cold PBS (10% $w/v$). The homogenate was then processed in refrigerated centrifuge, and the clear supernatant was collected and maintained at −20 °C for further biochemical investigations.

### 2.3. Assessment of Cardiac Biomarkers

A specific ELISA kit was used for measurement of serum cTnI levels, while CK-MB and LDH activities were evaluated spectrophotometrically with commercially available kits. All tests were carried out in accordance with the manufacturers' instructions.

### 2.4. Investigation of Oxidative Stress Markers and Antioxidants in the Heart

The malondialdehyde (MDA) contents were determined spectrophotometrically using thiobarbituric acid (TBA), as previously described [27]. In brief, the sample MDA was allowed to react with TBA in acidic medium at 95 °C for 30 min to form thiobarbituric acid reactive product, followed by measuring the absorbance of the resultant pink product at 532 nm. The level of protein carbonyl in the myocardial tissues was measured spectrophotometrically, as previously described [28]. The principle of the method depends on protein carbonyl interaction with 2,4-dinitrophenylhydrazine (DNPH) to generate the stable dinitrophenyl (DNP) hydrazine, which was analyzed spectrophotometrically at 375 nm. Nitric oxide (NO) level was determined spectrophotometrically by measurement of its more stable derivatives, nitrate and nitrite, according to the method described previously [29]. Briefly, nitrate was reduced first to nitrite followed by determination of total nitrite by a Griess reaction. This involved the formation of diazotize sulphanilamide from nitrite in an acidic medium, which is then coupled with N-(1–naphthyl) ethylenediamine to produce a chromophoric azo product measured with spectrophotometry at 540 nm.

The contents of reduced glutathione (GSH) were estimated using the method based on the continuous reduction of 5, 5 dithiobis (2-nitrobenzoic acid) by GSH to produce 5-thio-2-nitrobenzoic acid (TNB) which was spectrophotometrically quantified at 412 nm [30]. The superoxide dismutase (SOD) activity was estimated using an assay relying on the ability of the enzyme to inhibit the phenazine methosulphate-mediated reduction of nitro blue tetrazolium dye as previously described [31]. Catalase (CAT) activity was estimated using a method based on its ability to decompose hydrogen peroxide ($H_2O_2$) into water and oxygen [32]. The rate of $H_2O_2$ decomposition was assessed upon measurement of unconverted $H_2O_2$ by reacting it with 3,5-dichloro-2-hydroxybenzenesulfonic acid (DHBS) and 4-aminoantipyrine in the presence of horseradish peroxidase (HRP) to form a product measured via spectrophotometry at 570 nm. Furthermore, a specific ELISA kit was used for the measurement of heart HO-1 content according to the protocol provided.

### 2.5. Investigation of Pro-Inflammatory Cytokines in the Heart

Measurement of cardiac levels of TNF-$\alpha$ and IL-1$\beta$ were performed by following the manufacturer's protocol for the specific ELISA kits.

### 2.6. Histopathological and Immunohistochemical Analysis

Following a standard procedure, the NBF preserved cardiac tissue was paraffin embedded, 5-microne slices were prepared, and parts of the deparaffinized sections were stained with hematoxylin and eosin (H&E) [33]. The stained sections were then examined under a light microscope by a histopathologist.

The remaining deparaffinized slices were subjected to immunohistochemistry (IHC) staining [34]. Antigen retrieval was performed by successively treating cardiac tissues with 50 mM citrate buffer (pH 6.8) and 0.3% $H_2O_2$. Then, normal serum was added for 20 min to block nonspecific antigen–antibody binding. After washing the tissues with PBS, anti-NF-κB p65, anti-caspase-3, anti-Bax, anti-Bcl-2, and anti-Nrf2, antibodies were applied overnight at 4 °C. Unbound primary antibodies were washed with PBS followed by incubation of sections with the secondary antibodies. The substrate 3,3-Diaminobenzidine (DAB) was added for color development before the slides were counterstained with Mayer's hematoxylin and evaluated under a light microscope. The staining labelling indices of the anti-Bax, anti-caspase-3, and anti-NF-κB p65 antibodies were presented as a percentage of positive expression in a total of 1000 cells. By using ImageJ analysis software (NIH, Bethesda, MD, USA), the immunostaining intensity of anti-Bcl-2 and anti-Nrf2 antibodies was determined through the area of positive expression.

### *2.7. Analysis of Data*

Data are presented as the mean ± standard error of the mean (SEM). GraphPad Prism 7 software (San Diego, CA, USA) was employed to perform the statistical comparison between groups using one-way analysis of variance (ANOVA) followed by a Tukey's post-hoc test. Results were considered statistically significant when the $p$ value < 0.05.

## 3. Results

### *3.1. TAX Prevents ISO-Triggered Myocardial Injury in Mice*

To assess the protective effect of TAX on ISO-induced cardiac injury, we investigated serum cardiac biomarkers levels (Figure 2A–C) and histological changes (Figure 3) in both TAX-treated and untreated mice. The ISO-treated mice showed a significant ($p < 0.05$) increase in serum CK-MB (Figure 2A), cTnI (Figure 2B), and LDH (Figure 2C) as compared to control mice. Pretreatment of ISO-intoxicated mice with TAX markedly ($p < 0.05$) attenuated CK-MB, cTnI, and LDH in serum. Normal mice that received TAX alone had no effects on these cardiac biomarkers in serum.

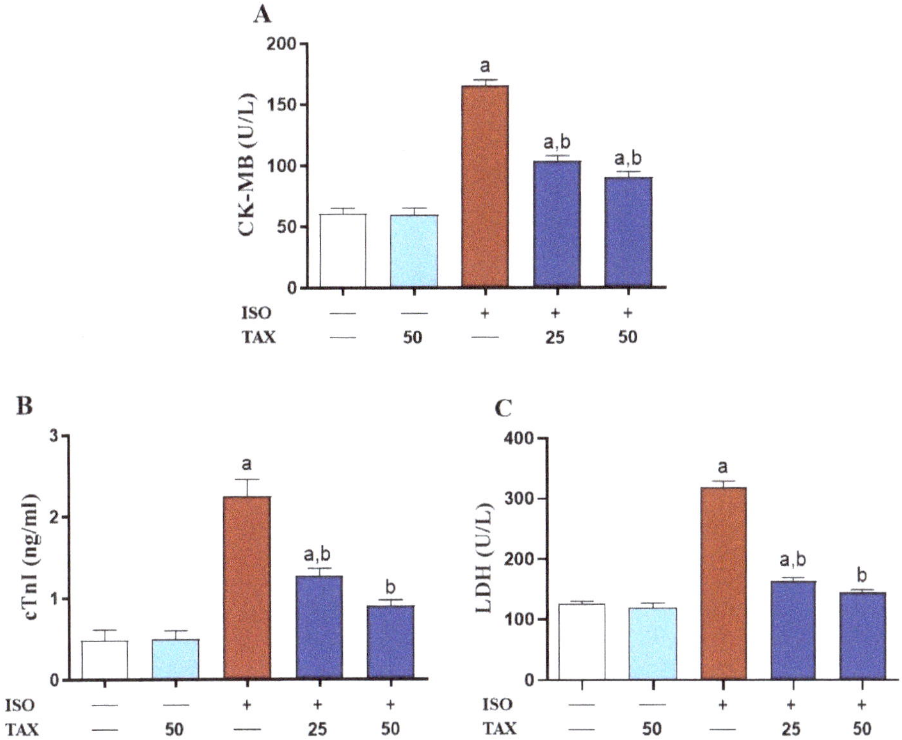


**Figure 2.** TAX attenuates the ISO-induced cardiac injury. Here, TAX reduced serum levels of (**A**) CK-MB, (**B**) cTnI, and (**C**) LDH in ISO-intoxicated animals. Data are expressed as mean ± SEM, ($n = 6$). Here, a indicates significant ($p < 0.05$) vs. control, while b indicates significant ($p < 0.05$) vs. ISO.

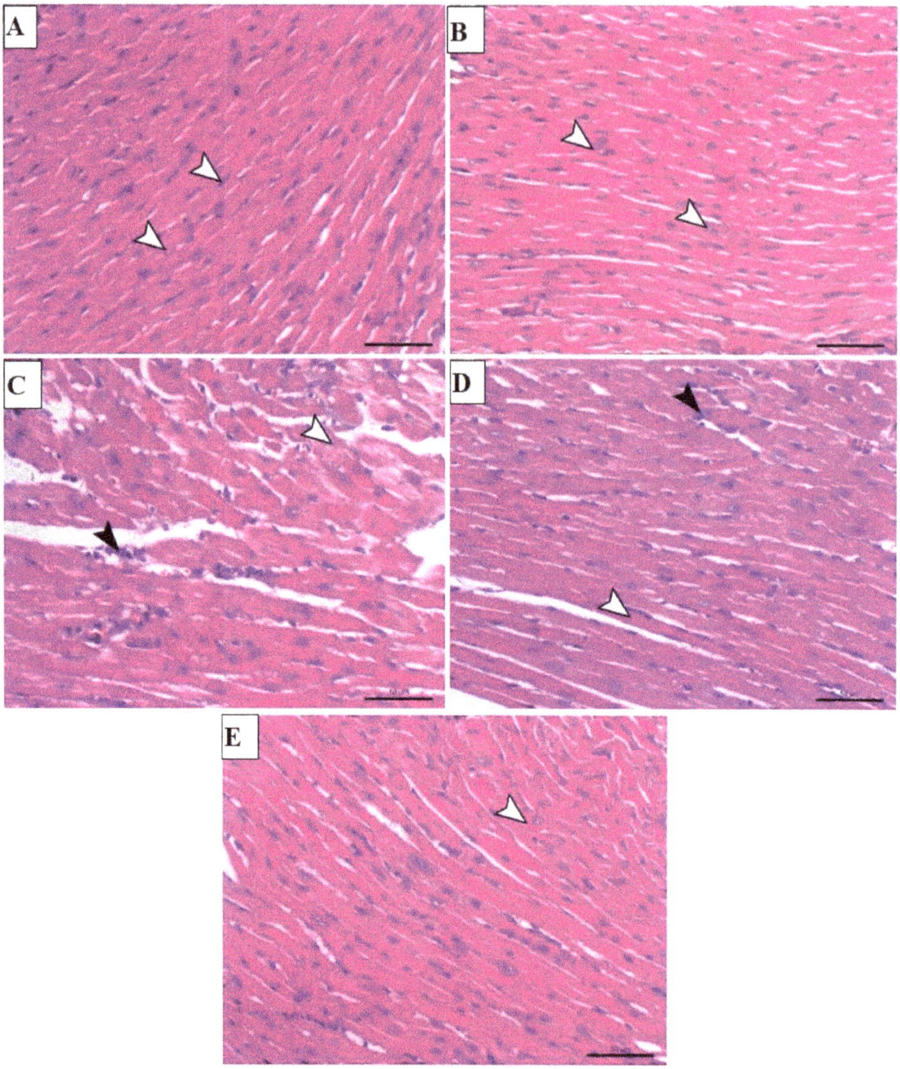


**Figure 3.** TAX prevents the ISO-induced histopathological damage in the heart. Images of the microscopic field of cardiac sections from (**A**) control and (**B**) TAX-treated mice showing normal branched connected myocardial fibers with a normal central nucleus (arrowheads); (**C**) ISO-treated animals demonstrating marked myolysis (white arrowhead), marked nuclear pyknosis, and infiltration of mononuclear inflammatory cells (black arrowhead); (**D**) ISO-injected mice pre-treated with 25 mg TAX demonstrating a reduction in myocardial degeneration of muscle fibers (white arrowhead) and mild infiltration of mononuclear inflammatory cells (black arrowhead); and (**E**) ISO-injected mice pre-treated with 50 mg TAX demonstrating a noticeable reduction in myocardial degeneration and with mild curving of cardiac muscle fibers (arrowhead) (H&E, X200, Scale bar = 50 µm).

Furthermore, the effect of TAX on ISO-induced cardiac injury was assessed by evaluating H&E stained heart sections of both TAX-treated and untreated mice. Microscopic observation of heart tissue slides from control and TAX-treated mice demonstrated normal branched connected myocardial fibers with a normal central nucleus (Figure 3A,B). A microscopic field of the hearts of ISO-treated mice demonstrated marked myolysis, marked nuclear pyknosis, and infiltration of mononuclear inflammatory cells (Figure 3C). These histopathological alterations were significantly ameliorated when ISO-intoxicated mice were treated with both doses of TAX (Figure 3D,E).

### 3.2. TAX Attenuates Oxidative Stress and Enhances Antioxidants in Heart of ISO-Injected Mice

Hearts of ISO-treated mice showed significantly ($p < 0.05$) increased MDA (Figure 4A), protein carbonyl (Figure 4B), and NO (Figure 3C) contents, along with significantly ($p < 0.05$) decreased GSH (Figure 4D) levels and SOD (Figure 4E) and CAT (Figure 4F) activities as compared to control mice. Pretreatment of ISO-injected mice with TAX resulted in a substantial ($p < 0.05$) reduction in MDA, protein carbonyl, and NO contents and an increase in antioxidants in the heart. The TAX alone did not affect these parameters in healthy mice.

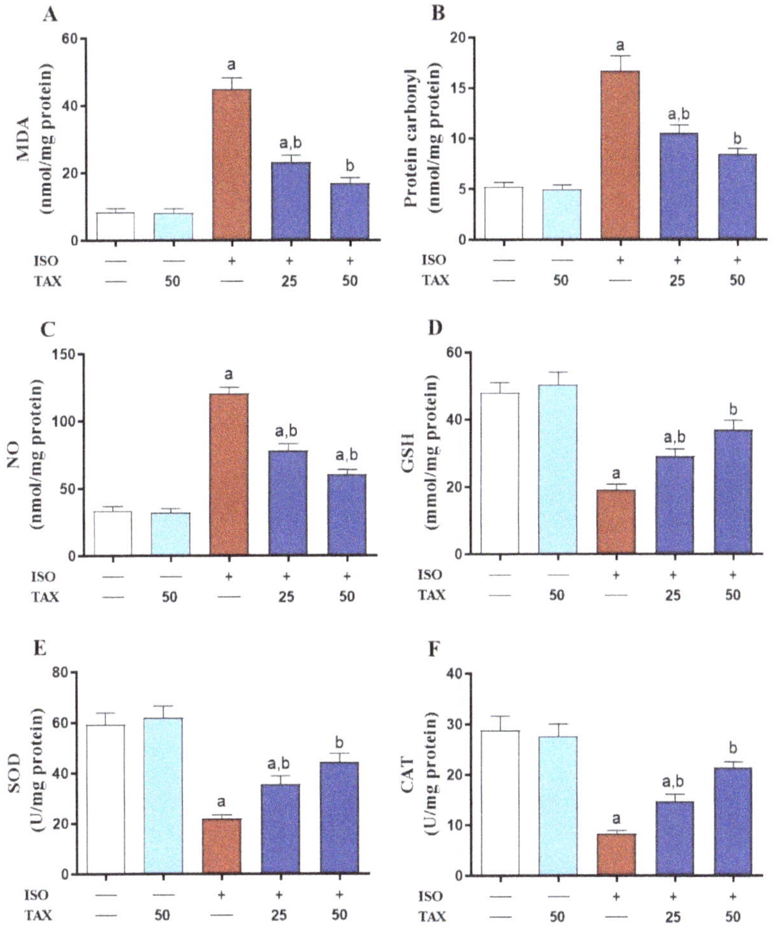


**Figure 4.** TAX attenuates the ISO-induced myocardial oxidative stress. Pre-treatment with TAX reduced cardiac levels of (**A**) MDA, (**B**) protein carbonyl, (**C**) NO, and elevated (**D**) GSH level, and activities of (**E**) SOD and (**F**) CAT in ISO-injected mice. Data are expressed as mean ± SEM, ($n = 6$). Here, a indicates significant ($p < 0.05$) vs. control, while b indicates significant ($p < 0.05$) vs. ISO.

### 3.3. TAX Mitigates Myocardial Inflammation in ISO-Injected Mice

The inflammatory response is known to fuel the major pathological processes in ISO-induced cardiac injury. Indeed, ISO induced a considerable ($p < 0.05$) rise in NF-κB p65 expression (Figure 5A–F) and myocardial levels of the pro-inflammatory cytokines, IL-1β (Figure 6A) and TNF-α (Figure 6B). The TAX treatment of ISO-injected mice significantly ($p < 0.05$) attenuated the myocardial contents of NF-κB p65, IL-1β and TNF-α. However, TAX alone did not affect the levels of the above-mentioned inflammatory proteins in the heart.

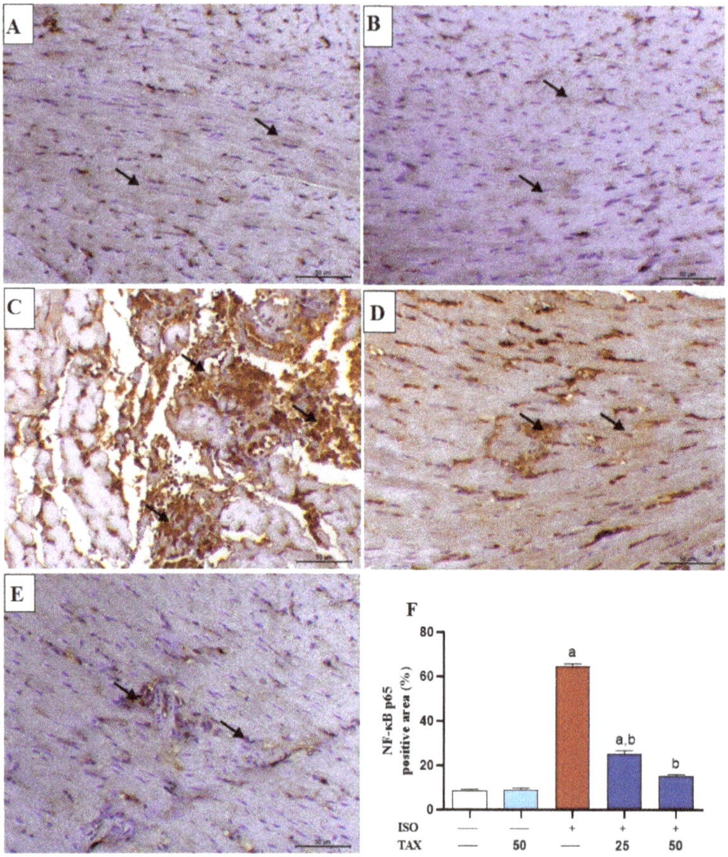


**Figure 5.** TAX attenuates ISO-induced myocardial NF-κB p65 upregulation. Images of microscopic field of IHC labeled cardiac sections from (**A**) control and (**B**) TAX-treated mice showing mild expression of myocardial NF-κB p65 (arrows designate positive immunostaining within the interstitial cells); (**C**) ISO-injected mice showing a marked increase in both cytoplasmic and nuclear levels of NF-κB p65 antibody within myocytes (arrows designate marked expression within both degenerated fibers and inflammatory cells); (**D**) ISO-administrated mice pre-treated with 25 mg TAX demonstrating a decreased level of the nuclear NF-κB p65 within the myocytes (arrows); and (**E**) ISO-administrated mice pre-treated with 50 mg TAX demonstrating a noticeably decreased expression of the nuclear NF-κB p65 within the myocardial fibers (arrows) (IHC, X200, Scale bar = 50 μm). (**F**) Analysis of the relative intensities of the images of IHC labeled myocardial NF-κB p65 showing a significant increase in ISO-injected mice and a significant reduction in mice pre-treated with both doses of TAX. Data are expressed as mean ± SEM, ($n = 6$). Here, a indicates significant ($p < 0.05$) vs. control, while b indicates significant ($p < 0.05$) vs. ISO.

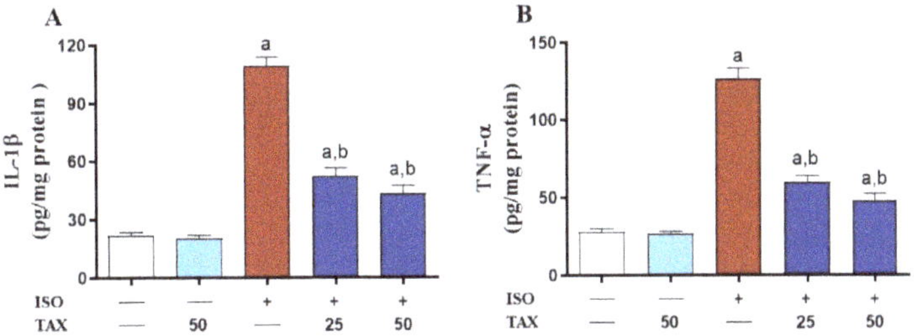


**Figure 6.** TAX reduced the ISO-induced increased myocardial (**A**) IL-1β and (**B**) TNF-α. Data are expressed as mean ± SEM, ($n$ = 6). Here, a indicates significant ($p < 0.05$) vs. control, while b indicates significant ($p < 0.05$) vs. ISO.

### 3.4. TAX Prevents Myocardial Apoptosis in ISO-Treated Mice

To further assess the protective impact of TAX treatment on ISO-triggered myocardial apoptosis, we evaluated Bax, Bcl-2, and caspase-3 expression levels in the myocardial tissues of both TAX-treated and untreated mice by IHC staining. The ISO resulted in a marked ($p < 0.05$) decrease in Bcl-2 expression level (Figure 7A,B), associated with a significant ($p < 0.05$) increase in Bax (Figure 7A,C) and caspase-3 (Figure 7A,D) expression in the heart. The ISO-induced imbalance in the expression of pro-apoptotic and anti-apoptotic proteins were remarkably ($p < 0.05$) attenuated when mice were pre-treated with both doses of TAX. Treatment of normal mice with TAX had no effects on the above-mentioned markers in the heart.

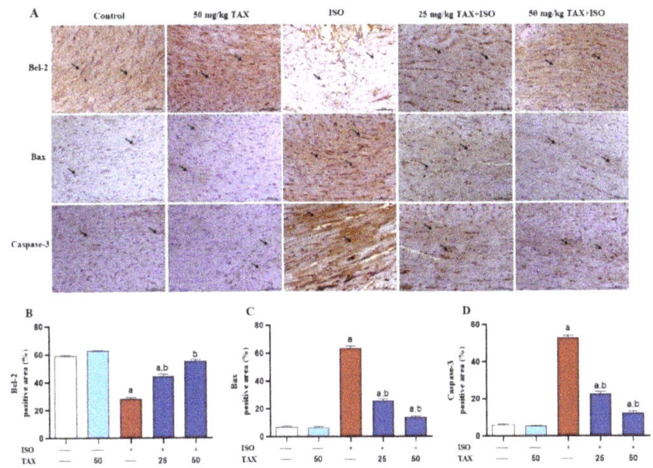


**Figure 7.** TAX attenuates the ISO-induced myocardial apoptosis in mice. (**A**) Images of the microscopic field of IHC labeled cardiac sections from control and TAX-treated mice, showing obvious expression of Bcl-2 within the myocardial fibers (arrows), along with mild expression of myocardial Bax and caspase-3 (arrows); ISO-injected mice demonstrating a noticeable reduction in the myocardial level of Bcl-2 (arrows), along with remarkably increased myocardial Bax and caspase-3 (arrows); and mice treated with 25 and 50 mg/kg TAX before ISO injection demonstrating a noticeable increase in myocardial expression of Bcl-2 (arrows), along with a marked reduction in the expression of Bax and caspase-3 in the myocardial tissues (arrows). (IHC, X200, Scale bar = 50 μm). (**B–D**) Analysis of the relative intensities of the images of IHC-labeled myocardial Bcl-2, Bax, and caspase-3, showing a significantly increased Bcl-2 and decreased Bax and caspase-3 in mice pre-treated with both doses of TAX. Data are expressed as mean ± SEM, ($n$ = 6). Here, a indicates significant ($p < 0.05$) vs. control, while b indicates significant ($p < 0.05$) vs. ISO.

### 3.5. TAX Activates Cardiac Nrf2/HO-1 in ISO-Injected Mice

Since Nrf2 is known to suppress surplus ROS and attenuate oxidative tissue injury, changes in its expression (Figure 8A–F) and HO-1 levels (Figure 8G) in the heart of both TAX-treated and untreated mice were estimated. The ISO-treated mice showed significantly ($p < 0.05$) downregulated Nrf2 expression and HO-1 levels in the heart as compared to the control mice. The ISO-induced downregulation of myocardial Nrf2 and HO-1 was significantly ($p < 0.05$) ameliorated by TAX treatment (Figure 8A–G). Treatment of normal mice with TAX alone did not affect Nrf2 and HO-1 levels in the heart.

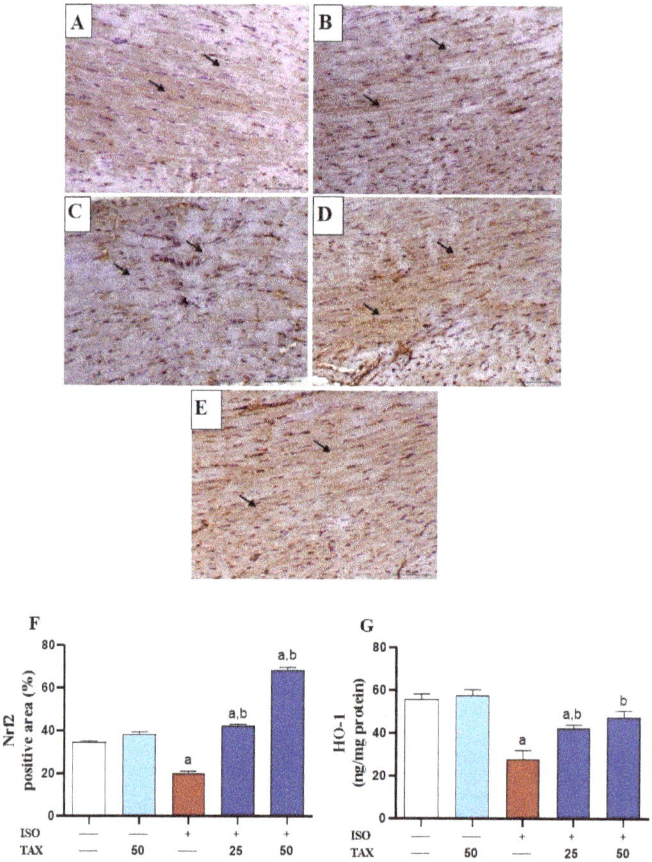


**Figure 8.** TAX upregulates Nrf2/HO-1 signaling in ISO-administrated mice. Images of the microscopic field of IHC labeled cardiac sections from (**A**) control and (**B**) TAX-treated mice, indicating obvious cytoplasmic immunostaining of Nrf2 within the myocardial fibers (arrows); (**C**) ISO-administered mice showing focal myocardial expression of Nrf2 (arrows); (**D**) ISO-injected mice pre-treated with 25 mg TAX demonstrating increased Nrf2 expression within the myocardial fibers (arrows); and (**E**) ISO-injected mice pre-treated with 50 mg demonstrating a noticeable rise in the myocardial expression of Nrf2 (arrows) (IHC, X200, Scale bar = 50 µm). (**F**) Analysis of the relative intensities of the images of IHC labeled myocardial Nrf2 demonstrating a significant increase in Nrf2 in the myocardium of mice treated with both doses of TAX before ISO injection. (**G**) TAX significantly attenuates myocardial levels of HO-1 in ISO-injected mice. Data are expressed as mean ± SEM, ($n$ = 6). Here, a indicates significant ($p < 0.05$) vs. control, while b indicates significant ($p < 0.05$) vs. ISO.

## 4. Discussion

As CVDs are the leading cause of mortality and morbidity worldwide, there is an urgent need to develop effective interventions to treat CVDs, principally MI. Since oxidative tissue injury and inflammation are key components implicated in the pathological processes of MI, novel compounds with antioxidant and anti-inflammatory effects could exert beneficial actions against the development of cardiovascular dysfunction in MI. Herein, we have assessed the cardioprotective effects of TAX, a flavononol with promising biological activities, on ISO-induced myocardial injury in mice. Our findings demonstrate that TAX prevented ISO-induced oxidative stress, NO formation, inflammation, and apoptosis in the heart, and consequently attenuated myocardial tissue damage.

Repeated injection of ISO offers a non-invasive, low-mortality animal model of acute myocardial injury that mimics acute MI in humans [35]. Consequently, the experimental model of ISO-induced myocardial damage has been employed to assess the effects of various treatments on cardiac dysfunctions. This ISO-induced myocardial damage has been linked to high circulating cardiac biomarkers, as well as the development of numerous histopathological alterations, such as myocyte loss, fibrosis, and inflammatory cell infiltration [11,12,36]. Likewise, our findings demonstrated that ISO-intoxicated mice had substantially higher serum CK-MB, cTnI, and LDH levels when compared to the control group, which was associated with morphological aberrations including necrotic features, misaligned myofibrils, and extensive neutrophil infiltration. It is well known that myocyte apoptosis and necrosis are preceded by membrane disruption, resulting in leakage of cardiac biomarkers and, hence, an increase in their circulating levels [37]. Thus, an increase in serum cardiac biomarkers is considered as a sensitive indicator of altered myocyte permeability under cardiac stress conditions [38]. On the other hand, TAX demonstrated potent cardioprotective actions in ISO-induced mice, which was evidenced by reducing circulating cardiac biomarkers and attenuating histological alterations. Consistently, TAX has shown cardioprotective actions and prevented cardiac dysfunction under various cardiac injurious conditions including ischemia–reperfusion cardiac injury [39] and diabetic cardiomyopathy [40].

The involvement of oxidative stress in the cardio–pathogenic action of ISO has been reported by several preclinical studies. This ISO-induced oxidative stress is triggered by disturbed mitochondrial membrane causing leakage of ROS and other free radicals. as well as catecholamine auto-oxidation [41–43]. Consequently, excessive production of oxidative agents, along with impairment of cardiac antioxidants, mediate oxidative damage of myocyte contents manifested by elevated cardiac lipid peroxidation (LPO) and protein carbonyl, as well as decreased activities of antioxidant mechanism, thereby aggravating myocardial damage and cardiac dysfunction [6,41]. In the present study, oxidative stress in hearts of ISO-intoxicated mice was demonstrated by a significant rise in myocardial MDA, protein carbonyl, and NO levels, in addition to a decrease in GSH and antioxidant enzyme activities. This is in line with previous reports indicating that injection of ISO increased myocardial LPO, protein carbonyl, and NO, while decreasing antioxidants [7,44–46]. The LPO disrupts the membrane fluidity and permeability in addition to deactivating membrane-bound enzymes, leading to cell injury. Additionally, cell destruction is provoked by the covalently modified proteins, which is exacerbated by permeated LPO-derived aldehydes [47]. Furthermore, oxidative alteration of proteins causes protein breakdown via promoting protein aggregation and/or fragmentation, as well as the inactivation of enzymes and transport proteins, which all end in cell death. Furthermore, the interaction of superoxide anions with NO generates peroxynitrite, a strong oxidant agent that exacerbates oxidative damage by further oxidizing cellular components, resulting in cell death [48].

Therefore, enhancing antioxidant capacity and mitigating oxidative damage are crucial strategies for protecting the myocardium from ISO-induced oxidative injury. In this context, the current study showed that pre-treatment of mice with TAX, the potent antioxidant flavononol, markedly attenuated oxidative stress and boosted the antioxidant defense in the myocardium exposed to ISO. This was evidenced by considerably lower MDA, protein carbonyl, and NO levels, while elevating SOD, CAT, and GSH in cardiac tissues of mice pre-treated with TAX prior to ISO administration. In accordance with our results, several investigations have shown that the cardioprotective action of TAX is associated with reduced oxidative markers and elevated antioxidant components, resulting in ameliorated oxidative injury to the myocardium [39,40,49]. Furthermore, TAX exhibited tissue protective effect against chemically-induced oxidative damage in various experimental models, such as acrylamide-induced heart [21] and kidney injury [50], alcohol [25] and carbon tetrachloride ($CCl_4$)-induced liver injury [25], and methanol-induced oxidative optic nerve damage [51].

A substantial amount of evidence demonstrates that enhanced ROS production is a critical factor in the activation of pro-inflammatory and stress signaling pathways, facilitating apoptosis in cardiomyocytes and endothelial cells [11,16,52–54]. The ISO-induced ROS production facilitates myocardial inflammation through NF-κB activation and the release of pro-inflammatory mediators, subsequently culminating in cardiac apoptosis and injury [55,56]. Consistent with several studies [11,12,18], hearts of ISO-intoxicated mice showed decreased Bcl-2 expression along with increased NF-κB p65, Bax, and caspase-3 expression, as well as TNF-α and IL-6 levels. Increased pro-inflammatory cytokines, including TNF-α and IL-6, can affect the myocardium directly or indirectly through changes in hemodynamic loading conditions, which trigger additional oxidative stress, myocyte contractility and viability alterations, endothelial and myocardial dysfunction, and myocardial necrosis [6,57–59]. The ISO-induced myocardial apoptosis is seemingly trigged by persistent ROS generation that facilitates the dissipation of mitochondrial membrane potential and the release of cytochrome c, leading to the execution phase of caspase-3-dependent apoptotic cell death [11,12,60,61]. Thus, the inflammatory response and consequent apoptosis are considered one of the key therapeutically important components associated with cardiac injury and dysfunction.

In the present study, TAX mitigated NF-κB and decreased TNF-α, and IL-1β in the myocardium of ISO-injected mice, indicating its anti-inflammatory action. Moreover, TAX attenuated ISO-induced myocardial apoptosis, as indicated by the diminished Bax and caspase-3 and enhanced Bcl-2 expressions. Accordingly, TAX prevented cadmium-induced renal inflammation and apoptosis via the suppression of NF-κB, TNF-α, IL-1β, Bax, and caspase-3, and increased Bcl-2 expression [26]. Furthermore, TAX attenuated alcoholic liver disease in mice via mitigation of the NF-κB-enhanced inflammatory response and the regulation of Bax, Bcl-2, and caspase-3 expression in the liver [25]. Moreover, TA was found to inhibit inflammatory response and apoptosis through attenuation of TNF-α, IL-1β, IL-6, Bax, Bcl-2, and caspase-3 in $CCl_4$-induced liver injury in mice [25]. Additionally, TAX attenuated myocardial apoptosis through upregulation of Bcl-2 protein expression and downregulation of protein expression levels of Bax, cytochrome c, and caspase-3 and 9 in a rat model of myocardial ischemia/reperfusion injury [23]. It has also been reported that inhibition of the NF-κB signaling pathway by TAX alleviated sepsis-induced acute lung injury in a mouse model of cecal ligation and puncture-induced sepsis [62]. The suppressive effect of TAX on ISO-induced myocardial inflammation and apoptosis appears to be attributable, at least in part, to its potential inhibitory effects on ROS production in the heart.

Since activation of Nrf2/HO-1 signaling plays a key role in protecting against ISO-induced acute myocardial injury [9,15,63], we assessed its involvement in the cardioprotective effect of TAX. Consistent with several studies [11,12,17], our findings showed ISO administration induced downregulation of Nrf2 and HO-1 in the heart. Indeed, Nrf2 is a crucial nuclear transcriptional factor that orchestrates the cellular defense against toxic and oxidative insults through regulation of the expression of an array of genes encoding multiple antioxidant and cytoprotective enzymes [63]. Additionally, Nrf2/HO-1 activation has been shown to regulate the inflammatory responses via suppressing NFκB signaling and upregulating anti-inflammatory cytokines [64,65]. Therefore, therapeutic approaches targeting Nrf2/HO-1 signaling can be considered as a potential therapeutic benefit for the treatment and prevention of CVDs where oxidative stress and inflammation are key factors. This assumption is supported by a previous study where Nrf2 knockdown resulted in increased mortality, decreased antioxidant genes expression in the heart, increased cardiac fibrosis, and apoptosis in mice after pathological pressure overload [66]. In this study, treatment of ISO-intoxicated mice with TAX largely upregulated the Nrf2/HO-1 signaling pathway in the myocardium. In accordance, TAX protected against cadmium-induced oxidative tissue injury, inflammation, and apoptosis in the kidney [26], attenuated oxidative stress-induced cellular damage, and apoptosis in human RPE [24], and prevented lipopolysaccharide (LPS)-induced inflammatory injury in mice [67] by stimulating the Nrf2/HO-1 signaling pathway. Thus, the TAX-mediated Nrf2/HO-1 activation is attributed, at least in part, to its antioxidant and anti-inflammatory properties against ISO-induced cardiac injury.

## 5. Conclusions

This study suggests that TAX can be of significant therapeutic benefit against ISO-induced cardiac injury by ameliorating oxidative stress, inflammatory response, and apoptosis in the heart. These beneficial effects of TAX were associated with increased Nrf2 expression and improved antioxidant defenses. Therefore, TAX could be suggested as a promising new protective strategy against ISO-induced myocardial injury and, perhaps, other CVDs, which undoubtedly deserves further research.

**Author Contributions:** Conceptualization, H.M.O., S.H.A. and M.H.A.; methodology, F.A., M.A.A., A.F.A., A.I.A., B.A., O.Y.A. and M.H.A.; validation, A.A.A., B.A. and M.H.A.; formal analysis, B.A., A.I.A. and M.H.A.; investigation, A.A.A., M.A.A., H.K.A., A.F.A., O.Y.A. and M.H.A.; resources, A.A.A., H.K.A., A.F.A. and M.A.A.; data curation, H.M.O., H.K.A., B.A. and M.H.A.; writing—original draft preparation, O.Y.A. and H.M.O.; writing—review and editing, O.Y.A., S.H.A. and M.H.A.; visualization, S.H.A. and O.Y.A.; supervision, M.H.A. and O.Y.A.; Funding acquisition, F.A. All authors have read and agreed to the published version of the manuscript.

**Funding:** This research was funded by the Researchers Supporting Project number (RSP2022R235), King Saud University, Riyadh, Saudi Arabia.

**Institutional Review Board Statement:** The animals handling, and experimental protocol conformed to the guidelines of the National Institutes of Health (NIH publication No. 85-23, revised 2011) and approved by ethical committee of animal experiments at Al-Hussein Bin Talal University.

**Informed Consent Statement:** Not applicable.

**Data Availability Statement:** Data analyzed or generated during this study are included in this manuscript.

**Conflicts of Interest:** The authors declare no conflict of interest.

## References

1. Kaptoge, S.; Pennells, L.; De Bacquer, D.; Cooney, M.T.; Kavousi, M.; Stevens, G.; Riley, L.M.; Savin, S.; Khan, T.; Altay, S. World Health Organization cardiovascular disease risk charts: Revised models to estimate risk in 21 global regions. *Lancet Glob. Health* **2019**, *7*, e1332–e1345. [CrossRef]
2. Saglietto, A.; Manfredi, R.; Elia, E.; D'Ascenzo, F.; GM, D.F.; Munzel, T. Cardiovascular disease burden: Italian and global perspectives. *Minerva Cardiol. Angiol.* **2021**, *69*, 231–240. [CrossRef] [PubMed]
3. Chang, X.; Lochner, A.; Wang, H.-H.; Wang, S.; Zhu, H.; Ren, J.; Zhou, H. Coronary microvascular injury in myocardial infarction: Perception and knowledge for mitochondrial quality control. *Theranostics* **2021**, *11*, 6766. [CrossRef] [PubMed]
4. Frangogiannis, N.G. Pathophysiology of myocardial infarction. *Compr. Physiol.* **2011**, *5*, 1841–1875.
5. Timercan, T.; Şveţ, I.; Pantea, V.; Ambros, A.; Lîsîi, L. Advanced glycation end products in isoproterenol-induced acute myocardial infarction. *Med. Pharm. Rep.* **2019**, *92*, 235. [CrossRef] [PubMed]
6. Wong, Z.W.; Thanikachalam, P.V.; Ramamurthy, S. Molecular understanding of the protective role of natural products on isoproterenol-induced myocardial infarction: A review. *Biomed. Pharmacother.* **2017**, *94*, 1145–1166. [CrossRef]
7. Upaganlawar, A.; Gandhi, H.; Balaraman, R. Isoproterenol induced myocardial infarction: Protective role of natural products. *J. Pharmacol. Toxicol.* **2011**, *6*, 1–17. [CrossRef]
8. Von Harsdorf, R.; Li, P.F.; Dietz, R. Signaling pathways in reactive oxygen species-induced cardiomyocyte apoptosis. *Circulation* **1999**, *99*, 2934–2941. [CrossRef] [PubMed]

9. Ouyang, B.; Li, Z.; Ji, X.; Huang, J.; Zhang, H.; Jiang, C. The protective role of lutein on isoproterenol-induced cardiac failure rat model through improving cardiac morphology, antioxidant status via positively regulating Nrf2/HO-1 signalling pathway. *Pharm. Biol.* **2019**, *57*, 529–535. [CrossRef]
10. Abdelzaher, W.Y.; Ahmed, S.M.; Welson, N.N.; Alsharif, K.F.; Batiha, G.E.-S.; Labib, D.A.A. Dapsone ameliorates isoproterenol-induced myocardial infarction via Nrf2/HO-1; TLR4/TNF-α signaling pathways and the suppression of oxidative stress, inflammation, and apoptosis in rats. *Front. Pharmacol.* **2021**, *12*, 669679. [CrossRef]
11. Abukhalil, M.H.; Hussein, O.E.; Aladaileh, S.H.; Althunibat, O.Y.; Al-Amarat, W.; Saghir, S.A.; Alfwuaires, M.A.; Algefare, A.I.; Alanazi, K.M.; Al-Swailmi, F.K. Visnagin prevents isoproterenol-induced myocardial injury by attenuating oxidative stress and inflammation and upregulating Nrf2 signaling in rats. *J. Biochem. Mol. Toxicol.* **2021**, *35*, e22906. [CrossRef]
12. Althunibat, O.Y.; Abduh, M.S.; Abukhalil, M.H.; Aladaileh, S.H.; Hanieh, H.; Mahmoud, A.M. Umbelliferone prevents isoproterenol-induced myocardial injury by upregulating Nrf2/HO-1 signaling, and attenuating oxidative stress, inflammation, and cell death in rats. *Biomed. Pharmacother.* **2022**, *149*, 112900. [CrossRef]
13. Velusamy, P.; Mohan, T.; Ravi, D.B.; Kishore Kumar, S.; Srinivasan, A.; Chakrapani, L.N.; Singh, A.; Varadharaj, S.; Kalaiselvi, P. Targeting the Nrf2/ARE signalling pathway to mitigate isoproterenol-induced cardiac hypertrophy: Plausible role of hesperetin in redox homeostasis. *Oxid. Med. Cell. Longev.* **2020**, *2020*, 9568278. [CrossRef]
14. Shanmugam, G.; Challa, A.K.; Devarajan, A.; Athmanathan, B.; Litovsky, S.H.; Krishnamurthy, P.; Davidson, C.J.; Rajasekaran, N.S. Exercise mediated Nrf2 signaling protects the myocardium from isoproterenol-induced pathological remodeling. *Front. Cardiovasc. Med.* **2019**, *6*, 68. [CrossRef] [PubMed]
15. Shanmugam, G.; Challa, A.K.; Litovsky, S.H.; Devarajan, A.; Wang, D.; Jones, D.P.; Darley-Usmar, V.M.; Rajasekaran, N.S. Enhanced Keap1-Nrf2 signaling protects the myocardium from isoproterenol-induced pathological remodeling in mice. *Redox Biol.* **2019**, *27*, 101212. [CrossRef] [PubMed]
16. Allam, M.A.M.; Khowailed, A.A.; Elattar, S.; Mahmoud, A.M. Umbelliferone ameliorates oxidative stress and testicular injury, improves steroidogenesis and upregulates peroxisome proliferator-activated receptor gamma in type 2 diabetic rats. *J. Pharm. Pharmacol.* **2021**, *74*, 573–584. [CrossRef] [PubMed]
17. Zhang, L.; Guo, Z.; Wang, Y.; Geng, J.; Han, S. The protective effect of kaempferol on heart via the regulation of Nrf2, NF-κβ, and PI3K/Akt/GSK-3β signaling pathways in isoproterenol—Induced heart failure in diabetic rats. *Drug Dev. Res.* **2019**, *80*, 294–309. [CrossRef] [PubMed]
18. Zhang, X.-J.; Cui, Z.-H.; Zhao, Y.-X.; He, T.-T.; Wang, L.; Liang, X.-W. Ferulic acid ameliorates isoproterenol-induced heart failure by decreasing oxidative stress and inhibiting cardiocyte apoptosis via activating Nrf2 signaling pathway in rats. *Biol. Pharm. Bull.* **2021**, *44*, 396–403. [CrossRef] [PubMed]
19. Das, A.; Baidya, R.; Chakraborty, T.; Samanta, A.K.; Roy, S. Pharmacological basis and new insights of taxifolin: A comprehensive review. *Biomed. Pharmacother.* **2021**, *142*, 112004. [CrossRef]
20. Sunil, C.; Xu, B. An insight into the health-promoting effects of taxifolin (dihydroquercetin). *Phytochemistry* **2019**, *166*, 112066. [CrossRef]
21. Coşgun, M.S.; Çoşkun, R.; Celik, A.I. The preventive effect of taxifolin on acrylamide-induced heart damage in rats. *Rev. Nutr.* **2022**, *35*. [CrossRef]
22. Ren, L.; Guo, H.-N.; Yang, J.; Guo, X.-Y.; Wei, Y.-S.; Yang, Z. Dissecting efficacy and metabolic characteristic mechanism of taxifolin on renal fibrosis by multivariate approach and ultra-performance liquid chromatography coupled with mass spectrometry-based metabolomics strategy. *Front. Pharmacol.* **2021**, *11*, 608511. [CrossRef] [PubMed]
23. Tang, Z.; Yang, C.; Zuo, B.; Zhang, Y.; Wu, G.; Wang, Y.; Wang, Z. Taxifolin protects rat against myocardial ischemia/reperfusion injury by modulating the mitochondrial apoptosis pathway. *PeerJ* **2019**, *7*, e6383. [CrossRef] [PubMed]
24. Xie, X.; Feng, J.; Kang, Z.; Zhang, S.; Zhang, L.; Zhang, Y.; Li, X.; Tang, Y. Taxifolin protects RPE cells against oxidative stress-induced apoptosis. *Mol. Vis.* **2017**, *23*, 520. [PubMed]
25. Liu, X.; Liu, W.; Ding, C.; Zhao, Y.; Chen, X.; Ling, D.; Zheng, Y.; Cheng, Z. Taxifolin, Extracted from Waste Larix olgensis Roots, Attenuates CCl4-Induced Liver Fibrosis by Regulating the PI3K/AKT/mTOR and TGF-β1/Smads Signaling Pathways. *Drug Des. Devel. Ther.* **2021**, *15*, 871. [CrossRef]
26. Algefare, A.I. Renoprotective and Oxidative Stress-Modulating Effects of Taxifolin against Cadmium-Induced Nephrotoxicity in Mice. *Life* **2022**, *12*, 1150. [CrossRef]
27. Ohkawa, H.; Ohishi, N.; Yagi, K. Assay for lipid peroxides in animal tissues by thiobarbituric acid reaction. *Anal. Biochem.* **1979**, *95*, 351–358. [CrossRef]
28. Levine, R.L.; Garland, D.; Oliver, C.N.; Amici, A.; Climent, I.; Lenz, A.-G.; Ahn, B.-W.; Shaltiel, S.; Stadtman, E.R. [49] Determination of carbonyl content in oxidatively modified proteins. In *Methods in Enzymology*; Elsevier: Amsterdam, The Netherlands, 1990; Volume 186, pp. 464–478.
29. Grisham, M.B.; Johnson, G.G.; Lancaster, J.R., Jr. Quantitation of nitrate and nitrite in extracellular fluids. In *Methods in Enzymology*; Elsevier: Amsterdam, The Netherlands, 1996; Volume 268, pp. 237–246.
30. Griffith, O.W. Determination of glutathione and glutathione disulfide using glutathione reductase and 2-vinylpyridine. *Anal. Biochem.* **1980**, *106*, 207–212. [CrossRef]
31. Nishikimi, M.; Rao, N.A.; Yagi, K. The occurrence of superoxide anion in the reaction of reduced phenazine methosulfate and molecular oxygen. *Biochem. Biophys. Res. Commun.* **1972**, *46*, 849–854. [CrossRef]

32. Aebi, H. [13] Catalase in vitro. In *Methods in Enzymology*; Academic Press: Cambridge, MA, USA, 1984; Volume 105, pp. 121–126.
33. Bancroft, J.D.; Gamble, M. *Theory and Practice of Histological Techniques*; Elsevier Health Sciences: Amsterdam, The Netherlands, 2008.
34. Kim, S.-W.; Roh, J.; Park, C.-S. Immunohistochemistry for pathologists: Protocols, pitfalls, and tips. *J. Pathol. Transl. Med.* **2016**, *50*, 411–418. [CrossRef]
35. Brooks, W.W.; Conrad, C.H. Isoproterenol-induced myocardial injury and diastolic dysfunction in mice: Structural and functional correlates. *Comp. Med.* **2009**, *59*, 339–343. [PubMed]
36. Li, X.; Zhou, R.; Zheng, P.; Yan, L.; Wu, Y.; Xiao, X.; Dai, G. Cardioprotective effect of matrine on isoproterenol-induced cardiotoxicity in rats. *J. Pharm. Pharmacol.* **2010**, *62*, 514–520. [CrossRef]
37. Nigam, P. Biochemical markers of myocardial injury. *Indian J. Clin. Biochem.* **2007**, *22*, 10–17. [CrossRef] [PubMed]
38. Patel, V.; Upaganlawar, A.; Zalawadia, R.; Balaraman, R. Cardioprotective effect of melatonin against isoproterenol induced myocardial infarction in rats: A biochemical, electrocardiographic and histoarchitectural evaluation. *Eur. J. Pharmacol.* **2010**, *644*, 160–168. [CrossRef]
39. Shu, Z.; Yang, Y.; Yang, L.; Jiang, H.; Yu, X.; Wang, Y. Cardioprotective effects of dihydroquercetin against ischemia reperfusion injury by inhibiting oxidative stress and endoplasmic reticulum stress-induced apoptosis via the PI3K/Akt pathway. *Food Funct.* **2019**, *10*, 203–215. [CrossRef] [PubMed]
40. Sun, X.; Chen, R.-C.; Yang, Z.-H.; Sun, G.-B.; Wang, M.; Ma, X.-J.; Yang, L.-J.; Sun, X.-B. Taxifolin prevents diabetic cardiomyopathy in vivo and in vitro by inhibition of oxidative stress and cell apoptosis. *Food Chem. Toxicol.* **2014**, *63*, 221–232. [CrossRef] [PubMed]
41. Singal, P. Potential oxidative pathways of catecholamines in the formation of lipid peroxides and genesis of heart disease. *Adv. Exp. Med. Biol.* **1980**, *191*, 421–427.
42. Kondo, T.; Ogawa, Y.; Sugiyama, S.; Ito, T.; Satake, T.; Ozawa, T. Mechanism of isoproterenol induced myocardial damage. *Cardiovasc. Res.* **1987**, *21*, 248–254. [CrossRef]
43. Ramakrishna, K.; Krishnamurthy, S. Indole-3-carbinol ameliorated the isoproterenol-induced myocardial infarction via multi-modal mechanisms in Wistar rats. *Nat. Prod. Res.* **2022**, 1–6. [CrossRef]
44. Khalil, M.; Ahmmed, I.; Ahmed, R.; Tanvir, E.; Afroz, R.; Paul, S.; Gan, S.H.; Alam, N. Amelioration of isoproterenol-induced oxidative damage in rat myocardium by Withania somnifera leaf extract. *BioMed Res. Int.* **2015**, *2015*, 1–10.
45. Othman, A.I.; Elkomy, M.M.; El-Missiry, M.; Dardor, M. Epigallocatechin-3-gallate prevents cardiac apoptosis by modulating the intrinsic apoptotic pathway in isoproterenol-induced myocardial infarction. *Eur. J. Pharmacol.* **2017**, *794*, 27–36. [CrossRef]
46. Zhou, Q.; Song, J.; Wang, Y.; Lin, T. Remifentanil attenuates cardiac dysfunction, lipid peroxidation and immune disorder in rats with isoproterenol-induced myocardial injury via JNK/NF-KB p65 inhibition. *Ann. Transl. Med.* **2020**, *8*, 551. [CrossRef] [PubMed]
47. Wang, Y.; Yang, J.; Yi, J. Redox sensing by proteins: Oxidative modifications on cysteines and the consequent events. *Antioxid. Redox Signal.* **2012**, *16*, 649–657. [CrossRef] [PubMed]
48. Pacher, P.; Beckman, J.S.; Liaudet, L. Nitric oxide and peroxynitrite in health and disease. *Physiol. Rev.* **2007**, *87*, 315–424. [CrossRef]
49. Najeb, S.M.; Jaccob, A.A.; Al-Moziel, M.S.; Abdulhameed, H.M. Cardioprotective and antioxidant effects of taxifolin and vitamin C against diazinone-induced myocardial injury in rats. *Environ. Anal. Health Toxicol.* **2022**, *37*, e2022002. [CrossRef]
50. Bedir, F.; Kocatürk, H.; Yapanoğlu, T.; Gürsul, C.; Arslan, R.; Mammadov, R.; Çoban, A.; Altuner, D.; Suleyman, H. Protective effect of taxifolin against prooxidant and proinflammatory kidney damage associated with acrylamide in rats. *Biomed. Pharmacother.* **2021**, *139*, 111660. [CrossRef] [PubMed]
51. Ahiskali, I.; Pinar, C.L.; Kiki, M.; Cankaya, M.; Kunak, C.S.; Altuner, D. Effect of taxifolin on methanol-induced oxidative and inflammatory optic nerve damage in rats. *Cutan. Ocul. Toxicol.* **2019**, *38*, 384–389. [CrossRef]
52. Allawadhi, P.; Khurana, A.; Sayed, N.; Kumari, P.; Godugu, C. Isoproterenol—Induced cardiac ischemia and fibrosis: Plant—Based approaches for intervention. *Phytother. Res.* **2018**, *32*, 1908–1932. [CrossRef]
53. Prabhu, S.D.; Frangogiannis, N.G. The biological basis for cardiac repair after myocardial infarction: From inflammation to fibrosis. *Circ. Res.* **2016**, *119*, 91–112. [CrossRef]
54. Senoner, T.; Dichtl, W. Oxidative stress in cardiovascular diseases: Still a therapeutic target? *Nutrients* **2019**, *11*, 2090. [CrossRef]
55. Sukoyan, G.; Golovach, S.; Dolidze, N.; Kezeli, T.; Gongadze, N. Hypertrophic and Inflammatory Markers in Isoproterenol-Induced Cardiac Hypertrophy and its Pharmacological Correction. *Cardiovasc. Pharm. Open Access* **2017**, *6*, 225–230.
56. Li, Y.; Feng, L.; Li, G.; An, J.; Zhang, S.; Li, J.; Liu, J.; Ren, J.; Yang, L.; Qi, Z. Resveratrol prevents ISO-induced myocardial remodeling associated with regulating polarization of macrophages through VEGF-B/AMPK/NF-kB pathway. *Int. Immunopharmacol.* **2020**, *84*, 106508. [CrossRef]
57. Freund, C.; Schmidt-Ullrich, R.; Baurand, A.; Dunger, S.; Schneider, W.; Loser, P.; El-Jamali, A.; Dietz, R.; Scheidereit, C.; Bergmann, M.W. Requirement of nuclear factor-κB in angiotensin II–and isoproterenol-induced cardiac hypertrophy in vivo. *Circulation* **2005**, *111*, 2319–2325. [CrossRef] [PubMed]
58. Verma, V.K.; Malik, S.; Narayanan, S.P.; Mutneja, E.; Sahu, A.K.; Bhatia, J.; Arya, D.S. Role of MAPK/NF-κB pathway in cardioprotective effect of Morin in isoproterenol induced myocardial injury in rats. *Mol. Biol. Rep.* **2019**, *46*, 1139–1148. [CrossRef] [PubMed]
59. Zhazykbayeva, S.; Pabel, S.; Mügge, A.; Sossalla, S.; Hamdani, N. The molecular mechanisms associated with the physiological responses to inflammation and oxidative stress in cardiovascular diseases. *Biophys. Rev.* **2020**, *12*, 947–968. [CrossRef]

60. Prince, P.S.M.; Hemalatha, K.L. A molecular mechanism on the antiapoptotic effects of zingerone in isoproterenol induced myocardial infarcted rats. *Eur. J. Pharmacol.* **2018**, *821*, 105–111. [CrossRef]
61. Jin, H.; Liu, A.D.; Holmberg, L.; Zhao, M.; Chen, S.; Yang, J.; Sun, Y.; Chen, S.; Tang, C.; Du, J. The role of sulfur dioxide in the regulation of mitochondrion-related cardiomyocyte apoptosis in rats with isopropylarterenol-induced myocardial injury. *Int. J. Mol. Sci.* **2013**, *14*, 10465–10482. [CrossRef]
62. Chen, W.; Deng, W.; Chen, S. Inactivation of Nf-kb Pathway by Taxifolin Attenuates Sepsis-Induced Acute Lung Injury. *Curr. Top. Nutraceutical Res.* **2020**, *18*, 176–182.
63. Satta, S.; Mahmoud, A.M.; Wilkinson, F.L.; Yvonne Alexander, M.; White, S.J. The role of Nrf2 in cardiovascular function and disease. *Oxidative Med. Cell. Longev.* **2017**, *2017*, 9237263. [CrossRef]
64. Wardyn, J.D.; Ponsford, A.H.; Sanderson, C.M. Dissecting molecular cross-talk between Nrf2 and NF-κB response pathways. *Biochem. Soc. Trans.* **2015**, *43*, 621–626. [CrossRef]
65. Ahmed, S.M.U.; Luo, L.; Namani, A.; Wang, X.J.; Tang, X. Nrf2 signaling pathway: Pivotal roles in inflammation. *Biochim. Biophys. Acta (BBA)-Mol. Basis Dis.* **2017**, *1863*, 585–597. [CrossRef] [PubMed]
66. Li, J.; Ichikawa, T.; Villacorta, L.; Janicki, J.S.; Brower, G.L.; Yamamoto, M.; Cui, T. Nrf2 protects against maladaptive cardiac responses to hemodynamic stress. *Arterioscler. Thromb. Vasc. Biol.* **2009**, *29*, 1843–1850. [CrossRef] [PubMed]
67. Lei, L.; Chai, Y.; Lin, H.; Chen, C.; Zhao, M.; Xiong, W.; Zhuang, J.; Fan, X. Dihydroquercetin activates AMPK/Nrf2/HO-1 signaling in macrophages and attenuates inflammation in LPS-induced endotoxemic mice. *Front. Pharmacol.* **2020**, *11*, 662. [CrossRef] [PubMed]

biomolecules

MDPI

*Article*

# Early Resveratrol Treatment Mitigates Joint Degeneration and Dampens Pain in a Mouse Model of Pseudoachondroplasia (PSACH)

Jacqueline T. Hecht [1], Alka C. Veerisetty [1], Debabrata Patra [2], Mohammad G. Hossain [1], Frankie Chiu [1], Claire Mobed [3], Francis H. Gannon [4] and Karen L. Posey [1,*]

1 Department of Pediatrics, McGovern Medical School, The University of Texas Health Science Center at Houston (UTHealth), Houston, TX 77030, USA; jacqueline.t.hecht@uth.tmc.edu (J.T.H.); alka.veerisetty@uth.tmc.edu (A.C.V.); mohammad.g.hossain@uth.tmc.edu (M.G.H.); frankie.chiu@uth.tmc.edu (F.C.)
2 Department of Developmental Biology, Washington University School of Medicine, St. Louis, MO 63110, USA; debabratapatra@wustl.edu
3 Department of Biology, Rice University, Houston, TX 77005, USA; com27@case.edu
4 Departments of Pathology and Immunology and Orthopedic Surgery, Baylor College of Medicine, Houston, TX 77030, USA; fgannon@bcm.edu
* Correspondence: karen.posey@uth.tmc.edu

**Abstract:** Pseudoachondroplasia (PSACH), a severe dwarfing condition associated with early-onset joint degeneration and lifelong joint pain, is caused by mutations in cartilage oligomeric matrix protein (COMP). The mechanisms underlying the mutant-COMP pathology have been defined using the MT-COMP mouse model of PSACH that has the common D469del mutation. Mutant-COMP protein does not fold properly, and it is retained in the rough endoplasmic reticulum (rER) of chondrocytes rather than being exported to the extracellular matrix (ECM), driving ER stress that stimulates oxidative stress and inflammation, driving a self-perpetuating cycle. CHOP (ER stress signaling protein) and TNFα inflammation drive high levels of mTORC1 signaling, shutting down autophagy and blocking ER clearance, resulting in premature loss of chondrocytes that negatively impacts linear growth and causes early joint degeneration in MT-COMP mice and PSACH. Previously, we have shown that resveratrol treatment from birth to 20 weeks prevents joint degeneration and decreases the pathological processes in articular chondrocytes. Resveratrol's therapeutic mechanism of action in the mutant-COMP pathology was shown to act by primarily stimulating autophagy and reducing inflammation. Importantly, we demonstrated that MT-COMP mice experience pain consistent with PSACH joint pain. Here, we show, in the MT-COMP mouse, that resveratrol treatment must begin within 4 weeks to preserve joint health and reduce pain. Resveratrol treatment started at 6 or 8 weeks (to 20 weeks) was not effective in preventing joint degeneration. Collectively, our findings in MT-COMP mice show that there is a postnatal resveratrol treatment window wherein the inevitable mutant-COMP joint degeneration and pain can be prevented.

**Keywords:** cartilage oligomeric matrix protein; COMP; autophagy; resveratrol; dwarfism; chondrocyte; articular cartilage; joint degeneration; joint pain

## 1. Introduction

COMP is a large pentameric, matricellular protein that binds to ECM proteins including collagen types I, II, IX, XII, XIV, matrilin-3, aggrecan, sparc, and fibronectin, providing an interaction hub [1–3] that contributes to cartilage homeostasis [1–4]. Mechanical loading increases COMP levels in tendons while aging or overuse decreases content, suggesting that COMP plays a role in the mechanical strength of ECM tissues [2,5]. Chondrogenesis and chondrocyte proliferation are stimulated by COMP, demonstrating that it has multiple functions [6,7]. In contrast, mutations in COMP cause pseudoachondroplasia (PSACH), a severe dwarfing condition characterized by disproportionately short stature, short limbs, joint laxity, pain, and early-onset joint degeneration [3,8–18]. PSACH birth parameters are within the normal range, and the first noticeable change is linear growth deceleration starting around one year of life and a waddling gait that appears after two years [19]. Characteristic rhizomelic shortening of long bones, small abnormal and underossified epiphyses, widened and irregular metaphyses, and platyspondyly are the radiograph findings that confirm the diagnosis [13,17,19]. Genetic testing is now used to confirm the clinical diagnosis. While the loss of linear growth is the most apparent feature of PSACH, joint dysfunction and pain are the most debilitating and long-term complications that diminish quality of life [20]. Significant pain begins in childhood and is most likely caused by inflammation related to the growth plate chondrocyte pathology [9,13,21,22]. Pain in adulthood is related to degenerative changes in the joints that necessitate hip replacement between the second and third decades of life [11,19]. Non-surgical treatment is critically needed since all joints are affected, especially hips, elbows, and shoulders, but not all joints are replaceable [13,20,23].

Massive retention of a lamellar material was first identified in electron microscopy (EM) images of PSACH growth plate chondrocytes from children aged 5 to 12 years, and after gene discovery, it was proven to be COMP [9,13,17]. Subsequent studies showed that an intracellular matrix was present in the enlarged ER cisternae, composed of COMP, types 2 and 9 collagens, and matrilin-3 [22]. Intracellular retention of COMP results from abnormal protein folding related to the loss of the calcium scaffold essential for proper 3-dimensional conformation [19]. To define the mechanisms that underlie the PSACH pathology, we generated the MT-COMP mouse that expresses mutant human D469del-COMP in tissues expressing type 2 collagen with doxycycline (DOX) administration that recapitulates both the clinical phenotype and PSACH chondrocyte pathology [19]. Using the MT-COMP mouse, we showed that the accumulation of mutant-COMP in the rER is toxic to the growth plate and articular chondrocytes [13,22,24]. The retention of mutant-COMP induces ER stress along with oxidative stress and inflammation, creating a self-perpetuating stress loop that leads to DNA damage and a blockage of autophagy [25]. The autophagy blockade is stimulated by increases in mTORC1 signaling driven by TNF$\alpha$ and CHOP [25]. mTORC1, a master regulator of growth, responds to nutritional status, cellular stress, and growth factors regulating general protein translation and autophagy [26]. Elevated mTORC1 signaling favors general protein synthesis at the expense of autophagy, which may exacerbate ER stress and directly inhibit autophagy clearance of the ER in chondrocytes [25,27]. The cumulative stress coupled with the block of ER clearance drives a senescent phenotype in articular chondrocytes that likely propagates degenerative changes to nearby cells and tissues of the joint [28]. Previously, we showed that resveratrol treatment from birth prevents mutant-COMP accumulation by stimulating autophagy and reducing inflammation and chondrocyte death, avoiding premature joint degeneration [27,29]. This work builds on the successful resveratrol joint-sparing treatment in the MT-COMP mice by defining the resveratrol treatment window that prevents MT-COMP joint degeneration.

## 2. Materials and Methods

### 2.1. Bigenic Mice

MT-COMP mice are a bigenic inducible mouse line that contains two plasmids: pTRE-COMP [22] (human COMPD469del + FLAG tag coding sequence driven by the tetracycline responsive element promoter) and pTET-On-Col II (rtTA coding sequence driven by a type II collagen promoter). These two constructs together express mutant-COMP with the D469del mutation in the cartilage of mice in the presence of DOX [22,30]. All mice were PCR genotyped to verify the presence of both transgenes [22]. DOX (500 ng/mL) was administered to mice at birth, through mother's milk, and collected in their drinking water. Human mutant-COMP is expressed in addition to endogenous mouse COMP. This study complied with the Guide for the Care and Use of Laboratory Animals, eighth edition (ISBN-10, 0-309-15396-4), was approved by the Animal Welfare Committee at the University of Texas Medical School at Houston, and complies with NIH guidelines. Importantly, transgenic mice overexpressing wild-type COMP are similar to C57BL/6 control mice [30] and therefore are used as controls. To control for untoward outcomes related to DOX, C57BL/6 mice (control) on DOX are included in all experiments [28,30].

### 2.2. Resveratrol Administration

Resveratrol (250 mg/L [19,27,31]) was administered through drinking water starting at the following time points: birth, 4, 6, 8, 16 to 20 weeks of age (Nature's Answer Alcohol-Free Resveratrol Reserve, Hauppauge, NY, USA). Prior to weaning, from birth to 3 weeks, resveratrol was transferred to pups through mother's milk. Mass spectroscopy confirmed the resveratrol concentration.

### 2.3. Ibuprofen Administration

Ibuprofen gel capsule (200 mg) was diluted into 1 L of water with 500 ng/mL DOX for a final concentration of 0.2 mg/mL [32]. Water consumption was monitored to ensure that mice consumed sufficient amounts of liquid prior to using ibuprofen DOX water in experiments. Mice were provided ibuprofen DOX water 24 h prior to grooming assessment.

### 2.4. Immunohistochemistry

Hind limbs from male and female MT-COMP and C57BL\6 control mice were collected and articular cartilage analyzed, as previously described [22,30]. Briefly, hind limbs were fixed in 4% PFA, and heat-induced epitope retrieval antigen retrieval was performed with sodium citrate buffer for pS6 (Cell Signaling Technology, Danvers, MA, USA; 2215S rabbit polyclonal, 1:200), tumor necrosis factor α (TNFα) (Abcam, Cambridge, United Kingdom; ab6671, 1:200), and MMP13 (Abcam ab39012, 1:50) antibodies. With IL-6 (Bioss Woburn, MA bs-0781R, 1:200), SIRT1 (Abcam 32441, 1:100), TRAIL (Abcam ab42243, 1:200), p16 INK4a (Abcam ab189034, 1:200), MID1 (Abcam ab-70770, 1:200), CHOP (Santa Cruz SC-575; 1:100), and human COMP (Thermofisher, Waltham, MA, USA; 1-20221, 1:100) antibodies, the hindlimbs were fixed in 95% ethanol, and pepsin (1 mg/mL in 0.1 N HCl) was used for antigen retrieval. Species-specific biotinylated secondary antibodies were used for 1 h. at RT. Sagittal sections of the same thickness (5 μm) were then washed and incubated with streptavidin, horseradish peroxidase (HRP), and DAB as the chromogen. The sections were dehydrated and mounted with cytoseal 60 (Thermofisher) and then visualized under a BX51 inverted microscope (Olympus America, Center Valley, PA, USA). Limbs were fixed in 10% wt/vol formalin for terminal deoxynucleotidyl transferase-mediated deoxyuridine triphosphate-biotin nick end labeling (TUNEL) staining using the Promega G3250 kit following the manufacturer's protocol. For proteoglycan stains, samples were deparaffinized, hydrated in distilled water, and stained with safranin-O (Spectrum Chemical, New Brunswick, NJ, USA, 477-73-6) according to the manufacturer's protocol. All assays were performed on 10 animals in each group.

### 2.5. Joint Degeneration Scoring

A joint scoring system that quantifies early degenerative changes in the proteoglycan content of the articular cartilage of the femur and tibia, the degree of synovitis, and bone/cartilage damage was developed and employed in this study [29]. This method is more sensitive to early joint changes than the OARSI scoring system, which best describes end-stage joint damage (Figure S1). Joint degeneration scoring was performed on 5 μm sagittal sections from 10 different mice. Only sections that contained both menisci were scored in order to ensure sections were obtained from the same area of the joint. While OARSI scoring covers a wide range of OA pathology, in this study, joint degeneration scoring was modified to optimize the evaluation of early joint pathology. Four areas—synovium, bone/cartilage, tibial, and femoral articular cartilage—were scored from 0 to 3 on each Safranin-O-stained section. This early-stage joint pathology scoring system was developed specifically after examining the pathology of MT-COMP mice to quantify the early degenerative characteristics. To show the utility of this scoring system, control and

MT-COMP mice were scored using OARSI scoring, OARSI scoring + meniscus/synovium component, and the early joint pathology score described here (Figure S1). All scoring systems showed a trend towards a higher level of damage for the MT-COMP joints. To show the utility of this scoring system, control and MT-COMP mice were scored using OARSI scoring, OARSI scoring + meniscus/synovium component, and the early joint pathology score described here (Figure S1). All scoring systems showed a trend towards a higher level of damage for the MT-COMP joints compared to controls, but only the early scoring system showed a significant difference. A score of 0 indicated normal or no damage, 1 = mild damage, 2 = moderate damage, and 3 = severe damage. Synovitis, bone/cartilage damage, proteoglycan of the tibia, and proteoglycan of the femur were scored individually, and all scores were summed with a maximal damage score of 12. Synovitis was defined as a mild—increase in thickness of synovial lining and increase in stromal area; a moderate—increase in stromal density; or a severe—thickening of synovial lining with further increase of stromal cellular density. Bone/cartilage damage was defined as: normal—the surface was smooth; mild—minor erosion of the surface; moderate—the presence of remodeling with minor erosion; or severe—major erosion. Proteoglycans of the articular cartilage of the tibia and femur were classified as: normal—if staining was even through to the subchondral bone; mild—when staining was thinned; moderate—thinning of the proteoglycan-stained layer and absence of staining in some areas; or severe—the widespread loss of proteoglycan staining. Ten mice per experimental group were used for each time point, providing 80–90% power to detect a minimal difference of 2 or 3 units. All scoring was performed blindly. Limb section depth, thickness, fixation, and decalcification conditions were identical. A *t*-test was used to evaluate the total joint degeneration score across six experiment groups, comparing MT-COMP to all other groups. The Kruskal-Wallis test was used to evaluate the distribution of individual scored features across six experiment groups, and the post-hoc Dunn's test compared MT-COMP to controls.

## 2.6. Voluntary Running

Mice were acclimated to the running wheel for 3 days before data collection. Running activity was collected on male mice for 7 days and nights. The cumulative running distance at night was calculated and compared to controls using a *t*-test. The standard deviation is represented by error bars in Figure 1.

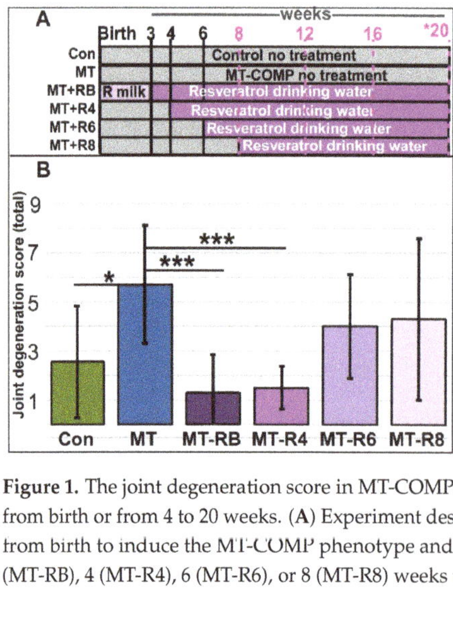


**Figure 1.** The joint degeneration score in MT-COMP mice was decreased with resveratrol treatment from birth or from 4 to 20 weeks. (**A**) Experiment design graphic. Male mice were administered DOX from birth to induce the MT-COMP phenotype and treated with resveratrol starting either at birth (MT-RB), 4 (MT-R4), 6 (MT-R6), or 8 (MT-R8) weeks until 20 weeks and analyzed. Timepoints noted

in pink denote grooming evaluations. (**B**) Joint degeneration scoring total is the sum of four scores based on scoring each area from 0–3 (synovitis, femoral proteoglycans, tibial proteoglycans, and cartilage/bone damage). Results were compared to untreated MT-COMP (MT) mice at 20 weeks. MT-COMP mice treated from birth to 4 weeks had significantly improved joint degeneration scores. Ages shown in purple text denote grooming timepoints and asterisk * denotes age at which joint degeneration was evaluated. (Abbreviations: R milk = resveratrol through mother's milk; Control = Con; MT-COMP = MT; MT-COMP mice treated with resveratrol at birth = M+RB; 4–20 weeks = M+R4; MT-COMP mice treated with resveratrol 6–20 weeks = M+R6). $n = 10$, * $p < 0.05$; *** $p < 0.0005$ (Kruskal-Wallis).

### *2.7. Grooming Assay*

Fluorescent dye (50 µL) was applied on the back of the neck equidistant between ears at time 0, and grooming efficiency was assessed 4 h later in male mice. Animals were imaged and compared to the scoring rubric previously published, with maximal grooming scoring at five [33]. The average grooming scores were analyzed with Kruskal-Wallis with a post-hoc Dwass-Steel-Critchlow-Fligner (DSCF) pairwise test. The standard deviation is represented by error bars in Figure 1.

## 3. Results

### *3.1. Resveratrol Treatment Starting at 4 Weeks Preserves Joint Health*

To identify the resveratrol postnatal joint degeneration prevention window, MT-COMP mice were administered resveratrol from birth at 4, 6, 8, and 12 to 20 weeks. Joint degeneration was quantified using a 0–3 scoring system, signifying no damage to severe damage in four regions: (1) cartilage/bone degeneration, (2) proteoglycan loss in the tibia, (3) proteoglycan loss in the femur, and (4) the presence of synovitis. The joint degeneration score is the sum of the four scored regions, with a maximal score of 12. In the knee joint, early degenerative changes include synovitis, meniscus damage, and the loss of proteoglycans. This early scoring system was developed to measure degenerative changes observed in safranin-O-stained sections of MT-COMP mice knee joints at 20 weeks that are not measurable with OARSI [34] or OARSI with synovitis scoring systems (see Section 2; Supplemental Figure S1).

MT-COMP mice administered resveratrol for 12–20 weeks were indistinguishable from untreated MT-COMP mice and were not further evaluated. The total joint degeneration score of 1.5 ± sd 0.9 (±standard deviation = sd) from MT-COMP mice treated with resveratrol from 4 to 20 weeks was significantly lower than the score of 5.7 ± 2.4 from untreated MT-COMP. This was also lower than the score of 4.0 ± sd 2.1 for resveratrol-treated MT-COMP mice from 6–20 weeks or a score of 4.3 ± sd 3.3 from 8–20-week mice (Figure 1 and Supplemental Figure S2). Proteoglycan loss, an early indicator of joint damage, is a part of this score [35–37]. Proteoglycan loss in the articular cartilage of femur and tibia MT-COMP mice treated with resveratrol from 4 to 20 weeks was significantly less (femur 0.3 ± 0.6 and tibia 0.1 ± 0.3) compared to untreated MT-COMP mice (1.8 ± 0.9; 1.2 ± 0.6). Synovitis, which makes up a substantial portion of the total score of untreated MT-COMP mice (1.8 ± 0.6 out of a total score of 5.7 ± 2.4), was significantly reduced in MT-COMP mice treated with resveratrol from birth to 20 weeks (0.5 ± 0.5) or 4 to 20 weeks (0.4 ± 0.6). These results suggest that both synovitis and proteoglycan loss are valuable markers for the analysis of PSACH joint pathology.

### *3.2. Pain Observed in MT-COMP Mice Was Mitigated by Resveratrol Treatment*

Murine pain was assessed using proxy assays evaluating alterations in instinctive behaviors [33], such as changes in physical activity (voluntary running) and grooming. Changes in voluntary running and grooming were evaluated based on the premise that pain will reduce the number of wheel rotations, or distance run, and grooming efficiency [38]. Each mouse was housed in a cage with a running wheel that measured the distance it ran. Since mice are nocturnal, voluntary running was measured over seven nights after a 3-night acclimation period. Running was assessed at 12, 16, 20, 24, 30, and 36 weeks. Figure 2 and Supplemental Figure S3 show that MT-COMP mice ran significantly less than control mice at each time point, suggesting the presence of pain.

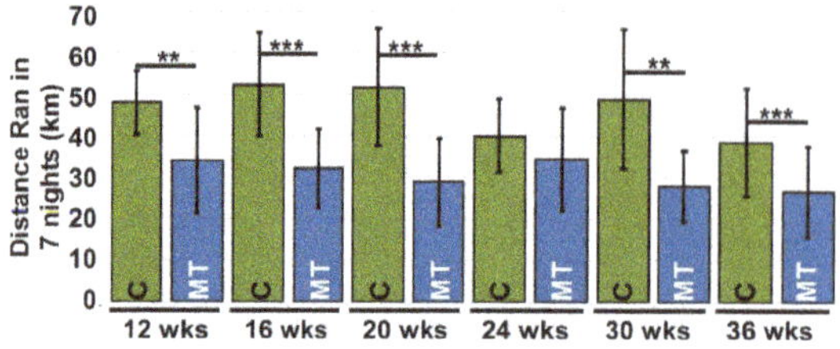


**Figure 2.** Voluntary running is reduced in MT-COMP mice, suggesting pain. All male mice were administered DOX from birth to induce mutant-COMP expression until the completion of the study. Voluntary running was used as a proxy for pain. Voluntary running data was collected at 12, 16, 20, 24, 30, and 36 weeks of age for both control C57BL\6 (C) and MT-COMP (MT) mice ($n$ = 10). All mice had a 3-night acclimation period to adjust to the running wheel. MT-COMP mice ran significantly less than controls at 12, 16, 20, 30, and 36 weeks. ** $p < 0.005$; *** $p < 0.0005$ (Kruskal Wallis).

Grooming, a natural activity, has been shown to be dampened or less efficient in the presence of pain [33]. To analyze this, we used a previously validated assay [33] where fluorescent dye was applied to the back of the neck equidistant between the ears, and the efficiency of dye removal by grooming was analyzed 4 h later. Figure 3 shows that MT-COMP mice consistently groom less than controls. To validate that grooming is associated with pain in MT-COMP mice, an analgesic (ibuprofen) was administered for 7 days, and then grooming was tested. Ibuprofen administration (0.2 mg/mL in drinking water) 7 days prior to grooming resulted in MT-COMP grooming scores comparable to those of control mice at 16 weeks (Supplemental Figure S4). Similarly, in the absence of DOX, which induces mutant-COMP expression, grooming scores were equivalent to controls (Figure 3 and Supplemental Figure S5). Importantly, MT-COMP mice administered resveratrol beginning at birth, 4, 6, and 8 weeks were groomed as efficiently as controls at all time points and better than untreated MT-COMP mice, despite some joint degeneration detected (Figures 1 and 3). These results suggest that pain is not directly linked to joint degeneration levels in MT-COMP mice (Figures 1–3).

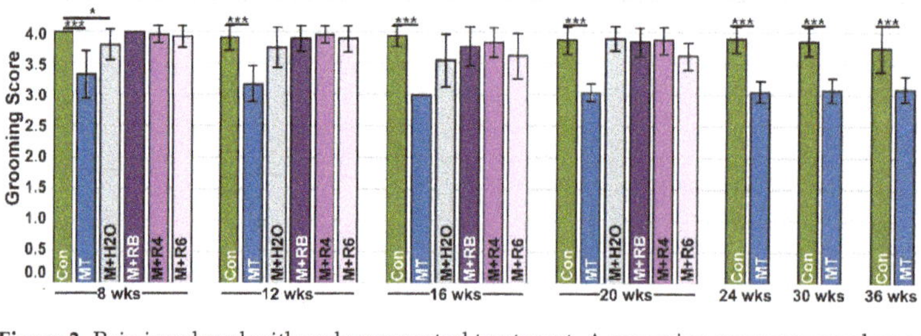


**Figure 3.** Pain is reduced with early resveratrol treatment. A grooming assay was used as a proxy for pain. The grooming assay measures the efficiency of the removal of a fluorescent dye from the fur, with a higher score indicating more effective elimination of dye (maximum score = 5). All male mice were administered DOX from birth to collection, except MT-COMP+H2O (M+H2O). Grooming was assessed at ages 8, 12, 16, 20, 24, 30, and 36 weeks in control C57BL\6 (Control) mice and MT-COMP mice, and MT-COMP mice treated with resveratrol beginning at birth (M+RB), 4 (M+4), 6 (M+6) weeks, and MT-COMP without DOX (M+H2O) ($n \geq 12$). MT-COMP mice have a significantly lower grooming

score than controls at all ages. Resveratrol treatment normalizes grooming at 8, 12, 16, and 20 weeks. Pairwise comparisons (Kruskal-Wallis) between control and all other groups are shown with asterisks. Importantly, MT-COMP grooming scores were lower than MT-COMP+H2O (MT+H2O) in the absence of the induction of mutant-COMP ($p < 0.005$ at 16 and 20 wks; $p < 0.0005$ at 8 and 12 wks). Moreover, pairwise comparisons show grooming scores from all resveratrol treatments are significantly higher than scores from untreated MT-COMP ($p < 0.0005$ at all ages tested). (Abbreviations: weeks = wks; R = resveratrol; Control = Con; MT-COMP = MT; MT-COMP mice treated with resveratrol at birth = M+RB; 4–20 weeks = M+R4; MT-COMP mice treated with resveratrol 6–20 weeks = M+R6). * $p < 0.05$; *** $p < 0.0005$.

### 3.3. *Early Resveratrol Treatment Is Associated with a Reduction in Intracellular Comp, ER Stress, and Inflammation*

Previously, we showed that mutant-COMP retention and inflammation in articular chondrocytes start at 4 weeks, and ER stress, autophagy blockage, and articular chondrocyte death were observed at 8 and 16 weeks, respectively [28]. Our published findings demonstrated that resveratrol treatment from birth to 20 weeks eliminates mutant-COMP protein retention, ER stress, inflammation, and autophagy blockage [28]. To assess resveratrol treatment at different intervals, MT-COMP mice were administered resveratrol from birth–20, 4–20, 6–20, and 8–20 weeks and compared to no treatment and control mice. Figure 4 showed that intracellular COMP accumulation (A–F), CHOP, ER stress marker (G–L), MMP13, a cartilage degradative enzyme (Y–AD), and articular chondrocyte death (AK–AP) were not observed in MT-COMP mice treated with resveratrol from birth to 20 and 4–20 weeks but were present in those treated from 6–20 and 8–20 weeks. This data again indicates a direct correlation between pathologies and mutant-COMP accumulation. pS6, a marker of mTORC1 activity, was minimally present in all treated animals (Figure 4M–R), while MID1, a driver of mTORC1 signaling, was markedly reduced in mice treated at birth, 4, and 6 weeks (Figure 4S–X). Senescence marker p16INK4a was observed at minimal levels in a few articular chondrocytes in MT-COMP mice treated at birth, 4 weeks, and 6 weeks, and more intensely in those treated at 8 weeks (Figure 4AE–AJ). Generally, ER stress, mutant-COMP protein accumulation, and chondrocyte death occur before degeneration markers are present, suggesting that ER stress driven by mutant-COMP protein accumulation in the ER and the associated chondrocyte death is the key to stimulating MT-COMP joint degeneration processes (MMP13, p16INK4a) [25,27].

### 3.4. *Resveratrol Treatment of MT-COMP Mice from 6–20 or 8–20 Weeks Correlates with the Presence of Inflammation*

We showed previously that MT-COMP mice have robust inflammatory processes and reduced levels of the pro-survival marker SIRT1 in the growth plate at 4 weeks [28,29] that were dampened by resveratrol treatment [27]. Based on those findings, inflammation was evaluated for MT-COMP mice treated with resveratrol from birth–20, 4–20, 6–20, and 8–20 weeks and compared to no treatment and control mice. Early resveratrol treatment beginning at birth or 4 weeks decreases TNF$\alpha$, TRAIL, and IL6 and increases the SIRT1 signal (Figure 5A–D,G–J,M–P,S–V). Despite resveratrol treatment for 6–20 and 8–20 weeks, the MT-COMP articular cartilage had active inflammation (TNF$\alpha$ and TRAIL) and repressed SIRT1, a pro-survival protein, with some reduction in IL-6 (Figure 5). These findings, along with joint scoring and pain outcomes, suggest that early resveratrol treatment results in a substantial reduction in inflammatory processes, an improvement in joint/articular chondrocyte health, and a reduction in pain in MT-COMP mice.

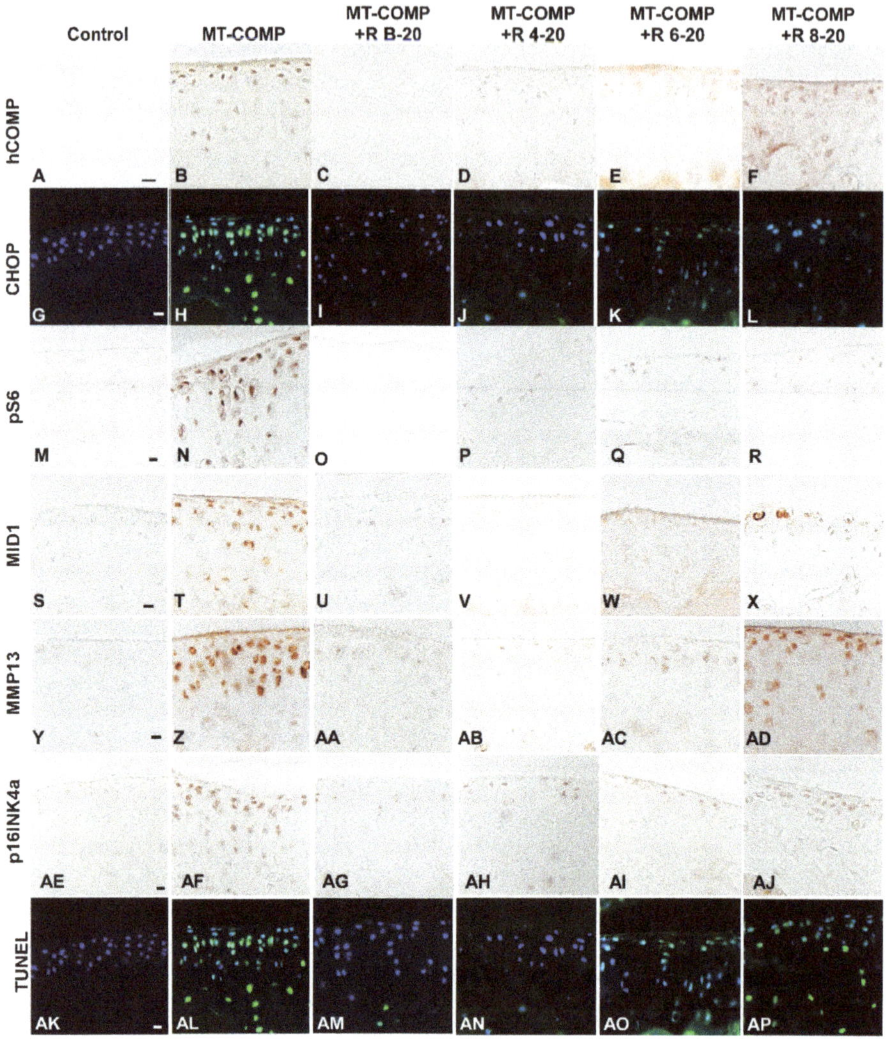


**Figure 4.** Early resveratrol treatment reduced mutant-COMP pathology in the articular cartilage of MT-COMP mice. All mice (both male and female) were administered DOX from birth to collection for analysis at 20 weeks. Resveratrol was administered from birth to 20 weeks (+R B-20), or 4 to 20 weeks (+R 4–20), 6 to 20 weeks (+R 6–20), or 8 to 20 weeks (+R 8–20). Tibial articular cartilage from control, MT-COMP, and resveratrol-treated MT-COMP mice from each group was immunostained (brown or green signal) for human COMP hCOMP (**A–F**), CHOP (**G–L**), pS6 (**M–R**), MID1 (**S–X**), MMP13 (**Y–AD**), pINK4a (**AE–AJ**), and TUNEL (**AK–AP**). DAPI (blue signal) shows nuclei in fluorescence images. MT-COMP articular chondrocytes show substantial intracellular hCOMP, CHOP (ER stress marker), pS6 (mTORC1 activity = blocked autophagy), MID1 (increases mTORC1 activity), MMP13 (degeneration marker), and pINK4a (senescence) signals. Resveratrol treatment for 4–20 weeks reduced MT-COMP pathology. Representative images from the analysis of 10 mice in each study group are shown. Bar (all panels) = 50 μm.

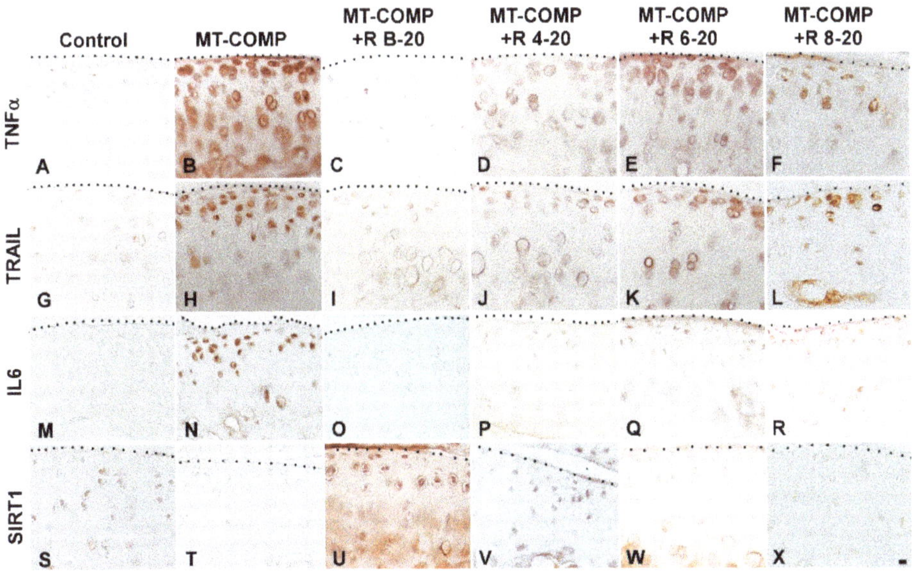


**Figure 5.** Early resveratrol treatment reduces inflammation and supports chondrocyte survival in the articular cartilage of MT-COMP mice. All mice (both male and female) were administered DOX from birth to collection for analysis at 20 weeks. Resveratrol (R) treatment was given from birth to 20 weeks (+R B-20), or 4 to 20 weeks (+R 4–20), 6 to 20 weeks (+R 6–20), or 8 to 20 weeks (+R 8–20). Tibial articular cartilage from control, MT-COMP, and MT-COMP treated with resveratrol was immunostained (brown) for TNFα (**A–F**), TRAIL (**G–L**), IL-6 (**M–R**), and SIRT1 (**S–X**). Representative images from the analysis of 10 mice in each study group are shown. Bar (all panels) = 50 μm.

## 4. Discussion

In this study, we show that there is a defined window of treatment, preferably immediately at birth to 4 weeks, in MT-COMP mice, after which treatment has little effect on abating the mutant-COMP pathology. Previously, we demonstrated that MT-COMP mice develop joint degeneration by 20 weeks of age, and resveratrol treatment from birth to 20 weeks prevents this pathology [28]. To define the effective postnatal resveratrol intervention window, MT-COMP mice were administered resveratrol from birth–20, 4–20, 6–20, and 8–20 weeks. Total joint degeneration scores from controls and MT-COMP mice treated with resveratrol from birth to 20 weeks were significantly lower than those from untreated MT-COMP mice (Figure 1) [27]. Resveratrol treatment that was started before retention of intracellular mutant-COMP (autophagy blockage) began and prior to the appearance of the inflammatory marker TNFα was the most effective [28]. This early resveratrol treatment at 4 weeks reduced ER stress (CHOP), mTORC1 activation (pS6/MID1), and MMP13. Resveratrol dampens articular chondrocyte stress, improves joint health as evidenced by less synovitis and bone/cartilage damage, diminishes the loss of proteoglycans in the articular cartilage, and, most importantly, reduces pain (Figures 1–3). These important findings provide the first indication that there is a postnatal treatment window during which resveratrol or other treatments can mitigate joint degeneration and pain. This has important implications for the treatment of PSACH.

Our work defined a complex stress process responsible for mutant-COMP accumulation in chondrocytes that becomes cytotoxic to the cells, resulting in cell death [19,22,25,27–31]. COMP mutations cause protein misfolding, with 97% of the COMP pentamers expected to have at least one mutant monomer, causing massive retention of COMP in the ER [25]. The ER quality control mechanisms that should refold or degrade misfolded proteins fail, and this is evident as mutant-COMP continuously accumulates in the ER [39]. Two major pathways, autophagy and proteasomal degradation, are responsible for clearing misfolded proteins from the ER [40]. In MT-COMP mice, both ER stress and TNF$\alpha$ drive high levels of mTORC1 signaling [31]. Furthermore, chronic inflammation leads to reduced levels of SIRT1 in MT-COMP chondrocytes, which, in turn, exacerbates mTORC1 activity [41–45]. Notably, the mTORC1 signaling in MT-COMP is found to be independent of AMPK [27]. This elevated mTORC1 signaling represses autophagy, preventing mutant-COMP clearance [25]. Mutant-COMP induces ER stress through CHOP [19,30] which drives oxidative and inflammation processes that sustain a self-perpetuating cycle of stress. IL-1$\beta$ and TNF$\alpha$ stimulate the synthesis of matrix metalloproteinases (MMPs), enzymes responsible for cleaving collagens and other proteoglycans in the extracellular matrix [46]. The elevated levels of MMP13 observed in MT-COMP articular chondrocytes are likely a consequence of chronic inflammation [28]. Oxidative stress associated with mutant-COMP generates reactive oxygen species, causing DNA damage and necroptotic chondrocyte death [19]. Chondrocyte death inhibits both long bone growth and severely compromises joint health. We previously showed that resveratrol treatment increased the number of autophagic vesicles (LC3-II positive vesicles), indicating that part of resveratrol's therapeutic mechanism of action is the activation of autophagy, which promotes the clearance of mutant-COMP [27]. The stimulation of autophagy by resveratrol results in reduced inflammation, a decrease in cartilage-degrading enzyme MMP13 (thus mitigating degeneration), normalization of chondrocyte function by alleviating ER stress (CHOP), and relief from joint degeneration and associated pain caused by mutant-COMP. Clearance of mutant-COMP restores homeostasis and alleviates the multiple associated chondrocyte stresses.

In previous studies, we showed that resveratrol treatment from birth in MT-COMP mice dampened inflammation (TNF$\alpha$, IL-1$\beta$, IL-6, and IL-18), oxidative stress, degradative enzymes (multiple MMPs), and joint damage [27,29,47,48]. IL6, TNF$\alpha$, and MMP13, all of which are associated with osteoarthritis (OA) and compromised joint health, have been observed in MT-COMP joints [28,31]. IL6 is a pro-inflammatory interleukin that is intimately involved in OA joint degeneration [47,49,50]. MMP13, a degradative enzyme, plays an important role in extracellular matrix degeneration in the articular cartilage [46,51,52] and both IL-6 and TNF$\alpha$ stimulate MMP13 expression [53,54]. Notably, these markers of degeneration were reduced by resveratrol treatment starting at birth or 4 weeks (Figures 4 and 5). In contrast, six or eight weeks of unmitigated accumulation of mutant-COMP was sufficient to drive joint degeneration in MT-COMP mice, although improved grooming scores were observed. Resveratrol treatment of MT-COMP mice from 6–20 and 8–20 weeks reduced the autophagy blockage; however, the presence of some mutant-COMP in articular chondrocytes, along with inflammation (TNF$\alpha$, TRAIL), MMP13, and senescence, was sufficient to drive some joint degeneration, despite improvements in pain measurements (Figures 1, 4 and 5). Given that multiple pathological events in the articular cartilage downstream of mutant-COMP protein accumulation occur prior to 6 or 8 weeks of age, it is difficult to identify the precise pathological event that plays the biggest role in joint degeneration. Based on the presence of inflammatory markers and their link to degenerative enzymes, we propose that inflammation may be one driver of MT-COMP joint degeneration. Importantly, early resveratrol treatment may limit degradation through downregulation of inflammation by restraining MMP13 degradation of the extracellular matrix [46,51,52] (Figure 4).

These findings also suggest that senescent articular chondrocytes generate a senescence-associated secretory phenotype (SASP), which likely creates a pro-degenerative environment in MT-COMP joints, including the presence of MMP13 [46]. Our finding that MT-COMP mice have synovitis and damage outside the articular cartilage supports the idea that SASP may contribute to mutant-COMP joint degeneration, leading to the destruction of tissues near the articular cartilage. Resveratrol treatment from birth to 20 weeks or 4–20 weeks substantially reduced synovitis and joint pain, suggesting that reducing stress in articular chondrocytes improves overall joint health, not solely the articular cartilage. Other studies have shown that resveratrol delays joint degeneration when administered simultaneously with an OA-inducing agent [55,56], which aligns with our observation that resveratrol treatment from birth to 20 or 4–20 weeks preserves joint health. Joint degeneration after 6 or 8 weeks of mutant-COMP expression suggests that either early events predetermine joint degeneration due to the poor quality of articular cartilage or early events set in motion destructive processes that cannot be resolved after a particular threshold is exceeded, leading to articular chondrocyte death.

Several aspects of the mutant-COMP pathology, including inflammation and senescence, are particularly important to joint degeneration beyond the articular cartilage. Senescent articular chondrocytes have been observed in MT-COMP mice along with IL6 and MMP13, degenerative components of SASP [28]. These findings suggest that senescent articular chondrocytes generate SASP, which likely creates a pro-degenerative environment in MT-COMP joints [28]. Importantly, the presence of synovitis, which causes damage outside the articular cartilage, supports the idea that SASP may contribute to MT-COMP joint degeneration, leading to the destruction of tissues near the articular cartilage.

Resveratrol treatment from birth to 20 or 4–20 weeks substantially reduces synovitis and joint pain, suggesting that dampening of articular chondrocyte stress improves overall joint health. Others have shown that resveratrol delays joint degeneration when administered simultaneously with OA-induction [57,58], which aligns with our observation that resveratrol from birth to 20 or 4–20 weeks preserves joint health. Joint degeneration from 6 or 8 weeks of mutant-COMP expression suggested that early events predetermine joint degeneration, perhaps due to the poor quality of articular cartilage or early events (mutant-COMP accumulation) that set-in motion destructive processes that cannot be resolved after a particular threshold leading to articular chondrocyte death. These findings indicate that some features of mutant-COMP joint degeneration may be dampened with later resveratrol therapy, but ultimately, early intervention is essential to preserving joint health.

In PSACH, joint pain is the most debilitating complication, and similarly, MT-COMP mice show evidence of pain (Figures 2 and 3). Since behavioral assays can be difficult to interpret, assessment of pain in MT-COMP mice was performed using two validated proxy assays, voluntary running [59–62] and grooming [63–65], both associated with instinctive behaviors. Importantly, we show resveratrol treatment reduces pain as measured by grooming. Similarly, in a clinical trial, resveratrol decreased pain in postmenopausal women [55,66], and resveratrol, when added to knee OA meloxicam therapy (anti-inflammatory), decreased pain more than meloxicam alone [56,67]. Pain is a critical aspect of modeling PSACH because it is a significant problem that often leads to joint replacements [20] and is closely intertwined with disability and decreased quality of life. Pain from joint degeneration comes either directly from damaged tissues or indirectly from inflammation or changes in tissue that alter function. Direct joint pain is associated with synovitis, meniscal damage, subchondral bone remodeling, abnormal ligament insertion, joint capsule thickening, and osteophytes [55,56]. In contrast, inflammation, cartilage degeneration, damage to the inner meniscus, and ligament laxity are indirect causes of degenerative joint pain [55,56]. Given the relatively high level of joint degeneration in the absence or decrease of pain in MT-COMP mice treated with resveratrol from 6–20 and 8–20 weeks, it is likely that resveratrol suppresses mutant-COMP-induced pain through indirect sources rather than direct ones. Any PSACH therapy must address pain. Since grooming assays indicate that resveratrol reduces pain associated with mutant-COMP pathology, this suggests that resveratrol or resveratrol in conjunction with other pain control measures could be beneficial for PSACH pain control, and early treatment may preserve joint health.

**Supplementary Materials:** The following supporting information can be downloaded at: https://www.mdpi.com/article/10.3390/biom13101553/s1, Figure S1: MT-COMP mouse joint degeneration scoring. All male mice were administered DOX from birth to collection at 20 weeks. Control (Con) and MT-COMP (MT) at 20 weeks were scored using OARSI score [28,34], OARSI score plus synovium/meniscus (syn/meni) component or the "early" OA scoring system described in this manuscript and previously used to evaluate MT-COMP mice [28,29]. Only the early joint degeneration scoring system quantified the early mild joint degeneration in MT-COMP mice, while the OARSI scoring system provided better quantification of late-stage severe degeneration. * $p < 0.05$. Figure S2: Joint degeneration score in MT-COMP mice was decreased with resveratrol treatment from birth or from 4 to 20 weeks. Male mice were administered with DOX from birth to induce the MT-COMP phenotype and treated with resveratrol starting either at: birth (MT-RB), or 4 (MT-R4), 6 (MT-R6) or 8 (MT-R8) weeks until 20 weeks and analyzed. Joint degeneration scoring total is the sum of four scores based on scoring each area from 0–3 (synovitis, femoral proteoglycans, tibial proteoglycans and cartilage/bone damage). Results were compared to untreated MT-COMP (MT) mice at 20 weeks. MT-COMP mice treated from birth and 4 weeks had significantly improved joint degeneration scores. (Abbreviations: Control = Con; MT-COMP = MT; MT-COMP mice treated with resveratrol at birth = M+RB; 4–20 weeks = M+R4; MT-COMP mice treated with resveratrol 6–20 weeks = M+R6). $N = 10$, * $p < 0.05$; *** $p < 0.0005$ (Kruskal Wallis). Figure S3: Voluntary running is reduced in MT-COMP mice suggesting pain. All male mice were administered DOX from birth to induce mutant-COMP expression until completion of study. Voluntary running was used as a proxy for pain. Voluntary running data was collected at 12, 16, 20, 24, 30 and 36 weeks of age for both controls C57BL\6 (C) and MT-COMP (MT) mice ($n = 10$). All mice had a 3-night acclimation period to adjust to running wheel. MT-COMP mice ran significantly less than controls at 12, 16, 20, 30 and 36 weeks. ** $p < 0.005$; *** $p < 0.0005$ (Kruskal Wallis). Figure S4: Ibuprofen improves grooming score of MT-COMP mice. All male mice were administered DOX from birth to analysis. Grooming was scored from control (green bar) and MT-COMP (blue bar) at 16 weeks. Another group of mice ($n = 10$) grooming was measured at 15 weeks (light blue bar MT-COMP 15 wks) ibuprofen 0.2 mg/mL [32] was administered for one week and grooming was remeasured (orange bar MT-COMP 16 wks ibup). Kruskal Wallis test was used for comparison *** $p < 0.0005$. Figure S5: Pain is reduced with early resveratrol treatment. A grooming assay was used as a proxy for pain. The grooming assay measures the efficiency of removal of a fluorescent dye from the fur, with higher score indicating more effective elimination of dye (maximal score = 5). All male mice were administered DOX from birth to collection except MT-COMP-H2O (M-H2O). Grooming was assessed at ages: 8, 12, 16, 20, 24, 30 and 36 weeks in control C57BL\6 (Control) mice and MT-COMP mice, and MT-COMP mice treated with resveratrol beginning at birth (M+RB), 4 (M+4), 6 (M+6) weeks and MT-COMP without DOX (M+H2O) ($n \geq 12$). MT-COMP mice have a significantly lower grooming score than controls at all ages. Resveratrol treatment normalizes grooming at 8, 12, 16 and 20 weeks. Pairwise comparisons (Kruskal Wallis) between control and all other groups are shown with asterisks. Importantly, MT-COMP grooming scores were lower than MT-COMP+H2O (MT+H2O) in the absence of the induction of mutant-COMP ($p < 0.005$ at 16 and 20 wks; $p < 0.0005$ at 8 and 12 wks). Moreover, pairwise comparisons show grooming scores from all resveratrol treatments are significantly higher than scores from untreated MT-COMP ($p < 0.0005$ at all ages tested). The number of mice in each group that scored 4.0 are shown in red, 3.5 are shown in brown, and 3.0 are shown in pink. (Abbreviations: weeks = wks; R = resveratrol, Control = Con; MT-COMP = MT; MT-COMP mice treated with resveratrol at birth = M+RB; 4–20 weeks = M+R4; MT-COMP mice treated with resveratrol 6–20 weeks = M+R6. * $p < 0.05$; *** $p < 0.0005$

**Author Contributions:** J.T.H. oversaw the entire project and prepared the manuscript; A.C.V., D.P., M.G.H., C.M. and F.C. performed research and collected data; F.H.G. developed the early joint degeneration scoring system; K.L.P. managed the project, designed research, analyzed data, and prepared the manuscript; K.L.P. is the guarantor of this work and, as such, had full access to all of the data in the study and takes responsibility for the integrity of the data and the accuracy of the data analysis. All authors have read and agreed to the published version of the manuscript.

**Funding:** This work was supported in part by the Leah Lewis Family Foundation (J.T.H.) and the National Institute of Arthritis and Musculoskeletal and Skin Diseases of the National Institutes of Health (NIAMS) Award 5R01AR057117-10 (J.T.H. and K.L.P.). The content is solely the responsibility of the authors and does not necessarily represent the official views of the NIH.

**Institutional Review Board Statement:** This study complied with the Guide for the Care and Use of Laboratory Animals, eighth edition (ISBN-10, 0-309-15396-4), was approved by the Animal Welfare Committee at the University of Texas Health Sciences Center at Houston, and complies with NIH guidelines.

**Informed Consent Statement:** Not applicable.

**Data Availability Statement:** Data is available upon request by writing to karen.posey@uth.tmc.edu. K.L.P. is the first and senior author.

**Acknowledgments:** We thank Kaleigh L. Riggs for statistical data analysis.

**Conflicts of Interest:** The authors declare no conflict of interest.

## References

1. Holden, P.; Meadows, R.S.; Chapman, K.L.; Grant, M.E.; Kadler, K.E.; Briggs, M.D. Cartilage oligomeric matrix protein interacts with type IX collagen, and disruptions to these interactions identify a pathogenetic mechanism in a bone dysplasia family. *J. Biol. Chem.* **2001**, *276*, 6046–6055. [CrossRef]
2. Mann, H.H.; Ozbek, S.; Engel, J.; Paulsson, M.; Wagener, R. Interactions between the cartilage oligomeric matrix protein and matrilins. Implications for matrix assembly and the pathogenesis of chondrodysplasias. *J. Biol. Chem.* **2004**, *279*, 25294–25298. [CrossRef]
3. Thur, J.; Rosenberg, K.; Nitsche, D.P.; Pihlajamaa, T.; Ala-Kokko, L.; Heinegard, D.; Paulsson, M.; Maurer, P. Mutations in cartilage oligomeric matrix protein causing pseudoachondroplasia and multiple epiphyseal dysplasia affect binding of calcium and collagen I, II, and IX. *J. Biol. Chem.* **2001**, *276*, 6083–6092. [CrossRef]
4. Di Cesare, P.E.; Chen, F.S.; Moergelin, M.; Carlson, C.S.; Leslie, M.P.; Perris, R.; Fang, C. Matrix-matrix interaction of cartilage oligomeric matrix protein and fibronectin. *Matrix Biol.* **2002**, *21*, 461–470. [CrossRef]
5. Smith, R.K.; Gerard, M.; Dowling, B.; Dart, A.J.; Birch, H.L.; Goodship, A.E. Correlation of cartilage oligomeric matrix protein (COMP) levels in equine tendon with mechanical properties: A proposed role for COMP in determining function-specific mechanical characteristics of locomotor tendons. *Equine Vet. J. Suppl.* **2002**, *34*, 241–244. [CrossRef]
6. Kipnes, J.; Carlberg, A.L.; Loredo, G.A.; Lawler, J.; Tuan, R.S.; Hall, D.J. Effect of cartilage oligomeric matrix protein on mesenchymal chondrogenesis in vitro. *Osteoarthr. Cartil.* **2003**, *11*, 442–454. [CrossRef]
7. Xu, K.; Zhang, Y.; Ilalov, K.; Carlson, C.S.; Feng, J.Q.; Di Cesare, P.E.; Liu, C.J. Cartilage oligomeric matrix protein associates with granulin-epithelin precursor (GEP) and potentiates GEP-stimulated chondrocyte proliferation. *J. Biol. Chem.* **2007**, *282*, 11347–11355. [CrossRef]
8. Briggs, M.D.; Chapman, K.L. Pseudoachondroplasia and multiple epiphyseal dysplasia: Mutation review, molecular interactions, and genotype to phenotype correlations. *Hum. Mutat.* **2002**, *19*, 465–478. [CrossRef]
9. Briggs, M.D.; Hoffman, S.M.G.; King, L.M.; Olsen, A.S.; Mohrenweiser, H.; Leroy, J.G.; Mortier, G.R.; Rimoin, D.L.; Lachman, R.S.; Gaines, E.S.; et al. Pseudoachondroplasia and multiple epiphyseal dysplasia due to mutations in the cartilage oligomeric matrix protein gene. *Nat. Genet.* **1995**, *10*, 330–336. [CrossRef]
10. Briggs, M.D.; Brock, J.; Ramsden, S.C.; Bell, P.A. Genotype to phenotype correlations in cartilage oligomeric matrix protein associated chondrodysplasias. *Eur. J. Hum. Genet.* **2014**, *22*, 1278–1282. [CrossRef]
11. Kung, L.H.W.; Mullan, L.; Soul, J.; Wang, P.; Mori, K.; Bateman, J.F.; Briggs, M.D.; Boot-Handford, R.P. Cartilage endoplasmic reticulum stress may influence the onset but not the progression of experimental osteoarthritis. *Arthritis Res. Ther.* **2019**, *21*, 206. [CrossRef] [PubMed]
12. DiCesare, P.E.; Morgelin, M.; Carlson, C.S.; Pasumarti, S.; Paulsson, M. Cartilage oligomeric matrix protein: Isolation and characterization from human articular cartilage. *J. Orthop. Res.* **1995**, *13*, 422–428. [CrossRef] [PubMed]
13. Cooper, R.R.; Ponseti, I.V.; Maynard, J.A. Pseudoachondroplasia dwarfism. A rough-surfaced endoplasmic reticulum disorder. *J. Bone Jt. Surg. Am.* **1973**, *55A*, 475–484. [CrossRef]
14. Dinser, R.; Zaucke, F.; Kreppel, F.; Hultenby, K.; Kochanek, S.; Paulsson, M.; Maurer, P. Pseudoachondroplasia is caused through both intra- and extracellular pathogenic pathways. *J. Clin. Invest.* **2002**, *110*, 505–513. [CrossRef]
15. Ikegawa, S.; Ohashi, H.; Nishimura, G.; Kim, K.C.; Sannohe, A.; Kimizuka, M.; Fukushima, Y.; Nagai, T.; Nakamura, Y. Novel and recurrent COMP (cartilage oligomeric matrix protein) mutations in pseudoachondroplasia and multiple epiphyseal dysplasia. *Hum. Genet.* **1998**, *103*, 633–638. [CrossRef]
16. Briggs, M.D.; Mortier, G.R.; Cole, W.G.; King, L.M.; Golik, S.S.; Bonaventure, J.; Nuytinck, L.; De Paepe, A.; Leroy, J.G.; Biesecker, L.; et al. Diverse mutations in the gene for cartilage oligomeric matrix protein in the pseudoachondroplasia-multiple epiphyseal dysplasia disease spectrum. *Am. J. Hum. Genet.* **1998**, *62*, 311–319. [CrossRef]
17. Hecht, J.T.; Nelson, L.D.; Crowder, E.; Wang, Y.; Elder, F.F.; Harrison, W.R.; Francomano, C.A.; Prange, C.K.; Lennon, G.G.; Deere, M.; et al. Mutations in exon 17B of cartilage oligomeric matrix protein (COMP) cause pseudoachondroplasia. *Nat. Genet.* **1995**, *10*, 325–329. [CrossRef]
18. Bonafe, L.; Cormier-Daire, V.; Hall, C.; Lachman, R.; Mortier, G.; Mundlos, S.; Nishimura, G.; Sangiorgi, L.; Savarirayan, R.; Sillence, D.; et al. Nosology and classification of genetic skeletal disorders: 2015 revision. *Am. J. Med. Genet. A* **2015**, *167A*, 2869–2892. [CrossRef]
19. Hecht, J.T.; Chiu, F.; Veerisetty, A.; Hossain, M.; Posey, K.L. Health consequences of mutant cartilage oligomeric matrix protein and its relationship to abnormal growth and joint degeneration. *Matrix Biol.* **2023**, *119*, 101–111. [CrossRef]
20. Gamble, C.; Nguyen, J.; Hashmi, S.S.; Hecht, J.T. Pseudoachondroplasia and painful sequelae. *Am. J. Med. Genet. A* **2015**, *167*, 2618–2622. [CrossRef]

21. Maynard, J.A.; Cooper, R.R.; Ponseti, I.V. A unique rough surfaced endoplasmic reticulum inclusion in pseudoachondroplasia. *Lab. Invest.* **1972**, *26*, 40–44.
22. Posey, K.L.; Veerisetty, A.C.; Liu, P.; Wang, H.R.; Poindexter, B.J.; Bick, R.; Alcorn, J.L.; Hecht, J.T. An inducible cartilage oligomeric matrix protein mouse model recapitulates human pseudoachondroplasia phenotype. *Am. J. Pathol.* **2009**, *175*, 1555–1563. [CrossRef] [PubMed]
23. Hall, J.G. Pseudoachondroplasia. *Birth Defects Orig. Artic. Ser.* **1975**, *11*, 187–202.
24. Kung, L.H.; Rajpar, M.H.; Preziosi, R.; Briggs, M.D.; Boot-Handford, R.P. Increased classical endoplasmic reticulum stress is sufficient to reduce chondrocyte proliferation rate in the growth plate and decrease bone growth. *PLoS ONE* **2015**, *10*, e0117016. [CrossRef]
25. Posey, K.L.; Coustry, F.; Veerisetty, A.C.; Hossain, M.G.; Gambello, M.J.; Hecht, J.T. Novel mTORC1 Mechanism Suggests Therapeutic Targets for COMPopathies. *Am. J. Pathol.* **2019**, *189*, 132–146. [CrossRef]
26. Laplante, M.; Sabatini, D.M. mTOR Signaling. *Cold Spring Harb. Perspect. Biol.* **2012**, *4*, a011593. [CrossRef]
27. Hecht, J.T.; Coustry, F.; Veerisetty, A.C.; Hossain, M.G.; Posey, K.L. Resveratrol Reduces COMPopathy in Mice Through Activation of Autophagy. *JBMR Plus* **2021**, *5*, e10456. [CrossRef] [PubMed]
28. Hecht, J.T.; Veerisetty, A.C.; Hossain, M.G.; Patra, D.; Chiu, F.; Coustry, F.; Posey, K.L. Joint Degeneration in a Mouse Model of Pseudoachondroplasia: ER Stress, Inflammation, and Block of Autophagy. *Int. J. Mol. Sci.* **2021**, *22*, 9239. [CrossRef] [PubMed]
29. Hecht, J.T.; Veerisetty, A.C.; Wu, J.; Coustry, F.; Hossain, M.G.; Chiu, F.; Gannon, F.H.; Posey, K.L. Primary Osteoarthritis Early Joint Degeneration Induced by Endoplasmic Reticulum Stress Is Mitigated by Resveratrol. *Am. J. Pathol.* **2021**, *191*, 1624–1637. [CrossRef] [PubMed]
30. Posey, K.L.; Coustry, F.; Veerisetty, A.C.; Liu, P.; Alcorn, J.L.; Hecht, J.T. Chondrocyte-specific pathology during skeletal growth and therapeutics in a murine model of pseudoachondroplasia. *J. Bone Miner. Res.* **2014**, *29*, 1258–1268. [CrossRef] [PubMed]
31. Posey, K.L.; Coustry, F.; Veerisetty, A.C.; Hossain, M.; Alcorn, J.L.; Hecht, J.T. Antioxidant and anti-inflammatory agents mitigate pathology in a mouse model of pseudoachondroplasia. *Hum. Mol. Genet.* **2015**, *24*, 3918–3928. [CrossRef]
32. Salama, R.A.M.; El Gayar, N.H.; Georgy, S.S.; Hamza, M. Equivalent intraperitoneal doses of ibuprofen supplemented in drinking water or in diet: A behavioral and biochemical assay using antinociceptive and thromboxane inhibitory dose-response curves in mice. *PeerJ* **2016**, *4*, e2239. [CrossRef] [PubMed]
33. Deuis, J.R.; Dvorakova, L.S.; Vetter, I. Methods Used to Evaluate Pain Behaviors in Rodents. *Front. Mol. Neurosci.* **2017**, *10*, 284. [CrossRef] [PubMed]
34. Glasson, S.S.; Chambers, M.G.; Van Den Berg, W.B.; Little, C.B. The OARSI histopathology initiative—Recommendations for histological assessments of osteoarthritis in the mouse. *Osteoarthr. Cartil.* **2010**, *18* (Suppl. S3), S17–S23. [CrossRef] [PubMed]
35. Knudson, C.B.; Knudson, W. Cartilage proteoglycans. *Semin. Cell Dev. Biol.* **2001**, *12*, 69–78. [CrossRef]
36. Martin, J.A.; Buckwalter, J.A. Aging, articular cartilage chondrocyte senescence and osteoarthritis. *Biogerontology* **2002**, *3*, 257–264. [CrossRef]
37. Wheaton, A.J.; Borthakur, A.; Shapiro, E.M.; Regatte, R.R.; Akella, S.V.; Kneeland, J.B.; Reddy, R. Proteoglycan loss in human knee cartilage: Quantitation with sodium MR imaging--feasibility study. *Radiology* **2004**, *231*, 900–905. [CrossRef]
38. Sheahan, T.D.; Copits, B.A.; Golden, J.P.; Gereau, R.W. Voluntary Exercise Training: Analysis of Mice in Uninjured, Inflammatory, and Nerve-Injured Pain States. *PLoS ONE* **2015**, *10*, e0133191. [CrossRef]
39. Malhotra, J.D.; Kaufman, R.J. The endoplasmic reticulum and the unfolded protein response. *Semin. Cell Dev. Biol.* **2007**, *18*, 716–731. [CrossRef]
40. Kadowaki, H.; Nishitoh, H. Endoplasmic reticulum quality control by garbage disposal. *FEBS J.* **2019**, *286*, 232–240. [CrossRef]
41. Liu, T.F.; McCall, C.E. Deacetylation by SIRT1 Reprograms Inflammation and Cancer. *Genes. Cancer* **2013**, *4*, 135–147. [CrossRef] [PubMed]
42. Yoshino, J.; Mills, K.F.; Yoon, M.J.; Imai, S. Nicotinamide mononucleotide, a key NAD(+) intermediate, treats the pathophysiology of diet- and age-induced diabetes in mice. *Cell Metab.* **2011**, *14*, 528–536. [CrossRef]
43. Kim, H.J.; Kim, J.H.; Noh, S.; Hur, H.J.; Sung, M.J.; Hwang, J.T.; Park, J.H.; Yang, H.J.; Kim, M.S.; Kwon, D.Y.; et al. Metabolomic analysis of livers and serum from high-fat diet induced obese mice. *J. Proteome Res.* **2011**, *10*, 722–731. [CrossRef]
44. Kendrick, A.A.; Choudhury, M.; Rahman, S.M.; McCurdy, C.E.; Friederich, M.; Van Hove, J.L.; Watson, P.A.; Birdsey, N.; Bao, J.; Gius, D.; et al. Fatty liver is associated with reduced SIRT3 activity and mitochondrial protein hyperacetylation. *Biochem. J.* **2011**, *433*, 505–514. [CrossRef]
45. Tao, R.; Wei, D.; Gao, H.; Liu, Y.; DePinho, R.A.; Dong, X.C. Hepatic FoxOs regulate lipid metabolism via modulation of expression of the nicotinamide phosphoribosyltransferase gene. *J. Biol. Chem.* **2011**, *286*, 14681–14690. [CrossRef] [PubMed]
46. Burrage, P.S.; Mix, K.S.; Brinckerhoff, C.E. Matrix metalloproteinases: Role in arthritis. *Front. Biosci.* **2006**, *11*, 529–543. [CrossRef]
47. Wei, Y.; Jia, J.; Jin, X.; Tong, W.; Tian, H. Resveratrol ameliorates inflammatory damage and protects against osteoarthritis in a rat model of osteoarthritis. *Mol. Med. Rep.* **2018**, *17*, 1493–1498. [CrossRef] [PubMed]
48. Yuce, P.; Hosgor, H.; Rencber, S.F.; Yazir, Y. Effects of Intra-Articular Resveratrol Injections on Cartilage Destruction and Synovial Inflammation in Experimental Temporomandibular Joint Osteoarthritis. *J. Oral. Maxillofac. Surg.* **2020**, *79*, 344-e1. [CrossRef]
49. Toegel, S.; Wu, S.Q.; Otero, M.; Goldring, M.B.; Leelapornpisid, P.; Chiari, C.; Kolb, A.; Unger, F.M.; Windhager, R.; Viernstein, H. Caesalpinia sappan extract inhibits IL1beta-mediated overexpression of matrix metalloproteinases in human chondrocytes. *Genes. Nutr.* **2012**, *7*, 307–318. [CrossRef]

50. Elsaid, K.A.; Ubhe, A.; Shaman, Z.; D'Souza, G. Intra-articular interleukin-1 receptor antagonist (IL1-ra) microspheres for posttraumatic osteoarthritis: In vitro biological activity and in vivo disease modifying effect. *J. Exp. Orthop.* **2016**, *3*, 18. [CrossRef]
51. Goldring, M.B.; Otero, M. Inflammation in osteoarthritis. *Curr. Opin. Rheumatol.* **2011**, *23*, 471–478. [CrossRef] [PubMed]
52. Goldring, M.B. Articular cartilage degradation in osteoarthritis. *HSS J.* **2012**, *8*, 7–9. [CrossRef] [PubMed]
53. Tang, C.H.; Chen, C.F.; Chen, W.M.; Fong, Y.C. IL-6 increases MMP-13 expression and motility in human chondrosarcoma cells. *J. Biol. Chem.* **2011**, *286*, 11056–11066. [CrossRef]
54. Liacini, A.; Sylvester, J.; Li, W.Q.; Huang, W.; Dehnade, F.; Ahmad, M.; Zafarullah, M. Induction of matrix metalloproteinase-13 gene expression by TNF-alpha is mediated by MAP kinases, AP-1, and NF-kappaB transcription factors in articular chondrocytes. *Exp. Cell Res.* **2003**, *288*, 208–217. [CrossRef]
55. Wong, R.H.X.; Evans, H.M.; Howe, P.R.C. Resveratrol supplementation reduces pain experience by postmenopausal women. *Menopause* **2017**, *24*, 916–922. [CrossRef] [PubMed]
56. Hussain, S.A.; Marouf, B.H.; Ali, Z.S.; Ahmmad, R.S. Efficacy and safety of co-administration of resveratrol with meloxicam in patients with knee osteoarthritis: A pilot interventional study. *Clin. Interv. Aging* **2018**, *13*, 1621–1630. [CrossRef]
57. Gu, H.; Li, K.; Li, X.; Yu, X.; Wang, W.; Ding, L.; Liu, L. Oral Resveratrol Prevents Osteoarthritis Progression in C57BL/6J Mice Fed a High-Fat Diet. *Nutrients* **2016**, *8*, 233. [CrossRef]
58. Jiang, M.; Li, X.; Yu, X.; Liu, X.; Xu, X.; He, J.; Gu, H.; Liu, L. Oral Administration of Resveratrol Alleviates Osteoarthritis Pathology in C57BL/6J Mice Model Induced by a High-Fat Diet. *Mediators Inflamm.* **2017**, *2017*, 7659023. [CrossRef]
59. Cobos, E.J.; Ghasemlou, N.; Araldi, D.; Segal, D.; Duong, K.; Woolf, C.J. Inflammation-induced decrease in voluntary wheel running in mice: A nonreflexive test for evaluating inflammatory pain and analgesia. *Pain.* **2012**, *153*, 876–884. [CrossRef]
60. Grace, P.M.; Strand, K.A.; Maier, S.F.; Watkins, L.R. Suppression of voluntary wheel running in rats is dependent on the site of inflammation: Evidence for voluntary running as a measure of hind paw-evoked pain. *J. Pain.* **2014**, *15*, 121–128. [CrossRef]
61. Contreras, K.M.; Caillaud, M.; Neddenriep, B.; Bagdas, D.; Roberts, J.L.; Ulker, E.; White, A.B.; Aboulhosn, R.; Toma, W.; Khalefa, T.; et al. Deficit in voluntary wheel running in chronic inflammatory and neuropathic pain models in mice: Impact of sex and genotype. *Behav. Brain Res.* **2021**, *399*, 113009. [CrossRef] [PubMed]
62. Oto, Y.; Takahashi, Y.; Kurosaka, D.; Kato, F. Alterations of voluntary behavior in the course of disease progress and pharmacotherapy in mice with collagen-induced arthritis. *Arthritis Res. Ther.* **2019**, *21*, 284. [CrossRef] [PubMed]
63. Roughan, J.V.; Flecknell, P.A. Effects of surgery and analgesic administration on spontaneous behaviour in singly housed rats. *Res. Vet. Sci.* **2000**, *69*, 283–288. [CrossRef] [PubMed]
64. Oliver, V.L.; Thurston, S.E.; Lofgren, J.L. Using Cageside Measures to Evaluate Analgesic Efficacy in Mice (Mus musculus) after Surgery. *J. Am. Assoc. Lab. Anim. Sci.* **2018**, *57*, 186–201. [PubMed]
65. Wright, F.L.; Rodgers, R.J. Low dose naloxone attenuates the pruritic but not anorectic response to rimonabant in male rats. *Psychopharmacology* **2013**, *226*, 415–431. [CrossRef]
66. Thaung Zaw, J.J.; Howe, P.R.C.; Wong, R.H.X. Long-term resveratrol supplementation improves pain perception, menopausal symptoms, and overall well-being in postmenopausal women: Findings from a 24-month randomized, controlled, crossover trial. *Menopause* **2020**, *28*, 40–49. [CrossRef]
67. Marouf, B.H.; Hussain, S.A.; Ali, Z.S.; Ahmmad, R.S. Resveratrol Supplementation Reduces Pain and Inflammation in Knee Osteoarthritis Patients Treated with Meloxicam: A Randomized Placebo-Controlled Study. *J. Med. Food* **2018**, *21*, 1253–1259. [CrossRef]

*biomolecules*

MDPI

*Article*

# Tanshinone IIA and Cryptotanshinone Counteract Inflammation by Regulating Gene and miRNA Expression in Human SGBS Adipocytes

**Sara Carpi [1,2,3,†], Stefano Quarta [4,†], Stefano Doccini [5], Anella Saviano [6], Noemi Marigliano [6], Beatrice Polini [3,7], Marika Massaro [8], Maria Annunziata Carluccio [8], Nadia Calabriso [8], Martin Wabitsch [9], Filippo Maria Santorelli [5], Marco Cecchini [2], Francesco Maione [6], Paola Nieri [3] and Egeria Scoditti [8,*]**

1 Science of Health Department, Magna Græcia University, 88100 Catanzaro, Italy; sara.carpi@unicz.it
2 NEST, Istituto Nanoscienze-CNR and Scuola Normale Superiore, 56100 Pisa, Italy; marco.cecchini@nano.cnr.it
3 Department of Pharmacy, University of Pisa, 56100 Pisa, Italy; beatrice.polini@farm.unipi.it (B.P.); paola.nieri@unipi.it (P.N.)
4 Department of Biological and Environmental Sciences and Technologies (DISTEBA), University of Salento, 73100 Lecce, Italy; stefano.quarta3@unisalento.it
5 IRCCS Fondazione Stella Maris, Calambrone, 56128 Pisa, Italy; stefano.doccini@fsm.unipi.it (S.D.); filippo.santorelli@fsm.unipi.it (F.M.S.)
6 ImmunoPharmaLab, Department of Pharmacy, School of Medicine and Surgery, University of Naples Federico II, 80131 Naples, Italy; anella.saviano@unina.it (A.S.); noemi.marigliano@outlook.com (N.M.); francesco.maione@unina.it (F.M.)
7 Department of Pathology, University of Pisa, 56100 Pisa, Italy
8 National Research Council (CNR), Institute of Clinical Physiology (IFC), 73100 Lecce, Italy; marika.massaro@ifc.cnr.it (M.M.); maria.carluccio@ifc.cnr.it (M.A.C.); nadia.calabriso@ifc.cnr.it (N.C.)
9 Division of Pediatric Endocrinology, Diabetes and Obesity, Department of Pediatrics and Adolescent Medicine, University of Ulm, 89075 Ulm, Germany; martin.wabitsch@uniklinik-ulm.de
* Correspondence: egeria.scoditti@ifc.cnr.it
† These authors contributed equally to this work.

**Abstract:** Inflammation of the adipose tissue contributes to the onset and progression of several chronic obesity-related diseases. The two most important lipophilic diterpenoid compounds found in the root of *Salvia milthorrhiza* Bunge (also called Danshen), tanshinone IIA (TIIA) and cryptotanshinone (CRY), have many favorable pharmacological effects. However, their roles in obesity-associated adipocyte inflammation and related sub-networks have not been fully elucidated. In the present study, we investigated the gene, miRNAs and protein expression profile of prototypical obesity-associated dysfunction markers in inflamed human adipocytes treated with TIIA and CRY. The results showed that TIIA and CRY prevented tumor necrosis factor (TNF)-α induced inflammatory response in adipocytes, by counter-regulating the pattern of secreted cytokines/chemokines associated with adipocyte inflammation (CCL2/MCP-1, CXCL10/IP-10, CCL5/RANTES, CXCL1/GRO-α, IL-6, IL-8, MIF and PAI-1/Serpin E1) via the modulation of gene expression (as demonstrated for CCL2/MCP-1, CXCL10/IP-10, CCL5/RANTES, CXCL1/GRO-α, and IL-8), as well as related miRNA expression (miR-126-3p, miR-223-3p, miR-124-3p, miR-155-5p, and miR-132-3p), and by attenuating monocyte recruitment. This is the first demonstration of a beneficial effect by TIIA and CRY on adipocyte dysfunction associated with obesity development and complications, offering a new outlook for the prevention and/or treatment of metabolic diseases.

**Keywords:** adipocyte; chemokine; cryptotanshinone; cytokine; Danshen; gene expression; inflammation; miRNA; obesity; *Salvia milthorrhiza* bunge; tanshinone-IIA

## 1. Introduction

Fat accumulation in the visceral adipose depot and consequent obesity are important risk factors for type 2 diabetes, non-alcoholic fatty liver disease (NAFLD), dyslipidemia, hypertension, cardiovascular disease, and cancer [1–3]. Far from being simply the result of an energy imbalance between calorie intake and expenditure, obesity involves a host of metabolic alterations, including mitochondrial dysfunction, oxidative stress, immune response, and chronic low-grade inflammation, which are significant correlates and possible determinants of the increased cardiometabolic risk associated with visceral obesity [1]. Under prolonged positive energy balance conditions, chronic fat accumulation in white adipose tissue leads to hypertrophic expansion of adipocytes, cell stress, and eventually death, along with impaired adipogenesis, altered tissue perfusion, and fibrosis [4]. As a consequence, adipocytes become metabolically dysfunctional, acquiring a pro-lipolytic and insulin-resistant phenotype, with consequent free fatty acid (FFA) spillover into the circulation and ectopic lipid accumulation, such as in the liver, skeletal muscle, and heart, which promotes local inflammation, insulin resistance, and organ damage [4]. Furthermore, adipose tissue expansion is associated with a shift of the transcriptional program of hypertrophic adipocytes toward a pro-inflammatory, diabetogenic, and proatherogenic pattern of adipokines, i.e., adipose tissue secretes bioactive factors [4–6]. In fact, the secretion of anti-inflammatory and insulin-sensitizing adipokines, such as adiponectin, is downregulated, while the secreted levels of pro-inflammatory mediators, such as interleukin (IL)-6, IL-8, monocyte chemoattractant protein (MCP)-1, chemokine (C-C motif) ligand 5 (CCL-5) and tumor necrosis factor (TNF)-α, increase [7]. In particular, chemokines production by adipocytes promotes the recruitment and infiltration of immune cells, mostly macrophages but also neutrophils, T cells, and B2 cells, in the adipose tissue [8]. Here, the dysmetabolic microenvironment of obese adipose tissue leads to pro-inflammatory activation of recruited immune cells; indeed, recruited adipose tissue macrophages in obesity differentiate into a M1 proinflammatory state, thus causing an imbalance between M1 and anti-inflammatory M2 macrophages [9,10]. Increased production of inflammatory chemokines and cytokines from these cells amplifies the inflamed status of the adipose tissue in a vicious cycle and inhibits the insulin signaling cascade [9,10]. Several studies have shown that decreased adipose tissue macrophage recruitment is associated with reduced adipose tissue inflammation and insulin resistance in obesity [11–13]. The altered pattern of adipose tissue-released cytokines/adipokines, lipids, and microRNA (miRNA)-containing exosomal vesicles, which travel through the circulation to influence distant organs, contributes to establishing a state of systemic chronic low-grade inflammation that is a pathogenic link between obesity and cardiometabolic complications [7].

Inflammation of adipocytes and immune cell activation/interactions in adipose tissue offer an attractive therapeutic target for treating obesity-related metabolic and cardiovascular conditions [14]. Neutralizing antibodies against cytokines, including TNF-α and IL-1β, have been shown to provide anti-diabetic effects in clinical studies [14]. Moreover, genetic deletion [15,16] or pharmacological antagonism of inflammatory pathways, for example, by using a high dose of salicylates [17] or chemokine antagonists [18], have been shown to alleviate obesity-induced inflammation and insulin resistance.

Currently, there are several pharmacological strategies, surgical approaches, and lifestyle regimens (e.g., diet and exercise) for the treatment of obesity, although they provide only limited long-term success. Research on natural compounds in different plants provides a growing arsenal of potential pharmaceutical agents using ethnopharmacological information. Indeed, a large percentage of marketed therapeutics (about 40%) rely on non-modified or semi-synthetic natural compounds with biological activity [19]. Consequently, in the past few decades, researchers have paid much attention to the active compounds present in medicinal and/or edible plants and their effects on human diseases, including obesity management [20,21].

In this context, a novel potential natural compound-based approach could be found in the root of *Salvia milthorrhiza* Bunge (also referred to as Danshen), a perennial herb belonging to the *Labiatae* family [22]. The dry root and/or rhizome of *Salvia milthorrhiza* Bunge is officially listed in the Chines Pharmacopoeia [23] and has been one of the main commercial medicinal herbs in China for over a thousand years. The main uses are to treat inflammatory [24,25], cardiovascular [26,27], cerebrovascular [28,29] disorders and other pathological conditions [30–32]. Moreover, extracts of *S. milthorrhiza* roots named "Compound Danshen Dripping pill" are investigated in six different clinical trials in China and the United States in conditions such as angina pectoris, coronary heart disease, diabetes and diabetic retinopathy [33]. The favorable pharmacological effects are mainly attributed to lipophilic diterpenoid compounds, such as tanshinone IIA (TIIA) and cryptotanshinone (CRY) (Figure 1) [34].

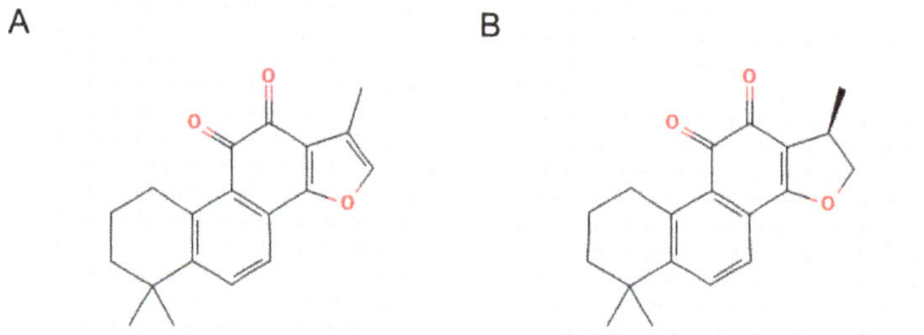

**Figure 1.** Chemical structure of (**A**) tanshinone IIA (TIIA) and (**B**) cryptotanshinone (CRY).

Indeed, TIIA has been shown to reduce inflammatory cytokine production induced by TNF-α in rheumatoid arthritis fibroblast-like synoviocytes and to ameliorate arthritis severity in adjuvant-induced arthritis mice models [35]. Moreover, TIIA inhibited key inflammatory factors, such as IL-1β, in lipopolysaccharide (LPS)-stimulated RAW264.7 macrophages [36,37] and has been shown to have immune-regulating and anti-inflammatory roles in murine models [38,39]. Similarly, CRY has been shown to suppress the production of pro-inflammatory cytokines both in in vitro and in vivo models [40] and to inhibit the nuclear factor-κB (NF-κB) pathway, a pro-inflammatory transcription factor, in immune cells, including THP-1 monocytes and RAW264.7 macrophages [41,42].

Despite the potential TIIA and CRY as anti-inflammatory agents, no experimental study has been conducted to determine their effects on human adipocyte models of inflammation and obesity. Against this background, we hypothesized that TIIA and CRY might modulate the obesity-associated adipocyte inflammatory response. We, therefore, evaluated the effects of TIIA and CRY on the expression of genes, proteins and miRNAs involved in TNF-α-induced inflammatory phenotype of human adipocytes and explored underlying molecular mechanisms.

We here demonstrate that TIIA and CRY can improve adipocyte dysfunction by beneficially regulating the expression of mRNAs, proteins and miRNAs in inflamed adipocytes and attenuating monocyte recruitment.

## 2. Materials and Methods

### 2.1. Materials

TIIA and CRY (≥98% purity) were obtained from Sigma Aldrich (now under Merck, Darmstadt, Germany) and dissolved in dimethyl sulfoxide (DMSO). All other chemicals were obtained from Sigma Aldrich, unless otherwise indicated. The proteome profiler human cytokine array kit (Catalog #ARY005B) was purchased from R&D System (Milan, Italy). Unless otherwise stated, all the other reagents were from BioCell (Milan, Italy).

### 2.2. Cell Cultures and Treatments

We used human Simpson–Golabi–Behmel syndrome (SGBS) preadipocytes, a physiologically relevant cell model system resembling human adipose tissue [43]. These were a generous gift from our co-author, Prof. Martin Wabitsch (Division of Pediatric Endocrinology, Diabetes, and Obesity, Department of Pediatrics and Adolescent Medicine, University of Ulm, Ulm, Germany), and were cultured and differentiated into mature adipocytes as previously described [44]. On day 14 postdifferentiation, >90% of these cells undergo complete differentiation into mature adipocytes, as assessed using Oil Red-O lipid staining and the expression of adipocyte-specific mRNAs, such as lipoprotein lipase, adipocyte fatty acid binding protein (FABP4), peroxisome proliferator-activated receptor (PPAR), and the glucose transporter GLUT-4.

Adipose tissue inflammation was mimicked by exposing SGBS adipocytes to the pro-inflammatory cytokine TNF-α at 10 ng/mL for 18 h. Though this concentration is higher compared with those found in the circulation in obesity [45], potentially high concentrations could be found in the interstitial fluid of adipose tissue in obese subjects [46–49], thus making (patho)physiologically relevant the 10 ng/mL TNF-α concentration as used in the present study as well as in other in vitro studies [50–53]. Unstimulated controls were adipocytes incubated without TNF-α. For TIIA and CRY treatment, SGBS cells were incubated with 10 μM TIIA or CRY for 1 h before stimulation with TNF-α.

Human monocytoid THP-1 cells were obtained from the American Tissue Culture Collection (Rockville, MD, USA) and maintained in RPMI 1640 medium supplemented with 2 mmol/L glutamine, 100 mg/mL streptomycin, 100 IU/mL penicillin, and 10% FBS in a 5% $CO_2$ humidified atmosphere at 37 °C.

### 2.3. Cell Viability

Cell viability was determined by the 3(4,5dimethylthiazol2yl) 2,5diphenyltetrazolium bromide (MTT) assay, as previously described [44].

### 2.4. RNA Isolation and Real-Time Quantitative Polymerase Chain Reaction

Total RNA was isolated using the TRIzol reagent (Invitrogen, Carlsbad, CA, USA) according to the manufacturer's protocol. For real-time quantitative polymerase chain reaction (qPCR), total RNA (2 μg) was converted into first-strand cDNA by using the High Capacity cDNA Reverse Transcription Kit (Applied Biosystems, Monza, Italy). The qPCR was performed in a CFX Connect Real-Time PCR Detection System (Bio-Rad Laboratories, Milan, Italy) by using primer sequences (Table 1) for the indicated adipokines. All reactions were performed in triplicate, and the amount of mRNA was calculated by the comparative critical threshold (CT) method. To account for possible variations related to cDNA input or the presence of PCR inhibitors, the endogenous reference gene ribosomal 18S was simultaneously quantified for each sample, and the data were normalized accordingly. Results are expressed as a fold increase relative to the unstimulated control (=1).

**Table 1.** Primer sequences used for qPCR analysis.

| Gene | Forward Primer | Reverse Primer |
|---|---|---|
| CCL2/MCP-1 | 5′-CCCCAGTCACCTGCTGTTAT-3′ | 5′-TCCTGAACCCACTTCTGCTT-3′ |
| CXCL1/GRO-α | 5′-CCCCAAGAACATCCAAAGTG-3′ | 5′-TGGATTTGTCACTGTTCAGCA-3′ |
| CXCL10/IP-10 | 5′-CAAGGATGGACCACACAGAG-3′ | 5′-GCAGGGTCAGAACATCCACT-3′ |
| CCL5/RANTES | 5′-CGCTGTCATCCTCATTGCTA-3′ | 5′-GAGCACTTGCCACTGGTGTA-3′ |
| IL-8 | 5′-GTGCAGTTTTGCCAAGGAGT-3′ | 5′-CTCTGCACCCAGTTTTCCTT-3′ |
| 18S | 5′-AAACGGCTACCACATCCAAG-3′ | 5′-CCTCCAATGGATCCTCGTTA-3′ |

MCP-1/CCL-2: monocyte chemoattractant protein-1/chemokine (C-C motif) ligand 2; CXCL-1/GRO- α; chemokine (C-X-C motif) ligand 1/growth regulated protein α; CXCL-10/IP-10: chemokine (C-X-C motif) ligand 10/interferon γ-induced protein-10; CCL-5/RANTES: chemokine (C-C motif) ligand-5/regulated on activation, normal T cell expressed and secreted; IL-8: interleukin-8.

### 2.5. Cytokines and Chemokines Protein Array

According to the manufacturer's instructions, equal volumes (1.5 mL) of the pulled cell supernatants from all experimental conditions were then incubated with the pre-coated proteome profiler array membranes. Dot plots were detected using the enhanced chemiluminescence detection kit and Image Quant 400 GE Healthcare software (GE Healthcare,

Italy) and successively quantified using GS 800 imaging densitometer software (Biorad, Italy) [54,55].

### 2.6. Evaluation of miRNAs Expression in Adipocytes

Total microRNAs (miRNAs) were extracted and purified from frozen adipocytes by using miRNeasy Mini Kit (Qiagen, Germany). Reverse transcription of the extracted miRNAs was performed by the miScript Reverse Transcription Kit (Qiagen, Germany). cDNA was diluted 1:3 in RNase-free water and then qPCRs were performed in triplicate using the miScript SYBR-Green PCR kit (Qiagen, Germany) on the MiniOpticon CFX 48 real-time PCR Detection System (Bio-Rad, Hercules, CA, USA) [56,57]. MiScript Primer Assays specific for hsa-miR-126-3p, hsa-miR-223-3p, hsa-miR-124-3p, hsa-miR-155-5p and hsa-miR-132-3p and hsa-RNU6 were obtained from Qiagen, and the probe sequences are proprietary. miRNAs expression was calculated using the Ct method and normalized to the expression of housekeeping RNU6. Mature miRNA sequences and miRBase Accession are reported in Table 2.

**Table 2.** Mature miRNAs sequences and miRBase Accession number.

| miRNA Name | Mature miRNA Sequences | miRBase Accession |
|---|---|---|
| hsa-miR-126-3p | 5′-UCGUACCGUGAGUAAUAAUGCG-3′ | MIMAT0000445 |
| hsa-miR-223-3p | 5′-UGUCAGUUUGUCAAAUACCCCA-3′ | MIMAT0000280 |
| hsa-miR-124-3p | 5′-UAAGGCACGCGGUGAAUGCC-3′ | MIMAT0000422 |
| hsa-miR-155-5p | 5′-UUAAUGCUAAUCGUGAUAGGGGU-3′ | MIMAT0000646 |
| hsa-miR-132-3p | 5′-UAACAGUCUACAGCCAUGGUCG-3 | MIMAT0000426 |

### 2.7. miRNA Target Prediction and Pathway Analysis

The bioinformatic survey was carried out using the Ingenuity Pathway Analysis software platform (IPA Winter Release–December 2022; QIAGEN, Hilden, Germany) [58] to recognize meaningful biological processes, disease annotations, and molecular interactions. Interaction networks were generated by selecting relations experimentally observed (maximum confidence) and functional annotations based on the number of interactions with the analyzed miRNA or mRNA targets (5) and their significance. P-value, which was ascertained by right-tailed Fisher's Exact Test following Benjamini and Hochberg (B-H) correction, indicating the robustness of correlations. Moreover, the IPA algorithm estimated the predicted activation or inhibition of a given connecting node or biological function, simulating the directional consequences associated with the modulated miRNAs or mRNAs.

### 2.8. Leukocyte-Adipocyte Adhesion Assay

SGBS cells were treated with TIIA or CRY at the indicated concentrations for 1 h and then stimulated with 10 ng/mL TNF-$\alpha$ for an additional 18 h. THP-1 cells ($10^6$ cells/mL) were fluorescently labeled by incubation with calcein-AM (5 ng/mL) for 30 min and then washed twice in RPMI medium. Suspended THP-1 cells were then added to the SGBS monolayers. Nonadherent cells were then removed, and monolayers were fixed with 1% paraformaldehyde. Images of SGBS and adherent calcein-labeled THP-1 cells were visualized and captured with a stereomicroscope (Nikon, Minato, Tokyo, Japan) equipped with the Nikon NIS-Elements D at 40× magnification. Finally, adherent monocytes were counted using the ImageJ program (http://imagej.nih.gov/ij/, accessed on 2 November 2022).

### 2.9. Statistical Analysis

Statistical analysis complies with international recommendations on experimental design and analysis in pharmacology [59–61]. All data are presented as means ± SD, and were analyzed using one/two-way ANOVA followed by Dunnett's for multiple comparisons. GraphPad Prism 8.0 software (San Diego, CA, USA) was used for analysis.

Differences among groups were considered significant when a value of $p \leq 0.05$ was achieved.

## 3. Results

### *3.1. Effect of TIIA and CRY on SGBS Adipocytes Viability*

Preliminary experiments were conducted to test the effects of TIIA and CRY on cell viability. Fully differentiated SGBS adipocytes were pre-treated with TIIA or CRY at the concentration range of 0.1–50 μM (based on literature studies [40,62]) for 1 h, and then stimulated with TNF-α 10 ng/mL for 18 h. After treatments, cell viability was determined by the MTT assay. 1–h pretreatment time was chosen as the most effective treatment time for both compounds in pilot time-course studies (6 h, 3 h, 1 h and 0 h pretreatment before TNF-α) evaluating their anti-inflammatory action against MCP-1 gene expression.

As shown in Figure 2A, TIIA at concentrations tested in the absence or presence of TNF-α did not significantly affect cell viability. Similar results were seen morphologically or observed at the protein level. CRY reduced adipocyte viability at concentrations ≥25 μM, in the absence or presence of TNF-α (Figure 2B). Thus, we used 10 μM as the concentration of TIIA and CRY for further experiments.

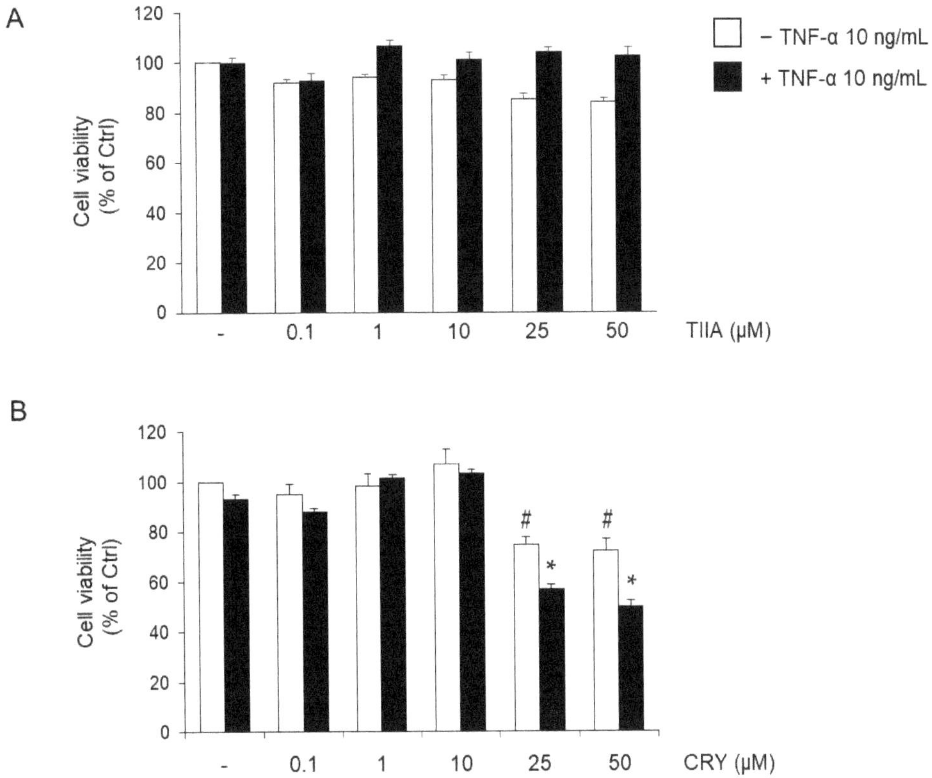


**Figure 2.** The effect of TIIA and CRY treatment on cell viability. SGBS adipocytes were treated with TIIA (**A**) or CRY (**B**) for 1 h at the concentrations indicated, and then either treated with 10 ng/mL TNF-α (black filled bars), or left untreated (open white bars) for 18 h. Cell viability was assessed by the MTT assay, and expressed as a percent of unstimulated control (Ctrl). Data are means ± SD ($n$ = 3). # $p < 0.05$ versus Ctrl. * $p < 0.05$ versus TNF-α alone.

### 3.2. TIIA and CRY Decrease the Release of Cyto/Chemokines after TNF−α Stimulation

The effect of TIIA and CRY on TNF−α-induced secretion of a panel of pro-inflammatory cyto/chemokine was evaluated (panel reported in Figure 3A). Among the anti-/pro-inflammatory cyto/chemokines tested, cell supernatants obtained from TNF−α (10 ng/mL) group (Figure 3C) displayed a large increase of pro-inflammatory cyto/chemokines compared to Ctrl (Figure 3B). Interestingly, when comparing cell supernatants from TIIA- and CRY-treated groups (both at 10 μM) with TNF−α (alone), we observed a significant decrease in several mediators (Figure 3D,E). Densitometric analysis, presented as a heatmap, revealed a significant reduction of the following TNF−α-stimulated pro-inflammatory factors: MCP-1/CCL2 ($p \leq 0.0001$), CCL5/RANTES ($p \leq 0.0001$), (CXCL1/GRO-α ($p \leq 0.0001$), CXCL10/IP-10 ($p \leq 0.0001$), IL-6 ($p \leq 0.05$), IL-8/CXCL8 ($p \leq 0.0001$, only for CRY), macrophage migration inhibitory factor (MIF) ($p \leq 0.001$ and $p \leq 0.0001$ for TIIA and CRY, respectively) and plasminogen activator inhibitor 1 (PAI-1/Serpin E1) ($p \leq 0.0001$) (Figure 3F).

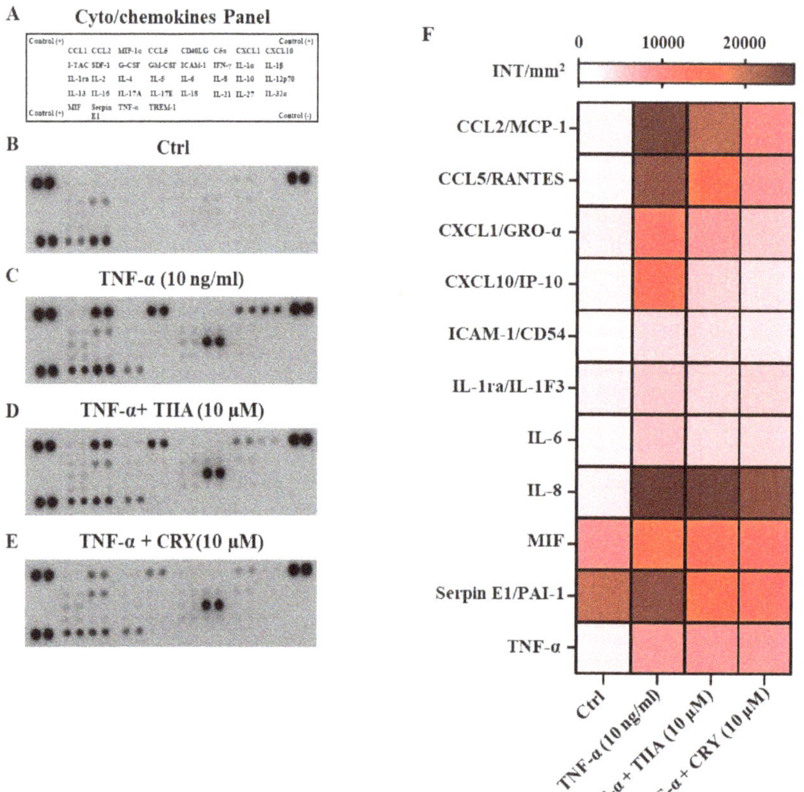


| | Δ TNFα − Ctrl | Δ TIIA − TNF-α | Δ CRY − TNF-α |
|---|---|---|---|
| CCL2/MCP-1 | 22,526 | -4,518 | -15,348.5 |
| CCL5/RANTES | 20,861.5 | -6,117 | -14,631 |
| CXCL1/GRO-α | 9,252 | -3,993.5 | -6,921 |
| CXCL10/IP-10 | 12,154 | -9,814 | -10,953.5 |
| ICAM-1/CD54 | 1,907.5 | -64 | -660.5 |
| IL-1ra/IL-1F3 | 2,360 | -770 | -603 |
| IL-6 | 3,098.5 | -1,189.5 | -1,178 |
| IL-8 | 23,230.5 | -398 | -2,291 |
| MIF | 6,256.5 | -1,651.5 | -2,823.5 |
| Serpin E1/PAI-1 | 3,341.5 | -7,135.5 | -10,820 |
| TNF-α | 6,191 | 248.5 | 51.5 |

**Figure 3.** Cell supernatants were assayed using a proteome profiler cytokine array (**A**) for Ctrl (**B**), TNF−α (10 ng/mL) (**C**), TNF−α + TIIA 10 μM (**D**) and TNF−α + CRY 10 μM (**E**) groups. Densitometric analysis is presented as a heatmap, expressed as INT(intensity)/mm$^2$ (**F**). On the right panel of the heatmap is reported a table list with Δ values (Δ TNF−α−Ctrl; Δ TIIA−TNF-α; Δ CRY−TNF−α) of all pro-inflammatory cyto/chemokines. Data are presented as means ± SD of positive spots of three separate independent experiments run each with $n = 3$ per group pooled (**F**). ElisaSpot assay statistical analysis (reported in the text) was performed by using two-way ANOVA followed by Dunnett's for multiple comparisons.

### 3.3. TIIA and CRY Attenuate Monocyte Adhesion to Inflamed Adipocytes

As a functional counterpart of the modulation of adipocyte secretion profiles by TIIA and CRY, we assessed the effect of both compounds on the adhesion of monocytes to inflamed adipocytes. As shown in Figure 4, TNF-α stimulated the adhesion of THP-1 monocytes on adipocytes as a consequence of the induced secretion of cytokines and chemokines. Contrarily, TIIA and CRY treatment reduced the adhesion of THP-1 monocytes on TNF-α-stimulated adipocytes, in line with the regulation of adipokines by TIIA and CRY toward an anti-inflammatory profile.

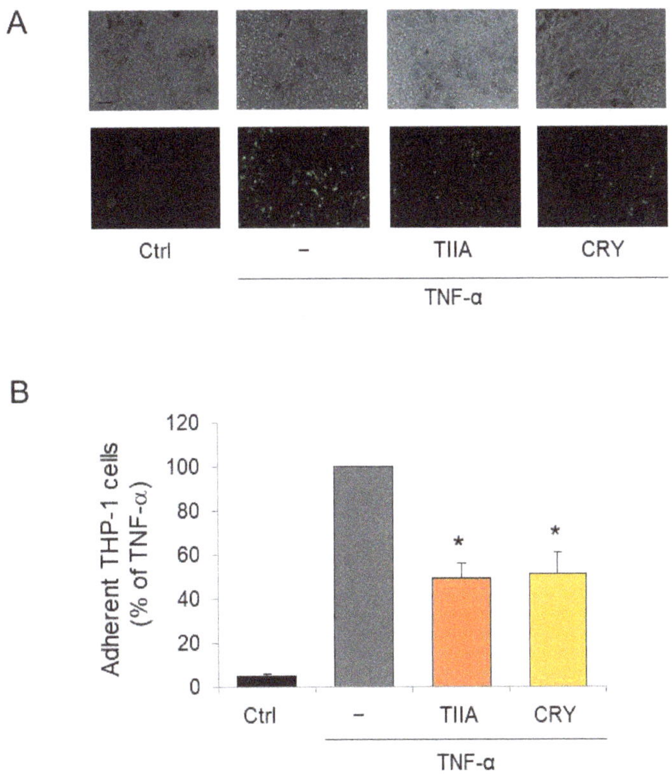


**Figure 4.** Modulation by TIIA and CRY of THP−1 monocytes adhesion to inflamed adipocytes. Fluorescently labeled THP−1 ($10^6$ cells/mL) were added to SGBS monolayers pretreated with 10 μM TIIA or CRY before stimulation with 10 ng/mL TNF−α for 18 h. After 1 h, non-adhering cells were removed and (**A**) images of SGBS and adherent calcein-labeled THP−1 cells were visualized and captured with a phase contrast microscope (upper panels) and a fluorescent microscope (lower panels). (**B**) The number of adherent THP−1 cells is expressed as a percent of TNF-α. Scale bar = 50 μm. * $p < 0.05$ versus TNF-α alone.

### 3.4. TIIA and CRY attenuate TNF-α–Mediated Inflammatory Gene Expression in Human Adipocytes

To evaluate the modulation of secretory proteins by TIIA and CRY at the transcriptional level, mRNA expression levels of MCP-1, CXCL-10, CXCL-1, CCL-5, and IL-8 were determined by qPCR. As shown in Figure 5, both compounds were able to significantly downregulate the TNF-α-stimulated increase in the mRNA expression of the tested chemokines and cytokines, with a more potent inhibitory effect of CRY compared with TIIA, in accordance with the modulation of corresponding protein release.

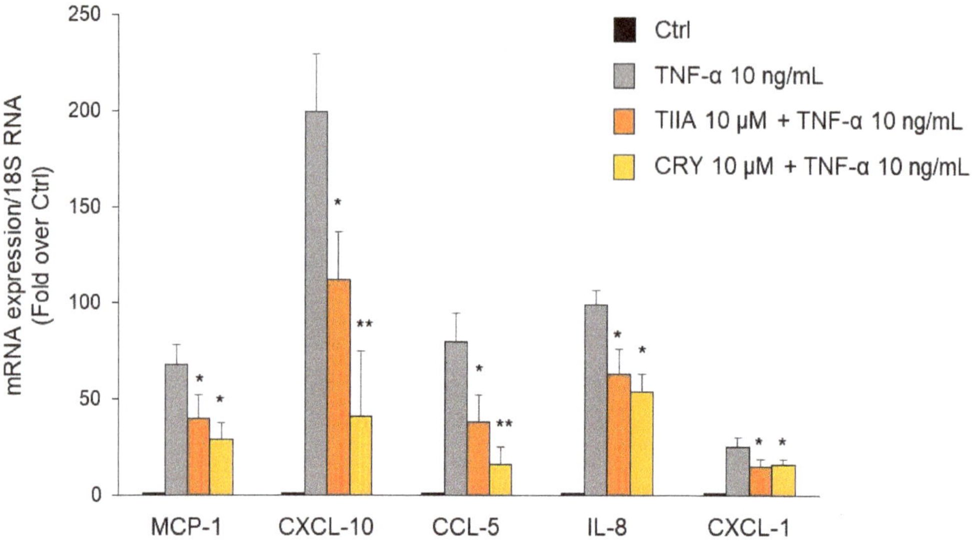


**Figure 5.** Modulation by TIIA and CRY of mRNA expression levels of genes associated with adipocyte inflammation. SGBS adipocytes were pretreated with 10 μM TIIA or CRY (1 h) at the concentrations indicated, and then treated with 10 ng/mL TNF-α for 18 h. Total RNA was extracted from cells, and mRNA levels of the indicated genes were measured by qPCR using specific primers and probes and normalized to 18S RNA. Data (means ± SD, $n = 3$) are expressed as fold induction over unstimulated control (Ctrl). * $p < 0.05$ versus TNF-α alone. ** $p < 0.01$ versus TNF-α alone.

### *3.5. TIIA and CRY Modulate TNF-α-Induced Inflammation-Linked miRNAs Expression in Adipocytes*

To evaluate the modulatory effect of the tested diterpenoids on miRNAs expression related to inflammation, SGBS adipocytes were pre-incubated with TIIA or CRY for 1 h and then stimulated with TNF-α for 18 h. Levels of miRNAs extracted from adipocytes were analyzed by qPCR. Under inflammatory conditions induced by TNF-α, we observed a decrease of miRNA known as anti-inflammatory i.e., miR-126-3p, miR-223-3p, and miR-124-3p (Figure 6A–C), and an increased expression of miRNAs known as pro-inflammatory such as miR-155-5p and miR-132-3p, compared to levels found in untreated cells (Figure 6D,E). Pre-treatment with TIIA or CRY significantly prevented the TNF-α-induced decreased expression of miR-126-3p, miR-223-3p, and miR-124-3p (Figure 6A–C). In parallel, treatment with TIIA or CRY significantly counteracted the higher expression of miR-132-3p and miR-155-5p observed under TNF-α alone (Figure 6D,E).

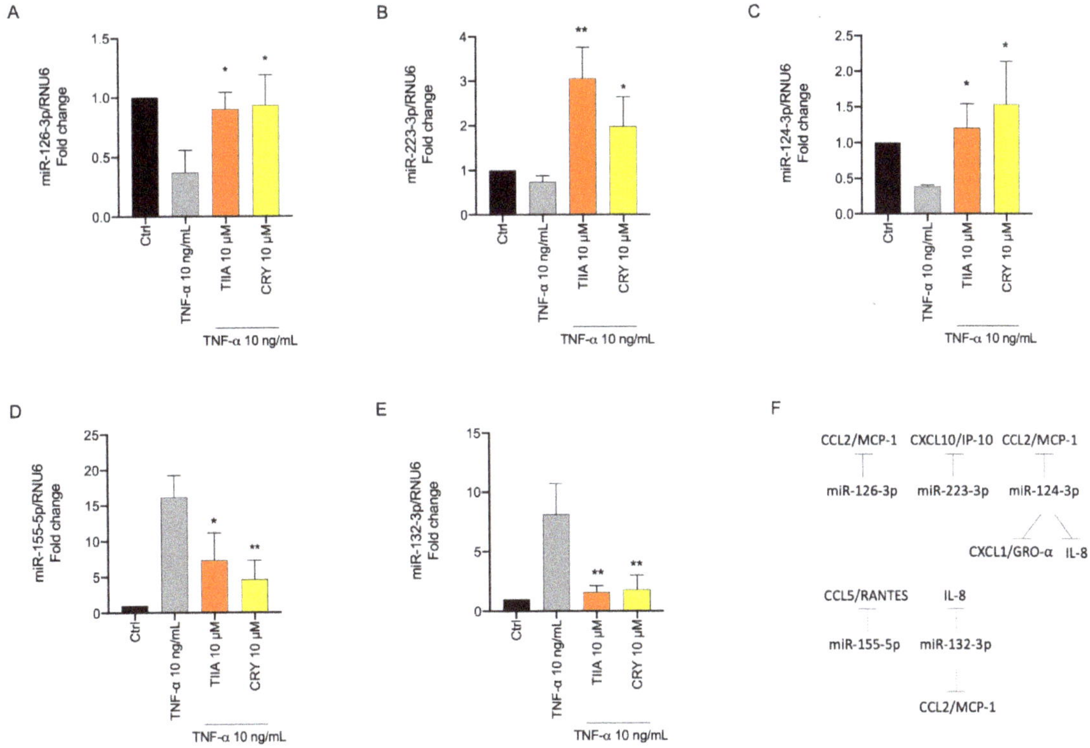


**Figure 6.** Modulation by TIIA and CRY of miRNA expression levels associated with inflammation. SGBS adipocytes were pretreated with 10 µM TIIA or CRY (1 h) and then treated with 10 ng/mL TNF−α for 18 h. Total miRNAs were extracted from cells, and miRNA levels were measured by qPCR and normalized to RNU6. (**A**) miR−126−3p, (**B**) miR−223−p, (**C**) miR−124−3p, (**D**) miR−155−5p, (**E**) miR−132−3p. (**F**) Schematic representation of the miRNA-mRNA targeting, arrow: direct inhibition, dotted arrow: indirect inhibition. Data (means ± SD, $n = 3$) are expressed as fold change (2–ΔΔCt) versus unstimulated control (Ctrl). * $p < 0.05$ versus TNF−α alone. ** $p < 0.01$ versus TNF-α alone.

### *3.6. Biological Processes Associated to miRNAs Modulated by TIIA and CRY in Inflamed Adipocytes*

A bioinformatic analysis of upregulated miR-124-3p, miR-126-3p, miR-223-3p and downregulated miR-155-5p, miR-132-3p, and the mRNAs for CCL2, CCL5, CXCL1, CXCL8, and CXCL10 were conducted. The miRNA–mRNA network showed that they participated in the regulation of immune response and inflammation mediated by other cytokines that are at least double-connected and predicted to be activated or inhibited by the modulation of miRNA/mRNA targets (Figure 7A). This approach allowed us to identify a set of relevant pro-inflammatory cytokines acting in connection with our miRNA/mRNA. Besides representing potential future targets, cytokines such as IL-1, IFN-γ, TNF-α, or IL-6 appear to be themselves influenced by TIIA and CRY treatment.

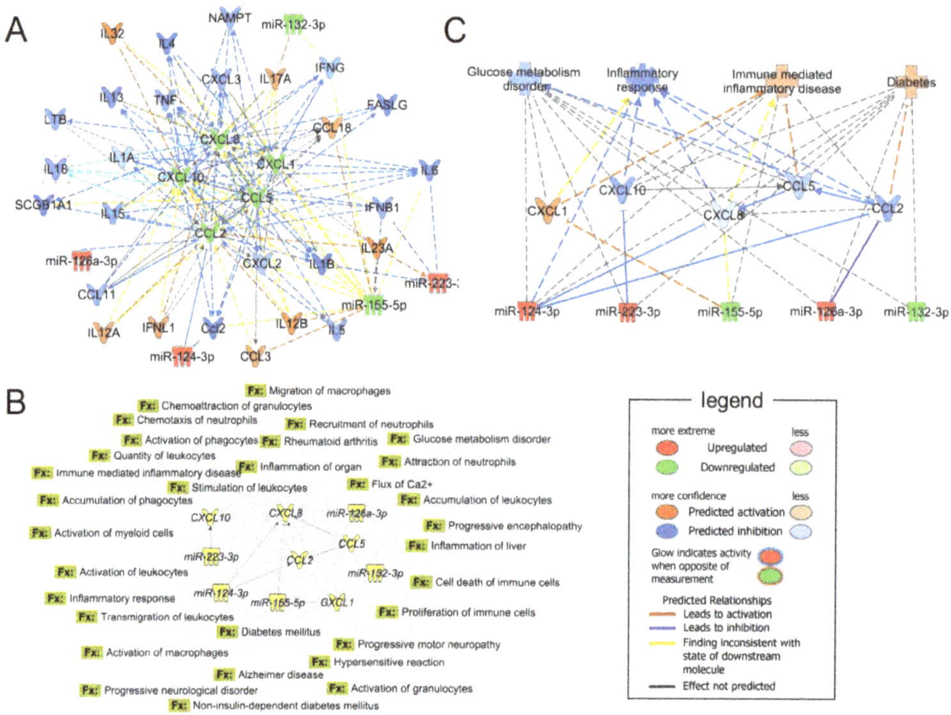


**Figure 7.** Bioinformatics survey of miRNA-mRNA modulated by TIIA and CRY. (**A**) Assessment of miRNA and genes regulation by a prediction network encompassing pro-inflammatory cytokines connected with at least two target miRNA/mRNA and bioinformatically predicted to be activated or inhibited by their modulation. (**B**) Functional association network generated according to the relationship of significant functions and the target miRNAs/genes. Significant molecular functions with less than 5 connections were not included. (**C**) Hierarchical network predicting the effect of miRNAs up/down-regulation on gene expression and the downstream effects on the top *Disease and Function* annotations.

Ingenuity knowledge base, a repository of biological interactions and functional annotations created from multiple individually modeled cellular relationships, was used to predict the possible functions significantly involved in the miRNA/mRNA modulation (Figure 7B). Downstream analysis, which scrutinized the main functions involving at least 5 out of 10 nodes with a $p$ value < 0.01, revealed the presence of five macro-categories of interest, including *Metabolic disorders*, *Neurological diseases*, *Inflammatory disease and response*, *Hematological system development and function* and *Cell signaling*.

Furthermore, we hierarchically connected miRNA, genes and the main *Disease and Function* with higher values of predicted activation/inhibition and lower $p$-value (Figure 7C). Also, in this case, annotations related to metabolic disorders and inflammation appeared to be the most involved in the modulation of our targets.

## 4. Discussion

In the present study, we investigated gene, miRNAs and protein expression profiles of prototypical obesity-associated dysfunction markers in inflamed human adipocytes treated with TIIA and CRY, the two most important lipophilic diterpenoid compounds of *Salvia milthorrhiza* Bunge. The results showed that TIIA and CRY prevented TNF-$\alpha$-induced inflammatory response in adipocytes, by counter-regulating the pattern of secreted cytokines/chemokines associated with adipocyte inflammation (CCL2/MCP-1, CXCL10/IP-10, CCL5/RANTES, CXCL1/GRO-α, IL-6, IL-8, MIF and PAI-1/Serpin E1) via the modulation of gene expression (as demonstrated for CCL2/MCP-1, CXCL10/IP-10, CCL5/RANTES, CXCL1/GRO-α, and IL-8), as well as related miRNA expression (miR-126-3p, miR-223-3p, miR-124-3p, miR-155-5p, and miR-132-3p). This is the first demonstration of a beneficial effect of TIIA and CRY on adipocyte dysfunction associated with obesity development and complications, thus opening new perspectives in pharmacological research for the prevention and/or treatment of metabolic diseases.

The altered pattern of adipokine production by hypertrophic adipocytes is a typical marker and a pathophysiological feature of visceral obesity [63]. Adipokines may exert both local and systemic roles: at the adipose tissue level, they regulate adipogenesis, insulin sensitivity, adipocyte metabolism and function, and immune cell recruitment and activation [63]. At the systemic level, they regulate appetite/satiety, energy metabolism, body fat distribution, insulin secretion and action, blood pressure, vascular function, and hemostasis, thus implying more widespread endocrine effects of adipose tissue inflammation. Obesity is associated with the upregulation of the expression and secretion of pro-inflammatory, diabetogenic and pro-atherogenic factors that therefore represent a pathogenic link–and a targetable component–between obesity and related metabolic and cardiovascular complications [63,64].

TNF-α is a prototypical pro-inflammatory adipocytokine whose expression, mostly by adipocytes as well as adipose tissue macrophages, increases during the development of obesity [46,49], and contributes to adipocyte dysfunction by inhibiting adipogenesis and insulin signaling and inducing lipolysis and inflammation [65,66]. Accordingly, we observed that adipocytes exposure to TNF-α led to increased levels of secreted adipokines i.e., CCL2/MCP-1, CXCL10/IP-10, CCL5/RANTES, CXCL1/GRO-α, IL-6, IL-8, MIF and PAI-1/Serpin E1, and corresponding mRNAs as surveyed for a subset of adipokines i.e., CCL2/MCP-1, CXCL10/IP-10, CCL5/RANTES, CXCL1/GRO-α, and IL-8. This was functionally accompanied, as occurs during obesity development, by the adhesion of THP-1 monocytes on inflamed adipocytes as a consequence of the induced adipocyte secretion of cytokines and chemokines.

Chemokines (or chemoattractant cytokines) play a pivotal role in the recruitment and inflammatory activation of immune cells at sites of chronic inflammation, including the expanding adipose tissue in obesity [67]. The contributory role of the chemokine CCL2/MCP-1 in macrophage recruitment in obesity in metabolic tissues and consequent insulin resistance has been shown early in animal models [11,12,68,69].

However, the genetic deficiency or inactivation of CCL2/MCP-1 or its receptor (CCR2) did not entirely restrain macrophage infiltration in obese adipose tissue, suggesting other chemokines may be involved [12]. In support of this, CCL5/RANTES has also been found to be overexpressed in obese adipose tissue [70,71], and implicated in macrophage accumulation and survival, inflammation and insulin resistance in the visceral adipose tissue during obesity [72,73]. Although CCL5 is thought to be mainly expressed by the stromavascular fraction of adipose tissue in obesity [74] and its deficiency led to a compensatory increase in T cell recruitment [75], adipocytes may also be an important source of CCL5 as demonstrated in previous findings [70], as well as in the present study. CXCL10/IP-10 is another potent chemotactic factor with higher expression in obese adipose tissue [70,76] and circulating levels significantly associated with visceral adipose tissue expansion, insulin resistance and metabolic syndrome [77]. Moreover, CXCL1/GRO-α and IL-8 also exhibited higher expression in the adipose tissue of obese subjects [70] as well as in the circulation of obese mice and humans in association with insulin resistance and type 2 diabetes [78,79]. Interestingly, several of these chemokines were shown to promote the recruitment of other immune cells besides macrophages, such as neutrophils (as for CXCL1 and IL-8) [80] and Th1 (pro-inflammatory) cells (as for CCL5 and CXCL-10/IP-10) [81], that also infiltrate and derange adipose tissue in response to obesity [81,82]. MIF is a pleiotropic cytokine exerting pro-inflammatory and dysmetabolic functions and exhibiting higher adipose tissue and plasma levels in obesity, insulin resistance, and type 2 diabetes [83,84]. PAI-1 is an adipokine that is highly expressed in adipose tissue in obesity, insulin resistance, and type 2 diabetes, having pro-inflammatory and pro-thrombotic functions that may play a role in impaired fibrinolysis associated with obesity [85,86]. IL-6 is another cytokine that can be produced by adipocytes and is crucially implicated in the pathogenesis of obesity and insulin resistance [87–89]. Circulating levels of IL-6 positively correlate with obesity in humans [88] and predict the risk of insulin resistance and type 2 diabetes [87]. Notably, IL-6 produced by visceral adipose depots is drained directly into the portal circulation and contributes to obesity-associated production of C reactive protein (CRP), a clinical index of cardiovascular risk [90].

We observed that TIIA and CRY were able to significantly downregulate the TNF-α-induced production of these cytokines and chemokines in adipocytes, along with a concordant inhibition of monocyte recruitment. Given the role of these adipokines in obesity and its adverse sequelae, we can hypothesize that the observed inhibitory effects by TIIA and CRY may attenuate the adipose tissue and systemic chronic inflammatory tone associated with obesity and the metabolic and vascular consequences [91].

Previous findings support the multifaceted pharmacological properties of the diterpenes TIIA and CRY. In preclinical studies, anti-cancer, cardioprotective, vasculoprotective, anti-platelet, hepatoprotective, neuroprotective, and antidiabetic effects have been attributed to both the active molecules of *S. milthorrhiza* [28,62,92,93]. Most of the benefits of both compounds are thought to derive from the ability to blunt inflammation as shown in several cell and animal models, for example by inhibiting: (1) the production of pro-inflammatory molecules, including some of those here investigated, adhesion molecules, pro-inflammatory or matrix remodeling enzymes; (2) the activation of signaling pathways such as mitogen-activated protein kinase (MAPK) and NLRP3 inflammasome, and of inflammatory transcription factors such as NF-κB and signal transducer and activator of transcription (STAT); and (3) by decreasing the overproduction of reactive oxygen species and/or inducing the cytoprotective antioxidant response via activation of Nrf2 [62,92,94].

Concerning metabolic diseases and especially adipocyte function, anti-obesity effects have been documented for TIIA and CRY. In animal models of obesity and NAFLD, TIIA attenuated weight gain, dyslipidemia, plasma levels of IL-6, TNF-α and IL-1β, and oxidative stress markers by inhibiting toll-like receptor (TLR)4/NF-κB signaling pathway and upregulating PPARγ [95]. Moreover, TIIA suppressed 3T3-L1 preadipocyte differentiation through antagonism of PPARγ transcriptional activities and decreased adipose mass, body weight, and improved glucose tolerance and blood lipid levels in high-fat diet-induced obese mice [96]. CRY promoted mitochondrial biogenesis and brown differentiation in mesenchymal stem cells [97]. This is in line with the in vivo [98] and in vitro [99] evidence of anti-obesity effect by CRY [98,99]. Beyond anti-adipogenic effects, insulin-sensitizing activity and anti-diabetic effects have been reported for TIIA [100–102], and CRY [103]. In accordance with these literature data, we here added novel evidence of potential anti-obesity effects through the modulation of the inflammatory response in adipocytes which is a pivotal process in the development of metabolic complications.

As a part of our investigation into the efficacy of TIIA and CRY on adipocyte inflammation, we evaluated their effects on epigenetic regulators of gene expression and specifically on miRNA, a small non-coding RNA molecule that functions to regulate gene expression by repressing translation or cleaving mRNAs [104,105]. There has been a growing interest in these molecules as potential disease biomarkers and therapeutic targets because a single miRNA can directly bind to several transcripts, affecting their expression and thereby simultaneously controlling a wide range of biological processes, such as adipogenesis, insulin signaling, adipokine expression, inflammation, and food intake [106–109].

The present study showed that TIIA and CRY can attenuate miRNA expression changes in human adipocytes induced by TNF-α. Indeed, TNF-α upregulated miR-155-5p and miR-132-3p and downregulated miR-126-3p, miR-223-3p and miR-124-3p. On the contrary, part of the anti-inflammatory effects of TIIA and CRY are likely to be explained by their ability to restore miRNAs expression in inflamed adipocytes. In detail, TIIA and CRY decreased the TNF-α-stimulated levels of CCL2/MCP-1 and CXCL-10/IP-10 mRNA and protein expression and, correspondingly, increased the levels of their inhibitory miRNAs, i.e., miR-126-3p [110], miR-223-3p [111] and miR-124-3p [112]. Furthermore, TIIA and CRY significantly decreased TNF-α-induced expression of IL-8 and CXCL-1/GRO-α and increased their experimentally validated inhibitory miR-124-3p [112]. TIIA and CRY also inhibited TNF-α-stimulated expression of miR-132-3p, indirectly reducing the expression of CCL2/MCP-1 and IL-8 [113]. Finally, both tested compounds attenuated TNF-α-stimulated expression of miR-155-5p, which may indirectly decrease CCL-5/RANTES expression in an NF-κB-dependent manner [114]. In line with our data, miR-155 is increased in obese adipose tissue [115] and is upregulated by TNF-α in adipocytes via NF-κB activation [116]. Interestingly, miR-155-5p is considered a master regulator of inflammation [117] and is involved in the pathogenesis of obesity and insulin resistance [118,119]. A similar modulatory effect by TIIA on miR-155-5p was observed in RAW264.7 macrophages incubated with LPS [36,37]. miR-155 inhibition may translate into reduced pro-inflammatory activation being miR-155 involved in pro-inflammatory M1 polarization of macrophages [120]. On the other hand, miR-223, which was upregulated by TIIA and CRY, was shown to protect against adipose tissue inflammation and systemic insulin resistance [121].

The regulation of miRNAs by the two diterpenes has been previously shown in a few studies concerning different pathophysiological settings, such as cardiovascular diseases [122–124] or cancer [125]. Our data, therefore, strengthen and expand evidence on potential molecular mechanisms of the anti-inflammatory action of TIIA and CRY, and indicate that TIIA and CRY exert their anti-inflammatory activity in adipocytes also by controlling miRNAs expression towards an anti-inflammatory profile that may improve adipocyte dysfunction. Notably, miRNAs can be packaged into exosomes that can regulate gene expression in a paracrine fashion and distant organs [118,126]. miRNAs, besides adipokines, may represent a significant and novel targetable way for the adipose tissue to cross-talk with other cells/tissues and contribute to local and systemic chronic inflammation in obesity.

Aiming at providing further functional insight into the effects of TIIA and CRY in adipocytes, the integrative bioinformatic analysis of the mRNA-miRNA networks revealed that other inflammation-related cytokines and chemokines were predicted to be affected by the diterpenes, thus potentially expanding the influence of these compounds on the networks of immune/inflammatory mediators. Moreover, several enriched functions, such as immune cell chemoattraction, migration and activation, or dysmetabolism, relevant to inflammatory and cardiometabolic diseases were identified among the most significantly modulated by TIIA and CRY. This analysis may allow a better understanding of the molecular mechanisms of action of TIIA and CRY [92] and provide avenues for identifying potential target genes and pathways for future studies.

In vivo studies conducted in mice reported the anti-inflammatory and anti-oxidant propriety of TIIA and CRY in a dosage range between 10–90 mg/kg [35,38–40,95,98,102]. Therefore, we estimated a human equivalent dose (HED) of 0.67–6.09 mg/kg and an estimated human dose of 47.4–426.3 mg/die in an adult of 70 kg [127]. These speculations could serve as the starting point for future studies aimed at supporting the clinical use of TIIA and CRY.

## 5. Conclusions

Our study demonstrates that TIIA and CRY ameliorate adipocyte dysfunction associated with obesity via the modulation of miRNA, gene, and protein expression toward an anti-inflammatory phenotype. Pending experimental studies to further extrapolate these effects in patients, these data recognize TIIA and CRY as natural compounds with attractive biological and pharmacological properties for the development of novel drugs/supplements for clinical use.

**Author Contributions:** Conceptualization, S.C., A.S., N.M. and E.S.; methodology, S.C., S.Q. and E.S.; software, S.Q. and S.D.; formal analysis, S.Q., E.S., B.P., S.C. and M.M.; investigation, S.Q., S.C. and E.S.; resources, M.W., M.M., M.A.C. and F.M.; data curation, E.S., S.Q., S.C., A.S. and N.M.; writing—original draft preparation, E.S., S.C., F.M. and A.S.; writing—review and editing, S.Q., B.P., M.M., M.A.C., N.C., P.N., S.D., M.W., M.C., F.M.S. and F.M.; supervision, S.C., F.M. and E.S. All authors have read and agreed to the published version of the manuscript.

**Funding:** This research was in part funded by MIUR (PRIN 2017; 2017, A95NCJ/2017A95NCJ_002, "Stolen molecules–Stealing natural products from the depot and reselling them as new drug candidates") and in part by Contributo di Ricerca Corrente RC 5X1000.

**Institutional Review Board Statement: Not** applicable.

**Informed Consent Statement:** Not applicable.

**Data Availability Statement:** The data presented in this study are available on request from the corresponding author.

**Acknowledgments:** A.S. is supported by Dompé farmaceutici S.p.A fellowship for PhD program in "Nutraceuticals, functional foods and human health" (University of Naples Federico II), whereas N.M. is supported by ABC Farmaceutici S.p.A. Scholarship (FARMA/BR/02/2023).

**Conflicts of Interest:** The authors declare no conflict of interest.

## References

1. Tchernof, A.; Després, J.-P. Pathophysiology of Human Visceral Obesity: An Update. *Physiol. Rev.* **2013**, *93*, 359–404. [CrossRef]
2. Berrington de Gonzalez, A.; Hartge, P.; Cerhan, J.R.; Flint, A.J.; Hannan, L.; MacInnis, R.J.; Moore, S.C.; Tobias, G.S.; Anton-Culver, H.; Freeman, L.B.; et al. Body-Mass Index and Mortality among 1.46 Million White Adults. *N. Engl. J. Med.* **2010**, *363*, 2211–2219. [CrossRef]
3. LeRoith, D.; Novosyadlyy, R.; Gallagher, E.; Lann, D.; Vijayakumar, A.; Yakar, S. Obesity and Type 2 Diabetes Are Associated with an Increased Risk of Developing Cancer and a Worse Prognosis; Epidemiological and Mechanistic Evidence. *Exp. Clin. Endocrinol. Diabetes* **2008**, *116*, S4–S6. [CrossRef] [PubMed]
4. Hammarstedt, A.; Gogg, S.; Hedjazifar, S.; Nerstedt, A.; Smith, U. Impaired Adipogenesis and Dysfunctional Adipose Tissue in Human Hypertrophic Obesity. *Physiol. Rev.* **2018**, *98*, 1911–1941. [CrossRef] [PubMed]
5. Van Gaal, L.F.; Mertens, I.L.; De Block, C.E. Mechanisms Linking Obesity with Cardiovascular Disease. *Nature* **2006**, *444*, 875–880. [CrossRef]
6. Scully, T.; Ettela, A.; LeRoith, D.; Gallagher, E.J. Obesity, Type 2 Diabetes, and Cancer Risk. *Front. Oncol.* **2021**, *10*, 615375. [CrossRef] [PubMed]
7. Lee, Y.S.; Wollam, J.; Olefsky, J.M. An Integrated View of Immunometabolism. *Cell* **2018**, *172*, 22–40. [CrossRef]
8. Chatzigeorgiou, A.; Karalis, K.P.; Bornstein, S.R.; Chavakis, T. Lymphocytes in Obesity-Related Adipose Tissue Inflammation. *Diabetologia* **2012**, *55*, 2583–2592. [CrossRef]
9. Lumeng, C.N.; Bodzin, J.L.; Saltiel, A.R. Obesity Induces a Phenotypic Switch in Adipose Tissue Macrophage Polarization. *J. Clin. Investig.* **2007**, *117*, 175–184. [CrossRef] [PubMed]
10. Xu, X.; Grijalva, A.; Skowronski, A.; van Eijk, M.; Serlie, M.J.; Ferrante, A.W. Obesity Activates a Program of Lysosomal-Dependent Lipid Metabolism in Adipose Tissue Macrophages Independently of Classic Activation. *Cell Metab.* **2013**, *18*, 816–830. [CrossRef]
11. Kanda, H. MCP-1 Contributes to Macrophage Infiltration into Adipose Tissue, Insulin Resistance, and Hepatic Steatosis in Obesity. *J. Clin. Investig.* **2006**, *116*, 1494–1505. [CrossRef] [PubMed]
12. Kamei, N.; Tobe, K.; Suzuki, R.; Ohsugi, M.; Watanabe, T.; Kubota, N.; Ohtsuka-Kowatari, N.; Kumagai, K.; Sakamoto, K.; Kobayashi, M.; et al. Overexpression of Monocyte Chemoattractant Protein-1 in Adipose Tissues Causes Macrophage Recruitment and Insulin Resistance. *J. Biol. Chem.* **2006**, *281*, 26602–26614. [CrossRef]
13. Weisberg, S.P.; McCann, D.; Desai, M.; Rosenbaum, M.; Leibel, R.L.; Ferrante, A.W. Obesity Is Associated with Macrophage Accumulation in Adipose Tissue. *J. Clin. Investig.* **2003**, *112*, 1796–1808. [CrossRef] [PubMed]
14. Lackey, D.E.; Olefsky, J.M. Regulation of Metabolism by the Innate Immune System. *Nat. Rev. Endocrinol.* **2016**, *12*, 15–28. [CrossRef] [PubMed]
15. Han, M.S.; Jung, D.Y.; Morel, C.; Lakhani, S.A.; Kim, J.K.; Flavell, R.A.; Davis, R.J. JNK Expression by Macrophages Promotes Obesity-Induced Insulin Resistance and Inflammation. *Science* **2013**, *339*, 218–222. [CrossRef] [PubMed]
16. Chiang, S.-H.; Bazuine, M.; Lumeng, C.N.; Geletka, L.M.; Mowers, J.; White, N.M.; Ma, J.-T.; Zhou, J.; Qi, N.; Westcott, D.; et al. The Protein Kinase IKKεRegulates Energy Balance in Obese Mice. *Cell* **2009**, *138*, 961–975. [CrossRef] [PubMed]
17. Goldfine, A.B.; Conlin, P.R.; Halperin, F.; Koska, J.; Permana, P.; Schwenke, D.; Shoelson, S.E.; Reaven, P.D. A Randomised Trial of Salsalate for Insulin Resistance and Cardiovascular Risk Factors in Persons with Abnormal Glucose Tolerance. *Diabetologia* **2013**, *56*, 714–723. [CrossRef] [PubMed]

18. Di Prospero, N.A.; Artis, E.; Andrade-Gordon, P.; Johnson, D.L.; Vaccaro, N.; Xi, L.; Rothenberg, P. CCR2 Antagonism in Patients with Type 2 Diabetes Mellitus: A Randomized, Placebo-Controlled Study. *Diabetes Obes. Metab.* **2014**, *16*, 1055–1064. [CrossRef]
19. Newman, D.J.; Cragg, G.M. Natural Products as Sources of New Drugs over the Nearly Four Decades from 01/1981 to 09/2019. *J. Nat. Prod.* **2020**, *83*, 770–803. [CrossRef]
20. Ohishi, T.; Fukutomi, R.; Shoji, Y.; Goto, S.; Isemura, M. The Beneficial Effects of Principal Polyphenols from Green Tea, Coffee, Wine, and Curry on Obesity. *Molecules* **2021**, *26*, 453. [CrossRef]
21. Bhardwaj, M.; Yadav, P.; Vashishth, D.; Sharma, K.; Kumar, A.; Chahal, J.; Dalal, S.; Kataria, S.K. A Review on Obesity Management through Natural Compounds and a Green Nanomedicine-Based Approach. *Molecules* **2021**, *26*, 3278. [CrossRef] [PubMed]
22. Li, X.W.; Ian, C.H. (Eds.) . *Flora of China*; Science Press China: Beijing, China, 1994; p. 213.
23. *Pharmacopoeia Committee of the People's Republic of China*; Chemical Industry Press: Beijing, China, 2005.
24. Zhang, S.; Huang, G.; Yuan, K.; Zhu, Q.; Sheng, H.; Yu, R.; Luo, G.; Xu, A. Tanshinone IIA Ameliorates Chronic Arthritis in Mice by Modulating Neutrophil Activities. *Clin. Exp. Immunol.* **2017**, *190*, 29–39. [CrossRef] [PubMed]
25. Feng, Z.; Zheng, W.; Li, X.; Lin, J.; Xie, C.; Li, H.; Cheng, L.; Wu, A.; Ni, W. Cryptotanshinone Protects against IL-1β-Induced Inflammation in Human Osteoarthritis Chondrocytes and Ameliorates the Progression of Osteoarthritis in Mice. *Int. Immunopharmacol* **2017**, *50*, 161–167. [CrossRef]
26. Bonito, M.C.; Cicala, C.; Marcotullio, M.C.; Maione, F.; Mascolo, N. Biological Activity of Bicyclic and Tricyclic Diterpenoids from Salvia Species of Immediate Pharmacological and Pharmaceutical Interest. *Nat. Prod. Commun.* **2011**, *6*, 1205–1215. [CrossRef]
27. Maione, F.; De Feo, V.; Caiazzo, E.; De Martino, L.; Cicala, C.; Mascolo, N. Tanshinone IIA, a Major Component of *Salvia Milthorriza* Bunge, Inhibits Platelet Activation via Erk-2 Signaling Pathway. *J. Ethnopharmacol.* **2014**, *155*, 1236–1242. [CrossRef] [PubMed]
28. Maione, F.; Piccolo, M.; de Vita, S.; Chini, M.G.; Cristiano, C.; de Caro, C.; Lippiello, P.; Miniaci, M.C.; Santamaria, R.; Irace, C.; et al. Down Regulation of Pro-Inflammatory Pathways by Tanshinone IIA and Cryptotanshinone in a Non-Genetic Mouse Model of Alzheimer's Disease. *Pharm. Res.* **2018**, *129*, 482–490. [CrossRef]
29. Zhang, X.-Z.; Qian, S.-S.; Zhang, Y.-J.; Wang, R.-Q. *Salvia Miltiorrhiza*: A Source for Anti-Alzheimer's Disease Drugs. *Pharm. Biol.* **2016**, *54*, 18–24. [CrossRef]
30. Chen, X.; Guo, J.; Bao, J.; Lu, J.; Wang, Y. The Anticancer Properties of *Salvia Miltiorrhiza* Bunge (Danshen): A Systematic Review. *Med. Res. Rev.* **2014**, *34*, 768–794. [CrossRef]
31. Maione, F.; Mascolo, N. Danshen and the Cardiovascular System: New Advances for an Old Remedy. *Semin. Thromb. Hemost.* **2016**, *42*, 321–322. [CrossRef]
32. Maione, F.; Cantone, V.; Chini, M.G.; de Feo, V.; Mascolo, N.; Bifulco, G. Molecular Mechanism of Tanshinone IIA and Cryptotanshinone in Platelet Anti-Aggregating Effects: An Integrated Study of Pharmacology and Computational Analysis. *Fitoterapia* **2015**, *100*, 174–178. [CrossRef]
33. Available online: https://Clinicaltrials.Gov/Ct2/Results?Cond=&term=Danshen+Dripping+Pills&cntry=&state=&city=&dist= (accessed on 11 May 2023).
34. Li, Z.; Xu, S.; Liu, P. Salvia MiltiorrhizaBurge (Danshen): A Golden Herbal Medicine in Cardiovascular Therapeutics. *Acta Pharm. Sin.* **2018**, *39*, 802–824. [CrossRef] [PubMed]
35. Du, H.; Wang, Y.; Zeng, Y.; Huang, X.; Liu, D.; Ye, L.; Li, Y.; Chen, X.; Liu, T.; Li, H.; et al. Tanshinone IIA Suppresses Proliferation and Inflammatory Cytokine Production of Synovial Fibroblasts from Rheumatoid Arthritis Patients Induced by TNF-α and Attenuates the Inflammatory Response in AIA Mice. *Front. Pharm.* **2020**, *11*, 568. [CrossRef] [PubMed]
36. Gao, S.; Wang, Y.; Li, D.; Guo, Y.; Zhu, M.; Xu, S.; Mao, J.; Fan, G. TanshinoneIIA Alleviates Inflammatory Response and Directs Macrophage Polarization in Lipopolysaccharide-Stimulated RAW264.7 Cells. *Inflammation* **2019**, *42*, 264–275. [CrossRef] [PubMed]
37. Fan, G.; Jiang, X.; Wu, X.; Fordjour, P.A.; Miao, L.; Zhang, H.; Zhu, Y.; Gao, X. Anti-Inflammatory Activity of Tanshinone IIA in LPS-Stimulated RAW264.7 Macrophages via MiRNAs and TLR4–NF-KB Pathway. *Inflammation* **2016**, *39*, 375–384. [CrossRef]
38. Chen, Z.; Gao, X.; Jiao, Y.; Qiu, Y.; Wang, A.; Yu, M.; Che, F.; Li, S.; Liu, J.; Li, J.; et al. Tanshinone IIA Exerts Anti-Inflammatory and Immune-Regulating Effects on Vulnerable Atherosclerotic Plaque Partially via the TLR4/MyD88/NF-KB Signal Pathway. *Front. Pharm.* **2019**, *10*, 850. [CrossRef] [PubMed]
39. Lu, T.-C.; Wu, Y.-H.; Chen, W.-Y.; Hung, Y.-C. Targeting Oxidative Stress and Endothelial Dysfunction Using Tanshinone IIA for the Treatment of Tissue Inflammation and Fibrosis. *Oxid. Med. Cell Longev.* **2022**, *2022*, 2811789. [CrossRef]
40. Wu, Y.H.; Wu, Y.R.; Li, B.; Yan, Z.Y. Cryptotanshinone: A Review of Its Pharmacology Activities and Molecular Mechanisms. *Fitoterapia* **2020**, *145*, 104633. [CrossRef]
41. Ma, S.; Zhang, D.; Lou, H.; Sun, L.; Ji, J. Evaluation of the Anti-Inflammatory Activities of Tanshinones Isolated from *Salvia Miltiorrhiza* Var. Alba Roots in THP-1 Macrophages. *J. Ethnopharmacol.* **2016**, *188*, 193–199. [CrossRef]
42. Tang, S.; Shen, X.-Y.; Huang, H.-Q.; Xu, S.-W.; Yu, Y.; Zhou, C.-H.; Chen, S.-R.; Le, K.; Wang, Y.-H.; Liu, P.-Q. Cryptotanshinone Suppressed Inflammatory Cytokines Secretion in RAW264.7 Macrophages through Inhibition of the NF-KB and MAPK Signaling Pathways. *Inflammation* **2011**, *34*, 111–118. [CrossRef]
43. Wabitsch, M.; Brenner, R.; Melzner, I.; Braun, M.; Möller, P.; Heinze, E.; Debatin, K.-M.; Hauner, H. Characterization of a Human Preadipocyte Cell Strain with High Capacity for Adipose Differentiation. *Int. J. Obes.* **2001**, *25*, 8–15. [CrossRef]

44. Scoditti, E.; Massaro, M.; Carluccio, M.A.; Pellegrino, M.; Wabitsch, M.; Calabriso, N.; Storelli, C.; De Caterina, R. Additive Regulation of Adiponectin Expression by the Mediterranean Diet Olive Oil Components Oleic Acid and Hydroxytyrosol in Human Adipocytes. *PLoS ONE* **2015**, *10*, e0128218. [CrossRef]
45. Quarta, S.; Massaro, M.; Carluccio, M.A.; Calabriso, N.; Bravo, L.; Sarria, B.; García-Conesa, M.-T. An Exploratory Critical Review on TNF-α as a Potential Inflammatory Biomarker Responsive to Dietary Intervention with Bioactive Foods and Derived Products. *Foods* **2022**, *11*, 2524. [CrossRef]
46. Hotamisligil, G.S.; Shargill, N.S.; Spiegelman, B.M. Adipose Expression of Tumor Necrosis Factor-α: Direct Role in Obesity-Linked Insulin Resistance. *Science* **1993**, *259*, 87–91. [CrossRef]
47. Kern, P.A.; Ranganathan, S.; Li, C.; Wood, L.; Ranganathan, G. Adipose Tissue Tumor Necrosis Factor and Interleukin-6 Expression in Human Obesity and Insulin Resistance. *Am. J. Physiol. Endocrinol. Metab.* **2001**, *280*, E745–E751. [CrossRef] [PubMed]
48. Kern, P.A.; Saghizadeh, M.; Ong, J.M.; Bosch, R.J.; Deem, R.; Simsolo, R.B. The Expression of Tumor Necrosis Factor in Human Adipose Tissue. Regulation by Obesity, Weight Loss, and Relationship to Lipoprotein Lipase. *J. Clin. Investig.* **1995**, *95*, 2111–2119. [CrossRef] [PubMed]
49. Hotamisligil, G.S.; Arner, P.; Caro, J.F.; Atkinson, R.L.; Spiegelman, B.M. Increased Adipose Tissue Expression of Tumor Necrosis Factor-Alpha in Human Obesity and Insulin Resistance. *J. Clin. Investig.* **1995**, *95*, 2409–2415. [CrossRef]
50. Souza, S.C.; Palmer, H.J.; Kang, Y.-H.; Yamamoto, M.T.; Muliro, K.V.; Eric Paulson, K.; Greenberg, A.S. TNF-? Induction of Lipolysis Is Mediated through Activation of the Extracellular Signal Related Kinase Pathway in 3T3-L1 Adipocytes. *J. Cell. Biochem.* **2003**, *89*, 1077–1086. [CrossRef]
51. Jack, B.U.; Mamushi, M.; Viraragavan, A.; Dias, S.; Pheiffer, C. Comparing the Effects of Tumor Necrosis Factor Alpha, Lipopolysaccharide and Palmitic Acid on Lipid Metabolism and Inflammation in Murine 3T3-L1 Adipocytes. *Life Sci.* **2022**, *297*, 120422. [CrossRef] [PubMed]
52. Hauner, H.; Petruschke, T.; Russ, M.; Röhrig, K.; Eckel, J. Effects of Tumour Necrosis Factor Alpha (TNFα) on Glucose Transport and Lipid Metabolism of Newly-Differentiated Human Fat Cells in Cell Culture. *Diabetologia* **1995**, *38*, 764–771. [CrossRef] [PubMed]
53. Zhang, H.H.; Halbleib, M.; Ahmad, F.; Manganiello, V.C.; Greenberg, A.S. Tumor Necrosis Factor-α Stimulates Lipolysis in Differentiated Human Adipocytes Through Activation of Extracellular Signal-Related Kinase and Elevation of Intracellular CAMP. *Diabetes* **2002**, *51*, 2929–2935. [CrossRef]
54. Caso, F.; Saviano, A.; Tasso, M.; Raucci, F.; Marigliano, N.; Passavanti, S.; Frallonardo, P.; Ramonda, R.; Brancaleone, V.; Bucci, M.; et al. Analysis of Rheumatoid- vs Psoriatic Arthritis Synovial Fluid Reveals Differential Macrophage (CCR2) and T Helper Subsets (STAT3/4 and FOXP3) Activation. *Autoimmun. Rev.* **2022**, *21*, 103207. [CrossRef]
55. Vellecco, V.; Saviano, A.; Raucci, F.; Casillo, G.M.; Mansour, A.A.; Panza, E.; Mitidieri, E.; Femminella, G.D.; Ferrara, N.; Cirino, G.; et al. Interleukin-17 (IL-17) Triggers Systemic Inflammation, Peripheral Vascular Dysfunction, and Related Prothrombotic State in a Mouse Model of Alzheimer's Disease. *Pharm. Res.* **2023**, *187*, 106595. [CrossRef]
56. Carpi, S.; Polini, B.; Poli, G.; Alcantara Barata, G.; Fogli, S.; Romanini, A.; Tuccinardi, T.; Guella, G.; Frontini, F.; Nieri, P.; et al. Anticancer Activity of Euplotin C, Isolated from the Marine Ciliate Euplotes Crassus, Against Human Melanoma Cells. *Mar. Drugs* **2018**, *16*, 166. [CrossRef] [PubMed]
57. Carpi, S.; Polini, B.; Manera, C.; Digiacomo, M.; Salsano, J.E.; Macchia, M.; Scoditti, E.; Nieri, P. MiRNA Modulation and Antitumor Activity by the Extra-Virgin Olive Oil Polyphenol Oleacein in Human Melanoma Cells. *Front. Pharm.* **2020**, *11*, 574317. [CrossRef] [PubMed]
58. Krämer, A.; Green, J.; Pollard, J.; Tugendreich, S. Causal Analysis Approaches in Ingenuity Pathway Analysis. *Bioinformatics* **2014**, *30*, 523–530. [CrossRef] [PubMed]
59. Curtis, M.J.; Alexander, S.; Cirino, G.; Docherty, J.R.; George, C.H.; Giembycz, M.A.; Hoyer, D.; Insel, P.A.; Izzo, A.A.; Ji, Y.; et al. Experimental Design and Analysis and Their Reporting II: Updated and Simplified Guidance for Authors and Peer Reviewers. *Br. J. Pharm.* **2018**, *175*, 987–993. [CrossRef]
60. Alexander, S.P.H.; Roberts, R.E.; Broughton, B.R.S.; Sobey, C.G.; George, C.H.; Stanford, S.C.; Cirino, G.; Docherty, J.R.; Giembycz, M.A.; Hoyer, D.; et al. Goals and Practicalities of Immunoblotting and Immunohistochemistry: A Guide for Submission to the British Journal of Pharmacology. *Br. J. Pharm.* **2018**, *175*, 407–411. [CrossRef] [PubMed]
61. George, C.H.; Stanford, S.C.; Alexander, S.; Cirino, G.; Docherty, J.R.; Giembycz, M.A.; Hoyer, D.; Insel, P.A.; Izzo, A.A.; Ji, Y.; et al. Updating the Guidelines for Data Transparency in the British Journal of Pharmacology—Data Sharing and the Use of Scatter Plots Instead of Bar Charts. *Br. J. Pharm.* **2017**, *174*, 2801–2804. [CrossRef]
62. Ansari, M.A.; Khan, F.B.; Safdari, H.A.; Almatroudi, A.; Alzohairy, M.A.; Safdari, M.; Amirizadeh, M.; Rehman, S.; Equbal, M.J.; Hoque, M. Prospective Therapeutic Potential of Tanshinone IIA: An Updated Overview. *Pharm. Res.* **2021**, *164*, 105364. [CrossRef]
63. Zorena, K.; Jachimowicz-Duda, O.; Ślęzak, D.; Robakowska, M.; Mrugacz, M. Adipokines and Obesity. Potential Link to Metabolic Disorders and Chronic Complications. *Int. J. Mol. Sci.* **2020**, *21*, 3570. [CrossRef]
64. Blüher, M. Adipokines—Removing Road Blocks to Obesity and Diabetes Therapy. *Mol. Metab.* **2014**, *3*, 230–240. [CrossRef] [PubMed]
65. Gustafson, B.; Gogg, S.; Hedjazifar, S.; Jenndahl, L.; Hammarstedt, A.; Smith, U. Inflammation and Impaired Adipogenesis in Hypertrophic Obesity in Man. *Am. J. Physiol. Endocrinol. Metab.* **2009**, *297*, E999–E1003. [CrossRef] [PubMed]

66. Laurencikiene, J.; Van Harmelen, V.; Nordström, E.A.; Dicker, A.; Blomqvist, L.; Näslund, E.; Langin, D.; Arner, P.; Rydén, M. NF-KB Is Important for TNF-α-Induced Lipolysis in Human Adipocytes. *J. Lipid Res.* **2007**, *48*, 1069–1077. [CrossRef] [PubMed]
67. Turner, M.D.; Nedjai, B.; Hurst, T.; Pennington, D.J. Cytokines and Chemokines: At the Crossroads of Cell Signalling and Inflammatory Disease. *Biochim. Biophys. Acta (BBA) Mol. Cell Res.* **2014**, *1843*, 2563–2582. [CrossRef]
68. Kim, J.; Chung, K.; Choi, C.; Beloor, J.; Ullah, I.; Kim, N.; Lee, K.Y.; Lee, S.-K.; Kumar, P. Silencing CCR2 in Macrophages Alleviates Adipose Tissue Inflammation and the Associated Metabolic Syndrome in Dietary Obese Mice. *Mol. Ther. Nucleic Acids* **2016**, *5*, e280. [CrossRef]
69. Ohman, M.K.; Wright, A.P.; Wickenheiser, K.J.; Luo, W.; Russo, H.M.; Eitzman, D.T. Monocyte Chemoattractant Protein-1 Deficiency Protects Against Visceral Fat-Induced Atherosclerosis. *Arter. Thromb. Vasc. Biol.* **2010**, *30*, 1151–1158. [CrossRef]
70. Tourniaire, F.; Romier-Crouzet, B.; Lee, J.H.; Marcotorchino, J.; Gouranton, E.; Salles, J.; Malezet, C.; Astier, J.; Darmon, P.; Blouin, E.; et al. Chemokine Expression in Inflamed Adipose Tissue Is Mainly Mediated by NF-KB. *PLoS ONE* **2013**, *8*, e66515. [CrossRef] [PubMed]
71. Huber, J.; Kiefer, F.W.; Zeyda, M.; Ludvik, B.; Silberhumer, G.R.; Prager, G.; Zlabinger, G.J.; Stulnig, T.M. CC Chemokine and CC Chemokine Receptor Profiles in Visceral and Subcutaneous Adipose Tissue Are Altered in Human Obesity. *J. Clin. Endocrinol. Metab.* **2008**, *93*, 3215–3221. [CrossRef] [PubMed]
72. Keophiphath, M.; Rouault, C.; Divoux, A.; Clément, K.; Lacasa, D. CCL5 Promotes Macrophage Recruitment and Survival in Human Adipose Tissue. *Arter. Thromb. Vasc. Biol.* **2010**, *30*, 39–45. [CrossRef]
73. Chan, P.-C.; Lu, C.-H.; Chien, H.-C.; Tian, Y.-F.; Hsieh, P.-S. Adipose Tissue-Derived CCL5 Enhances Local Pro-Inflammatory Monocytic MDSCs Accumulation and Inflammation via CCR5 Receptor in High-Fat Diet-Fed Mice. *Int. J. Mol. Sci.* **2022**, *23*, 14226. [CrossRef]
74. Matter, C.M.; Handschin, C. RANTES (Regulated on Activation, Normal T Cell Expressed and Secreted), Inflammation, Obesity, and the Metabolic Syndrome. *Circulation* **2007**, *115*, 946–948. [CrossRef]
75. Zhou, H.; Liao, X.; Zeng, Q.; Zhang, H.; Song, J.; Hu, W.; Sun, X.; Ding, Y.; Wang, D.; Xiao, Y.; et al. Metabolic Effects of CCL5 Deficiency in Lean and Obese Mice. *Front. Immunol.* **2023**, *13*, 1059687. [CrossRef] [PubMed]
76. Faber, D.R.; Moll, F.L.; Vink, A.; van der Waal, C.; Kalkhoven, E.; Schipper, H.S.; Hajer, G.R.; Monajemi, H.; Visseren, F.L.J. Adipose Tissue Quantity and Composition Contribute to Adipokine Concentrations in the Subclavian Vein and the Inferior Mesenteric Vein. *Int. J. Obes.* **2012**, *36*, 1078–1085. [CrossRef]
77. Faber, D.R.; van der Graaf, Y.; Westerink, J.; Kanhai, D.A.; Monajemi, H.; Visseren, F.L.J. Hepatocyte Growth Factor and Interferon-γ Inducible Protein-10 Are Related to Visceral Adiposity. *Eur. J. Clin. Investig.* **2013**, *43*, 369–378. [CrossRef] [PubMed]
78. Nunemaker, C.S.; Chung, H.G.; Verrilli, G.M.; Corbin, K.L.; Upadhye, A.; Sharma, P.R. Increased Serum CXCL1 and CXCL5 Are Linked to Obesity, Hyperglycemia, and Impaired Islet Function. *J. Endocrinol.* **2014**, *222*, 267–276. [CrossRef] [PubMed]
79. Straczkowski, M.; Kowalska, I.; Nikolajuk, A.; Dzienis-Straczkowska, S.; Szelachowska, M.; Kinalska, I. Plasma Interleukin 8 Concentrations in Obese Subjects with Impaired Glucose Tolerance. *Cardiovasc. Diabetol.* **2003**, *2*, 5. [CrossRef] [PubMed]
80. Viola, A.; Luster, A.D. Chemokines and Their Receptors: Drug Targets in Immunity and Inflammation. *Annu. Rev. Pharm. Toxicol.* **2008**, *48*, 171–197. [CrossRef]
81. Wu, H.; Ghosh, S.; Perrard, X.D.; Feng, L.; Garcia, G.E.; Perrard, J.L.; Sweeney, J.F.; Peterson, L.E.; Chan, L.; Smith, C.W.; et al. T-Cell Accumulation and Regulated on Activation, Normal T Cell Expressed and Secreted Upregulation in Adipose Tissue in Obesity. *Circulation* **2007**, *115*, 1029–1038. [CrossRef]
82. Elgazar-Carmon, V.; Rudich, A.; Hadad, N.; Levy, R. Neutrophils Transiently Infiltrate Intra-Abdominal Fat Early in the Course of High-Fat Feeding. *J. Lipid Res.* **2008**, *49*, 1894–1903. [CrossRef]
83. Morrison, M.C.; Kleemann, R. Role of Macrophage Migration Inhibitory Factor in Obesity, Insulin Resistance, Type 2 Diabetes, and Associated Hepatic Co-Morbidities: A Comprehensive Review of Human and Rodent Studies. *Front. Immunol.* **2015**, *6*, 308. [CrossRef]
84. Finucane, O.M.; Reynolds, C.M.; McGillicuddy, F.C.; Harford, K.A.; Morrison, M.; Baugh, J.; Roche, H.M. Macrophage Migration Inhibitory Factor Deficiency Ameliorates High-Fat Diet Induced Insulin Resistance in Mice with Reduced Adipose Inflammation and Hepatic Steatosis. *PLoS ONE* **2014**, *9*, e113369. [CrossRef]
85. Skurk, T.; Hauner, H. Obesity and Impaired Fibrinolysis: Role of Adipose Production of Plasminogen Activator Inhibitor-1. *Int. J. Obes.* **2004**, *28*, 1357–1364. [CrossRef] [PubMed]
86. Ma, L.-J.; Mao, S.-L.; Taylor, K.L.; Kanjanabuch, T.; Guan, Y.; Zhang, Y.; Brown, N.J.; Swift, L.L.; McGuinness, O.P.; Wasserman, D.H.; et al. Prevention of Obesity and Insulin Resistance in Mice Lacking Plasminogen Activator Inhibitor 1. *Diabetes* **2004**, *53*, 336–346. [CrossRef] [PubMed]
87. Spranger, J.; Kroke, A.; Möhlig, M.; Hoffmann, K.; Bergmann, M.M.; Ristow, M.; Boeing, H.; Pfeiffer, A.F.H. Inflammatory Cytokines and the Risk to Develop Type 2 Diabetes. *Diabetes* **2003**, *52*, 812–817. [CrossRef] [PubMed]
88. Bastard, J.-P.; Jardel, C.; Bruckert, E.; Blondy, P.; Capeau, J.; Laville, M.; Vidal, H.; Hainque, B. Elevated Levels of Interleukin 6 Are Reduced in Serum and Subcutaneous Adipose Tissue of Obese Women after Weight Loss. *J. Clin. Endocrinol. Metab.* **2000**, *85*, 3338–3342. [CrossRef]
89. Franckhauser, S.; Elias, I.; Rotter Sopasakis, V.; Ferré, T.; Nagaev, I.; Andersson, C.X.; Agudo, J.; Ruberte, J.; Bosch, F.; Smith, U. Overexpression of Il6 Leads to Hyperinsulinaemia, Liver Inflammation and Reduced Body Weight in Mice. *Diabetologia* **2008**, *51*, 1306–1316. [CrossRef]

90. Mortensen, R.F. C-Reactive Protein, Inflammation, and Innate Immunity. *Immunol. Res.* **2001**, *24*, 163–176. [CrossRef]
91. Charo, I.F.; Ransohoff, R.M. The Many Roles of Chemokines and Chemokine Receptors in Inflammation. *N. Engl. J. Med.* **2006**, *354*, 610–621. [CrossRef]
92. Li, H.; Gao, C.; Liu, C.; Liu, L.; Zhuang, J.; Yang, J.; Zhou, C.; Feng, F.; Sun, C.; Wu, J. A Review of the Biological Activity and Pharmacology of Cryptotanshinone, an Important Active Constituent in Danshen. *Biomed. Pharmacother.* **2021**, *137*, 111332. [CrossRef]
93. Su, C.-Y.; Ming, Q.-L.; Rahman, K.; Han, T.; Qin, L.-P. Salvia Miltiorrhiza: Traditional Medicinal Uses, Chemistry, and Pharmacology. *Chin. J. Nat. Med.* **2015**, *13*, 163–182. [CrossRef] [PubMed]
94. Gao, S.; Liu, Z.; Li, H.; Little, P.J.; Liu, P.; Xu, S. Cardiovascular Actions and Therapeutic Potential of Tanshinone IIA. *Atherosclerosis* **2012**, *220*, 3–10. [CrossRef]
95. Huang, L.; Ding, W.; Wang, M.-Q.; Wang, Z.-G.; Chen, H.-H.; Chen, W.; Yang, Q.; Lu, T.-N.; Yang, Q.; He, J.-M. Tanshinone IIA Ameliorates Non-Alcoholic Fatty Liver Disease through Targeting Peroxisome Proliferator-Activated Receptor Gamma and Toll-like Receptor 4. *J. Int. Med. Res.* **2019**, *47*, 5239–5255. [CrossRef]
96. Gong, Z.; Huang, C.; Sheng, X.; Zhang, Y.; Li, Q.; Wang, M.-W.; Peng, L.; Zang, Y.Q. The Role of Tanshinone IIA in the Treatment of Obesity through Peroxisome Proliferator-Activated Receptor $\gamma$ Antagonism. *Endocrinology* **2009**, *150*, 104–113. [CrossRef] [PubMed]
97. Imran, K.M.; Rahman, N.; Yoon, D.; Jeon, M.; Lee, B.-T.; Kim, Y.-S. Cryptotanshinone Promotes Commitment to the Brown Adipocyte Lineage and Mitochondrial Biogenesis in C3H10T1/2 Mesenchymal Stem Cells via AMPK and P38-MAPK Signaling. *Biochim. Biophys. Acta BBA Mol. Cell Biol. Lipids* **2017**, *1862*, 1110–1120. [CrossRef]
98. Kim, E.J.; Jung, S.-N.; Son, K.H.; Kim, S.R.; Ha, T.Y.; Park, M.G.; Jo, I.G.; Park, J.G.; Choe, W.; Kim, S.-S.; et al. Antidiabetes and Antiobesity Effect of Cryptotanshinone via Activation of AMP-Activated Protein Kinase. *Mol. Pharm.* **2007**, *72*, 62–72. [CrossRef] [PubMed]
99. Rahman, N.; Jeon, M.; Song, H.-Y.; Kim, Y.-S. Cryptotanshinone, a Compound of *Salvia Miltiorrhiza* Inhibits Pre-Adipocytes Differentiation by Regulation of Adipogenesis-Related Genes Expression via STAT3 Signaling. *Phytomedicine* **2016**, *23*, 58–67. [CrossRef]
100. Hwang, S.-L.; Yang, J.H.; Jeong, Y.-T.; Kim, Y.D.; Li, X.; Lu, Y.; Chang, Y.-C.; Son, K.H.; Chang, H.W. Tanshinone IIA Improves Endoplasmic Reticulum Stress-Induced Insulin Resistance through AMP-Activated Protein Kinase. *Biochem. Biophys. Res. Commun.* **2013**, *430*, 1246–1252. [CrossRef]
101. Jung, S.H.; Seol, H.J.; Jeon, S.J.; Son, K.H.; Lee, J.R. Insulin-Sensitizing Activities of Tanshinones, Diterpene Compounds of the Root of Salvia Miltiorrhiza Bunge. *Phytomedicine* **2009**, *16*, 327–335. [CrossRef] [PubMed]
102. Yuan, F.; Zhang, M.; Xu, P.; Xu, D.; Chen, P.; Ren, M.; Sun, Q.; Chen, J.; Du, J.; Tang, X. Tanshinone IIA Improves Diabetes Mellitus via the NF-κB-induced AMPK Signal Pathway. *Exp. Ther. Med.* **2018**, *16*, 4225–4231. [CrossRef] [PubMed]
103. Huang, Y.; Li, W.; Wang, C.C.; Wu, X.; Zheng, J. Cryptotanshinone Reverses Ovarian Insulin Resistance in Mice through Activation of Insulin Signaling and the Regulation of Glucose Transporters and Hormone Synthesizing Enzymes. *Fertil. Steril.* **2014**, *102*, 589–596.e4. [CrossRef]
104. Gebert, L.F.R.; MacRae, I.J. Regulation of MicroRNA Function in Animals. *Nat. Rev. Mol. Cell Biol.* **2019**, *20*, 21–37. [CrossRef]
105. Scoditti, E.; Naccarati, A.; Carpi, S.; Polini, B.; Ebada, S.S.; Nieri, P. Editorial: Non-Coding RNAs as Mediators of the Activity of Natural Compounds. *Front. Pharm.* **2021**, *12*, 751956. [CrossRef]
106. Deiuliis, J.A. MicroRNAs as Regulators of Metabolic Disease: Pathophysiologic Significance and Emerging Role as Biomarkers and Therapeutics. *Int. J. Obes.* **2016**, *40*, 88–101. [CrossRef]
107. Ferrero, G.; Carpi, S.; Polini, B.; Pardini, B.; Nieri, P.; Impeduglia, A.; Grioni, S.; Tarallo, S.; Naccarati, A. Intake of Natural Compounds and Circulating MicroRNA Expression Levels: Their Relationship Investigated in Healthy Subjects with Different Dietary Habits. *Front. Pharm.* **2021**, *11*, 619200. [CrossRef]
108. Castaño, C.; Kalko, S.; Novials, A.; Párrizas, M. Obesity-Associated Exosomal MiRNAs Modulate Glucose and Lipid Metabolism in Mice. *Proc. Natl. Acad. Sci. USA* **2018**, *115*, 12158–12163. [CrossRef] [PubMed]
109. van Rooij, E.; Purcell, A.L.; Levin, A.A. Developing MicroRNA Therapeutics. *Circ. Res.* **2012**, *110*, 496–507. [CrossRef] [PubMed]
110. Cerutti, C.; Edwards, L.J.; de Vries, H.E.; Sharrack, B.; Male, D.K.; Romero, I.A. MiR-126 and MiR-126* Regulate Shear-Resistant Firm Leukocyte Adhesion to Human Brain Endothelium. *Sci. Rep.* **2017**, *7*, 45284. [CrossRef]
111. He, Y.; Hwang, S.; Cai, Y.; Kim, S.; Xu, M.; Yang, D.; Guillot, A.; Feng, D.; Seo, W.; Hou, X.; et al. MicroRNA-223 Ameliorates Nonalcoholic Steatohepatitis and Cancer by Targeting Multiple Inflammatory and Oncogenic Genes in Hepatocytes. *Hepatology* **2019**, *70*, 1150–1167. [CrossRef]
112. Available online: https://Mirtarbase.Cuhk.Edu.Cn/~miRTarBase/MiRTarBase_2022/Php/Search.Php?Org=hsa&opt=mirna_id&kw=miR-124-3p) (accessed on 11 May 2023).
113. Strum, J.C.; Johnson, J.H.; Ward, J.; Xie, H.; Feild, J.; Hester, A.; Alford, A.; Waters, K.M. MicroRNA 132 Regulates Nutritional Stress-Induced Chemokine Production through Repression of SirT1. *Mol. Endocrinol.* **2009**, *23*, 1876–1884. [CrossRef] [PubMed]
114. Wang, K.; Deng, P.; Sun, Y.; Ye, P.; Zhang, A.; Wu, C.; Yue, Z.; Chen, Z.; Xia, J. MicroRNA-155 Promotes Neointimal Hyperplasia through Smooth Muscle-like Cell-Derived RANTES in Arteriovenous Fistulas. *J. Vasc. Surg.* **2018**, *67*, 933–944.e3. [CrossRef]

115. Ortega, F.J.; Moreno, M.; Mercader, J.M.; Moreno-Navarrete, J.M.; Fuentes-Batllevell, N.; Sabater, M.; Ricart, W.; Fernández-Real, J.M. Inflammation Triggers Specific MicroRNA Profiles in Human Adipocytes and Macrophages and in Their Supernatants. *Clin. Epigenetics* **2015**, *7*, 49. [CrossRef]
116. Karkeni, E.; Astier, J.; Tourniaire, F.; el Abed, M.; Romier, B.; Gouranton, E.; Wan, L.; Borel, P.; Salles, J.; Walrand, S.; et al. Obesity-Associated Inflammation Induces MicroRNA-155 Expression in Adipocytes and Adipose Tissue: Outcome on Adipocyte Function. *J. Clin. Endocrinol. Metab.* **2016**, *101*, 1615–1626. [CrossRef]
117. Mahesh, G.; Biswas, R. MicroRNA-155: A Master Regulator of Inflammation. *J. Interferon Cytokine Res.* **2019**, *39*, 321–330. [CrossRef]
118. Ying, W.; Riopel, M.; Bandyopadhyay, G.; Dong, Y.; Birmingham, A.; Seo, J.B.; Ofrecio, J.M.; Wollam, J.; Hernandez-Carretero, A.; Fu, W.; et al. Adipose Tissue Macrophage-Derived Exosomal MiRNAs Can Modulate In Vivo and In Vitro Insulin Sensitivity. *Cell* **2017**, *171*, 372–384.e12. [CrossRef]
119. Gaudet, A.D.; Fonken, L.K.; Gushchina, L.V.; Aubrecht, T.G.; Maurya, S.K.; Periasamy, M.; Nelson, R.J.; Popovich, P.G. MiR-155 Deletion in Female Mice Prevents Diet-Induced Obesity. *Sci. Rep.* **2016**, *6*, 22862. [CrossRef]
120. Zhang, Y.; Mei, H.; Chang, X.; Chen, F.; Zhu, Y.; Han, X. Adipocyte-Derived Microvesicles from Obese Mice Induce M1 Macrophage Phenotype through Secreted MiR-155. *J. Mol. Cell Biol.* **2016**, *8*, 505–517. [CrossRef] [PubMed]
121. Zhuang, G.; Meng, C.; Guo, X.; Cheruku, P.S.; Shi, L.; Xu, H.; Li, H.; Wang, G.; Evans, A.R.; Safe, S.; et al. A Novel Regulator of Macrophage Activation. *Circulation* **2012**, *125*, 2892–2903. [CrossRef] [PubMed]
122. Zhou, B.; Li, L.-H.; Tan, L.-M.; Luo, W.-B.; Xiong, H.; Lu, X.-L.; Liu, D.; Li, W.-Y.; Guo, Y.-X.; Tang, Z.; et al. Tanshinone IIA Ameliorates Inflammation Response in Osteoarthritis via Inhibition of MiR-155/FOXO3 Axis. *Pharmacology* **2021**, *106*, 20–28. [CrossRef]
123. Xu, H.; Li, H.; Zhu, P.; Liu, Y.; Zhou, M.; Chen, A. Tanshinone IIA Ameliorates Progression of CAD Through Regulating Cardiac H9c2 Cells Proliferation and Apoptosis by MiR-133a-3p/EGFR Axis. *Drug Des. Dev. Ther.* **2020**, *14*, 2853–2863. [CrossRef]
124. Du, X.; Wang, X.; Cui, K.; Chen, Y.; Zhang, C.; Yao, K.; Hao, Y.; Chen, Y. Tanshinone IIA and Astragaloside IV Inhibit MiR-223/JAK2/STAT1 Signalling Pathway to Alleviate Lipopolysaccharide-Induced Damage in Nucleus Pulposus Cells. *Dis. Mrk.* **2021**, *2021*, 6554480. [CrossRef] [PubMed]
125. Wang, H.; Zhang, Y.; Zhang, Y.; Liu, W.; Wang, J. Cryptotanshinone Inhibits Lung Cancer Invasion via MicroRNA-133a/Matrix Metalloproteinase 14 Regulation. *Oncol. Lett.* **2019**, *18*, 2554–2559. [CrossRef] [PubMed]
126. Thomou, T.; Mori, M.A.; Dreyfuss, J.M.; Konishi, M.; Sakaguchi, M.; Wolfrum, C.; Rao, T.N.; Winnay, J.N.; Garcia-Martin, R.; Grinspoon, S.K.; et al. Adipose-Derived Circulating MiRNAs Regulate Gene Expression in Other Tissues. *Nature* **2017**, *542*, 450–455. [CrossRef] [PubMed]
127. Nair, A.; Jacob, S. A Simple Practice Guide for Dose Conversion between Animals and Human. *J. Basic Clin. Pharm.* **2016**, *7*, 27. [CrossRef] [PubMed]

*biomolecules*

*Review*

# The Benefits of Anthocyanins against Obesity-Induced Inflammation

**Chanya Ngamsamer [1], Jintana Sirivarasai [2] and Nareerat Sutjarit [2,*]**

[1] Doctor of Philosophy Program in Nutrition, Faculty of Medicine, Ramathibodi Hospital and Institute of Nutrition, Mahidol University, Bangkok, 10400, Thailand; chanya.ngm@student.mahidol.edu
[2] Graduate Program in Nutrition, Faculty of Medicine, Ramathibodi Hospital, Mahidol University, Bangkok, 10400, Thailand; jintana.sir@mahidol.ac.th
* Correspondence: nareerat.sut@mahidol.ac.th

**Abstract:** Obesity has become a serious public health epidemic because of its associations with chronic conditions such as type 2 diabetes mellitus, hypertension, cardiovascular disease, and cancer. Obesity triggers inflammation marked by the secretion of low-grade inflammatory cytokines including interleukin-6, C-reactive protein, and tumor necrosis factor-α, leading to a condition known as "meta-inflammation". Currently, there is great interest in studying the treatment of obesity with food-derived bioactive compounds, which have low toxicity and no severe adverse events compared with pharmacotherapeutic agents. Here, we reviewed the beneficial effects of the bioactive compounds known as anthocyanins on obesity-induced inflammation. Foods rich in anthocyanins include tart cherries, red raspberries, black soybeans, blueberries, sweet cherries, strawberries and Queen Garnet plums. These anthocyanin-rich foods have been evaluated in cell culture, animal, and clinical studies, and found to be beneficial for health, reportedly reducing inflammatory markers. One factor in the development of obesity-related inflammation may be dysbiosis of the gut microbiome. Therefore, we focused this review on the in vitro and in vivo effects of anthocyanins on inflammation and the gut microbiota in obesity.

**Keywords:** obesity; inflammation; meta-inflammation; anthocyanin; antioxidant; anti-inflammation

## 1. Introduction

Obesity, which has been rising in prevalence worldwide over the past several decades, is now considered a public health epidemic. Obesity for adults is defined as a body mass index of $\geq$30 kg/m$^2$ in Western populations, while the current Asia-Pacific guidelines recommend using a lower body mass index of $\geq$25 kg/m$^2$ [1]. Obesity is linked to metabolic conditions such as hyperglycemia and dyslipidemia, which are well-known risk factors for developing chronic noncommunicable diseases, including hypertension, type II diabetes mellitus (T2DM), cancer, and cardiovascular disease (CVD) [2]. Individuals with obesity exhibit chronic low-grade systemic inflammation, which is characterized by increased secretion of pro-inflammatory cytokines from adipose tissue, made dysfunctional by the excessive accumulation of fat into the circulation; these cytokines include interleukin-6 (IL-6), C-reactive protein (CRP) and tumor necrosis factor-α (TNF-α) [3].

Generally, treatments for obesity include lifestyle modification and pharmacotherapy aimed at reducing body weight and alleviating inflammation [4]. Clinical guidelines suggest a combination of lifestyle modification and pharmacotherapy [5]. However, such medicines are associated with severe adverse events such as mental disorders and stroke, and they are quite expensive [6,7]. Currently, there are many studies aimed at treating obesity through dietary lifestyle modifications. Bioactive chemicals, particularly those that are plant or animal-based compounds, have the potential to diminish the prevalence of chronic disorders [8]. Some bioactive compounds are also classified as phytochemicals. Phytochemicals such as polyphenols have antioxidant, anti-inflammatory, anti-obesity, and anti-aging properties [9,10].

## 2. Human Obesity: Its Causes and Consequences

The World Health Organization reports that "obesity has almost tripled globally since 1975", highlighting that obesity has become a global epidemic. In 2016, more than 1.9 billion of the world's population were overweight; of these, over 650 million individuals were classified as obese [11]. More importantly, the obesity epidemic is increasingly affecting children, with an estimated 38.2 million children under the age of 5 years suffering from obesity in 2019 [12]. The cornerstone of the disease of obesity is an energy imbalance that occurs when energy consumption, particularly excessive fat and sugar intake, exceeds energy expenditure. Currently, many individuals with extreme obesity lead a sedentary lifestyle and eventually become bedridden. In fact, the continuous increase in noncommunicable diseases such as T2DM, CVD, hepatic steatosis, neurodegenerative diseases, biliary diseases, and certain cancers are directly linked to the increased prevalence of human obesity [13]. Importantly, these obesity-related diseases also result in a shorter life expectancy and premature death.

## 3. Obesity-Induced Inflammation

Obesity is associated with inflammation through increased pro-inflammatory cytokine secretion resulting from inflammatory responses and altered metabolic homeostasis. "Meta-inflammation" is a term used to describe chronic low-grade inflammation as a response to obesity [13]. In obesity, excessive accumulation of fat in adipocytes causes adipose tissue dysfunction [14]. Compared with non-obese individuals, the dysfunctional adipose tissues in rodent models of obesity and humans with obesity secrete lower amounts of adiponectin, which exerts an anti-inflammatory effect on the liver, skeletal muscle, and adipose tissue. These dysfunctional tissues simultaneously secrete increased levels of pro-inflammatory cytokines such as CRP, IL-6 and TNF-α [15], which disrupt the metabolic state required to maintain immune homeostasis. Innate immune cells secrete several cytokines and acute-phase proteins to the sites of inflammation [3,16]. In people with obesity, there is a phenotypic switching from M2 macrophages to M1 macrophages in the adipose tissues. M2 macrophages, which are anti-inflammatory, play a role in the maintenance of tissue and are typical in the adipose tissues of lean individuals. By contrast, M1 macrophages produce pro-inflammatory cytokines and thus contribute to the development of insulin resistance [17]. An increase in M1 macrophages, which form a structure around adipocytes and release pro-inflammatory mediators such as IL-1, IL-6, IL-12, TNF-α, and chemokines, is part of the alteration in immune cell profile that is characteristic of obesity (Figure 1). As a result, tissue damage to the liver, colon, and arterial walls may occur. One possible underlying mechanism of obesity-induced inflammation is the activation of toll-like receptors (TLRs) of the innate immune system in obese patients, particularly TLR4, which involves the activation of transcription factors such as nuclear factor kappa B (NF-κB) and the production of pro-inflammatory markers [18,19].

In patients with prolonged obesity, there may be a switch from the innate immune response to the adaptive immune response, leading to the onset of obesity-associated chronic diseases. Thus, obesity-related chronic diseases can be prevented or delayed by preventing long-term obesity [16]. Below we review the nutraceutical and pharmaceutical effects of anthocyanins that might be beneficial for treating obesity-related inflammation.

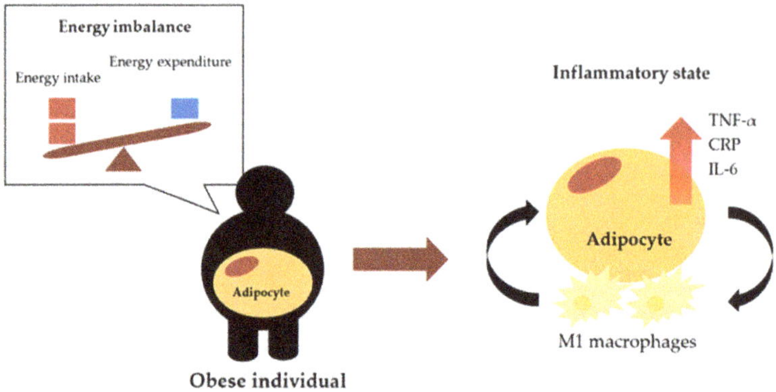


**Figure 1.** Hallmark of meta-inflammation. The major cause of obesity is a positive energy balance between energy intake and energy expenditure leading to meta-inflammation, a chronic low-grade inflammatory condition.

## 4. Anthocyanins

Anthocyanins are water-soluble pigments within the phenol class of compounds. They are abundant in nature, and can be found in vegetables, fruits, and flowers. Anthocyanin pigments in plants occur in the form of glycosides. All anthocyanins play a significant role in preventing chronic noncommunicable diseases.

### *4.1. Chemistry of Anthocyanins*

Anthocyanins are considered to be flavonoids, although they have a positive charge at the oxygen atom of the C-ring of the fundamental flavonoid structure. Such a compound is also called the flavylium (2-phenylchromenylium) ion. The general molecular structure of anthocyanins is shown in Figure 2. Anthocyanins are derived from flavanols, but with a flavylium ion lacking a ketone oxygen at the 4-position (Figure 3) [20]. The empirical formula for the flavylium ion of anthocyanin is $C_{15}H_{11}O^{+}$ with a molecular weight of 207.24724 g/mol [21].

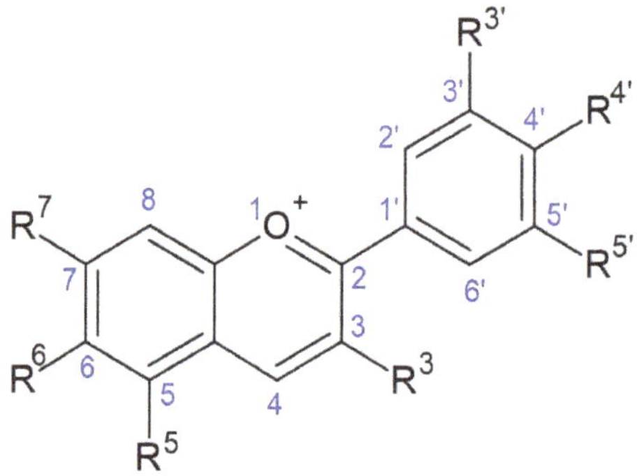

**Figure 2.** Fundamental structure of anthocyanin. (Drawn using the tool ACD/ChemSketch, version 2021.2.1, Advanced Chemistry Development, Inc., Toronto, ON, Canada).

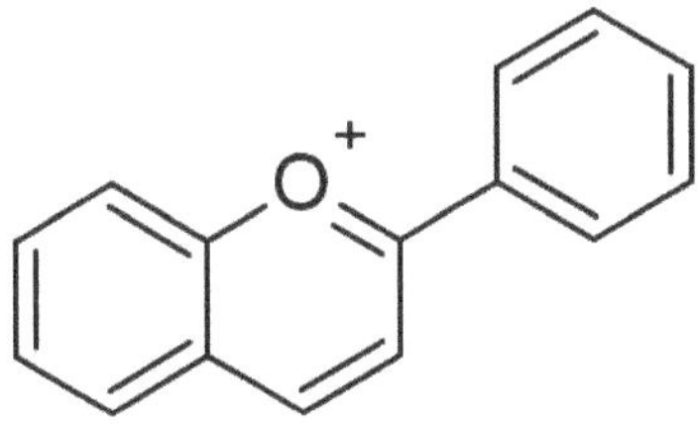

**Figure 3.** Two-dimensional structure of a flavylium ion. (Drawn using the tool ACD/ChemSketch, version 2021.2.1, Advanced Chemistry Development, Inc., Toronto, ON, Canada).

### 4.2. Bioavailability of Anthocyanins

The majority of bioactive compounds in functional foods have been linked to lower risks of chronic conditions including CVD and T2DM [8]. Some bioactive compounds, such as polyphenols, are also phytochemicals. Polyphenolic compounds have antioxidant, anti-inflammatory, anti-obesity, and anti-aging properties [9,10]. Apart from the overall content and biological functions of bioactive components, bioavailability is also an important factor. The term "bioavailability" is defined by the US Food and Drug Administration as the rate and extent to which an active ingredient is absorbed from a drug product and becomes available at the site of action, and includes metabolism and excretion [22]. The anthocyanins have exhibited low bioavailability in animal studies, as indicated by low systemic concentrations [23–25]. Examinations of plasma after intake of anthocyanin-rich foods such as blackcurrant juice, red wine, and strawberries, have revealed that anthocyanins have limited bioavailability in humans as well [26–28]. Previous research in both humans and animals has established that the low bioavailability of anthocyanins is attributable to their limited absorption into the circulation and significant elimination in urine and feces.

In contrast, a study on the phytochemical uptake following human consumption of the Montmorency tart cherry, which is rich in anthocyanins, has shown that anthocyanin metabolites are most bioavailable in the plasma at 1–2 h post consumption, indicating that anthocyanins might be rapidly absorbed [29]. Moreover, a literature survey on the bioavailability of anthocyanins and anthocyanin-containing foods among humans by Manach et al. (2005), found that the bioavailability of anthocyanins is generally poor, although they may be rapidly absorbed at first. Multiple authors, as cited by Manach, et al. (2005), note that anthocyanins are absorbed rapidly in a manner consistent with stomach absorption, but the absorption efficiencies were very poor with plasma concentrations approximately six orders of magnitude lower than the ingested dose. Furthermore this small absorbed fraction was rapidly eliminated via the urine [30]. However, there are novel technologies that can be utilized to increase anthocyanin bioavailability, as discussed below.

### 4.3. Stability of Anthocyanins

The stability of anthocyanin pigments are affected by pH, light, temperature, and structure [20]. Anthocyanins change color in response to different pH levels. Under acidic conditions, anthocyanins may appear as the red-colored flavylium ion, while under alkaline conditions, they change to the blueish color of quinoidal bases [31]. Temperature also affects anthocyanin color. In higher solution temperatures, anthocyanins can become less stable due to coupled oxidation reactions of peroxidase and hydrogen peroxide ($H_2O_2$) [32]. The food industry offers many different fruit and vegetable products that are rich in anthocyanins, such as juices and supplements. However, industrial anthocyanin-containing products are often subjected to thermal processing and other common food processing methods that can decrease stability and total anthocyanin content [33]. In the last decade, novel food-processing technologies such as high pressure processing (HPP), high hydrostatic pressure (HHP), and pulsed electric fields (PEF) have been developed that may help prevent the loss of color, total anthocyanin content, and potential health benefits of anthocyanins [34]. To prevent the loss of potential health benefits specifically, we suggest the application of the following novel technologies in food processing: A previous study reported that HPP treatment of blueberry juice led to minor changes in ascorbic acid, total phenolics, anthocyanin stability, and total antioxidant capacity compared with fresh blueberry juice because the pressure–time conditions were maximized at a maximum temperature of 42 °C [35]. Encapsulation technology can also protect a bioactive molecule such as anthocyanin from the effects of oxygen, light, or other factors. Spray-drying is a common technique used for the encapsulation of compounds [36]. An extract of cultivated purple flesh potato (PFPE) was encapsulated by spray-drying with maltodextrin (MD) as the encapsulating agent. The constant degradation rate and bioavailability of anthocyanins within an in vitro gastrointestinal digestion model are significantly higher in PFPE-MD than in non-encapsulated PFPE because powder encapsulation protects the stable colorant and health benefits of anthocyanins [37]. Kanokpanont et al. (2018) demonstrated that anthocyanin-encapsulated alginate/chitosan beads derived from spray-dried mulberries were stable under gastric conditions that simulated gastric fluid conditions or enhanced the bioavailability of anthocyanins [38]. In conclusion, novel technologies such as microencapsulation and HPP treatment are an effective strategy to preserve the biological functions and nutrient content of anthocyanins in food processing.

Anthocyanins comprise a subclass of phenolic phytochemicals. They are water-soluble compounds with orange, red, purple, and blue colors, and are present in fruits and vegetables such as pomegranates, berries, red grapes, purple tomatoes, and red cabbage. The most common types of anthocyanins are pelargonidin, cyanidin, peonidin, petunidin, delphinidin, and malvidin (Table 1) [21,39,40]. Anthocyanins have been investigated as bio-functional molecules possessing anti-inflammatory, antioxidant, and chemoprotective properties, and they play significant preventive roles in chronic diseases [41].

**Table 1.** Colors and sources of the six most well-studied anthocyanins [21,39,40].

| Main Anthocyanins | Color | pH Ranges | Sources |
|---|---|---|---|
| Pelargonidin | Red, orange | Low pH (pH < 3) | Radish, pomegranate, red potato, ripe raspberry |
| Cyanidin | Red, reddish-purple | Low to neutral pH (pH 3–7) | Blackberry, red sweet potato, purple corn, tart and sweet cherry |
| Peonidin | Purplish-red | Neutral pH (pH 6–7) | Sweet potato, cranberry, grape, purple corn |
| Petunidin | Purple, dark red | Low to high pH (pH 3–8) | Blackcurrant, black bean, red berry |
| Delphinidin | Purple, blue-reddish | Neutral to high pH (pH 7–11) | Pomegranate, black bean, purple tomato |
| Malvidin | Purple | Neutral pH (pH 7–8) | Blueberry, red wine, bilberry, mulberry |

## 5. Effects of Anthocyanins in Obesity-Associated Inflammation

The abundant natural anthocyanins are well-characterized antioxidants that have been shown to eliminate reactive oxygen species (ROS) in cells, animals, and clinical studies [42–48].

In cell culture studies, strawberries were found to be rich in anthocyanins and contain a high antioxidant capacity. Heo and Lee (2005) showed that strawberries, as compared with bananas and oranges, dramatically reduced oxidative-stress-induced neurotoxicity in PC12 cells treated with $H_2O_2$ [42].

In animal studies, the purple sweet potato color (PSPC) protected against ROS production and restored glutathione content in high-fat diet (HFD)-induced mice [43,44]. Blackberry and blueberry anthocyanin (BLA and BBA) supplementation reduced oxidative stress and inflammation by increasing first-line defense antioxidants such as superoxide dismutase (SOD) and glutathione peroxidase (GPx). Furthermore, supplementation of BLA and BBA for 12 weeks prevented weight gain in HFD-induced obesity in C57BL/6 mice [45]. Mulberry and cherry anthocyanins reduced body weight and improved SOD and GPx activities in HFD-fed mice [46]. In clinical studies, healthy volunteers who supplemented their daily diet with 500 g of strawberries rich in anthocyanins for 1 month lowered their risk for developing CVD, as assessed by improvements in lipid profiles and antioxidant activity [47]. Li et al. (2015) studied participants with T2DM who received purified anthocyanin supplements for 24 weeks and found that anthocyanin improved lipid profiles and enhanced antioxidant capacity, as well as insulin sensitivity [48].

In summary, anthocyanins play a role as antioxidants that can eliminate ROS and help prevent chronic diseases including CVD and T2DM. Additionally, anthocyanins can reduce pro-inflammatory markers associated with obesity, such as CRP, IL-6, and TNF-α (Table 2).

**Table 2.** Effects of anthocyanins on obesity-associated inflammation [49–63].

| Food Source | Bioactive of Anthocyanins | Effects | Study Group | Reference |
|---|---|---|---|---|
| Tart cherries | • Cyanidin-3-*O*-glucoside<br>• Cyanidin-3-*O*-glucosyl-rutinoside<br>• Cyanidin-3-*O*-rutinoside | Reduced IL-6 level | Adipose stem cells | [49] |
| Red raspberries | Identified anthocyanins N/A | • Reduced NO, COX-2, IL-1β, IL-6 expression<br>• Suppressed NF-kB pathway | RAW264.7 macrophages | [50] |
| Black soybeans | • Cyanidin-3-*O*-glucoside<br>• Pelargonidin 3-glucoside<br>• Delphinidin-3-glucoside | • Reduced TNF-α secretion<br>• Increased adiponectin and insulin sensitivity | 3T3-L1 cells | [51] |
| Sweet cherry | • Cyanidin 3-*O*-rutinoside | • Reduced NO level<br>• Decreased COX-2 and iNOS expression | RAW 264.7 macrophages | [52] |
| Blueberry supplementation | Phenolics and anthocyanin | • Improved TNF-α and IL-1β gene expression<br>• Improved gut microbiota | Male Wistar rats fed HFD | [53] |
| Tart cherry extract | • Cyanidin-3-*O*-glucoside<br>• Cyanidin-3-*O*-glucosyl-rutinoside<br>• Cyanidin-3-*O*-rutinoside | Decreased IL-6 andleptin levels | Obese mice | [54] |
| Tart cherry seed powder and tart cherry juice | Identified anthocyanins N/A | • Reduced TNF-α, IL-1β and IL-6 expression<br>• Suppressed NF-kB pathway | Male Wistar rats fed HFD | [55] |
| Sweet cherry | • Cyanidin-3-*O*-glucosyl-rutinoside<br>• Cyanidin-3-*O*-rutinoside<br>• Pelargonidin 3-rutinoside | • Decreased body weight<br>• Decreased IL-6 and TNF-α gene expression in WAT | Male C57BL/6 mice fed HFD | [56] |
| Pomegranate peel extract | Ellagitannins and anthocyanins | • Reduced COX-2 in the colon and visceral adipose tissue<br>• Reduced mRNA levels of IL-1β in the visceral adipose tissue<br>• Reduced IL-6 in the colon | Balb/c obese mice | [57] |
| Strawberrybeverage | • Pelargonidin 3-glucoside<br>• Pelargonidin 3-malonylglucoside<br>• Pelargonidin 3-rutinoside | Improved hs-CRPand IL-6 | Overweight adults | [58] |

**Table 2.** *Cont.*

| | | | | |
|---|---|---|---|---|
| Black soybean testa extracts | • Cyanidin-3-glucoside<br>• Delphinidin-3-glucoside<br>• Petunidin-3-glucoside | • Improved visceral fat<br>• Improved plasma<br>• Lipid profiles<br>• Reduced TNF-α and MCP-1 | Overweight and obese adults | [59] |
| Queen garnet plum | Cyanidin 3-*O*-β-D-glucoside | Reduced LDL level, plasma glucose, insulin, C-peptide, leptin and GLP-1 concentrations | Mildly hypertensive obese/overweight adults | [60] |
| Authentic tart cherry juice | • Cyanidin-3-2G-glucosylrutinoside<br>• Cyanidin-3-glucoside<br>• Cyanidin-3-rutinoside<br>• Peonidin-3-rutinoside | Reduced TNF-α and MCP-1 | Overweight and obese adults | [61] |
| Strawberry | Identified anthocyanins N/A | No effect on inflammatory markers (e.g., IL-6, IL-1β, TNF-α) | Obese healthy males and females | [62] |
| Commercially available red orange juice | Identified anthocyanins N/A | No effect on body weight and plasma inflammatory markers | Overweight or obese females | [63] |

NO = nitric oxide; iNOS = inducible nitric oxide synthase; COX-2 = cyclooxygenase-2; IL = interleukin; NF-kB = nuclear factor-kB; TNF = tumor necrosis factor; HFD = high-fat diet; N/A = not applicable; hs-CRP = high-sensitivity C-reactive protein; MCP-1 = monocyte chemoattractant protein-1; LDL = low-density lipoprotein; GLP-1 = glucagon-like peptide 1.

### 5.1. Anti-Inflammatory Effect

Obesity has been linked to chronic low-grade systemic inflammation. Secretion of inflammatory cytokines such as IL-6 and TNF-α by adipocytes and macrophages in adipose tissue have been shown to activate the inflammatory response [64]. Alleviating inflammation can be achieved by decreasing the release of pro-inflammatory cytokines. Several studies have been conducted using cell cultures and animal models, as well as clinical trials in humans, to examine the impact of anthocyanin-rich foods on obesity-induced inflammation.

In vitro studies have shown that anthocyanin-rich fruit extracts can moderate inflammatory cytokines associated with obesity. Tart cherries rich in anthocyanins significantly reduced the expression of the IL-6 gene in lipopolysaccharide (LPS)-induced adipose stem cells [49]. Consistent with other studies, RAW264.7 mouse macrophage cells that were induced with both *Escherichia coli* LPS and recombinant interferon-c, and treated with anthocyanin-rich fractions of red raspberries at concentrations of 150 and 200 μg/mL, showed reduced nitric oxide (NO), cyclooxygenase-2 (COX-2), IL-1β, and IL-6 expression, as well as inhibition of the NF-kB inflammatory pathway [50]. Likewise, treatment with black soybean extract decreased TNF-α production, but increased adiponectin secretion and insulin sensitivity, in 3T3-L1 adipocytes [51]. In another study, sweet cherry phenolic-rich extract significantly reduced NO levels and decreased COX-2 and inducible nitric oxide synthase (iNOS) expression in LPS-induced adipose stem cells [52].

Among the animal studies, administration of a blueberry supplement containing phenolics and anthocyanins also decreased the expression of the TNF-α and IL-1β genes in HFD–fed male rats [53]. Anthocyanin-rich tart cherry extract effectively improved the pro-inflammatory cytokine profile by reducing IL-6 and leptin levels in an obese mouse model [54]. Similarly, tart cherry reduced TNF-α, IL-1β, and IL-6 expression and suppressed NF-kB inflammatory expression in HFD-induced obese rats after a 17-week intervention with tart cherry seed powder and tart cherry juice [55]. Therefore, tart cherries have the potential to prevent obesity-induced inflammation. Purified sweet cherry anthocyanins at 200 mg/kg in mice reduced body weight by approximately 11.2% and decreased the expression of the IL-6 and TNF-α genes in white adipose tissue (WAT), slowing down the progression of obesity in these mice [56]. In another study, a pomegranate peel extract (PPE) with high polyphenol content decreased COX-2 and IL-1 mRNA levels in visceral adipose tissue [57].

Anthocyanins have been shown in animal studies and cell cultures to reduce obesity-related inflammation. Therefore, many studies have aimed to evaluate the effect of anthocyanins on inflammation associated with obesity in humans. Strawberry antioxidants containing 81.65 mg anthocyanins/10 g of freeze-dried powder in a beverage significantly attenuated the postprandial inflammatory response by decreasing high-sensitivity CRP and IL-6) in an overweight population, following the consumption of a high-carbohydrate, moderate-fat meal (HCFM) [58]. Black soybean testa (BBT) extracts rich in three major anthocyanins reduced visceral fat, and improved plasma lipid profiles and inflammatory markers including TNF-α and monocyte chemoattractant protein-1 in overweight/obese adults [59]. Bhaswant et al. (2019) discovered that the high anthocyanin content in Queen Garnet plum juice significantly reduced IL-2, IL-6, IL-13, and TNF-α levels in mildly hypertensive overweight or obese people. Queen Garnet plum juice also reduced the levels of low-density lipoprotein, plasma glucose, insulin, C-peptide, leptin, and glucagon-like peptide-1 [60]. Authentic tart cherry juice (TCJ) at 240 mL of 100% TCJ (equivalent to 50 tart cherries) reduced TNF-α and monocyte chemoattractant protein-1 (MCP-1) levels compared to placebo after 4 weeks of consumption in overweight and obese individuals [61].

Both in vitro and in vivo studies have reported that foods rich in anthocyanins reduced inflammation associated with obesity, including tart cherries, PPE, and strawberry beverages. By contrast, some clinical studies found no anti-inflammatory effect of anthocyanins in people with obesity [62,63]. Zunino et al. (2012) showed that people consuming 80 g/serving of freeze-dried strawberry powder mixed with food and drinks for 3 weeks had no effect on inflammatory markers (IL-6, IL-1β, TNF-α) [62]. Similarly, consumption of commercially available red orange juice (250 mg anthocyanins/day) for 12 weeks did not show any effect on body weight or plasma inflammatory markers [63]. Possible reasons for the lack of beneficial effects of anthocyanins on obesity-induced inflammation in these studies are the use of relatively low dosages and the short length of the intervention.

### 5.2. Regulating the Gut Microbiota

Microbes living in the gut play a role in the onset of obesity [65]. Obesity is associated with changes in the gut microbiota that lead to gut dysbiosis. People with obesity have lower proportions of bacteria from the phylum Bacteroidetes, such as *Bacteroides* spp., and higher proportions from the phylum Firmicutes, such as *Bacillus* spp. and *Clostridium*. Firmicutes bacteria are distinguished by higher endotoxic activity of membrane-bound LPS compared with other gram-negative bacteria, thereby inducing more systemic inflammation. Furthermore, obesity promotes the initiation of numerous pro-inflammatory pathways by causing TLR4-mediated inflammatory responses that involve activation of the transcription factor NF-κB and the production of pro-inflammatory mediators such as IL-6, IL-1, and TNF-α [65–68]. A previous study reported that pomegranate peel extract (PPE) is rich in anthocyanins. Treatment with a PPE was shown to modulate the gut microbiota in obese mice by lowering COX-2 levels in the colon and visceral adipose tissue, as well as mRNA levels of IL-1 and IL-6 in the colon [57]. Another study in mice demonstrated that cranberry extract at 200 mg/kg reduced the abundance of *Firmicutes* but enhanced that of *Bacteroidetes*, and concluded that anthocyanin-rich cranberry extract protects dysbiosis of the microbiota in mice with diet-induced obesity [69].

### 5.3. Molecular Pathway for Effects of Anthocyanin on Obesity-Associated Inflammation

NF-κB is a family of inducible transcription factors that regulate genes involved in inflammatory responses and includes NF-κB1 (p50), NF-κB2 (p52), RelA, RelB, and c-Rel [70,71]. Under normal conditions, an NF-κB heterodimer (p50-RelA) is retained in an active form through interaction with inhibitory kappa B (IκB). The NF-κB pathway can be activated by extracellular stimuli such as ROS and a HFD. Upon stimulation, IκB is phosphorylated by the IκB kinase complex, then subsequently ubiquitinated and degraded by the 26S proteasome, allowing the p50-RelA heterodimer to translocate into the nucleus, where it promotes the expression of proinflammatory genes including TNF-α, IL-1α, and IL-1β. These genes are pathogenic in a variety of inflammatory disorders, such as obesity, T2DM, and CVD [71–73]. Vendramea et al. (2013) reported that an 8-week freeze-dried wild-blueberry-powder-enriched diet containing 1.5% w/w of total anthocyanin content significantly reduced the expression of genes encoding the inflammatory mediators NF-κB, TNF-α, IL-6, and CRP, both in the liver and abdominal adipose tissues, in obese Zucker rats [74]. This study concluded that anthocyanins modulated the molecular pathway of the inflammatory process, i.e., the expression of transcription factor NF-κB.

TLRs are a class of pattern recognition receptors that play a key role in the innate immune system [75]. The binding of LPS to the TLR4 receptor promotes the activation of M1 macrophages, which release pro-inflammatory cytokines [76]. Upon binding of TLR4 ligands, TLR4 recruits the adaptor molecule myeloid differentiation factor 88 (MyD88), which then recruits IL-1 receptor-associated kinase-4 (IRAK-4). Subsequently, IRAK-4 recruits TNF receptor-associated factor 6 (TRNF6) to the receptor complex. When IRAK-4 is phosphorylated, the IRAK-4–TRAF6 complex separates from the receptor complex to conjugate with transforming growth factor-β-activated kinase 1 (TAK1) [77]. Upon activation, TAK1 causes IκB to be phosphorylated and ubiquitinated, leading to degradation of IκB. As a result, the NF-κB signaling pathway is activated and inflammatory mediators are released [78]. Karunarathne et al. (2020) studied anthocyanins isolated from *Hibiscus syriacus* L. in RAW 264.7 macrophages and found that anthocyanin inhibits TLR4 in LPS-induced cells. Consequently, phosphorylation of MyD88 and IRAK4 is decreased, leading to the inhibition of the NF-κB signaling pathway [79]. Anthocyanins from black soybeans were shown to inhibit IκB phosphorylation, thereby inhibiting NF-κB nuclear translocation and activation in *Helicobacter-pylori*-induced inflammation in human gastric epithelial cells. As a result, NO and COX-2 gene expression was significantly attenuated [80]. In pro-inflammatory conditions, excess NO is converted to a pro-oxidative role, while COX-2 leads to the formation of pro-inflammatory mediators such as prostaglandin E2 [81]. These mechanisms of NF-κB pathways act as a link between oxidative stress and inflammation. Anthocyanins prevent NF-κB activation, thus suppressing the entire downstream cascade, including proinflammatory cytokines, chemokines, adhesion molecules, NO, and COX-2.

**SUMMARY POINTS**

- Obesity is associated with meta-inflammation, a newly coined term for chronic low-grade inflammation as a response to obesity.
- The state of inflammation in obesity is associated with adipose tissue dysfunction, which increases the secretion of pro-inflammatory cytokines such as TNF-α, CRP, and IL-6. Additionally, increased numbers of M1 macrophages in adipose tissue can release pro-inflammatory cytokines.
- Anthocyanins decrease the production of inflammatory cytokines in obese animal models, in studies conducted both in vitro and in vivo.
- Anthocyanins might have potential to be a treatment for obesity-related inflammation and chronic diseases.
- Be aware that the bioavailability, including metabolism and excretion, of anthocyanins is low. Therefore, future research is needed to increase the bioavailability of anthocyanins to improve their beneficial anti-inflammatory effects in people with obesity.

## 6. Conclusions

We reviewed the benefits of anthocyanin-rich foods such as sweet cherries, black soybeans, and strawberries on obesity prevention, including antioxidant and anti-inflammatory effects, and regulation of the gut microbiota in cell cultures, animal models, and human clinical trials. As shown in Figure 4, dietary anthocyanins may have anti-obesity effects and may reduce the risks of the most common chronic noncommunicable diseases.

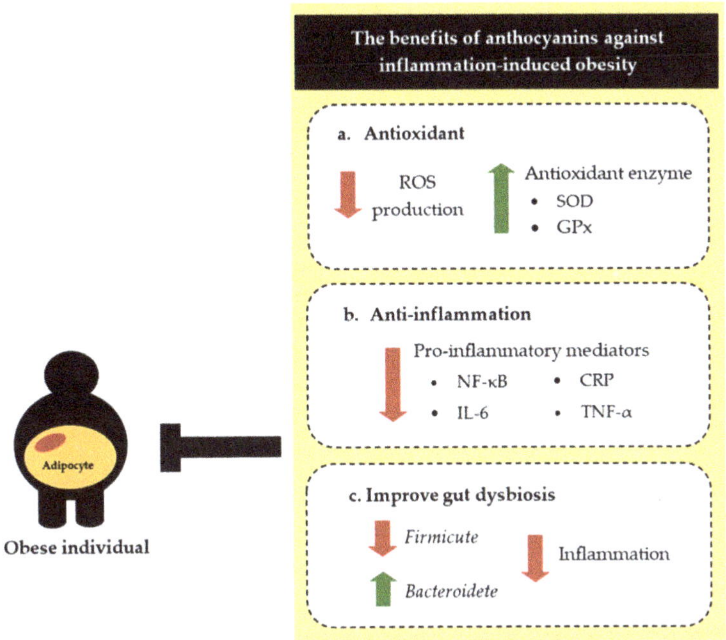


**Figure 4.** The anti-obesity effects of anthocyanins. (**a**) Anthocyanins are antioxidants that eliminate reactive oxygen species (ROS) by increasing antioxidant enzymes such as superoxide dismutase (SOD) and glutathione peroxidase (GPx). (**b**) Anthocyanins prevent nuclear factor kappa B (NF-κB) activation, thus decreasing the entire downstream cascade of pro-inflammatory mediators such as C-reactive protein (CRP), interleukin (IL)-6, and tumor necrosis factor (TNF)-α. (**c**) Anthocyanins also improve gut dysbiosis, restoring a balanced gut microbiota.

**Author Contributions:** Conceptualization: C.N. and N.S.; writing—original draft preparation: C.N.; writing—review and editing, C.N., N.S. and J.S. All authors have read and agreed to the published version of the manuscript.

**Funding:** This research received no external funding.

**Institutional Review Board Statement:** Not applicable.

**Informed Consent Statement:** Not applicable.

**Conflicts of Interest:** The authors declare no conflict of interest.

## References

1. WHO Expert Consultation. Appropriate body-mass index for Asian populations and its implications for policy and intervention strategies. *Lancet.* **2004**, *363*, 157–163. [CrossRef]
2. Finucane, M.M.; Stevens, G.A.; Cowan, M.J.; Danaei, G.; Lin, J.K.; Paciorek, C.J.; Singh, G.M.; Gutierrez, H.R.; Lu, Y.; Bahalim, A.N.; et al. National, regional, and global trends in body-mass index since 1980: Systematic analysis of health examination surveys and epidemiological studies with 960 country-years and 91 million participants. *Lancet* **2011**, *377*, 557–567. [CrossRef]
3. Gregor, M.F.; Hotamisligil, G.S. Inflammatory mechanisms in obesity. *Annu. Rev. Immunol.* **2011**, *29*, 415–445. [CrossRef] [PubMed]
4. Forsythe, L.K.; Wallace, J.M.; Livingstone, M.B. Obesity and inflammation: The effects of weight loss. *Nutr. Res. Rev.* **2008**, *21*, 117–133. [CrossRef] [PubMed]
5. Garvey, W.T.; Mechanick, J.I.; Brett, E.M.; Garber, A.J.; Hurley, D.L.; Jastreboff, A.M.; Nadolsky, K.; Pessah-Pollack, R.; Plodkowski, R.; Reviewers of the AACE/ACE Obesity Clinical Practice Guidelines. American Association of Clinical Endocrinologists and American College of Endocrinology Comprehensive Clinical Practice Guidelines for Medical Care of Patients with Obesity. *Endocr Pract.* **2016**, *22* (Suppl. S3), 1–203. [CrossRef] [PubMed]

6. Bessesen, D.H.; Van Gaal, L.F. Progress and challenges in anti-obesity pharmacotherapy. *Lancet Diabetes Endocrinol.* **2018**, *6*, 237–248. [CrossRef]
7. Kang, J.G.; Park, C.Y. Anti-obesity drugs: A review about their effects and safety. *Diabetes Metab. J.* **2012**, *36*, 13–25. [CrossRef]
8. Konstantinidi, M.; Koutelidakis, A.E. Functional foods and bioactive compounds: A review of its possible role on weight management and obesity's metabolic consequences. *Medicines* **2019**, *6*, 94. [CrossRef]
9. Wang, S.; Moustaid-Moussa, N.; Chen, L.; Mo, H.; Shastri, A.; Su, R.; Bapat, P.; Kwun, I.; Shen, C.-L. Novel insights of dietary polyphenols and obesity. *J. Nutr. Biochem.* **2014**, *25*, 1–18. [CrossRef]
10. Joseph, S.V.; Edirisinghe, I.; Burton-Freeman, B.M. Fruit polyphenols: A review of anti-inflammatory effects in humans. *Crit. Rev. Food Sci. Nutr.* **2016**, *56*, 419–444. [CrossRef]
11. WHO. WHO Fact Files: Obesity and overweight. Geneva: WHO 2020. Available online: https://www.who.int/news-room/fact-sheets/detail/obesity-and-overweight (accessed on 12 December 2021).
12. Ogden, C.L.; Carroll, M.D.; Curtin, L.R.; Lamb, M.M.; Flegal, K.M. Prevalence of high body mass index in US children and adolescents, 2007–2008. *JAMA.* **2010**, *303*, 242–249. [CrossRef] [PubMed]
13. Hotamisligil, G.S. Inflammation and metabolic disorders. *Nature* **2006**, *444*, 860–867. [CrossRef] [PubMed]
14. Longo, M.; Zatterale, F.; Naderi, J.; Parrillo, L.; Formisano, P.; Raciti, G.A.; Beguinot, F.; Miele, C. Adipose tissue dysfunction as determinant of obesity-associated metabolic complications. *Int. J. Mol. Sci.* **2019**, *20*, 2358. [CrossRef] [PubMed]
15. Shoelson, S.E.; Lee, J.; Goldfine, A.B. Inflammation and insulin resistance. *J. Clin. Investig.* **2006**, *116*, 1793–1801. [CrossRef]
16. Lee, B.C.; Lee, J. Cellular and molecular players in adipose tissue inflammation in the development of obesity-induced insulin resistance. *Biochim. Biophys. Acta* **2014**, *1842*, 446–462. [CrossRef] [PubMed]
17. Schipper, H.S.; Prakken, B.; Kalkhoven, E.; Boes, M. Adipose tissue-resident immune cells: Key players in immunometabolism. *Trends Endocrinol. Metab.* **2012**, *23*, 407–415. [CrossRef]
18. Odegaard, J.I.; Chawla, A. The immune system as a sensor of the metabolic state. *Immunity* **2013**, *38*, 644–654. [CrossRef]
19. Li, C.; Xu, M.M.; Wang, K.; Adler, A.J.; Vella, A.T.; Zhou, B. Macrophage polarization and meta-inflammation. *Transl. Res.* **2018**, *191*, 29–44. [CrossRef]
20. Lahleh, G.H.; Frydoonfar, H.; Heidary, R.; Jamei, R.; Zare, S. The effect of light, temperature, pH and species on stability of anthocyanin pigments in four berberis species. *Pak. J. Nutr.* **2006**, *5*, 90–92.
21. Khoo, H.E.; Azlan, A.; Tang, S.T.; Lim, S.M. Anthocyanidins and anthocyanins: Colored pigments as food, pharmaceutical ingredients, and the potential health benefits. *Food Nutr. Res.* **2017**, *61*, 1361779. [CrossRef]
22. Fang, J. Bioavailability of anthocyanins. *Drug Metab. Rev.* **2014**, *46*, 508–520. [CrossRef] [PubMed]
23. Borges, G.; Roowi, S.; Rouanet, J.M.; Duthie, G.G.; Lean, M.E.; Crozier, A. The bioavailability of raspberry anthocyanins and ellagitannins in rats. *Mol. Nutr. Food Res.* **2007**, *51*, 714–725. [CrossRef] [PubMed]
24. Felgines, C.; Texier, O.; Besson, C.; Fraisse, D.; Lamaison, J.L.; Rémésy, C. Blackberry anthocyanins are slightly bioavailable in rats. *J. Nutr.* **2002**, *132*, 1249–1253. [CrossRef] [PubMed]
25. Ichiyanagi, T.; Shida, Y.; Rahman, M.M.; Hatano, Y.; Konishi, T. Bioavailability and tissue distribution of anthocyanins in bilberry (*Vaccinium myrtillus* L.) extract in rats. *J. Agric. Food Chem.* **2006**, *54*, 6578–6587. [CrossRef]
26. Rechner, A.R.; Kuhnle, G.; Hu, H.; Roedig-Penman, A.; van den Braak, M.H.; Moore, K.P.; Rice-Evans, C.A. The metabolism of dietary polyphenols and the relevance to circulating levels of conjugated metabolites. *Free Radic. Res.* **2002**, *36*, 1229–1241. [CrossRef]
27. Bub, A.; Watzl, B.; Heeb, D.; Rechkemmer, G.; Briviba, K. Malvidin-3-glucoside bioavailability in humans after ingestion of red wine, dealcoholized red wine and red grape juice. *Eur. J. Nutr.* **2001**, *40*, 113–120. [CrossRef]
28. Felgines, C.; Talavéra, S.; Gonthier, M.P.; Texier, O.; Scalbert, A.; Lamaison, J.L.; Rémésy, C. Strawberry anthocyanins are recovered in urine as glucuro- and sulfoconjugates in humans. *J. Nutr.* **2003**, *133*, 1296–1301. [CrossRef]
29. Keane, K.M.; Bell, P.G.; Lodge, J.K.; Constantinou, C.L.; Jenkinson, S.E.; Bass, R.; Howatson, G. Phytochemical uptake following human consumption of Montmorency tart cherry (L. *Prunus cerasus*) and influence of phenolic acids on vascular smooth muscle cells in vitro. *Eur. J. Nutr.* **2016**, *55*, 1695–1705.
30. Manach, C.; Williamson, G.; Morand, C.; Scalbert, A.; Rémésy, C. Bioavailability and bioefficacy of polyphenols in humans. I. Review of 97 bioavailability studies. *Am. J. Clin. Nutr.* **2005**, *81*, 230s–242s. [CrossRef]
31. Andersen, O.M.J. *Basic Anthocyanin Chemistry and Dietary Sources*; CRC Press: Boca Raton, FL, USA, 2014.
32. Rehman, R.N.U.; You, Y.; Zhang, L.; Goudia, B.D.; Khan, A.R.; Li, P.; Ma, F. High temperature induced anthocyanin inhibition and active degradation in Malus profusion. *Front. Plant Sci.* **2017**, *8*, 1401. [CrossRef]
33. Patras, A.; Brunton, N.P.; O'Donnell, C.; Tiwari, B.K. Effect of thermal processing on anthocyanin stability in foods, mechanisms and kinetics of degradation. *Trends Food Sci. Technol.* **2010**, *21*, 3–11. [CrossRef]
34. Tiwari, B.K.; O'Donnell, C.P.; Cullen, P.J. Effect of non-thermal processing technologies on the anthocyanin content of fruit juices. *Trends Food Sci. Technol.* **2009**, *20*, 137–145. [CrossRef]
35. Barba, F.J.; Esteve, M.J.; Frigola, A. Physicochemical and nutritional characteristics of blueberry juice after high pressure processing. *Food Res. Int.* **2013**, *50*, 545–549. [CrossRef]
36. Akhavan Mahdavi, S.; Jafari, S.M.; Assadpoor, E.; Dehnad, D. Microencapsulation optimization of natural anthocyanins with maltodextrin, gum arabic and gelatin. *Int. J. Biol. Macromol.* **2016**, *85*, 379–385. [CrossRef] [PubMed]
37. Vergara, C.; Pino, M.T.; Zamora, O.; Parada, J.; Pérez, R.; Uribe, M.; Kalazich, J. Microencapsulation of anthocyanin extracted from purple flesh cultivated potatoes by spray drying and its effects on in vitro gastrointestinal digestion. *Molecules* **2020**, *25*, 722. [CrossRef]

38. Kanokpanont, S.; Yamdech, R.; Aramwit, P. Stability enhancement of mulberry-extracted anthocyanin using alginate/chitosan microencapsulation for food supplement application. *Artif. Cells Nanomed. Biotechnol.* **2018**, *46*, 773–782. [CrossRef]
39. Castañeda-Ovando, A.; Pacheco-Hernández, M.D.L.; Páez-Hernández, M.E.; Rodríguez, J.A.; Galán-Vidal, C.A. Chemical studies of anthocyanins: A review. *Food Chem.* **2009**, *113*, 859–871. [CrossRef]
40. He, J.; Giusti, M.M. Anthocyanins: Natural colorants with health-promoting properties. *Annu Rev Food Sci Technol.* **2010**, *1*, 163–187. [CrossRef]
41. Cianciosi, D.; Varela-Lopez, A.; Forbes-Hernandez, T.Y.; Gasparrini, M.; Afrin, S.; Reboredo-Rodriguez, P.; Zhang, J.; Quiles, J.L.; Nabavi, S.F.; Battino, M.; et al. Targeting molecular pathways in cancer stem cells by natural bioactive compounds. *Pharmacol. Res.* **2018**, *135*, 150–165. [CrossRef]
42. Heo, H.J.; Lee, C.Y. Strawberry and its anthocyanins reduce oxidative stress-induced apoptosis in PC12 cells. *J. Agric. Food Chem.* **2005**, *53*, 1984–1989. [CrossRef]
43. Zhang, Z.F.; Lu, J.; Zheng, Y.L.; Wu, D.M.; Hu, B.; Shan, Q.; Cheng, W.; Li, M.-Q.; Sun, Y.-Y. Purple sweet potato color attenuates hepatic insulin resistance via blocking oxidative stress and endoplasmic reticulum stress in high-fat-diet-treated mice. *J. Nutr. Biochem.* **2013**, *24*, 1008–1018. [CrossRef]
44. Esatbeyoglu, T.; Rodríguez-Werner, M.; Schlösser, A.; Winterhalter, P.; Rimbach, G. Fractionation, enzyme inhibitory and cellular antioxidant activity of bioactives from purple sweet potato (*Ipomoea batatas*). *Food Chem.* **2017**, *221*, 447–456. [CrossRef] [PubMed]
45. Wu, T.; Gao, Y.; Guo, X.; Zhang, M.; Gong, L. Blackberry and blueberry anthocyanin supplementation counteract high-fat-diet-induced obesity by alleviating oxidative stress and inflammation and accelerating energy expenditure. *Oxid. Med. Cell Longev.* **2018**, *2018*, 4051232. [CrossRef] [PubMed]
46. Wu, T.; Yin, J.; Zhang, G.; Long, H.; Zheng, X. Mulberry and cherry anthocyanin consumption prevents oxidative stress and inflammation in diet-induced obese mice. *Mol. Nutr. Food Res.* **2016**, *60*, 687–694. [CrossRef] [PubMed]
47. Alvarez-Suarez, J.M.; Giampieri, F.; Tulipani, S.; Casoli, T.; Di Stefano, G.; González-Paramás, A.M.; Santos-Buelga, C.; Busco, F.; Quiles, J.L.; Cordero, M.D.; et al. One-month strawberry-rich anthocyanin supplementation ameliorates cardiovascular risk, oxidative stress markers and platelet activation in humans. *J. Nutr. Biochem.* **2014**, *25*, 289–294. [CrossRef] [PubMed]
48. Li, D.; Zhang, Y.; Liu, Y.; Sun, R.; Xia, M. Purified anthocyanin supplementation reduces dyslipidemia, enhances antioxidant capacity, and prevents insulin resistance in diabetic patients. *J. Nutr.* **2015**, *145*, 742–748. [CrossRef] [PubMed]
49. Zhou, Z.; Nair, M.G.; Claycombe, K.J. Synergistic inhibition of interleukin-6 production in adipose stem cells by tart cherry anthocyanins and atorvastatin. *Phytomedicine* **2012**, *19*, 878–881. [CrossRef]
50. Li, L.; Wang, L.; Wu, Z.; Yao, L.; Wu, Y.; Huang, L.; Liu, K.; Zhou, X.; Gou, D. Anthocyanin-rich fractions from red raspberries attenuate inflammation in both RAW264.7 macrophages and a mouse model of colitis. *Sci. Rep.* **2014**, *4*, 6234. [CrossRef]
51. Matsukawa, T.; Inaguma, T.; Han, J.; Villareal, M.O.; Isoda, H. Cyanidin-3-glucoside derived from black soybeans ameliorate type 2 diabetes through the induction of differentiation of preadipocytes into smaller and insulin-sensitive adipocytes. *J. Nutr. Biochem.* **2015**, *26*, 860–867. [CrossRef]
52. Gonçalves, A.C.; Costa, A.R.; Flores-Félix, J.D.; Falcão, A.; Alves, G.; Silva, L.R. Anti-inflammatory and antiproliferative properties of sweet cherry phenolic-rich extracts. *Molecules* **2022**, *27*, 268. [CrossRef]
53. Lee, S.; Keirsey, K.I.; Kirkland, R.; Grunewald, Z.I.; Fischer, J.G.; de La Serre, C.B. Blueberry supplementation influences the gut microbiota, inflammation, and insulin resistance in high-fat-diet-fed rats. *J. Nutr.* **2018**, *148*, 209–219. [CrossRef] [PubMed]
54. Nemes, A.; Homoki, J.R.; Kiss, R.; Hegedűs, C.; Kovács, D.; Peitl, B.; Gál, F.; Stündl, L.; Szilvássy, Z.; Remenyik, J. Effect of anthocyanin-rich tart cherry extract on inflammatory mediators and adipokines involved in type 2 diabetes in a high fat diet induced obesity mouse model. *Nutrients* **2019**, *11*, 1966. [CrossRef] [PubMed]
55. Martinelli, I.; Tomassoni, D.; Bellitto, V.; Roy, P.; Di Bonaventura, M.V.M.; Amenta, F.; Amantini, C.; Cifani, C.; Tayebati, S.K. Anti-inflammatory and antioxidant properties of tart cherry consumption in the heart of obese rats. *Biology* **2022**, *11*, 646. [CrossRef] [PubMed]
56. Wu, T.; Tang, Q.; Yu, Z.; Gao, Z.; Hu, H.; Chen, W.; Zheng, X.; Yu, T. Inhibitory effects of sweet cherry anthocyanins on the obesity development in C57BL/6 mice. *Int. J. Food Sci. Nutr.* **2014**, *65*, 351–359. [CrossRef] [PubMed]
57. Neyrinck, A.M.; Van Hée, V.F.; Bindels, L.B.; De Backer, F.; Cani, P.D.; Delzenne, N.M. Polyphenol-rich extract of pomegranate peel alleviates tissue inflammation and hypercholesterolaemia in high-fat diet-induced obese mice: Potential implication of the gut microbiota. *Br. J. Nutr.* **2013**, *109*, 802–809. [CrossRef]
58. Edirisinghe, I.; Banaszewski, K.; Cappozzo, J.; Sandhya, K.; Ellis, C.L.; Tadapaneni, R.; Kappagoda, C.T.; Burton-Freeman, B.M. Strawberry anthocyanin and its association with postprandial inflammation and insulin. *Br. J. Nutr.* **2011**, *106*, 913–922. [CrossRef]
59. Lee, M.; Sorn, S.R.; Park, Y.; Park, H.K. Anthocyanin rich-black soybean testa improved visceral fat and plasma lipid profiles in overweight/obese Korean adults: A randomized controlled trial. *J. Med. Food* **2016**, *19*, 995–1003. [CrossRef]
60. Bhaswant, M.; Brown, L.; Mathai, M.L. Queen garnet plum juice and raspberry cordial in mildly hypertensive obese or overweight subjects: A randomized, double-blind study. *J. Funct. Foods* **2019**, *56*, 119–126. [CrossRef]
61. Martin, K.R.; Burrell, L.; Bopp, J. Authentic tart cherry juice reduces markers of inflammation in overweight and obese subjects: A randomized, crossover pilot study. *Food Funct.* **2018**, *9*, 5290–5300. [CrossRef]
62. Zunino, S.J.; Parelman, M.A.; Freytag, T.L.; Stephensen, C.B.; Kelley, D.S.; Mackey, B.E.; Woodhouse, L.R.; Bonnel, E.L. Effects of dietary strawberry powder on blood lipids and inflammatory markers in obese human subjects. *Br. J. Nutr.* **2012**, *108*, 900–909. [CrossRef]

63. Azzini, E.; Venneria, E.; Ciarapica, D.; Foddai, M.S.; Intorre, F.; Zaccaria, M.; Maiani, F.; Palomba, L.; Barnaba, L.; Tubili, C.; et al. Effect of red orange juice consumption on body composition and nutritional status in overweight/obese female: A pilot study. *Oxid. Med. Cell Longev.* **2017**, *2017*, 1672567. [CrossRef] [PubMed]
64. Bastard, J.P.; Maachi, M.; Lagathu, C.; Kim, M.J.; Caron, M.; Vidal, H.; Capeau, J.; Feve, B. Recent advances in the relationship between obesity, inflammation, and insulin resistance. *Eur. Cytokine Netw.* **2006**, *17*, 4–12. [PubMed]
65. John, G.K.; Mullin, G.E. The gut microbiome and obesity. *Curr. Oncol. Rep.* **2016**, *18*, 45. [CrossRef] [PubMed]
66. Ley, R.E.; Bäckhed, F.; Turnbaugh, P.; Lozupone, C.A.; Knight, R.D.; Gordon, J.I. Obesity alters gut microbial ecology. *Proc. Natl. Acad. Sci. USA* **2005**, *102*, 11070–11075. [CrossRef]
67. Green, M.; Arora, K.; Prakash, S. Microbial medicine: Prebiotic and probiotic functional foods to target obesity and metabolic syndrome. *Int. J. Mol. Sci.* **2020**, *21*, 2890. [CrossRef]
68. Lu, Y.C.; Yeh, W.C.; Ohashi, P.S. LPS/TLR4 signal transduction pathway. *Cytokine* **2008**, *42*, 145–151. [CrossRef]
69. Anhê, F.F.; Roy, D.; Pilon, G.; Dudonné, S.; Matamoros, S.; Varin, T.V.; Garofalo, C.; Moine, Q.; Desjardins, Y.; Levy, E.; et al. A polyphenol-rich cranberry extract protects from diet-induced obesity, insulin resistance and intestinal inflammation in association with increased *Akkermansia* spp. population in the gut microbiota of mice. *Gut* **2015**, *64*, 872–883. [CrossRef]
70. Sun, S.C.; Chang, J.H.; Jin, J. Regulation of nuclear factor-κB in autoimmunity. *Trends Immunol.* **2013**, *34*, 282–289. [CrossRef]
71. Liu, T.; Zhang, L.; Joo, D.; Sun, S.-C. NF-κB signaling in inflammation. *Signal Transduct. Target Ther.* **2017**, *2*, 17023. [CrossRef]
72. Hayden, M.S.; Ghosh, S. Shared principles in NF-kappaB signaling. *Cell* **2008**, *132*, 344–362. [CrossRef]
73. Singh, A.; Kumar, D.S.; Vinayak, M. Recent development in antihyperalgesic effect of phytochemicals: Anti-inflammatory and neuro-modulatory actions. *Inflamm. Res.* **2018**, *67*, 633–654. [CrossRef] [PubMed]
74. Vendrame, S.; Daugherty, A.; Kristo, A.S.; Riso, P.; Klimis-Zacas, D. Wild blueberry (*Vaccinium angustifolium*) consumption improves inflammatory status in the obese Zucker rat model of the metabolic syndrome. *J. Nutr. Biochem.* **2013**, *24*, 1508–1512. [CrossRef] [PubMed]
75. Vidya, M.K.; Kumar, V.G.; Sejian, V.; Bagath, M.; Krishnan, G.; Bhatta, R. Toll-like receptors: Significance, ligands, signaling pathways, and functions in mammals. *Int. Rev. Immunol.* **2018**, *37*, 20–36. [CrossRef] [PubMed]
76. Stephens, M.; Liao, S.; von der Weid, P.-Y. Ultra-purification of lipopolysaccharides reveals species-specific signalling bias of TLR4: Importance in macrophage function. *Sci. Rep.* **2021**, *11*, 1335. [CrossRef]
77. Kim, T.W.; Staschke, K.; Bulek, K.; Yao, J.; Peters, K.; Oh, K.H.; Vandenburg, Y.; Xiao, H.; Qian, W.; Hamilton, T.; et al. A critical role for IRAK4 kinase activity in Toll-like receptor-mediated innate immunity. *J. Exp. Med.* **2007**, *204*, 1025–1036. [CrossRef]
78. Sato, S.; Sanjo, H.; Takeda, K.; Ninomiya-Tsuji, J.; Yamamoto, M.; Kawai, T.; Matsumoto, K.; Takeuchi, O.; Akira, S. Essential function for the kinase TAK1 in innate and adaptive immune responses. *Nat. Immunol.* **2005**, *6*, 1087–1095. [CrossRef]
79. Karunarathne, W.; Lee, K.T.; Choi, Y.H.; Jin, C.Y.; Kim, G.Y. Anthocyanins isolated from *Hibiscus syriacus* L. attenuate lipopolysaccharide-induced inflammation and endotoxic shock by inhibiting the TLR4/MD2-mediated NF-κB signaling pathway. *Phytomedicine.* **2020**, *76*, 153237. [CrossRef]
80. Kim, J.-M.; Kim, K.-M.; Park, E.-H.; Seo, J.-H.; Song, J.-Y.; Shin, S.-C.; Kang, H.-Y.; Lee, W.-K.; Cho, M.-J.; Rhee, K.-H.; et al. Anthocyanins from black soybean inhibit *Helicobacter pylori*-induced inflammation in human gastric epithelial AGS cells. *Microbiol. Immunol.* **2013**, *57*, 366–373. [CrossRef]
81. Vendrame, S.; Klimis-Zacas, D. Anti-inflammatory effect of anthocyanins via modulation of nuclear factor-κB and mitogen-activated protein kinase signaling cascades. *Nutr. Rev.* **2015**, *73*, 348–358. [CrossRef]

*biomolecules*

MDPI

*Review*

# Role of Natural Compounds Modulating Heme Catabolic Pathway in Gut, Liver, Cardiovascular, and Brain Diseases

**Sri Jayanti [1,2,†], Libor Vitek [3,†], Camilla Dalla Verde [1,4,†], John Paul Llido [1,4,5], Caecilia Sukowati [1,2], Claudio Tiribelli [1] and Silvia Gazzin [1,*]**

[1] Liver brain Unit "Rita Moretti", Fondazione Italiana Fegato-Onlus, Bldg. Q, AREA Science Park, ss14, Km 163,5, Basovizza, 34149 Trieste, Italy; sri.jayanti@fegato.it or srij001@brin.go.id (S.J.); camilla.dallaverde@fegato.it (C.D.V.); johnpaul.llido@fegato.it (J.P.L.); caecilia.sukowati@fegato.it or caecilia.sukowati@brin.go.id (C.S.); ctliver@fegato.it (C.T.)

[2] Eijkman Research Centre for Molecular Biology, Research Organization for Health, National Research and Innovation Agency, Cibinong 16915, Indonesia

[3] Institute of Medical Biochemistry and Laboratory Diagnostics, and 4th Department of Internal Medicine, General University Hospital and 1st Faculty of Medicine, Charles University, 12000 Prague, Czech Republic; vitek@cesnet.cz

[4] Department of Life Sciences, University of Trieste, 34139 Trieste, Italy

[5] Department of Science and Technology, Philippine Council for Health Research and Development, Bicutan, Taguig City 1631, Philippines

* Correspondence: silvia.gazzin@fegato.it

† These authors contributed equally to this work.

**Abstract:** The crucial physiological process of heme breakdown yields biliverdin (BV) and bilirubin (BR) as byproducts. BV, BR, and the enzymes involved in their production (the "yellow players—YP") are increasingly documented as endogenous modulators of human health. Mildly elevated serum bilirubin concentration has been correlated with a reduced risk of multiple chronic pro-oxidant and pro-inflammatory diseases, especially in the elderly. BR and BV per se have been demonstrated to protect against neurodegenerative diseases, in which heme oxygenase (HMOX), the main enzyme in the production of pigments, is almost always altered. HMOX upregulation has been interpreted as a tentative defense against the ongoing pathologic mechanisms. With the demonstration that multiple cells possess YP, their propensity to be modulated, and their broad spectrum of activity on multiple signaling pathways, the YP have assumed the role of an adjustable system that can promote health in adults. Based on that, there is an ongoing effort to induce their activity as a therapeutic option, and natural compounds are an attractive alternative to the goal, possibly requiring only minimal changes in the life style. We review the most recent evidence of the potential of natural compounds in targeting the YP in the context of the most common pathologic condition of adult and elderly life.

**Keywords:** bilirubin; herbal medicine; NRF2; heme-oxygenase; MAFLD; neurodegeneration; nutraceuticals; Alzheimer's disease; cancer; Parkinson's disease

## 1. Introduction

Far from being only a diagnostic marker for liver diseases, mild elevation of the total serum bilirubin (TSB) concentration has been repeatedly correlated with a lower risk of developing chronic diseases typical of adult and elderly life [1–6]. The protection is mediated by bilirubin modulatory activity on the cellular redox state, immunity, cellular metabolism, body glucose, insulin balance, etc. Thus, bilirubin is now considered a biomarker of disease resistance, a predictor of all-cause mortality, and a molecule that can promote health in adults. A complex network links the enzymes involved in bilirubin production and multiple signaling pathways that may be targets of pharmacologic induction [2,4,5]. Based on that, there is an ongoing effort to induce their activity as a therapeutic option. Natural compounds are attractive alternatives to the goal, especially in chronic diseases, requiring a lifelong approach.

## 2. The Heme Catabolic Pathway and Metabolic Checkpoints

Bilirubin (BR) and its precursor biliverdin (BV) are entirely derived from heme degradation of hemoglobin (Figure 1), which is catabolized by the unique and ubiquitous microsomal enzyme heme oxygenase (HMOX) [7]. Simultaneously, in the reaction catalyzed by HMOX, a molecule of carbon monoxide (CO), a vasoactive and biologically important gaseous molecule, is released together with iron [8]. BV is not present in the circulation due to its rapid conversion to BR by biliverdin reductase (BLVR), a biologically potent enzyme with an important impact on cell function and metabolism [9]. Interestingly, BR has been shown to have even more versatile biological functions in all organs and compartments of the human body, including general antioxidant activities, immunosuppressive functions, and potent cell signaling properties [1,10–14]. HMOX, the key enzyme responsible for the initiation of heme catabolism, has two isoforms (HMOX1, OMIM 141250; and HMOX2, OMIM 141251) [15,16]. HMOX1 is considered the most inducible enzyme in the human body [17,18], and a variety of natural compounds are known as potent inducers of HMOX1 activity [3,19].

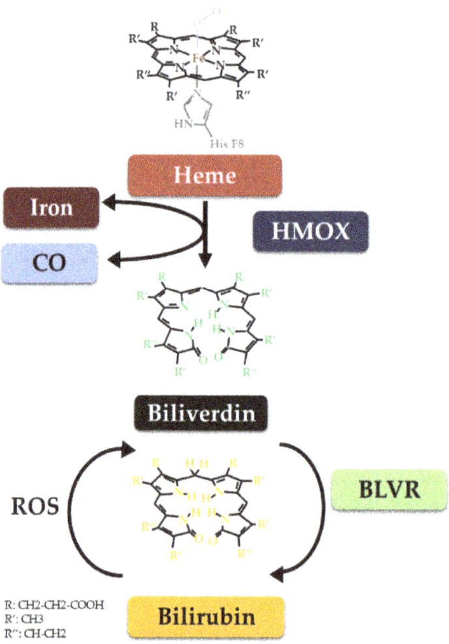


**Figure 1.** Bilirubin metabolism. CO: carbon monoxide, HMOX: heme oxygenase, ROS: reactive oxygen species, BLVR: biliverdin reductase.

The HMOX pathway, through its activation role in gene transcriptions, is closely associated with another evolutionarily conserved cell system, the multifunctional nuclear factor erythroid 2-related factor (NRF2). It is considered not only a regulator of cellular resistance to oxidants, inflammatory stimuli, and toxic xenobiotics, but also a potent modulator of longevity [20,21] and cardiovascular and metabolic diseases [22]. On the other hand, its aberrant activation can increase the risk of development of various diseases, such as diabetes or cancer [23]. Interestingly, also the NRF2 pathway is highly responsive to natural products (see Table 1). Indeed, both stimulatory and inhibitory effects of a wide array of natural products have been reported, with a possible deep impact on the prevention of various civilization diseases [23–29], including neurodegenerative conditions, such as

Alzheimer's disease (AD), Parkinson's disease (PD), schizophrenia, multiple sclerosis, amyotrophic lateral sclerosis, perinatal brain injuries, and Duchenne muscular dystrophy, as well as ischemic stroke, CNS trauma, cerebral neoplasms, etc. [5,30–40]. HMOX1 and its products are believed to protect astrocytes and microglia from increased oxidative stress, apoptosis, and inflammation [35,41–43] and to promote angiogenesis [44]. The induction of HMOX1 activity by various natural products has become an important therapeutic target for combating certain neurodegenerative and other autoimmune diseases [19,28,45–48].

**Table 1.** Metabolic checkpoints affected by natural compounds and also by the heme catabolic pathway. NRF2, nuclear factor erythroid 2—related factor 2; HMOX, heme oxygenase; PI3K, phosphoinositide 3-kinase; AKT, serine/threonine kinase; EGFR, epidermal growth factor receptor; ERK, extracellular signal-regulated kinase; ROS, reactive oxygen species; NQO1, NAD(P)H: Quinone Acceptor Oxidoreductase Type 1; NO, nitric oxide; TNFα, tumor necrosis factor alpha; IL, interleukin; SOD, superoxide dismutase; GSH, glutathione; PERK, protein kinase R (PKR)-like endoplasmic reticulum kinase; NFkB, nuclear factor k-light-chain-enhancer of activated B cells; TLR4, toll like receptor 4; MAPK, mitogen-activated protein kinases; PKC, protein kinase C; GST, glutathione g-transferases; p62, selective autophagy receptor p62; AKT, protein kinase B (PKB), also known as Akt; NADPH, nicotinamide adenine dinucleotide phosphate; PPAR, peroxisome proliferator-activated receptor; PGC1α, peroxisome proliferator-activated receptor-α coactivator-1α; CD, cluster of differentiation; CO, carbon monoxide; FGF21, fibroblast growth hormone 21; mTOR, mammalian target of rapamycin; SIRT1, NAD-dependent deacetylase sirtuin-1; AMPK, AMP-activated protein kinase; T2DM, type 2 diabetes; GSK3b, glycogen synthase kinase-3b; BLVRA, biliverdin reductase A.

| Metabolic Checkpoint | Heme Catabolic Pathway Modulator | Natural Compound (Some Examples) | Possible Clinical Impact |
|---|---|---|---|
| NRF2 | NRF2 activates HMOX1 [29,49,50]<br>Bilirubin activates NRF2 [51] | Sulphoraphan, curcumin, bixin, apigenin, cinnamaldehyde, withaferin A, luteolin, wogonin, chrysin... [23–28] | Regulator of cellular resistance to oxidants, inflammatory stimuli and toxic xenobiotics, modulator of longevity and cardiovascular and metabolic diseases. |
| | HMOX1 | 20C (bibenzyl compound isolated from Gastrodia elvata) [47] | Suppresses the pro-apoptotic effect of Rot by inhibition of Bax and suppress the accumulation of intracellular ROS and the collapse of the mitochondrial membrane potential. |
| | HMOX1 | (ABPK) achyrantes bidentata polypeptide K [43] | Neuroprotective agent inhibiting the neuroinflammation on BV2 microglia cell culture. |
| | HMOX1 | Coriolus versicolor, Hericium erinaceus [48] | Anti-inflammatory modulating the lipoxin A4 levels (LXA4), resolving neuroinflammation and limiting the motor and non-motor symptoms, typical of PD. |
| | HMOX1 | Hyperoside (quercetin 3-O-galactoside) [52] | Protects cultured dopaminergic neurons from death via ROS-dependent mechanisms. |
| | HMOX1 | Berberine (BBR) [53] | Binds specific DNA sequences triggering DNA repair process. |
| | HMOX1 | Breviscapine [54] | HMOX1 and NQO1 increases. |
| | HMOX1 via PI3K/AKT | Acerogin A [55] | Prevent glutamate-induced oxidative damage. |
| | HMOX1 via EGFR/ERK | Astragaloside IV+/− Panax notoginseng [56,57] | Reduction of the oxidative stress markers, inhibition inflammatory mediators (NO, TNFα, IL6) and increase of SOD and GSH level. |
| | HMOX1 and NFkB/TLR4 signaling cascade | S-allyl cysteine (SAC) from aged garlic extract [58] | Improve cognitive deficits by attenuation of oxidative stress and neuroinflammation. |
| | HMOX1 | Mangiferin [59] | Protects neurons and glia from the oxidative damage by increasing HMOX1 in AD. |
| | HMOX1 | Luteolin [60] | Increases cells' survival by preventing apoptosis and oxidative stress. |
| | - | Curcumin [61] | Inhibits the secretion of pro-neuroinflammatory mediators by increasing HMOX. |
| | - | Curcumin [62–65] | Protects neurons by ameliorating brain water content, oxidative stress, inflammation, and apoptosis, as well as reversal of depressive-like behaviors. |

**Table 1.** *Cont.*

| Metabolic Checkpoint | Heme Catabolic Pathway Modulator | Natural Compound (Some Examples) | Possible Clinical Impact |
|---|---|---|---|
| NRF2 | - | Quercetin, anthocyanins, tea polyphenols, kaempferol, hesperetin, icariin, and various forms of terpenoids [28] | Protect from glutamate neurotoxicity and rescue of impaired cognitive function by increasing antioxidant responses, improving cell viability, and decreasing pro-inflammatory mediators. |
| | - | Curcumin [66] | Improves motor deficits and morphological alterations through antioxidant activity in an in vivo model of quinolinic acid neurotoxicity. |
| | NRF2 and PERK pathway | Curcumin [67] | Improves motor, sensory, reflex, and balance through inhibition of oxidative stress and apoptotic process. |
| NFkB/TLR4 | NRF2 and NFkB | Curcumin [68] | Improves memory and behavior. |
| NFkB/STAT3/Ap-1 | | Luteolin [60] | Reduces neuroinflammation induced by astrocytes. |
| NFkB/MAPKs | NFkB/MAPK pathways | Curcumin [61] | Inhibits the secretion of pro-neuroinflammatory mediators by increasing Hmox. |
| NFkB | | Ellagic acid [69] | Promotes anti-inflammatory and anti-antioxidant effects in AD and PD. |
| PKC | PKC activates NRF2 | Curcumin [70] | Neurons are stimulated to increase antioxidant gene expression (GST-mu1, NQO1, and Hmox1), as well as p62, resulting in a positive feedback loop. |
| ERK | ERK modulate NRF2 anti-oxidant signaling | Curcumin [66] | Improve motor deficits and morphological alterations through antioxidant activity. |
| | AKT2/NRF2 pathways | Kaempferol<br>Ginsenoside rh2 [71] | Upregulation of the antioxidant enzyme thioredoxin linked to antidepressant mechanism. |
| | AKT/NRF2 | Curcumin [72] | Protects neurons and reduces infarct size in in vitro (oxygen and glucose deprivation/reoxygenation) and in vivo (middle cerebral artery occlusion) models of ischemic injury. |
| NADPH oxidase | Bilirubin, biliverdin | C-phycocyanin (C- PC) [73] | Protective in many neurodegenerative diseases and in COVID-19-induced neurologic damage. |
| PPARs | Bilirubin [74] | Resveratrol [75,76] | Beneficiary effects on glucose and adipose tissue metabolism. |
| PGC1a | Bilirubin [77] | Resveratrol, quercetin, curcumin, saponins, epigallocatechin-3-gallate (EGCG) [37] | Regulation of cellular energy metabolism with beneficiary effects on civilization diseases. |
| CD39 | Bilirubin [78]<br>CO [79] | Resveratrol [38]<br>Curcumin [39] | Control of inflammatory processes via purinergic signaling. |
| FGF21 | Bilirubin [80,81] | Coffee phytochemicals (chlorogenic/protocatechuic acid) [82]<br>Cocoa phytochemicals (theobromine/protocatechuic acid) [83] | Energy homeostasis, via FGF21 signaling, a late-acting fed and fasting-state hormone. |
| mTOR | Bilirubin [10,14]<br>biliverdin [84]<br>CO [85] | Curcumin, quercetin, apigenin [86] | Modulation of nutrient-sensing with impact on intermediary metabolism, aging processes, and overall life span. |
| SIRT1 | Bilirubin [87] | Resveratrol, butein, quercetin [88], astragaloside IV [40] | Control of fat and glucose metabolism, and energy expenditure.<br>Anti-inflammatory and antioxidant. Reducing infarct size in ischemic stroke. |
| AMPK | Bilirubin [77]<br>biliverdin [89]<br>CO [85] | Resveratrol<br>Berberine<br>Quercetin [90] | Prevention of cardiovascular and metabolic diseases (T2DM), energy homeostasis. |
| GSK3b | BLVRA/bilirubin [91] | Resveratrol, curcumin, berberine [92] | Modulation of cellular kinase, with >100 known targets affecting lipid and glucose metabolism, and cell proliferation. |

It should be mentioned that, after HMOX1 induction, iron is also released [93]. Especially in the brain, excessive iron levels can be toxic due to its pro-oxidant capacity (as example [5,6,44,94–96]). On the contrary, HMOX1, iron, cellular redox status, and inflammation regulate the increased vulnerability to ferroptosis in glioblastoma, the most recurrent brain tumor [97–103].

Also interesting is the observation that BV enhances the expression of CD36 [104], which is involved in fatty acid oxidation and diabetes control [105]. In this regard, it is interesting to note that CD36 has been identified as the aryl hydrocarbon receptor (AhR) target gene [106], with BR being a potent AhR activator [107]. In fact, almost 400 peroxisome proliferator-activated receptors (PPARs) α-dependent genes have been reported to have been significantly modulated by exposure of BV to HepG2 human hepatoblastoma cells [108]. BV also reduces the activation of c-Jun NH2 terminal kinase (JNK), protecting endothelial cells from undergoing apoptosis in vascular injury-induced intimal hyperplasia [109], and has been reported to be protective in cerebral infarction and cerebral ischemia-reperfusion [110,111]. BLVRA activity has significant implications in AD. Its reduction in AD models not only led to increased BACE1 (β-site amyloid precursor protein cleaving enzyme 1) phosphorylation, resulting in higher Aβ deposits, but it also induces insulin resistance by downregulating the insulin receptor (IR) and inhibiting insulin receptor substrate 1 (IRS1) [112]. Meanwhile, the increase of the BLVRA protein level and activity in the brain of the AD model, followed by the increase of UCB, shows a negative (protective) correlation with oxidative stress markers and cognition [113]. BLVRA, which also functions as a transcription factor, interacts with nuclear factor kappa B (NFκB) and leads to the halt of the cell cycle [114]. This results in the downregulation of BLVRA in brain tumors, specifically, meningiomas and gliomas, with implications not yet elucidated with antioxidant status and chemoresistance in these tumor types [6,115]. Moreover, NFkB activation during hypoxic-ischemic injury [116] may contribute to the underlying cause of cerebral palsy [117–120] and autism spectrum disorders [121–125].

Bilirubin UDP-glucuronosyltransferase (UGT1A1), a phylogenetically old bio-transforming enzyme conjugating bilirubin with glucuronic acid in the liver cell [7], is another important enzyme in the heme catabolic pathway, which is modulated by various natural products [3,126]. *UGT1A1* expression is substantially under the control of multifunctional nuclear receptors, including the constitutive androstane receptor (CAR), the pregnane X receptor (PXR), the glucocorticoid receptor (GR), the aryl hydrocarbon receptor (AhR), and the hepatocyte nuclear receptor 1α (HNF1α) [127], which regulate *UGT1A1* transcription via the phenobarbital-responsive enhancer module (PBREM) [127].

Interestingly, BR serves as a ligand for these nuclear receptors, which also contribute to glucose and lipid metabolism [12] and the pathogenesis of cardiovascular diseases and other diseases of civilization [128,129]. Meanwhile, a slightly elevated TSB is negatively correlated with multiple neurologic conditions (for a review on the correlation between TSB and neurologic conditions, see [6]). This strongly supports the idea that the nutraceuticals able to increase bilirubin production might help to prevent neurodegenerative damage [44]. Of relevance, BR has been recently demonstrated to prevent dopaminergic neuron (DOPAn) demise in an ex vivo model of Parkinson's disease by normalizing TNFα, the determinant in DOPAn death [130]. Remarkably, CAR/PXR/GR/AhR/HNF1α are substantially modulated by dietary components, including natural products [131,132]. It is also important to note that BR per se also serves as a ligand for other metabolic nuclear receptors, such as peroxisome proliferator-activated receptors (PPARs) α and γ, being considered master regulators of cellular and whole-body energy homeostasis, with important roles in the pathogenesis of obesity, diabetes mellitus, metabolic syndrome, and aging [12,133–135]. The direct activating effect of bilirubin on PPARα has recently been convincingly demonstrated in various experimental studies (reviewed in [74]). In fact, physiologically relevant BR concentrations were capable of activating PPARα with the same magnitude as that of fenofibrate, a clinically used agonist of this nuclear receptor. In addition, a recent in silico analysis revealed that bilirubin resembles structurally the molecule of fenofibrate [39]. Furthermore, acute treatment of mice with bilirubin resulted in increased expression of hepatic PPARα target genes, including fibroblast growth factor 21 (FGF21) [104]. BR also affects FGF21 [80,81], considered a late-acting fed and fasting-state hormone [136], but interestingly, also an insulin signaling pathway [137–140] (insulin being considered an immediately acting hormone). BR also affects the expression of PPARγ [91,137], another master regulator of adipogenesis and obesity [141]. Although not all available data are consistent, it seems that modulation of the PPARγ pathway is biologically and clinically relevant. It has been reported that PPARγ agonists activated the AMP-activated protein kinase (AMPK) [77,142–144], an important cellular energy sensor [145,146]. Increased AMPK phosphorylation has been reported in subjects with benign hyperbilirubinemia (Gilbert's syndrome) [77]. PPARγ activation of AMPK also led to inhibition of the mammalian target of rapamycin (mTOR signaling), an evolutionarily conserved nutrient-sensing protein kinase that regulates metabolism, aging processes, and overall life span [147,148], as well as dephosphorylation of a downstream factor, S6K [143]. Similar results on inhibition of S6K phosphorylation were also obtained in our study in human fibroblasts exposed to physiological concentrations of BR [14,149]. Interestingly, dysregulation of AMPK and mTOR hyperactivation was reported in BLVRA-deficient mice [84]; and in another study, BV inhibited TLR4 signaling in leukocytes and triggered phosphorylation of mTOR-specific targets, including Akt, PKC, and AMPK [89], suggesting the importance of BV and BR for modulation of these signaling pathways (Figure 2).

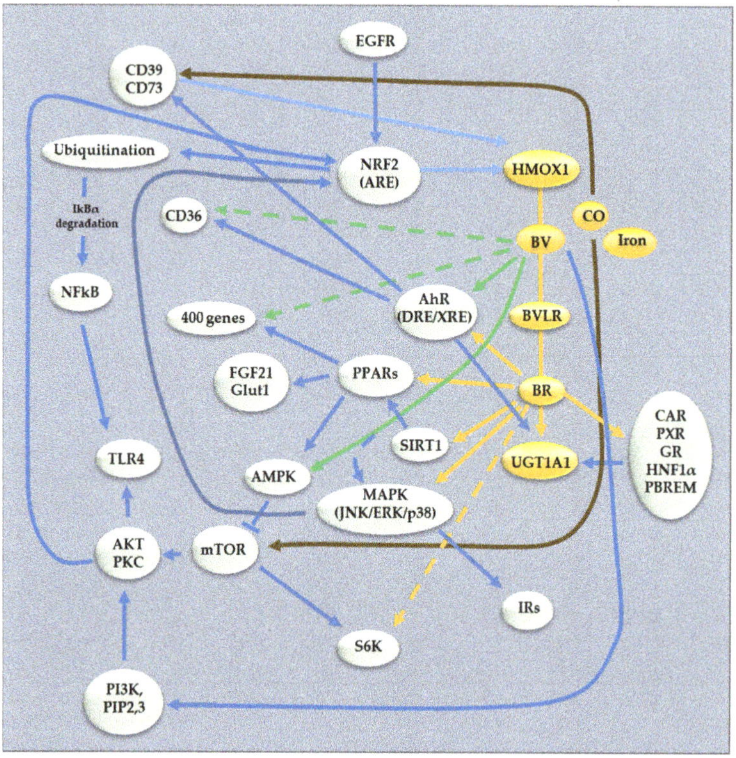


**Figure 2.** Interconnections between heme metabolism and signaling pathways. Solid line: experimental evidence of a direct link. Dash lines: no experimental evidence of a direct action. Green lines: biliverdin action; yellow lines: bilirubin actinin; brown lines: CO action; blue lines: all other actions/connections. Yellow EFGR: epidermal growth factor receptor; NRF2: nuclear factor erythroid 2–related factor 2; ARE: antioxidant response element; CD: cluster of differentiation; HMOX: heme oxygenase; BV: biliverdin; BVLR: biliverdin reductase; BR: bilirubin; UGT1A1: UDP-glucuronosyltransferase 1A1; CAR: constitutive active/androstane receptor; PXR: pregnane X receptor; GR: glucocorticoid receptor; AhR: aryl hydrocarbon receptor; DRE/XRE: drought/xenobiotic-responsive elements; HNF1: hepatocyte nuclear factor 1α; PBREM: phenobarbital (PB)-responsive

enhancer module; PPAR: peroxisome proliferator-activated receptor; SIRT: silent information regulator; MAPK: mitogen-activated protein kinase; JNK: Jun kinase; ERK: extracellular signal-regulated kinase; mTOR: mechanistic Target of Rapamycin; AMPK: AMP-activated protein kinase; AKT: protein kinase B; PKC: protein kinase C; TLR4: toll-Like receptor 4; NFkB: nuclear factor-κB; PIP: phosphatidylinositol biphosphate; PI3K: phosphatidylinositol-3-kinase; IRs: insulin receptor; S6K: protein S6 kinase; FGF: fibroblast growth factor; Glut1: glucose transporter type 1.

The same effects on the activation of AMPK and mTOR inhibition due to BR were also observed in mice with NAFLD treated with CO, another product of the heme catabolic pathway [85]. PPARγ coactivator- α (PGC1α) is a transcription coactivator of PPARγ that plays a central role in the regulation of cellular energy metabolism [150]. It has been shown that hyperbilirubinemic subjects with Gilbert's syndrome have substantially lower BMI and serum concentrations of glucose, insulin, C-peptide, and triacylglycerol, the activation of AMPK, PPARα/γ, and peroxisome proliferator-activated receptor-gamma coactivator 1α (PgC1α) being considered the most important factors responsible for these observations [77]. Sirtuins (SIRTs) are a family of signaling proteins involved in metabolic regulation [151]; SIRT1, a deacetylase modulating PPARγ and PGC1α, hence controlling fat and glucose metabolism, is metabolically one of the most important. SIRT1, PGC1α, together with AMPK form an energy-sensing network that controls energy expenditure [152]. Interestingly, BR was demonstrated to upregulate SIRT1 in an experimental NAFLD in vitro study [87]. Furthermore, SIRT1 influences the acetylation of microtubule-associated tau (MAPT) that contributes to the preservation of neuronal cytoskeletal stability important for neuroprotection in cerebral ischemia/reperfusion [153]. All BR-modulated pathways described above are interlinked, as demonstrated by the role of the CD39 ectonucleotidase/adenosine pathway in immunity and inflammation, which is under the control of HMOX1 [154,155], and also have important effects on glucose metabolism and insulin signaling [156], the AMPK pathway [157], and the pathogenesis of liver diseases, such as metabolic dysfunction-associated fatty liver disease (MAFLD) [158,159], indicating the complexity of BR-related modulation of cell signaling. It should be noted that some of the anti-inflammatory effects of CO are mediated via its CD39 ectonucleotidase/adenosine pathway [79]. All of these experimental data fit in with the clinical observation of low PPARα expression, as well as systemic BR concentrations in obese subjects [160]. In addition, BR has also been shown to ameliorate experimental colitis by upregulation of CD39 [78]. The effect of the heme catabolic pathway on cell metabolism is probably even more complex. As proven in a liver-specific BLVRA knockout mouse model [91], BLVRA (and hence bilirubin, its enzymatic product) regulates hepatic lipid metabolism by directly affecting the key enzymes implicated in lipid metabolism. In addition to AMPK, these also include acetyl-CoA carboxylase, an essential and rate-limiting enzyme in fatty acid metabolism [161] and glycogen synthase kinase 3b (GSK3b), one of the most active cellular kinases, with more than 100 known targets, involved in the regulation of multiple cellular functions, including lipid and glucose metabolism, among others [162,163]. Thus, BR appears to act as a multifunctional modulator at multiple cellular metabolic checkpoints, which are often mutually interrelated. This is exemplified by bilirubin-induced suppression of p38 MAPK (but also other MAPKs) [164] in ischemia/reperfusion after heart transplantation, known to be crucial for insulin signaling, as well as atherogenesis [165].

Several natural products have been reported to activate both PPARα and PPARγ nuclear receptors [75,76], but also AMPK [90]. Similarly, a variety of natural products, including curcumin, resveratrol, quercetin, or apigenin, to mention at least some of them, have been shown to inhibit mTOR signaling [86] or activate SIRT1 [88] (summarized in Table 1).

## 3. Natural Compounds with Demonstrated Effects on Bilirubin and Its Metabolic Enzymes

### *3.1. Natural Compounds*

#### 3.1.1. Flavonoids

There is strong clinical evidence indicating that treatment with silymarin (a seed extract of milk thistle, *Silybium marianum* (L.) Gartn.) [166] flavonolignans results in mild elevation of TSB concentrations, as reported in silybin-treated patients with prostate cancer [167], as well as hepatitis C [168]. It is important to note that some of the minor silymarin flavonolignans are even more biologically active, as evidenced in our recent experimental study, in which dehydrosilybin A and B were potent partial inhibitors of UGT1A1 activity [166]. Flavonoids are also promising chemoadjuvants for the treatment of neurodegenerative diseases [169]. Fisetin is one of the examples of flavonoid tested with positive results in the experimental model of amyotrophic lateral sclerosis by inducing HMOX1 [170]. A natural aglycone flavonoid from the Erigeron plant, known as breviscapine, decreases neuronal apoptosis by promoting the expression of NRF2, followed by the increase in antioxidant enzymes, including HMOX1, in traumatic brain injury [54]. Eriodictyol, a flavonoid compound that is commonly present in the rinds of citrus fruits and certain traditional Chinese herbal remedies, also exhibits anti-AD's effect. It ameliorates memory impairment, inhibits Aβ aggregation, and decreases Tau phosphorylation in the brains of AD mice by modulating the Nrf2/HMOX1 signaling pathway [171].

Another compound in this category, luteolin, has also been proposed as a phototherapeutic agent for PD due to its antioxidant, anti-inflammatory, and anti-apoptosis effects. It can enhance the HMOX1 expression and the link between Nrf2 and its antioxidant element, providing a good adaptive survival response against oxidative damage. Furthermore, this compound also inhibits the expression of various pro-inflammatory compounds, after the LPS stimulus, such as inducible NO synthase (iNOS), TNFα, cyclooxygenase 2 (COX2), interleukin 1β (1Lβ), nitric oxide (NO), and prostaglandin E2 [60].

#### 3.1.2. Curcumin

Curcumin is a major polyphenolic compound from the curcuminoid group of phytochemicals, originating from the plant *Curcuma longa*. Curcumin is a potent inducer of HMOX1 with many clinical consequences. Its use as a nutraceutical has been investigated in several studies focused on the prevention of neurotoxicity and neurodegeneration (reviewed in [172]). Despite the small number of participants, a clinical trial of a curcumin formulation with enhanced oral bioavailability showed an improvement in survival probability over a 12-month period in amyotrophic lateral sclerosis [173,174]. In an in vitro model of the disease (microglia cells that have been activated by LPS), curcumin reduces the production of iNOS and NO release by increasing the expression of NRF2 and HMOX1 and downregulating the NFkB signaling pathway, which inhibits the release of pro-inflammatory cytokines IL6, IL1, and TNFα [175]. NFkβ/MAPK pathways have also been involved in curcumin-mediated anti-inflammatory action in microglia [61]. Similarly, in primary astrocytes, curcumin activates NRF2 target genes, including HMOX1 and NQO1 (NAD(P)H dehydrogenase quinone 1); lowers the amount of intracellular ROS; and lessens oxidative damage and mitochondrial dysfunction [176], as well as curcumin; it also induces an improvement of mercury-induced cytotoxicity via the NRF2/ARE/PKCδ pathway [64]. The antioxidant properties of curcumin have also been shown in neurons, by a NRF2/PKCδ-mediated induction of p62 (ubiquitin sensor and selective autophagy receptor) phosphorylation [70]. Although the in vitro neuroprotective action of curcumin (and astragaloside) in depression has been suggested to be mediated by the tyrosine protein kinase (TRK) β/MAPK/PI3K/CREB signaling pathways-induced upregulation of BDNF (brain-derived neurotrophic factor) [177], in in vivo models, the antidepressant properties of curcumin have been found to be involved via AKT1, NRF2, and ARE signaling [65,71]. Furthermore, in vivo, dietary supplementation with curcumin has been shown to reverse the microglial and astrocytes activation via NRF2/TLR4/NFkB/RAGE (receptor for advance glycation end products) signaling in a rodent model ethanol-induced neurotoxicity [68], and acts as an antioxidant in intracerebral hemorrhage, traumatic brain injury, quinoline acid-induced glutamate neurotoxicity, and ischemic injury models via the NRF2/HMOX1/AKT pathways [49,62,63,66,67,72]. Moreover, curcumin also exhibits anti-cancer effects by enhancing HMOX1 expression and activating ferroptosis, a form of oxidative cell death, in thyroid cancer [178].

#### 3.1.3. Astragaloside

Astragaloside IV (AST) is one of the main active ingredients of *Astragalus* membranaceus var. mongholicus, or A. membranaceus, a traditional Chinese herb. AST was found to exhibit neuroprotection in rats with ischemic stroke by upregulating SIRT1 expression, which promotes anti-inflammation and antioxidant production [153]. Similarly, AST reduces LPS-mediated neuroinflammation (microglia) by acting on the activation of NRF2, and HMOX1 via the ERK pathway [57]. AST also improves endothelial dysfunction in cardiovascular disease by activating the NRF2 signaling pathway and promoting the transcription of antioxidant enzymes including HMOX1 [179]. The combination of AST and Panax notoginseng results in protection in cerebral ischemia-reperfusion mice by increasing the nuclear translocation of Nrf2, followed by the increase of HMOX1, with a decrease of oxidative stress markers and an increase of the SOD and GSH level [180].

#### 3.1.4. Vitamins

Vitamin C consumption in combination with vitamin E for one month among AD patients could reduce oxidative stress markers in plasma and CSF (cerebrospinal fluid) [181]. Vitamin C indeed promotes antioxidant activity by activating NRF2 and increasing HMOX1 expressions [182]. It also acts against inflammation by decreasing TNF$\alpha$ and IL6 and increasing IL10 [183]. There is growing evidence that vitamin D deficiency can affect brain processes, including cognition, and significantly increase the risk of AD [184], speeding up the aging process through the disruption of redox cell signaling pathways [184]. In line, maxacalcitol, a vitamin D analogue, substantially lowered neuroinflammation and enhanced expression of NRF2 and its downstream effectors (HMOX1 and GSH) in the animal model for AD [185].

#### 3.1.5. Madecassoside

The Indian Ayurvedic system of medicine has utilized *Centella asiatica* to improve neurocognitive functioning. The most abundant triterpenoid saponin isolated from the plant, madecassoside, protects the neurons from apoptosis due to free radicals by increasing the antioxidant activity in the ALS mouse model. It has also been reported to improve LPS-mediated neurotoxicity in rats with its anti-inflammatory activities and activation of the Keap1-Nrf2/HMOX1 signaling pathway [186].

#### 3.1.6. Green Tea

The clinical study by Arab et al. showed that drinking green tea (leaf of *Camellia sinensis*) 2 g/day for two months has a positive impact on cognitive performance in severe AD patients by increasing the total antioxidant capacity by around 20% [187]. Additionally, epigallocatechin-3-gallate (EGCG), a bioactive compound of green tea, has been demonstrated to stimulate antioxidant defense mechanisms by activating the Nrf2/ARE pathway and antioxidant enzymes by activating AKT and ERK1/2 pathways [188]. In particular, EGCG functions as an antioxidant by modulating neurodegenerative inflammatory processes, such as ferroptosis and microglia-induced cytotoxicity, and by activating signaling pathways important for neuronal survival [174].

#### 3.1.7. S-Allyl Cysteine

S-allyl cysteine, a water-soluble organosulfur compound containing an amino acid isolated from the garlic bulb (*Allium sativum*), ameliorates cognitive deficits in streptozotocin-diabetic rats via suppression of oxidative stress, inflammation acting on the NRF2- HMOX1, TLR4, and NFκB signaling cascade [58].

#### 3.1.8. 20C

20C, a bibenzyl compound isolated from *Gastrodia elata* (a commonly used traditional Chinese medicine for therapeutic applications, like epilepsy and vertigo treatment), protects PC12 neurons from rotenone-induced lesions mimicking PD in vitro by normalizing (increasing) HMOX1 and NQO1 mRNA and the protein expression level, and by reverting ROS accumulation, cytochrome c release, and apoptosis via NRF2 signaling [47].

#### 3.1.9. *Achyranthes bidentata*

The extract *Achyrantes bidentata* polypeptide K (ABPK) from the *Amaranthaceae* family, another Chinese herbal medicine, reduces inflammation in LPS-challenged microglia (BV2 cells, NO, IL6, TNFα, PGF2) via NRF2/HMOX1 signaling [43].

#### 3.1.10. *Coriolus versicolor* and *Hericium erinaceus*

Consumption of certain mushrooms affects neurocognitive functioning. Two of the widely studied medicinal mushrooms are *Coriolus versicolor* and *Hericium erinaceus*. The combined extracts of these two mushrooms lower LXA4, a metabolic product of arachidonic acid, which is considered an endogenous "stop signal" for inflammation, astrocytes and microglia activation, α-synuclein content, apoptosis, and dopaminergic neuron death, finally improving the motor abilities in a rodent model of PD [48].

#### 3.1.11. Hyperoside (Quercetin 3-O-galactoside)

Hyperoside (quercetin 3-O-galactoside) from *Acer tegmentosum*, a Korean traditional medicine, has demonstrated antioxidant action in cultured neurons serving as an in vitro model of PD via NRF2/HMOX1 [52].

#### 3.1.12. Acerogin A

Acerogenin A, a natural compound from *Acer nikoense* Maxim used in Japanese folk medicine, has been shown to prevent glutamate neurotoxicity in vitro by acting through NRF2/HMOX1 and PIK3/Akt signaling, with glutamate neurotoxicity being a common pathological mechanism in many neurodegenerative and neurologic conditions [55].

#### 3.1.13. Kaempferol, Ginsenoside rh2

Kaempferol and ginsenoside rh2, the most active principle of the Kaixinsan formula, used as an antidepressant in traditional Chinese medicine, are effective in reverting the pro-oxidant cellular status of SH-SY5Y cells exposed to $H_2O_2$ by increasing thioredoxin reductase activity via NRF2/AKT [71].

#### 3.1.14. Mangiferin

Mangiferin is a natural compound originating from multiple plants, including *Mangifera indica* L. Through recent studies, it has been demonstrated to have an important role in protecting neurons from degeneration. In fact, it has some crucial antioxidant properties by preventing the formation of hydroxyl radicals and ROS [59]. Mangiferin has protective effects on PD in vitro and in vivo models by enhancing antioxidant defense, including the expression of NRF2 and HMOX1 [189]. Mangiferin also can downregulate NFκB in breast cancer, followed by increased apoptosis as one of the consequential effects [190].

#### 3.1.15. Ellagic Acid

Ellagic acid is a chromene-dione derivative ($C_{14}H_6O_8$) present in many fruits and nuts, such as pomegranates, black raspberries, raspberries, strawberries, walnuts, and almonds. It is a phenolic antioxidant with higher antioxidant activity than vitamin E, succinate, and melatonin. It can also inhibit the major pro-inflammatory pathways, such as NFkB, MPAKs, and JAK/STAT, preventing inflammatory molecules' release (e.g., TNFα, IL1β, IL6, and iNOX) [69].

### 3.2. Natural Compounds Mimicking Products of the Heme Catabolic Pathway

#### 3.2.1. Tetrapyrroles from *Spirulina platensis*, Phycocyanin, and Phycocyanobilin

*Spirulina (Arthrospira) platensis* is a blue-green freshwater alga widely used as a dietary supplement. It is rich in proteins, carotenoids, essential fatty acids, vitamin B complex, vitamin E, and minerals, such as copper, manganese, magnesium, iron, selenium, and zinc. It is a source of potent antioxidants, including spirulans (sulfated polysaccharides), seleno compounds, phenolic compounds, and phycobiliproteins (C-phycocyanin and allophycocyanin) [191]. Numerous studies have demonstrated that dietary supplementation of *S. platensis* is helpful in the prevention and treatment of atherosclerosis, diabetes, and/or cancers (for review see [192]). C-phycocyanin is a light-harvesting biliprotein that is possibly implicated in the biological effects of *S. platensis*. C-phycocyanin contains a covalently linked chromophore called phycocyanobilin (PCB) [193], a linear tetrapyrrolic molecule structurally similar to BV that constitutes up to 1% of the dry weight of *Spirulina*. Interestingly, phycocyanobilin (PCB) can be metabolized by BLVR to phycocyanorubin, similarly as BV is reduced to BR in the human body [194]. Therefore, *S. platensis* appears to be a rich nutraceutical source of PCB [195], which is known to significantly inhibit nicotinamide adenine dinucleotide phosphate (NADPH) oxidase activity, activate the action of antioxidant enzymes, and have anti-inflammatory and cell signaling activities with an expected substantial impact on human health, including fighting the neurologic adverse effects of COVID-19 [73,196,197].

We have shown that *S. platensis* and PCB markedly increased Hmox1 in experimental models of atherosclerosis [198]. Therefore, activation of HMOX1 and the heme catabolic pathway appears to be an important mechanism of this food supplement for the reduction of atherosclerotic diseases. Due to their antioxidant effects, C-phycocyanin and PCB (but also BR and BV) protected diabetic db/db mice against albuminuria and renal mesangial expansion in db/db mice, and normalized tumor growth factor-β and fibronectin expression, suggesting a novel and feasible therapeutic approach to prevent diabetic nephropathy [199]. Importantly, antidiabetic effects of *S. platensis*, C-phycocyanin, and PCB were reported in diabetic male albino rats treated with streptozotocin, with decreases of blood glucose concentrations, improvement of insulin resistance and blood lipids levels, and restoration of pancreatic cell morphology [200].

The anticancer effects of *S. platensis* and its tetrapyrrolic components (PCB and chlorophyllin, a surrogate molecule for chlorophyll A) in biologically relevant doses were proved on experimental pancreatic cancer models in another study. These models included nude mice xenotransplanted with pancreatic cancer cells, suggesting a chemopreventive role of this edible alga, whose dietary supplementation with this alga might enhance the systemic pool of linear tetrapyrroles mimicking BR [201]. Similarly, in an experimental in vitro study by Hussein et al., C-phycocyanin and PCB treatment led to potent antiproliferative, and pro-apoptotic effects in MCF-7 breast cancer cells [202].

PCB also showed strong anti-inflammatory and general hepatoprotective effects in mice with CCl4-induced liver injury, with a marked improvement inf the survival rate of acute liver failure in mice injected with a lethal dose of CCl4 [203].

Importantly, *Spirulina alga* and its associated bioactive compounds appear to also be important in chemoprevention of neurodegenerative diseases [73,204]. Activated microglia, displaying NADPH oxidase activity, are believed to contribute substantially to the pathogenesis of many brain diseases, such as PD and AD, and multiple sclerosis, but also cerebral ischemia [205–207] or COVID-19-induced damage of the nervous system [73]. Although direct clinical evidence is lacking, experimental data suggest that *S. platensis* extracts could ameliorate the risks of these diseases [208]. A diet high in *Spirulina* ameliorates the loss of dopaminergic neurons in the mouse model of PD [209]. Furthermore, PCB improved the neurological outcomes in a mouse model of experimental autoimmune encephalitis (EAE) that mimics multiple sclerosis conditions, with significant inhibitory effects on pro-inflammatory cytokines. Interestingly, a reduction of demyelination, active microglia/macrophages density, and axonal damage was detected, along with an increase in oligodendrocyte precursor cells and mature oligodendrocytes, when assessing the spinal cords of EAE mice treated with PCB. Due to the increased treatment effects of PCB when used with IFNβ therapy, PCB was suggested for use with IFNβ as a disease-modifying agent for multiple sclerosis [210]. Similar results were found also in other EAE experimental studies, demonstrating a clear tendency for amelioration of the clinical severity of the disease promoted by the treatment with PCB in an EAE model with a reduction in the levels of the pro-inflammatory cytokines IL17, IFNγ, and IL-6 in the brains of animals treated with PCB. Similar observations were also obtained in animal models of cerebral ischemia [205,211]. The demyelination potential of C-phycocyanin and PCB with their possible clinical use in patients with demyelinization disorders has been reviewed recently [207].

PCB also was shown to have potential inhibitor activity against main protease (Mpro) and papain-like protease (PLpro) of human and animal coronaviruses, indicating broad-spectrum inhibitor activity of PCB. In addition, in silico studies with the Mpro and PLpro enzymes revealed that other tetrapyrrolic phycobilins, such as phycourobilin and phycoerythrobilin, will have similar binding affinity to SARS-CoV-2 Mpro and PLpro [212]. Hence, PCB and *S. platensis* were suggested as promising nutraceuticals against COVID-19-induced brain damage [73]. In addition, antiviral activities of *S. platensis*-derived compounds were shown in a study by Chen et al., who reported inhibitory effects of Spirulina extract against influenza virus replication and reduction of virus-induced mortality [213]. In in vitro studies, antiviral activities of *S. platensis* extract were also reported against herpetic viruses, measles, and mumps viruses, as well as human immunodeficiency virus (for a review, see [214]). Interestingly, in a small clinical study, *S. platensis* consumption for 12 months improved the immunological profile of HIV-infected patients [215]. It should be noted that in countries with high algae consumption, such as Japan, Korea, or Chad, the prevalence rate of HIV infection and AIDS remains substantially lower compared to other countries [214,216].

Due to the potent antioxidant, antibacterial, and other beneficial biological effects, extracts from *S. platensis* exhibits have been used in skincare formulations for the treatment of acne [217], sunscreens (*S. platensis* having a very high sun protecting factor, SPF) [218], face masks lip balms, and ointments for their wound-healing and anti-aging properties [219]. Therefore, the global market is estimated to be as much as USD 2000 million by 2026, with an overall *S. platensis* consumption of more than 321,000 tons [220].

#### 3.2.2. Artificial Bezoar

BR is present also in pulverized bovine gallstones (*Calculus bovis*, artificial bezoar, *Niu Huang* in Chinese) used for centuries for a variety of human diseases in traditional Chinese medicine [221,222]. Preparations of these pulverized gallstones have a high content of BR of no less than 25% by weight [222].

Neuroprotective effects of artificial bezoar were reported in an experimental study on male Sprague-Dawley rats with induced cerebral ischemia. Pretreatment with An-Gong-Niu-Huang Wan, a complex traditional Chinese medicine formulation containing Calculus bovis, significantly ameliorated ischemic damage to the brain in a dose-dependent manner, including a reduction in the neurological deficit score and infarct area [223]. Other reported activities of *Calculus bovis* on the nervous system include sedative, anticonvulsant, analgesic, antiepileptic, or even anti-schizophrenic (reviewed in [222]). Furthermore, *Calculus bovis* was found to also have protective effects on the heart, vessels, lungs, or liver [222], or even experimental cancer, as shown for experimental breast cancer, human hepatoblastoma [224].

3.2.3. Chlorophylls

Chlorophyll is a tetrapyrrolic compound structurally related to BR [7] that is likely to exert similar protective biological activities. Dietary intake of green leafy vegetables rich in chlorophyll is associated with protection against cancer and other civilization diseases, including neurodegenerative diseases [50,225]. Although their resorption from the gut lumen is not high, chlorophylls appear to be important for their potential systemic cancer-preventive effects [226], reaching biologically relevant concentrations even in peripheral tissues [227]. Anti-proliferative and antioxidant effects of chlorophylls (chlorophyll a/b, chlorophyllin, and pheophytin a) were reported recently in our experimental in vitro and in vivo pancreatic cancer study [228].

## 4. Natural Compounds Targeting the Heme Catabolic Pathway

As described in detail in our recent review [14], targeting the individual parts of the heme catabolic pathway represents an important way to increase tissue or systemic levels of BR.

### *4.1. Modulation of HMOX1*

As emphasized above, HMOX1 is inducible by a variety of natural products and is an important chemotherapeutic target for the prevention and treatment of many neurodegenerative and other autoimmune diseases [19,45,46]. These natural HMOX1 modulators include various polyphenolic compounds found in plants, such as curcuminoids; flavonoids, such as quercetin, EGCC, genistein, eriodictyol, or certain flavonolignans of silymarin complex, caffeic acid, and resveratrol; and natural coumarins, such as esculetin or fraxetin [3,229,230].

### *4.2. Modulation of BLVRA*

BLVRA induction seems always beneficial, while its enzymatic inactivation looks detrimental, possibly by reducing the final concentration of unconjugated bilirubin (UCB) inside the cell [9]. Interestingly, BLVRA expression is also inducible by natural products, leading to increased production of BR, as demonstrated for Korean red ginseng in an in vitro study in murine hippocampal astrocytes improving mitochondrial functions via the LKB1-SIRT1-ERR$\alpha$ axis [231].

### *4.3. Modulation of the Hepatic Transport of Bilirubin*

Inhibition of organic anion transporter 1B1 (OATP1B1), a transporter crucial for UCB uptake in the liver, by compounds or substances relying on this transporter for transport could lead to a substantial increase in bilirubin concentrations [3]. Basolateral uptake of bilirubin in liver cells is another druggable target to increase mild serum BR concentrations. In fact, interference with OATP1B1 bilirubin transporter was reported for many natural compounds commonly present in certain foods, food supplements, and herbs [3,232].

### *4.4. UGT1A1 Modulation*

As UGT1A1 is responsible for intrahepatic BR conjugation, its suppression will lead to an increase in BR level [3]. Many natural compounds, often used as nutraceuticals, have UGT1A1-suppressing activity resulting in mild elevation of systemic BR concentrations, including silymarin flavonolignans or EGCC, or many Japanese and Chinese herbs commonly used in traditional herbal medicine [3].

### *4.5. Gut Microbiome Modulation*

By affecting the enterohepatic and enterosystemic circulation of BR [233], BR-reducing bacteria in the gut lumen can affect systemic concentrations of the pigment [234]. Although no experimental and clinical data on possible modulation of gut microbiome metabolizing bilirubin are available today, likely, certain probiotics or other modulating agents capable of affecting intestinal metabolism of bilirubin with possible beneficial clinical consequences will become available [235].

## 5. Conclusions

The beneficial potential of bilirubin in human health is clear in large measure. The modulability of HMOX1 and its metabolic checkpoints by natural compounds makes this system a great and feasible target to combat various public health concerns and age-related diseases with minimal changes in the diet. Further studies, and collaborations with the food or pharma-nutraceutical industry, are needed to specify all this solid information in a varied and easy-to-take functionalized alimentary regimen to promote health day by day.

**Author Contributions:** Conceptualization, L.V., C.T. and S.G.; data curation, L.V., S.J., C.D.V., J.P.L., C.S., C.T. and S.G.; writing—original draft preparation, L.V., S.J., C.D.V., J.P.L. and S.G.; writing—review and editing, S.J., L.V., C.S., C.T. and S.G.; visualization, S.J. and S.G. All authors have read and agreed to the published version of the manuscript.

**Funding:** S.G., C.T. and C.D.V. have been funded by an internal grant from Fondazione Italiana Fegato (FIF). J.P.L. has been funded in part by an internal grant from Fondazione Italiana Fegato (FIF), in part by the Department of Science and Technology through the Philippine Council for Health Research and Development (DOST-PCHRD). S.J. has been funded by a 2023/2024 postdoctoral fellowship of Manajenen Talenta, Badan Riset dan Inovasi Nasional, Indonesia. C.S. has been funded in part by a 2023 grant of the Fondazione Veronesi, Milan, Italy, in part from the National Research and Innovation Agency of Indonesia. L.V. has been supported by grants MH CZ-DRO-VFN64165 from the Czech Ministry of Health, Cooperation Program, research area DIAG given by Charles University, and the project National Institute for Research of Metabolic and Cardiovascular Diseases (Programme EXCELES LX22NPO5104) funded by the European Union—Next Generation EU.

**Data Availability Statement:** All the data are inside the review.

**Conflicts of Interest:** The authors declare no conflicts of interest.

## References

1. Gazzin, S.; Vitek, L.; Watchko, J.; Shapiro, S.M.; Tiribelli, C. A Novel Perspective on the Biology of Bilirubin in Health and Disease. *Trends Mol. Med.* **2016**, *22*, 758–768. [CrossRef] [PubMed]
2. Wagner, K.-H.; Wallner, M.; Mölzer, C.; Gazzin, S.; Bulmer, A.C.; Tiribelli, C.; Vitek, L. Looking to the Horizon: The Role of Bilirubin in the Development and Prevention of Age-Related Chronic Diseases. *Clin. Sci.* **2015**, *129*, 1–25. [CrossRef] [PubMed]
3. Vitek, L.; Bellarosa, C.; Tiribelli, C. Induction of Mild Hyperbilirubinemia: Hype or Real Therapeutic Opportunity? *Clin. Pharmacol. Ther.* **2019**, *106*, 568–575. [CrossRef] [PubMed]
4. Creeden, J.F.; Gordon, D.M.; Stec, D.E.; Hinds, T.D. Bilirubin as a Metabolic Hormone: The Physiological Relevance of Low Levels. *Am. J. Physiol.-Endocrinol. Metab.* **2020**, *320*, E191–E207. [CrossRef]
5. Jayanti, S.; Dalla Verde, C.; Tiribelli, C.; Gazzin, S. Inflammation, Dopaminergic Brain and Bilirubin. *Int. J. Mol. Sci.* **2023**, *24*, 11478. [CrossRef] [PubMed]
6. Llido, J.P.; Jayanti, S.; Tiribelli, C.; Gazzin, S. Bilirubin and Redox Stress in Age-Related Brain Diseases. *Antioxidants* **2023**, *12*, 1525. [CrossRef] [PubMed]
7. Ostrow, J.D.; Vitek, L. Bilirubin Chemistry and Metabolism; Harmful and Protective Aspects. *Curr. Pharm. Design* **2009**, *15*, 2869–2883.
8. Ryter, S.W.; Otterbein, L.E. Carbon Monoxide in Biology and Medicine. *BioEssays* **2004**, *26*, 270–280. [CrossRef]
9. Wegiel, B.; Otterbein, L. Go Green: The Anti-Inflammatory Effects of Biliverdin Reductase. *Front. Pharmacol.* **2012**, *3*, 47. [CrossRef]
10. Gazzin, S.; Masutti, F.; Vítek, L.; Tiribelli, C. The Molecular Basis of Jaundice: An Old Symptom Revisited. *Liver Int.* **2016**, *37*, 1094–1102. [CrossRef]
11. Vítek, L. Bilirubin and Atherosclerotic Diseases. *Physiol. Res.* **2017**, *66*, S11–S20. [CrossRef] [PubMed]
12. Vítek, L. Bilirubin as a Signaling Molecule. *Med. Res. Rev.* **2020**, *40*, 1335–1351. [CrossRef] [PubMed]
13. Vítek, L. The Protective Role of the Heme Catabolic Pathway in Hepatic Disorders. *Antioxid. Redox Signal.* **2021**, *35*, 734–752. [CrossRef] [PubMed]
14. Vítek, L. Biliverdin and Bilirubin as Parallel Products of CO Formation. In *Carbon Monoxide in Drug Discovery*; John Wiley & Sons, Ltd.: Hoboken, NJ, USA, 2022; pp. 175–194. [CrossRef]
15. Ryter, S.W.; Alam, J.; Choi, A.M.K. Heme Oxygenase-1/Carbon Monoxide: From Basic Science to Therapeutic Applications. *Physiol. Rev.* **2006**, *86*, 583–650. [CrossRef] [PubMed]
16. Wilks, A.; Knör, G.; Wu, H.; Zheng, Y.; Liu, J.; Zhang, H.; Chen, H.; Buchberger, T.; Lamparter, T.; Estes, S.; et al. Heme Oxygenase: Evolution, Structure, and Mechanism. *Antioxid. Redox Signal.* **2002**, *4*, 603–614. [CrossRef] [PubMed]
17. Maines, M.D. THE HEME OXYGENASE SYSTEM: A Regulator of Second Messenger Gases. *Annu. Rev. Pharmacol. Toxicol.* **1997**, *37*, 517–554. [CrossRef] [PubMed]

18. Morse, D.; Choi, A.M.K. Heme Oxygenase-1. *Am. J. Respir. Cell Mol. Biol.* **2002**, *27*, 8–16. [CrossRef]
19. Funes, S.C.; Rios, M.; Fernández-Fierro, A.; Covián, C.; Bueno, S.M.; Riedel, C.A.; Mackern-Oberti, J.P.; Kalergis, A.M. Naturally Derived Heme-Oxygenase 1 Inducers and Their Therapeutic Application to Immune-Mediated Diseases. *Front. Immunol.* **2020**, *11*, 1467. [CrossRef]
20. Ma, Q. Role of Nrf2 in Oxidative Stress and Toxicity. *Annu. Rev. Pharmacol. Toxicol.* **2013**, *53*, 401–426. [CrossRef]
21. Loboda, A.; Damulewicz, M.; Pyza, E.; Jozkowicz, A.; Dulak, J. Role of Nrf2/HO-1 System in Development, Oxidative Stress Response and Diseases: An Evolutionarily Conserved Mechanism. *Cell. Mol. Life Sci.* **2016**, *73*, 3221–3247. [CrossRef]
22. da Costa, R.M.; Rodrigues, D.; Pereira, C.A.; Silva, J.F.; Alves, J.V.; Lobato, N.S.; Tostes, R.C. Nrf2 as a Potential Mediator of Cardiovascular Risk in Metabolic Diseases. *Front. Pharmacol.* **2019**, *10*, 382. [CrossRef] [PubMed]
23. Zhang, D.D.; Chapman, E. The Role of Natural Products in Revealing NRF2 Function. *Nat. Prod. Rep.* **2020**, *37*, 797–826. [CrossRef] [PubMed]
24. Qader, M.; Xu, J.; Yang, Y.; Liu, Y.; Cao, S. Natural Nrf2 Activators from Juices, Wines, Coffee, and Cocoa. *Beverages* **2020**, *6*, 68. [CrossRef]
25. Zhang, J.; Xu, H.-X.; Zhu, J.-Q.; Dou, Y.-X.; Xian, Y.-F.; Lin, Z.-X. Natural Nrf2 Inhibitors: A Review of Their Potential for Cancer Treatment. *Int. J. Biol. Sci.* **2023**, *19*, 3029–3041. [CrossRef] [PubMed]
26. Robledinos-Antón, N.; Fernández-Ginés, R.; Manda, G.; Cuadrado, A. Activators and Inhibitors of NRF2: A Review of Their Potential for Clinical Development. *Oxid. Med. Cell Longev.* **2019**, *2019*, 9372182. [CrossRef] [PubMed]
27. Singh, S.; Nagalakshmi, D.; Sharma, K.K.; Ravichandiran, V. Natural Antioxidants for Neuroinflammatory Disorders and Possible Involvement of Nrf2 Pathway: A Review. *Heliyon* **2021**, *7*, e06216. [CrossRef] [PubMed]
28. Moratilla-Rivera, I.; Sánchez, M.; Valdés-González, J.A.; Gómez-Serranillos, M.P. Natural Products as Modulators of Nrf2 Signaling Pathway in Neuroprotection. *Int. J. Mol. Sci.* **2023**, *24*, 3748. [CrossRef]
29. Puentes-Pardo, J.D.; Moreno-SanJuan, S.; Carazo, Á.; León, J. Heme Oxygenase-1 in Gastrointestinal Tract Health and Disease. *Antioxidants* **2020**, *9*, 1214. [CrossRef]
30. Song, W.; Zukor, H.; Lin, S.-H.; Hascalovici, J.; Liberman, A.; Tavitian, A.; Mui, J.; Vali, H.; Tong, X.-K.; Bhardwaj, S.K.; et al. Schizophrenia-Like Features in Transgenic Mice Overexpressing Human HO-1 in the Astrocytic Compartment. *J. Neurosci.* **2012**, *32*, 10841–10853. [CrossRef]
31. Laskaris, L.E.; Di Biase, M.A.; Everall, I.; Chana, G.; Christopoulos, A.; Skafidas, E.; Cropley, V.L.; Pantelis, C. Microglial Activation and Progressive Brain Changes in Schizophrenia. *Br. J. Pharmacol.* **2016**, *173*, 666–680. [CrossRef]
32. Zhuo, C.; Tian, H.; Song, X.; Jiang, D.; Chen, G.; Cai, Z.; Ping, J.; Cheng, L.; Zhou, C.; Chen, C. Microglia and Cognitive Impairment in Schizophrenia: Translating Scientific Progress into Novel Therapeutic Interventions. *Schizophrenia* **2023**, *9*, 42. [CrossRef] [PubMed]
33. Li, J.; Wang, Y.; Yuan, X.; Kang, Y.; Song, X. New Insight in the Cross-Talk between Microglia and Schizophrenia: From the Perspective of Neurodevelopment. *Front. Psychiatry* **2023**, *14*, 1126632. [CrossRef] [PubMed]
34. Fleiss, B.; Van Steenwinckel, J.; Bokobza, C.; Shearer, I.K.; Ross-Munro, E.; Gressens, P. Microglia-Mediated Neurodegeneration in Perinatal Brain Injuries. *Biomolecules* **2021**, *11*, 99. [CrossRef] [PubMed]
35. Schipper, H.M.; Song, W.; Tavitian, A.; Cressatti, M. The Sinister Face of Heme Oxygenase-1 in Brain Aging and Disease. *Progress. Neurobiol.* **2019**, *172*, 40–70. [CrossRef] [PubMed]
36. Bereczki, D.; Balla, J.; Bereczki, D. Heme Oxygenase-1: Clinical Relevance in Ischemic Stroke. *Curr. Pharm. Des.* **2018**, *24*, 2229–2235. [CrossRef] [PubMed]
37. Suntar, I.; Sureda, A.; Belwal, T.; Sanches Silva, A.; Vacca, R.A.; Tewari, D.; Sobarzo-Sánchez, E.; Nabavi, S.F.; Shirooie, S.; Dehpour, A.R.; et al. Natural Products, PGC-1 $\alpha$, and Duchenne Muscular Dystrophy. *Acta Pharm. Sin. B* **2020**, *10*, 734–745. [CrossRef] [PubMed]
38. Bottari, N.B.; Reichert, K.P.; Fracasso, M.; Dutra, A.; Assmann, C.E.; Ulrich, H.; Schetinger, M.R.C.; Morsch, V.M.; Da Silva, A.S. Neuroprotective Role of Resveratrol Mediated by Purinergic Signalling in Cerebral Cortex of Mice Infected by Toxoplasma Gondii. *Parasitol. Res.* **2020**, *119*, 2897–2905. [CrossRef]
39. Costa, P.; Gonçalves, J.; Baldissarelli, J.; Mann, T.; Abdalla, F.; Fiorenza, A.; Rosa, M.; Carvalho, F.; Gutierres, J.; Andrade, C.; et al. Curcumin Attenuates Memory Deficits and the Impairment of Cholinergic and Purinergic Signaling in Rats Chronically Exposed to Cadmium. *Environ. Toxicol.* **2015**, *32*, 70–83. [CrossRef]
40. Zhang, Y.; Zhang, Y.; Jin, X.; Zhou, X.; Dong, X.; Yu, W.; Gao, W. The Role of Astragaloside IV against Cerebral Ischemia/Reperfusion Injury: Suppression of Apoptosis via Promotion of P62-LC3-Autophagy. *Molecules* **2019**, *24*, 1838. [CrossRef]
41. Nitti, M.; Piras, S.; Brondolo, L.; Marinari, U.M.; Pronzato, M.A.; Furfaro, A.L. Heme Oxygenase 1 in the Nervous System: Does It Favor Neuronal Cell Survival or Induce Neurodegeneration? *Int. J. Mol. Sci.* **2018**, *19*, 2260. [CrossRef]
42. Dwyer, B.E.; Nishimura, R.N.; Lu, S.Y. Differential Expression of Heme Oxygenase-1 in Cultured Cortical Neurons and Astrocytes Determined by the Aid of a New Heme Oxygenase Antibody. Response to Oxidative Stress. *Brain Res. Mol. Brain Res.* **1995**, *30*, 37–47. [CrossRef] [PubMed]
43. Cheng, Q.; Shen, Y.; Cheng, Z.; Shao, Q.; Wang, C.; Sun, H.; Zhang, Q. Achyranthes Bidentata Polypeptide k Suppresses Neuroinflammation in BV2 Microglia through Nrf2-Dependent Mechanism. *Ann. Transl. Med.* **2019**, *7*, 575. [CrossRef]

44. Jayanti, S.; Vítek, L.; Tiribelli, C.; Gazzin, S. The Role of Bilirubin and the Other "Yellow Players" in Neurodegenerative Diseases. *Antioxidants* **2020**, *9*, 900. [CrossRef] [PubMed]
45. Sahebnasagh, A.; Eghbali, S.; Saghafi, F.; Sureda, A.; Avan, R. Neurohormetic Phytochemicals in the Pathogenesis of Neurodegenerative Diseases. *Immun. Ageing* **2022**, *19*, 36. [CrossRef] [PubMed]
46. Marino, A.; Battaglini, M.; Moles, N.; Ciofani, G. Natural Antioxidant Compounds as Potential Pharmaceutical Tools against Neurodegenerative Diseases. *ACS Omega* **2022**, *7*, 25974–25990. [CrossRef] [PubMed]
47. Huang, J.-Y.; Yuan, Y.-H.; Yan, J.-Q.; Wang, Y.-N.; Chu, S.-F.; Zhu, C.-G.; Guo, Q.-L.; Shi, J.-G.; Chen, N.-H. 20C, a Bibenzyl Compound Isolated from Gastrodia Elata, Protects PC12 Cells against Rotenone-Induced Apoptosis via Activation of the Nrf2/ARE/HO-1 Signaling Pathway. *Acta Pharmacol. Sin.* **2016**, *37*, 731–740. [CrossRef]
48. Cordaro, M.; Modafferi, S.; D'Amico, R.; Fusco, R.; Genovese, T.; Peritore, A.F.; Gugliandolo, E.; Crupi, R.; Interdonato, L.; Di Paola, D.; et al. Natural Compounds Such as *Hericium erinaceus* and *Coriolus versicolor* Modulate Neuroinflammation, Oxidative Stress and Lipoxin A4 Expression in Rotenone-Induced Parkinson's Disease in Mice. *Biomedicines* **2022**, *10*, 2505. [CrossRef]
49. Duan, C.; Wang, H.; Jiao, D.; Geng, Y.; Wu, Q.; Yan, H.; Li, C. Curcumin Restrains Oxidative Stress of After Intracerebral Hemorrhage in Rat by Activating the Nrf2/HO-1 Pathway. *Front. Pharmacol.* **2022**, *13*, 889226. [CrossRef]
50. Trock, B.; Lanza, E.; Greenwald, P. Dietary Fiber, Vegetables, and Colon Cancer: Critical Review and Meta-Analyses of the Epidemiologic Evidence. *J. Natl. Cancer Inst.* **1990**, *82*, 650–661. [CrossRef]
51. Qaisiya, M.; Coda Zabetta, C.D.; Bellarosa, C.; Tiribelli, C. Bilirubin Mediated Oxidative Stress Involves Antioxidant Response Activation via Nrf2 Pathway. *Cell. Signal.* **2014**, *26*, 512–520. [CrossRef]
52. Kwon, S.-H.; Lee, S.R.; Park, Y.J.; Ra, M.; Lee, Y.; Pang, C.; Kim, K.H. Suppression of 6-Hydroxydopamine-Induced Oxidative Stress by Hyperoside via Activation of Nrf2/HO-1 Signaling in Dopaminergic Neurons. *Int. J. Mol. Sci.* **2019**, *20*, 5832. [CrossRef] [PubMed]
53. Zhang, C.; Li, C.; Chen, S.; Li, Z.; Jia, X.; Wang, K.; Bao, J.; Liang, Y.; Wang, X.; Chen, M.; et al. Berberine Protects against 6-OHDA-Induced Neurotoxicity in PC12 Cells and Zebrafish through Hormetic Mechanisms Involving PI3K/AKT/Bcl-2 and Nrf2/HO-1 Pathways. *Redox Biol.* **2017**, *11*, 1–11. [CrossRef] [PubMed]
54. Li, F.; Wang, X.; Zhang, Z.; Gao, P.; Zhang, X. Breviscapine Provides a Neuroprotective Effect after Traumatic Brain Injury by Modulating the Nrf2 Signaling Pathway. *J. Cell Biochem.* **2019**, *120*, 14899–14907. [CrossRef] [PubMed]
55. Lee, D.-S.; Cha, B.-Y.; Woo, J.-T.; Kim, Y.-C.; Jang, J.-H. Acerogenin A from *Acer nikoense* Maxim Prevents Oxidative Stress-Induced Neuronal Cell Death through Nrf2-Mediated Heme Oxygenase-1 Expression in Mouse Hippocampal HT22 Cell Line. *Molecules* **2015**, *20*, 12545–12557. [CrossRef] [PubMed]
56. Gu, D.-M.; Lu, P.-H.; Zhang, K.; Wang, X.; Sun, M.; Chen, G.-Q.; Wang, Q. EGFR Mediates Astragaloside IV-Induced Nrf2 Activation to Protect Cortical Neurons against in vitro Ischemia/Reperfusion Damages. *Biochem. Biophys. Res. Commun.* **2015**, *457*, 391–397. [CrossRef]
57. Li, C.; Yang, F.; Liu, F.; Li, D.; Yang, T. NRF2/HO-1 Activation via ERK Pathway Involved in the Anti-Neuroinflammatory Effect of Astragaloside IV in LPS Induced Microglial Cells. *Neurosci. Lett.* **2018**, *666*, 104–110. [CrossRef] [PubMed]
58. Baluchnejadmojarad, T.; Kiasalari, Z.; Afshin-Majd, S.; Ghasemi, Z.; Roghani, M. S-Allyl Cysteine Ameliorates Cognitive Deficits in Streptozotocin-Diabetic Rats via Suppression of Oxidative Stress, Inflammation, and Acetylcholinesterase. *Eur. J. Pharmacol.* **2017**, *794*, 69–76. [CrossRef]
59. Feng, S.-T.; Wang, Z.-Z.; Yuan, Y.-H.; Sun, H.-M.; Chen, N.-H.; Zhang, Y. Mangiferin: A Multipotent Natural Product Preventing Neurodegeneration in Alzheimer's and Parkinson's Disease Models. *Pharmacol. Res.* **2019**, *146*, 104336. [CrossRef]
60. Siddique, Y.H. Role of Luteolin in Overcoming Parkinson's Disease. *Biofactors* **2021**, *47*, 198–206. [CrossRef]
61. Jin, M.; Park, S.Y.; Shen, Q.; Lai, Y.; Ou, X.; Mao, Z.; Lin, D.; Yu, Y.; Zhang, W. Anti-Neuroinflammatory Effect of Curcumin on Pam3CSK4-Stimulated Microglial Cells. *Int. J. Mol. Med.* **2018**, *41*, 521–530. [CrossRef]
62. Dai, W.; Wang, H.; Fang, J.; Zhu, Y.; Zhou, J.; Wang, X.; Zhou, Y.; Zhou, M. Curcumin Provides Neuroprotection in Model of Traumatic Brain Injury via the Nrf2-ARE Signaling Pathway. *Brain Res. Bull.* **2018**, *140*, 65–71. [CrossRef] [PubMed]
63. Dong, W.; Yang, B.; Wang, L.; Li, B.; Guo, X.; Zhang, M.; Jiang, Z.; Fu, J.; Pi, J.; Guan, D.; et al. Curcumin Plays Neuroprotective Roles against Traumatic Brain Injury Partly via Nrf2 Signaling. *Toxicol. Appl. Pharmacol.* **2018**, *346*, 28–36. [CrossRef] [PubMed]
64. Yang, B.; Yin, C.; Zhou, Y.; Wang, Q.; Jiang, Y.; Bai, Y.; Qian, H.; Xing, G.; Wang, S.; Li, F.; et al. Curcumin Protects against Methylmercury-Induced Cytotoxicity in Primary Rat Astrocytes by Activating the Nrf2/ARE Pathway Independently of PKCδ. *Toxicology* **2019**, *425*, 152248. [CrossRef]
65. Liao, D.; Lv, C.; Cao, L.; Yao, D.; Wu, Y.; Long, M.; Liu, N.; Jiang, P. Curcumin Attenuates Chronic Unpredictable Mild Stress-Induced Depressive-Like Behaviors via Restoring Changes in Oxidative Stress and the Activation of Nrf2 Signaling Pathway in Rats. *Oxidative Med. Cell. Longev.* **2020**, *2020*, e9268083. [CrossRef] [PubMed]
66. Santana-Martínez, R.A.; Silva-Islas, C.A.; Fernández-Orihuela, Y.Y.; Barrera-Oviedo, D.; Pedraza-Chaverri, J.; Hernández-Pando, R.; Maldonado, P.D. The Therapeutic Effect of Curcumin in Quinolinic Acid-Induced Neurotoxicity in Rats Is Associated with BDNF, ERK1/2, Nrf2, and Antioxidant Enzymes. *Antioxidants* **2019**, *8*, 388. [CrossRef] [PubMed]
67. Huang, T.; Zhao, J.; Guo, D.; Pang, H.; Zhao, Y.; Song, J. Curcumin Mitigates Axonal Injury and Neuronal Cell Apoptosis through the PERK/Nrf2 Signaling Pathway Following Diffuse Axonal Injury. *NeuroReport* **2018**, *29*, 661. [CrossRef] [PubMed]

68. Ikram, M.; Saeed, K.; Khan, A.; Muhammad, T.; Khan, M.S.; Jo, M.G.; Rehman, S.U.; Kim, M.O. Natural Dietary Supplementation of Curcumin Protects Mice Brains against Ethanol-Induced Oxidative Stress-Mediated Neurodegeneration and Memory Impairment via Nrf2/TLR4/RAGE Signaling. *Nutrients* **2019**, *11*, 1082. [CrossRef] [PubMed]
69. Zhu, H.; Yan, Y.; Jiang, Y.; Meng, X. Ellagic Acid and Its Anti-Aging Effects on Central Nervous System. *Int. J. Mol. Sci.* **2022**, *23*, 10937. [CrossRef]
70. Park, J.-Y.; Sohn, H.-Y.; Koh, Y.H.; Jo, C. Curcumin Activates Nrf2 through PKCδ-Mediated P62 Phosphorylation at Ser351. *Sci. Rep.* **2021**, *11*, 8430. [CrossRef]
71. Li, X.; Wang, Y.; Wang, C.; Jing, R.; Mu, L.; Liu, P.; Hu, Y. Antidepressant Mechanism of Kaixinsan and Its Active Compounds Based on Upregulation of Antioxidant Thioredoxin. *Evid. -Based Complement. Altern. Med.* **2022**, *2022*, e7302442. [CrossRef]
72. Wu, J.; Li, Q.; Wang, X.; Yu, S.; Li, L.; Wu, X.; Chen, Y.; Zhao, J.; Zhao, Y. Neuroprotection by Curcumin in Ischemic Brain Injury Involves the Akt/Nrf2 Pathway. *PLoS ONE* **2013**, *8*, e59843. [CrossRef] [PubMed]
73. Pentón-Rol, G.; Marín-Prida, J.; McCarty, M.F. C-Phycocyanin-Derived Phycocyanobilin as a Potential Nutraceutical Approach for Major Neurodegenerative Disorders and COVID-19-Induced Damage to the Nervous System. *Curr. Neuropharmacol.* **2021**, *19*, 2250–2275. [CrossRef] [PubMed]
74. Vitek, L.; Hinds, T.D.; Stec, D.E.; Tiribelli, C. The Physiology of Bilirubin: Health and Disease Equilibrium. *Trends Mol. Med.* **2023**, *29*, 315–328. [CrossRef] [PubMed]
75. Wang, L.; Waltenberger, B.; Pferschy-Wenzig, E.-M.; Blunder, M.; Liu, X.; Malainer, C.; Blazevic, T.; Schwaiger, S.; Rollinger, J.M.; Heiss, E.H.; et al. Natural Product Agonists of Peroxisome Proliferator-Activated Receptor Gamma (PPARγ): A Review. *Biochem. Pharmacol.* **2014**, *92*, 73–89. [CrossRef] [PubMed]
76. Rigano, D.; Sirignano, C.; Taglialatela-Scafati, O. The Potential of Natural Products for Targeting PPARα. *Acta Pharm. Sin. B* **2017**, *7*, 427–438. [CrossRef] [PubMed]
77. Mölzer, C.; Wallner, M.; Kern, C.; Tosevska, A.; Schwarz, U.; Zadnikar, R.; Doberer, D.; Marculescu, R.; Wagner, K.-H. Features of an Altered AMPK Metabolic Pathway in Gilbert's Syndrome, and Its Role in Metabolic Health. *Sci. Rep.* **2016**, *6*, 30051. [CrossRef]
78. Longhi, M.S.; Vuerich, M.; Kalbasi, A.; Kenison, J.E.; Yeste, A.; Csizmadia, E.; Vaughn, B.; Feldbrugge, L.; Mitsuhashi, S.; Wegiel, B.; et al. Bilirubin Suppresses Th17 Immunity in Colitis by Upregulating CD39. *JCI Insight* **2017**, *2*, e92791. [CrossRef]
79. Correa-Costa, M.; Gallo, D.; Csizmadia, E.; Gomperts, E.; Lieberum, J.-L.; Hauser, C.J.; Ji, X.; Wang, B.; Câmara, N.O.S.; Robson, S.C.; et al. Carbon Monoxide Protects the Kidney through the Central Circadian Clock and CD39. *Proc. Natl. Acad. Sci. USA* **2018**, *115*, E2302–E2310. [CrossRef]
80. McCarty, M.F. Practical Prospects for Boosting Hepatic Production of the "pro-Longevity" Hormone FGF21. *Horm. Mol. Biol. Clin. Investig.* **2017**, *30*, 20150057. [CrossRef]
81. Hinds, T.D.; Stec, D.E. Bilirubin, a Cardiometabolic Signaling Molecule. *Hypertension* **2018**, *72*, 788–795. [CrossRef]
82. Rebollo-Hernanz, M.; Aguilera, Y.; Martín-Cabrejas, M.A.; Gonzalez de Mejia, E. Activating Effects of the Bioactive Compounds From Coffee By-Products on FGF21 Signaling Modulate Hepatic Mitochondrial Bioenergetics and Energy Metabolism in vitro. *Front. Nutr.* **2022**, *9*, 866233. [CrossRef] [PubMed]
83. Rebollo-Hernanz, M.; Aguilera, Y.; Martin-Cabrejas, M.A.; Gonzalez de Mejia, E. Phytochemicals from the Cocoa Shell Modulate Mitochondrial Function, Lipid and Glucose Metabolism in Hepatocytes via Activation of FGF21/ERK, AKT, and mTOR Pathways. *Antioxidants* **2022**, *11*, 136. [CrossRef] [PubMed]
84. Lanzillotta, C.; Zuliani, I.; Vasavda, C.; Snyder, S.H.; Paul, B.D.; Perluigi, M.; Di Domenico, F.D.; Barone, E. BVR-A Deficiency Leads to Autophagy Impairment through the Dysregulation of AMPK/mTOR Axis in the Brain-Implications for Neurodegeneration. *Antioxidants* **2020**, *9*, 671. [CrossRef] [PubMed]
85. Kim, H.J.; Joe, Y.; Kim, S.-K.; Park, S.-U.; Park, J.; Chen, Y.; Kim, J.; Ryu, J.; Cho, G.J.; Surh, Y.-J.; et al. Carbon Monoxide Protects against Hepatic Steatosis in Mice by Inducing Sestrin-2 via the PERK-eIF2α-ATF4 Pathway. *Free Radic. Biol. Med.* **2017**, *110*, 81–91. [CrossRef] [PubMed]
86. Huang, S. Inhibition of PI3K/Akt/mTOR Signaling by Natural Products. *Anticancer. Agents Med. Chem.* **2013**, *13*, 967–970. [CrossRef] [PubMed]
87. Vakili, O.; Borji, M.; Saffari-Chaleshtori, J.; Shafiee, S.M. Ameliorative Effects of Bilirubin on Cell Culture Model of Non-Alcoholic Fatty Liver Disease. *Mol. Biol. Rep.* **2023**, *50*, 4411–4422. [CrossRef] [PubMed]
88. Alcaín, F.J.; Villalba, J.M. Sirtuin Activators. *Expert. Opin. Ther. Pat.* **2009**, *19*, 403–414. [CrossRef]
89. Zhang, Z.; Amorosa, L.F.; Petrova, A.; Coyle, S.; Macor, M.; Nair, M.; Lee, L.Y.; Haimovich, B. TLR4 Counteracts BVRA Signaling in Human Leukocytes via Differential Regulation of AMPK, mTORC1 and mTORC2. *Sci. Rep.* **2019**, *9*, 7020. [CrossRef]
90. Heidary Moghaddam, R.; Samimi, Z.; Asgary, S.; Mohammadi, P.; Hozeifi, S.; Hoseinzadeh-Chahkandak, F.; Xu, S.; Farzaei, M.H. Natural AMPK Activators in Cardiovascular Disease Prevention. *Front. Pharmacol.* **2021**, *12*, 738420. [CrossRef]
91. Hinds, T.D.; Burns, K.A.; Hosick, P.A.; McBeth, L.; Nestor-Kalinoski, A.; Drummond, H.A.; AlAmodi, A.A.; Hankins, M.W.; Vanden Heuvel, J.P.; Stec, D.E. Biliverdin Reductase A Attenuates Hepatic Steatosis by Inhibition of Glycogen Synthase Kinase (GSK) 3β Phosphorylation of Serine 73 of Peroxisome Proliferator-Activated Receptor (PPAR) α. *J. Biol. Chem.* **2016**, *291*, 25179–25191. [CrossRef]
92. Duda, P.; Akula, S.M.; Abrams, S.L.; Steelman, L.S.; Martelli, A.M.; Cocco, L.; Ratti, S.; Candido, S.; Libra, M.; Montalto, G.; et al. Targeting GSK3 and Associated Signaling Pathways Involved in Cancer. *Cells* **2020**, *9*, 1110. [CrossRef]

93. Song, X.; Long, D. Nrf2 and Ferroptosis: A New Research Direction for Neurodegenerative Diseases. *Front. Neurosci.* **2020**, *14*, 267. [CrossRef] [PubMed]
94. Gozzelino, R. The Pathophysiology of Heme in the Brain. *Curr. Alzheimer Res.* **2016**, *13*, 174–184. [CrossRef] [PubMed]
95. Schipper, H.M. Brain Iron Deposition and the Free Radical-Mitochondrial Theory of Ageing. *Ageing Res. Rev.* **2004**, *3*, 265–301. [CrossRef] [PubMed]
96. Ozen, M.; Kitase, Y.; Vasan, V.; Burkhardt, C.; Ramachandra, S.; Robinson, S.; Jantzie, L.L. Chorioamnionitis Precipitates Perinatal Alterations of Heme-Oxygenase-1 (HO-1) Homeostasis in the Developing Rat Brain. *Int. J. Mol. Sci.* **2021**, *22*, 5773. [CrossRef]
97. Kram, H.; Prokop, G.; Haller, B.; Gempt, J.; Wu, Y.; Schmidt-Graf, F.; Schlegel, J.; Conrad, M.; Liesche-Starnecker, F. Glioblastoma Relapses Show Increased Markers of Vulnerability to Ferroptosis. *Front. Oncol.* **2022**, *12*, 841418. [CrossRef]
98. Hara, E.; Takahashi, K.; Tominaga, T.; Kumabe, T.; Kayama, T.; Suzuki, H.; Fujita, H.; Yoshimoto, T.; Shirato, K.; Shibahara, S. Expression of Heme Oxygenase and Inducible Nitric Oxide Synthase mRNA in Human Brain Tumors. *Biochem. Biophys. Res. Commun.* **1996**, *224*, 153–158. [CrossRef]
99. Vandenbark, A.A.; Offner, H.; Matejuk, S.; Matejuk, A. Microglia and Astrocyte Involvement in Neurodegeneration and Brain Cancer. *J. Neuroinflammation* **2021**, *18*, 298. [CrossRef]
100. Maas, S.L.N.; Abels, E.R.; Van De Haar, L.L.; Zhang, X.; Morsett, L.; Sil, S.; Guedes, J.; Sen, P.; Prabhakar, S.; Hickman, S.E.; et al. Glioblastoma Hijacks Microglial Gene Expression to Support Tumor Growth. *J. Neuroinflammation* **2020**, *17*, 120. [CrossRef]
101. Catalano, M.; Serpe, C.; Limatola, C. Microglial Extracellular Vesicles as Modulators of Brain Microenvironment in Glioma. *Int. J. Mol. Sci.* **2022**, *23*, 13165. [CrossRef]
102. Lanza, M.; Casili, G.; Campolo, M.; Paterniti, I.; Colarossi, C.; Mare, M.; Giuffrida, R.; Caffo, M.; Esposito, E.; Cuzzocrea, S. Immunomodulatory Effect of Microglia-Released Cytokines in Gliomas. *Brain Sci.* **2021**, *11*, 466. [CrossRef] [PubMed]
103. Haghshenas, M.R.; Saffarian, A.; Khademolhosseini, A.; Dehghanian, A.; Ghaderi, A.; Sotoodeh Jahromi, A. Simultaneous Increase in Serum Levels of IL-37 and IL-18 Binding Protein In Low-Grade and High-Grade Brain Tumors. *Asian Pac. J. Cancer Prev.* **2022**, *23*, 2851–2856. [CrossRef] [PubMed]
104. Stec, D.E.; John, K.; Trabbic, C.J.; Luniwal, A.; Hankins, M.W.; Baum, J.; Hinds, T.D. Bilirubin Binding to PPARα Inhibits Lipid Accumulation. *PLoS ONE* **2016**, *11*, e0153427. [CrossRef] [PubMed]
105. Pepino, M.Y.; Kuda, O.; Samovski, D.; Abumrad, N.A. Structure-Function of CD36 and Importance of Fatty Acid Signal Transduction in Fat Metabolism. *Annu. Rev. Nutr.* **2014**, *34*, 281–303. [CrossRef]
106. Lee, J.H.; Wada, T.; Febbraio, M.; He, J.; Matsubara, T.; Lee, M.J.; Gonzalez, F.J.; Xie, W. A Novel Role for the Dioxin Receptor in Fatty Acid Metabolism and Hepatic Steatosis. *Gastroenterology* **2010**, *139*, 653–663. [CrossRef]
107. Phelan, D.; Winter, G.M.; Rogers, W.J.; Lam, J.C.; Denison, M.S. Activation of the Ah Receptor Signal Transduction Pathway by Bilirubin and Biliverdin. *Arch. Biochem. Biophys.* **1998**, *357*, 155–163. [CrossRef]
108. Gordon, D.M.; Blomquist, T.M.; Miruzzi, S.A.; McCullumsmith, R.; Stec, D.E.; Hinds, T.D. RNA Sequencing in Human HepG2 Hepatocytes Reveals PPAR-α Mediates Transcriptome Responsiveness of Bilirubin. *Physiol. Genom.* **2019**, *51*, 234–240. [CrossRef]
109. Nakao, A.; Murase, N.; Ho, C.; Toyokawa, H.; Billiar, T.R.; Kanno, S. Biliverdin Administration Prevents the Formation of Intimal Hyperplasia Induced by Vascular Injury. *Circulation* **2005**, *112*, 587–591. [CrossRef]
110. Deguchi, K.; Hayashi, T.; Nagotani, S.; Sehara, Y.; Zhang, H.; Tsuchiya, A.; Ohta, Y.; Tomiyama, K.; Morimoto, N.; Miyazaki, M.; et al. Reduction of Cerebral Infarction in Rats by Biliverdin Associated with Amelioration of Oxidative Stress. *Brain Res.* **2008**, *1188*, 1–8. [CrossRef]
111. Zou, Z.-Y.; Liu, J.; Chang, C.; Li, J.-J.; Luo, J.; Jin, Y.; Ma, Z.; Wang, T.-H.; Shao, J.-L. Biliverdin Administration Regulates the microRNA-mRNA Expressional Network Associated with Neuroprotection in Cerebral Ischemia Reperfusion Injury in Rats. *Int. J. Mol. Med.* **2019**, *43*, 1356–1372. [CrossRef]
112. Triani, F.; Tramutola, A.; Di Domenico, F.; Sharma, N.; Butterfield, D.A.; Head, E.; Perluigi, M.; Barone, E. Biliverdin Reductase-A Impairment Links Brain Insulin Resistance with Increased Aβ Production in an Animal Model of Aging: Implications for Alzheimer Disease. *Biochim. Et Biophys. Acta (BBA)—Mol. Basis Dis.* **2018**, *1864*, 3181–3194. [CrossRef] [PubMed]
113. Barone, E.; Mancuso, C.; Di Domenico, F.; Sultana, R.; Murphy, M.P.; Head, E.; Butterfield, D.A. Biliverdin Reductase-A: A Novel Drug Target for Atorvastatin in a Dog Pre-Clinical Model of Alzheimer Disease. *J. Neurochem.* **2012**, *120*, 135–146. [CrossRef] [PubMed]
114. Gibbs, P.E.M.; Maines, M.D. Biliverdin Inhibits Activation of NF-κB: Reversal of Inhibition by Human Biliverdin Reductase. *Int. J. Cancer* **2007**, *121*, 2567–2574. [CrossRef] [PubMed]
115. Atukeren, P.; Oner, S.; Baran, O.; Kemerdere, R.; Eren, B.; Cakatay, U.; Tanriverdi, T. Oxidant and Anti-Oxidant Status in Common Brain Tumors: Correlation to TP53 and Human Biliverdin Reductase. *Clin. Neurol. Neurosurg.* **2017**, *158*, 72–76. [CrossRef] [PubMed]
116. Zaghloul, N.; Kurepa, D.; Bader, M.Y.; Nagy, N.; Ahmed, M.N. Prophylactic Inhibition of NF-κB Expression in Microglia Leads to Attenuation of Hypoxic Ischemic Injury of the Immature Brain. *J. Neuroinflammation* **2020**, *17*, 365. [CrossRef] [PubMed]
117. Costa-De-Santana, B.J.R.; Manhães-De-Castro, R.; Gouveia, H.J.C.B.; Silva, E.R.; Araújo, M.A.d.S.; Lacerda, D.C.; Guzmán-Quevedo, O.; Torner, L.; Toscano, A.E. Motor Deficits Are Associated with Increased Glial Cell Activation in the Hypothalamus and Cerebellum of Young Rats Subjected to Cerebral Palsy. *Brain Res.* **2023**, *1814*, 148447. [CrossRef] [PubMed]
118. Mallard, C.; Davidson, J.O.; Tan, S.; Green, C.R.; Bennet, L.; Robertson, N.J.; Gunn, A.J. Astrocytes and Microglia in Acute Cerebral Injury Underlying Cerebral Palsy Associated with Preterm Birth. *Pediatr. Res.* **2014**, *75*, 234–240. [CrossRef]

119. Hu, C.; Li, H.; Li, J.; Luo, X.; Hao, Y. Microglia: Synaptic Modulator in Autism Spectrum Disorder. *Front. Psychiatry* **2022**, *13*, 958661. [CrossRef]
120. Brégère, C.; Schwendele, B.; Radanovic, B.; Guzman, R. Microglia and Stem-Cell Mediated Neuroprotection after Neonatal Hypoxia-Ischemia. *Stem Cell Rev. Rep.* **2022**, *18*, 474–522. [CrossRef]
121. Zhang, Y.; Xie, Y.; Cheng, Z.; Zhang, Y.; Wang, W.; Guo, B.; Wu, S. Mechanism of Action and Therapeutic Targeting of Microglia in Autism Spectrum Disorder. *Adv. Neurol.* **2022**, *1*, 167. [CrossRef]
122. Zhang, F.; Nance, E.; Alnasser, Y.; Kannan, R.; Kannan, S. Microglial Migration and Interactions with Dendrimer Nanoparticles Are Altered in the Presence of Neuroinflammation. *J. Neuroinflammation* **2016**, *13*, 65. [CrossRef] [PubMed]
123. Davoli-Ferreira, M.; Thomson, C.A.; McCoy, K.D. Microbiota and Microglia Interactions in ASD. *Front. Immunol.* **2021**, *12*, 676255. [CrossRef] [PubMed]
124. Koyama, R.; Ikegaya, Y. Microglia in the Pathogenesis of Autism Spectrum Disorders. *Neurosci. Res.* **2015**, *100*, 1–5. [CrossRef] [PubMed]
125. Tsilioni, I.; Patel, A.B.; Pantazopoulos, H.; Berretta, S.; Conti, P.; Leeman, S.E.; Theoharides, T.C. IL-37 Is Increased in Brains of Children with Autism Spectrum Disorder and Inhibits Human Microglia Stimulated by Neurotensin. *Proc. Natl. Acad. Sci. USA* **2019**, *116*, 21659–21665. [CrossRef]
126. Vítek, L.; Tiribelli, C. Gilbert's Syndrome Revisited. *J. Hepatol.* **2023**, *79*, 1049–1055. [CrossRef]
127. Sugatani, J.; Mizushima, K.; Osabe, M.; Yamakawa, K.; Kakizaki, S.; Takagi, H.; Mori, M.; Ikari, A.; Miwa, M. Transcriptional Regulation of Human UGT1A1 Gene Expression through Distal and Proximal Promoter Motifs: Implication of Defects in the UGT1A1 Gene Promoter. *Naunyn Schmiedebergs Arch. Pharmacol.* **2008**, *377*, 597–605. [CrossRef] [PubMed]
128. Bock, K.W.; Köhle, C. Contributions of the Ah Receptor to Bilirubin Homeostasis and Its Antioxidative and Atheroprotective Functions. *Biol. Chem.* **2010**, *391*, 645–653. [CrossRef]
129. Xiao, L.; Zhang, Z.; Luo, X. Roles of Xenobiotic Receptors in Vascular Pathophysiology. *Circ. J.* **2014**, *78*, 1520–1530. [CrossRef]
130. Jayanti, S.; Moretti, R.; Tiribelli, C.; Gazzin, S. Bilirubin Prevents the TH+ Dopaminergic Neuron Loss in a Parkinson's Disease Model by Acting on TNF-α. *Int. J. Mol. Sci.* **2022**, *23*, 14276. [CrossRef]
131. Hernandez, J.P.; Mota, L.C.; Baldwin, W.S. Activation of CAR and PXR by Dietary, Environmental and Occupational Chemicals Alters Drug Metabolism, Intermediary Metabolism, and Cell Proliferation. *Curr. Pharmacogenomics Person. Med.* **2009**, *7*, 81–105. [CrossRef]
132. Busbee, P.B.; Rouse, M.; Nagarkatti, M.; Nagarkatti, P.S. Use of Natural AhR Ligands as Potential Therapeutic Modalities against Inflammatory Disorders. *Nutr. Rev.* **2013**, *71*, 353–369. [CrossRef] [PubMed]
133. Hong, F.; Pan, S.; Guo, Y.; Xu, P.; Zhai, Y. PPARs as Nuclear Receptors for Nutrient and Energy Metabolism. *Molecules* **2019**, *24*, 2545. [CrossRef] [PubMed]
134. Bragt, M.C.E.; Popeijus, H.E. Peroxisome Proliferator-Activated Receptors and the Metabolic Syndrome. *Physiol. Behav.* **2008**, *94*, 187–197. [CrossRef] [PubMed]
135. Duszka, K.; Gregor, A.; Guillou, H.; König, J.; Wahli, W. Peroxisome Proliferator-Activated Receptors and Caloric Restriction—Common Pathways Affecting Metabolism, Health, and Longevity. *Cells* **2020**, *9*, 1708. [CrossRef] [PubMed]
136. Potthoff, M.J.; Kliewer, S.A.; Mangelsdorf, D.J. Endocrine Fibroblast Growth Factors 15/19 and 21: From Feast to Famine. *Genes. Dev.* **2012**, *26*, 312–324. [CrossRef]
137. Liu, J.; Dong, H.; Zhang, Y.; Cao, M.; Song, L.; Pan, Q.; Bulmer, A.; Adams, D.B.; Dong, X.; Wang, H. Bilirubin Increases Insulin Sensitivity by Regulating Cholesterol Metabolism, Adipokines and PPARγ Levels. *Sci. Rep.* **2015**, *5*, 9886. [CrossRef] [PubMed]
138. Dong, H.; Huang, H.; Yun, X.; Kim, D.; Yue, Y.; Wu, H.; Sutter, A.; Chavin, K.D.; Otterbein, L.E.; Adams, D.B.; et al. Bilirubin Increases Insulin Sensitivity in Leptin-Receptor Deficient and Diet-Induced Obese Mice through Suppression of ER Stress and Chronic Inflammation. *Endocrinology* **2014**, *155*, 818–828. [CrossRef]
139. Zhang, F.; Guan, W.; Fu, Z.; Zhou, L.; Guo, W.; Ma, Y.; Gong, Y.; Jiang, W.; Liang, H.; Zhou, H. Relationship between Serum Indirect Bilirubin Level and Insulin Sensitivity: Results from Two Independent Cohorts of Obese Patients with Impaired Glucose Regulation and Type 2 Diabetes Mellitus in China. *Int. J. Endocrinol.* **2020**, *2020*, 5681296. [CrossRef]
140. Lin, L.-Y.; Kuo, H.-K.; Hwang, J.-J.; Lai, L.-P.; Chiang, F.-T.; Tseng, C.-D.; Lin, J.-L. Serum Bilirubin Is Inversely Associated with Insulin Resistance and Metabolic Syndrome among Children and Adolescents. *Atherosclerosis* **2009**, *203*, 563–568. [CrossRef]
141. Shao, X.; Wang, M.; Wei, X.; Deng, S.; Fu, N.; Peng, Q.; Jiang, Y.; Ye, L.; Xie, J.; Lin, Y. Peroxisome Proliferator-Activated Receptor-γ: Master Regulator of Adipogenesis and Obesity. *Curr. Stem Cell Res. Ther.* **2016**, *11*, 282–289. [CrossRef]
142. Wang, L.; Yin, Y.; Hou, G.; Kang, J.; Wang, Q. Peroxisome Proliferator-Activated Receptor (PPARγ) Plays a Protective Role in Cigarette Smoking-Induced Inflammation via AMP-Activated Protein Kinase (AMPK) Signaling. *Med. Sci. Monit.* **2018**, *24*, 5168–5177. [CrossRef] [PubMed]
143. He, G.; Sung, Y.M.; Digiovanni, J.; Fischer, S.M. Thiazolidinediones Inhibit Insulin-like Growth Factor-i-Induced Activation of p70S6 Kinase and Suppress Insulin-like Growth Factor-I Tumor-Promoting Activity. *Cancer Res.* **2006**, *66*, 1873–1878. [CrossRef] [PubMed]
144. Hinds, T.D.; Hosick, P.A.; Chen, S.; Tukey, R.H.; Hankins, M.W.; Nestor-Kalinoski, A.; Stec, D.E.; Creeden, J.F.; Gordon, D.M.; Hipp, J.A.; et al. Mice with Hyperbilirubinemia Due to Gilbert's Syndrome Polymorphism Are Resistant to Hepatic Steatosis by Decreased Serine 73 Phosphorylation of PPARα. *Am. J. Physiol. Endocrinol. Metab.* **2017**, *312*, E244–E252. [CrossRef] [PubMed]

145. Viollet, B.; Guigas, B.; Leclerc, J.; Hébrard, S.; Lantier, L.; Mounier, R.; Andreelli, F.; Foretz, M. AMP-Activated Protein Kinase in the Regulation of Hepatic Energy Metabolism: From Physiology to Therapeutic Perspectives. *Acta Physiol.* **2009**, *196*, 81–98. [CrossRef] [PubMed]
146. Lin, S.-C.; Hardie, D.G. AMPK: Sensing Glucose as Well as Cellular Energy Status. *Cell Metab.* **2018**, *27*, 299–313. [CrossRef] [PubMed]
147. Stallone, G.; Infante, B.; Prisciandaro, C.; Grandaliano, G. mTOR and Aging: An Old Fashioned Dress. *Int. J. Mol. Sci.* **2019**, *20*, 2774. [CrossRef]
148. González, A.; Hall, M.N.; Lin, S.-C.; Hardie, D.G. AMPK and TOR: The Yin and Yang of Cellular Nutrient Sensing and Growth Control. *Cell Metab.* **2020**, *31*, 472–492. [CrossRef]
149. Zelenka, J.; Dvořák, A.; Alán, L.; Zadinová, M.; Haluzík, M.; Vítek, L. Hyperbilirubinemia Protects against Aging-Associated Inflammation and Metabolic Deterioration. *Oxid. Med. Cell Longev.* **2016**, *2016*, 6190609. [CrossRef]
150. Liang, H.; Ward, W.F.; Shute, R.J.; Heesch, M.W.; Zak, R.B.; Kreiling, J.L.; Slivka, D.R.; Sun, S.; Li, H.; Chen, J.; et al. PGC-1alpha: A Key Regulator of Energy Metabolism. *Adv. Physiol. Educ.* **2006**, *30*, 145–151. [CrossRef]
151. Jiang, W. Sirtuins: Novel Targets for Metabolic Disease in Drug Development. *Biochem. Biophys. Res. Commun.* **2008**, *373*, 341–344. [CrossRef]
152. Cantó, C.; Auwerx, J. PGC-1alpha, SIRT1 and AMPK, an Energy Sensing Network That Controls Energy Expenditure. *Curr. Opin. Lipidol.* **2009**, *20*, 98–105. [CrossRef] [PubMed]
153. Shi, Y.-H.; Zhang, X.-L.; Ying, P.-J.; Wu, Z.-Q.; Lin, L.-L.; Chen, W.; Zheng, G.-Q.; Zhu, W.-Z. Neuroprotective Effect of Astragaloside IV on Cerebral Ischemia/Reperfusion Injury Rats Through Sirt1/Mapt Pathway. *Front. Pharmacol.* **2021**, *12*, 639898. [CrossRef] [PubMed]
154. Antonioli, L.; Pacher, P.; Vizi, E.S.; Haskó, G. CD39 and CD73 in Immunity and Inflammation. *Trends Mol. Med.* **2013**, *19*, 355–367. [CrossRef] [PubMed]
155. Lee, G.R.; Shaefi, S.; Otterbein, L.E. HO-1 and CD39: It Takes Two to Protect the Realm. *Front. Immunol.* **2019**, *10*, 1765. [CrossRef] [PubMed]
156. Enjyoji, K.; Kotani, K.; Thukral, C.; Blumel, B.; Sun, X.; Wu, Y.; Imai, M.; Friedman, D.; Csizmadia, E.; Bleibel, W.; et al. Deletion of Cd39/Entpd1 Results in Hepatic Insulin Resistance. *Diabetes* **2008**, *57*, 2311–2320. [CrossRef] [PubMed]
157. da Silva, C.G.; Jarzyna, R.; Specht, A.; Kaczmarek, E. Extracellular Nucleotides and Adenosine Independently Activate AMP-Activated Protein Kinase in Endothelial Cells. *Circ. Res.* **2006**, *98*, e39–e47. [CrossRef] [PubMed]
158. Wang, P.; Jia, J.; Zhang, D. Purinergic Signalling in Liver Diseases: Pathological Functions and Therapeutic Opportunities. *JHEP Rep.* **2020**, *2*, 100165. [CrossRef]
159. Wang, S.; Gao, S.; Zhou, D.; Qian, X.; Luan, J.; Lv, X. The Role of the CD39-CD73-Adenosine Pathway in Liver Disease. *J. Cell Physiol.* **2021**, *236*, 851–862. [CrossRef]
160. Andersson, C.; Weeke, P.; Fosbøl, E.L.; Brendorp, B.; Køber, L.; Coutinho, W.; Sharma, A.M.; Van Gaal, L.; Finer, N.; James, W.P.T.; et al. Acute Effect of Weight Loss on Levels of Total Bilirubin in Obese, Cardiovascular High-Risk Patients: An Analysis from the Lead-in Period of the Sibutramine Cardiovascular Outcome Trial. *Metabolism* **2009**, *58*, 1109–1115. [CrossRef]
161. Chen, L.; Duan, Y.; Wei, H.; Ning, H.; Bi, C.; Zhao, Y.; Qin, Y.; Li, Y. Acetyl-CoA Carboxylase (ACC) as a Therapeutic Target for Metabolic Syndrome and Recent Developments in ACC1/2 Inhibitors. *Expert. Opin. Investig. Drugs* **2019**, *28*, 917–930. [CrossRef]
162. Ahmad, F.; Woodgett, J.R. Emerging Roles of GSK-3α in Pathophysiology: Emphasis on Cardio-Metabolic Disorders. *Biochim. Biophys. Acta Mol. Cell Res.* **2020**, *1867*, 118616. [CrossRef] [PubMed]
163. Beurel, E.; Grieco, S.F.; Jope, R.S. Glycogen Synthase Kinase-3 (GSK3): Regulation, Actions, and Diseases. *Pharmacol. Ther.* **2015**, *148*, 114–131. [CrossRef] [PubMed]
164. Bösch, F.; Thomas, M.; Kogler, P.; Oberhuber, R.; Sucher, R.; Aigner, F.; Semsroth, S.; Wiedemann, D.; Yamashita, K.; Troppmair, J.; et al. Bilirubin Rinse of the Graft Ameliorates Ischemia Reperfusion Injury in Heart Transplantation. *Transpl. Int.* **2014**, *27*, 504–513. [CrossRef] [PubMed]
165. Liu, Z.; Cao, W. P38 Mitogen-Activated Protein Kinase: A Critical Node Linking Insulin Resistance and Cardiovascular Diseases in Type 2 Diabetes Mellitus. *Endocr. Metab. Immune Disord. Drug Targets* **2009**, *9*, 38–46. [CrossRef] [PubMed]
166. Šuk, J.; Jašprová, J.; Biedermann, D.; Petrásková, L.; Valentová, K.; Křen, V.; Muchová, L.; Vítek, L. Isolated Silymarin Flavonoids Increase Systemic and Hepatic Bilirubin Concentrations and Lower Lipoperoxidation in Mice. *Oxidative Med. Cell. Longev.* **2019**, *2019*, e6026902. [CrossRef] [PubMed]
167. Flaig, T.W.; Gustafson, D.L.; Su, L.-J.; Zirrolli, J.A.; Crighton, F.; Harrison, G.S.; Pierson, A.S.; Agarwal, R.; Glodé, L.M. A Phase I and Pharmacokinetic Study of Silybin-Phytosome in Prostate Cancer Patients. *Investig. New Drugs* **2007**, *25*, 139–146. [CrossRef] [PubMed]
168. Mariño, Z.; Crespo, G.; D'Amato, M.; Brambilla, N.; Giacovelli, G.; Rovati, L.; Costa, J.; Navasa, M.; Forns, X. Intravenous Silibinin Monotherapy Shows Significant Antiviral Activity in HCV-Infected Patients in the Peri-Transplantation Period. *J. Hepatol.* **2013**, *58*, 415–420. [CrossRef]
169. Maher, P. The Potential of Flavonoids for the Treatment of Neurodegenerative Diseases. *Int. J. Mol. Sci.* **2019**, *20*, 3056. [CrossRef]
170. Wang, T.H.; Wang, S.Y.; Wang, X.D.; Jiang, H.Q.; Yang, Y.Q.; Wang, Y.; Cheng, J.L.; Zhang, C.T.; Liang, W.W.; Feng, H.L. Fisetin Exerts Antioxidant and Neuroprotective Effects in Multiple Mutant hSOD1 Models of Amyotrophic Lateral Sclerosis by Activating ERK. *Neuroscience* **2018**, *379*, 152–166. [CrossRef]

171. Li, L.; Li, W.-J.; Zheng, X.-R.; Liu, Q.-L.; Du, Q.; Lai, Y.-J.; Liu, S.-Q. Eriodictyol Ameliorates Cognitive Dysfunction in APP/PS1 Mice by Inhibiting Ferroptosis via Vitamin D Receptor-Mediated Nrf2 Activation. *Mol. Med.* **2022**, *28*, 11. [CrossRef]
172. Mhillaj, E.; Tarozzi, A.; Pruccoli, L.; Cuomo, V.; Trabace, L.; Mancuso, C. Curcumin and Heme Oxygenase: Neuroprotection and Beyond. *Int. J. Mol. Sci.* **2019**, *20*, 2419. [CrossRef] [PubMed]
173. Ahmadi, M.; Agah, E.; Nafissi, S.; Jaafari, M.R.; Harirchian, M.H.; Sarraf, P.; Faghihi-Kashani, S.; Hosseini, S.J.; Ghoreishi, A.; Aghamollaii, V.; et al. Safety and Efficacy of Nanocurcumin as Add-on Therapy to Riluzole in Patients with Amyotrophic Lateral Sclerosis: A Pilot Randomized Clinical Trial. *Neurotherapeutics* **2018**, *15*, 430–438. [CrossRef] [PubMed]
174. Valverde-Salazar, V.; Ruiz-Gabarre, D.; García-Escudero, V. Alzheimer's Disease and Green Tea: Epigallocatechin-3-Gallate as a Modulator of Inflammation and Oxidative Stress. *Antioxidants* **2023**, *12*, 1460. [CrossRef] [PubMed]
175. Mohi-Ud-Din, R.; Mir, R.H.; Shah, A.J.; Sabreen, S.; Wani, T.U.; Masoodi, M.H.; Akkol, E.K.; Bhat, Z.A.; Khan, H. Plant-Derived Natural Compounds for the Treatment of Amyotrophic Lateral Sclerosis: An Update. *Curr. Neuropharmacol.* **2022**, *20*, 179–193. [CrossRef] [PubMed]
176. Jiang, H.; Tian, X.; Guo, Y.; Duan, W.; Bu, H.; Li, C. Activation of Nuclear Factor Erythroid 2-Related Factor 2 Cytoprotective Signaling by Curcumin Protect Primary Spinal Cord Astrocytes against Oxidative Toxicity. *Biol. Pharm. Bull.* **2011**, *34*, 1194–1197. [CrossRef] [PubMed]
177. Wang, R.; Li, Y.-H.; Xu, Y.; Li, Y.-B.; Wu, H.-L.; Guo, H.; Zhang, J.-Z.; Zhang, J.-J.; Pan, X.-Y.; Li, X.-J. Curcumin Produces Neuroprotective Effects via Activating Brain-Derived Neurotrophic Factor/TrkB-Dependent MAPK and PI-3K Cascades in Rodent Cortical Neurons. *Prog. Neuropsychopharmacol. Biol. Psychiatry* **2010**, *34*, 147–153. [CrossRef] [PubMed]
178. Chen, H.; Li, Z.; Xu, J.; Zhang, N.; Chen, J.; Wang, G.; Zhao, Y. Curcumin Induces Ferroptosis in Follicular Thyroid Cancer by Upregulating HO-1 Expression. *Oxidative Med. Cell. Longev.* **2023**, *2023*, e6896790. [CrossRef]
179. Meng, P.; Yang, R.; Jiang, F.; Guo, J.; Lu, X.; Yang, T.; He, Q. Molecular Mechanism of Astragaloside IV in Improving Endothelial Dysfunction of Cardiovascular Diseases Mediated by Oxidative Stress. *Oxidative Med. Cell. Longev.* **2021**, *2021*, e1481236. [CrossRef]
180. Huang, X.-P.; Qiu, Y.-Y.; Wang, B.; Ding, H.; Tang, Y.-H.; Zeng, R.; Deng, C.-Q. Effects of Astragaloside IV Combined with the Active Components of Panax Notoginseng on Oxidative Stress Injury and Nuclear Factor-Erythroid 2-Related Factor 2/Heme Oxygenase-1 Signaling Pathway after Cerebral Ischemia-Reperfusion in Mice. *Pharmacogn. Mag.* **2014**, *10*, 402–409. [CrossRef]
181. Kontush, A.; Mann, U.; Arlt, S.; Ujeyl, A.; Lührs, C.; Müller-Thomsen, T.; Beisiegel, U. Influence of Vitamin E and C Supplementation on Lipoprotein Oxidation in Patients with Alzheimer's Disease. *Free Radic. Biol. Med.* **2001**, *31*, 345–354. [CrossRef]
182. Mudgal, R.; Sharma, S.; Singh, S.; Ravichandiran, V. The Neuroprotective Effect of Ascorbic Acid against Imidacloprid-Induced Neurotoxicity and the Role of HO-1 in Mice. *Front. Neurol.* **2023**, *14*, 1130575. [CrossRef] [PubMed]
183. Zhang, N.; Zhao, W.; Hu, Z.-J.; Ge, S.-M.; Huo, Y.; Liu, L.-X.; Gao, B.-L. Protective Effects and Mechanisms of High-Dose Vitamin C on Sepsis-Associated Cognitive Impairment in Rats. *Sci. Rep.* **2021**, *11*, 14511. [CrossRef] [PubMed]
184. Littlejohns, T.J.; Henley, W.E.; Lang, I.A.; Annweiler, C.; Beauchet, O.; Chaves, P.H.M.; Fried, L.; Kestenbaum, B.R.; Kuller, L.H.; Langa, K.M.; et al. Vitamin D and the Risk of Dementia and Alzheimer Disease. *Neurology* **2014**, *83*, 920–928. [CrossRef] [PubMed]
185. Saad El-Din, S.; Rashed, L.; Medhat, E.; Emad Aboulhoda, B.; Desoky Badawy, A.; Mohammed ShamsEldeen, A.; Abdelgwad, M. Active Form of Vitamin D Analogue Mitigates Neurodegenerative Changes in Alzheimer's Disease in Rats by Targeting Keap1/Nrf2 and MAPK-38p/ERK Signaling Pathways. *Steroids* **2020**, *156*, 108586. [CrossRef] [PubMed]
186. Liu, S.; Li, G.; Tang, H.; Pan, R.; Wang, H.; Jin, F.; Yan, X.; Xing, Y.; Chen, G.; Fu, Y.; et al. Madecassoside Ameliorates Lipopolysaccharide-Induced Neurotoxicity in Rats by Activating the Nrf2-HO-1 Pathway. *Neurosci. Lett.* **2019**, *709*, 134386. [CrossRef] [PubMed]
187. Arab, H.; Mahjoub, S.; Hajian-Tilaki, K.; Moghadasi, M. The Effect of Green Tea Consumption on Oxidative Stress Markers and Cognitive Function in Patients with Alzheimer's Disease: A Prospective Intervention Study. *Casp. J. Intern. Med.* **2016**, *7*, 188–194.
188. Na, H.-K.; Kim, E.-H.; Jung, J.-H.; Lee, H.-H.; Hyun, J.-W.; Surh, Y.-J. (−)-Epigallocatechin Gallate Induces Nrf2-Mediated Antioxidant Enzyme Expression via Activation of PI3K and ERK in Human Mammary Epithelial Cells. *Arch. Biochem. Biophys.* **2008**, *476*, 171–177. [CrossRef]
189. Zhou, H.; Mao, Z.; Zhang, X.; Li, R.; Yin, J.; Xu, Y. Neuroprotective Effect of Mangiferin against Parkinson's Disease through G-Protein-Coupled Receptor-Interacting Protein 1 (GIT1)-Mediated Antioxidant Defense. *ACS Chem. Neurosci.* **2023**, *14*, 1379–1387. [CrossRef]
190. Gold-Smith, F.; Fernandez, A.; Bishop, K. Mangiferin and Cancer: Mechanisms of Action. *Nutrients* **2016**, *8*, 396. [CrossRef]
191. Dillon, J.C.; Phuc, A.P.; Dubacq, J.P. Nutritional Value of the Alga Spirulina. *World Rev. Nutr. Diet.* **1995**, *77*, 32–46. [CrossRef]
192. Gershwin, M.E.; Amha, B. (Eds.) *Spirulina in Human Nutrition and Health*; CRC Press: Boca Raton, FL, USA, 2007. [CrossRef]
193. Padyana, A.K.; Bhat, V.B.; Madyastha, K.M.; Rajashankar, K.R.; Ramakumar, S. Crystal Structure of a Light-Harvesting Protein C-Phycocyanin from *Spirulina platensis*. *Biochem. Biophys. Res. Commun.* **2001**, *282*, 893–898. [CrossRef] [PubMed]
194. Terry, M.J.; Maines, M.D.; Lagarias, J.C. Inactivation of Phytochrome- and Phycobiliprotein-Chromophore Precursors by Rat Liver Biliverdin Reductase. *J. Biol. Chem.* **1993**, *268*, 26099–26106. [CrossRef] [PubMed]
195. McCarty, M.F. Clinical Potential of Spirulina as a Source of Phycocyanobilin. *J. Med. Food* **2007**, *10*, 566–570. [CrossRef] [PubMed]
196. McCarty, M.F.; Hendler, S.S.; Rorvik, D.M.; Inoguchi, T. Compositions for Inhibiting NADPH Oxidase Activity. US20100172971A1, 8 July 2010. Available online: https://patents.google.com/patent/US20100172971A1/en (accessed on 15 November 2023).
197. Li, Y. The Bioactivities of Phycocyanobilin from Spirulina. *J. Immunol. Res.* **2022**, *2022*, e4008991. [CrossRef] [PubMed]

198. Strasky, Z.; Zemankova, L.; Nemeckova, I.; Rathouska, J.; Wong, R.J.; Muchova, L.; Subhanova, I.; Vanikova, J.; Vanova, K.; Vitek, L.; et al. *Spirulina platensis* and Phycocyanobilin Activate Atheroprotective Heme Oxygenase-1: A Possible Implication for Atherogenesis. *Food Funct.* **2013**, *4*, 1586–1594. [CrossRef] [PubMed]
199. Zheng, J.; Inoguchi, T.; Sasaki, S.; Maeda, Y.; McCarty, M.F.; Fujii, M.; Ikeda, N.; Kobayashi, K.; Sonoda, N.; Takayanagi, R. Phycocyanin and Phycocyanobilin from *Spirulina platensis* Protect against Diabetic Nephropathy by Inhibiting Oxidative Stress. *Am. J. Physiol. -Regul. Integr. Comp. Physiol.* **2013**, *304*, R110–R120. [CrossRef]
200. El-Sayed, E.-S.M.; Hikal, M.S.; Abo El-Khair, B.E.; El-Ghobashy, R.E.; El-Assar, A.M. Hypoglycemic and Hypolipidemic Effects of *Spirulina platensis*, Phycocyanin, Phycocyanopeptide and Phycocyanobilin on Male Diabetic Rats. *Arab. Univ. J. Agric. Sci.* **2018**, *26* (Suppl. S2), 1121–1134. [CrossRef]
201. Koníčková, R.; Vaňková, K.; Vaníková, J.; Vánová, K.; Muchová, L.; Subhanová, I.; Zadinová, M.; Zelenka, J.; Dvořák, A.; Kolář, M.; et al. Anti-Cancer Effects of Blue-Green Alga *Spirulina platensis*, a Natural Source of Bilirubin-like Tetrapyrrolic Compounds. *Ann. Hepatol.* **2014**, *13*, 273–283. [CrossRef]
202. Hussein, N.; Ebied, S.; Saleh, M. Evaluation of the Anticancer Effect of Violacein, Phycocyanin and Phycocyanobilin on Apoptotic Genes Expression and Glycan Profiles in Breast Cancer Cells. *Int. J. Cancer Biomed. Res.* **2021**, *5*, 81–97. [CrossRef]
203. Liu, J.; Zhang, Q.-Y.; Yu, L.-M.; Liu, B.; Li, M.-Y.; Zhu, R.-Z. Phycocyanobilin Accelerates Liver Regeneration and Reduces Mortality Rate in Carbon Tetrachloride-Induced Liver Injury Mice. *World J. Gastroenterol.* **2015**, *21*, 5465–5472. [CrossRef]
204. Trotta, T.; Porro, C.; Cianciulli, A.; Panaro, M.A. Beneficial Effects of Spirulina Consumption on Brain Health. *Nutrients* **2022**, *14*, 676. [CrossRef] [PubMed]
205. Cervantes-Llanos, M.; Lagumersindez-Denis, N.; Marín-Prida, J.; Pavón-Fuentes, N.; Falcon-Cama, V.; Piniella-Matamoros, B.; Camacho-Rodríguez, H.; Fernández-Massó, J.R.; Valenzuela-Silva, C.; Raíces-Cruz, I.; et al. Beneficial Effects of Oral Administration of C-Phycocyanin and Phycocyanobilin in Rodent Models of Experimental Autoimmune Encephalomyelitis. *Life Sci.* **2018**, *194*, 130–138. [CrossRef] [PubMed]
206. Pavón-Fuentes, N.; Marín-Prida, J.; Llópiz-Arzuaga, A.; Falcón-Cama, V.; Campos-Mojena, R.; Cervantes-Llanos, M.; Piniella-Matamoros, B.; Pentón-Arias, E.; Pentón-Rol, G. Phycocyanobilin Reduces Brain Injury after Endothelin-1-Induced Focal Cerebral Ischaemia. *Clin. Exp. Pharmacol. Physiol.* **2020**, *47*, 383–392. [CrossRef] [PubMed]
207. Pentón-Rol, G.; Marín-Prida, J.; Falcón-Cama, V. C-Phycocyanin and Phycocyanobilin as Remyelination Therapies for Enhancing Recovery in Multiple Sclerosis and Ischemic Stroke: A Preclinical Perspective. *Behav. Sci.* **2018**, *8*, 15. [CrossRef] [PubMed]
208. McCarty, M.F.; Barroso-Aranda, J.; Contreras, F. Oral Phycocyanobilin May Diminish the Pathogenicity of Activated Brain Microglia in Neurodegenerative Disorders. *Med. Hypotheses* **2010**, *74*, 601–605. [CrossRef]
209. Chamorro, G.; Pérez-Albiter, M.; Serrano-García, N.; Mares-Sámano, J.J.; Rojas, P. Spirulina Maxima Pretreatment Partially Protects against 1-Methyl-4-Phenyl-1,2,3,6-Tetrahydropyridine Neurotoxicity. *Nutr. Neurosci.* **2006**, *9*, 207–212. [CrossRef]
210. Marín-Prida, J.; Pavón-Fuentes, N.; Lagumersindez-Denis, N.; Camacho-Rodríguez, H.; García-Soca, A.M.; Sarduy-Chávez, R.d.l.C.; Vieira, L.M.; Carvalho-Tavares, J.; Falcón-Cama, V.; Fernández-Massó, J.R.; et al. Anti-Inflammatory Mechanisms and Pharmacological Actions of Phycocyanobilin in a Mouse Model of Experimental Autoimmune Encephalomyelitis: A Therapeutic Promise for Multiple Sclerosis. *Front. Immunol.* **2022**, *13*, 1036200. [CrossRef]
211. Gardón, D.P.; Cervantes-Llanos, M.; Matamoros, B.P.; Rodríguez, H.C.; Tan, C.-Y.; Marín-Prida, J.; Falcón-Cama, V.; Pavón-Fuentes, N.; Lemus, J.G.; Ruiz, L.d.l.C.B.; et al. Positive Effects of Phycocyanobilin on Gene Expression in Glutamate-Induced Excitotoxicity in SH-SY5Y Cells and Animal Models of Multiple Sclerosis and Cerebral Ischemia. *Heliyon* **2022**, *8*, e09769. [CrossRef]
212. Pendyala, B.; Patras, A.; Dash, C. Phycobilins as Potent Food Bioactive Broad-Spectrum Inhibitors Against Proteases of SARS-CoV-2 and Other Coronaviruses: A Preliminary Study. *Front. Microbiol.* **2021**, *12*, 645713. [CrossRef]
213. Chen, Y.-H.; Chang, G.-K.; Kuo, S.-M.; Huang, S.-Y.; Hu, I.-C.; Lo, Y.-L.; Shih, S.-R. Well-Tolerated Spirulina Extract Inhibits Influenza Virus Replication and Reduces Virus-Induced Mortality. *Sci. Rep.* **2016**, *6*, 24253. [CrossRef]
214. Teas, J.; Hebert, J.R.; Fitton, J.H.; Zimba, P.V. Algae—A Poor Man's HAART? *Med. Hypotheses* **2004**, *62*, 507–510. [CrossRef] [PubMed]
215. Ngo-Matip, M.-E.; Pieme, C.A.; Azabji-Kenfack, M.; Moukette, B.M.; Korosky, E.; Stefanini, P.; Ngogang, J.Y.; Mbofung, C.M. Impact of Daily Supplementation of *Spirulina platensis* on the Immune System of Naïve HIV-1 Patients in Cameroon: A 12-Months Single Blind, Randomized, Multicenter Trial. *Nutr. J.* **2015**, *14*, 70. [CrossRef] [PubMed]
216. Teas, J.; Irhimeh, M.R. Dietary Algae and HIV/AIDS: Proof of Concept Clinical Data. *J. Appl. Phycol.* **2012**, *24*, 575–582. [CrossRef] [PubMed]
217. Nihal, B.; Gupta, N.V.; Gowda, D.V.; Manohar, M. Formulation and Development of Topical Anti Acne Formulation of Spirulina Extract. *Int. J. Appl. Pharm.* **2018**, *10*, 229–233. [CrossRef]
218. Mapoung, S.; Arjsri, P.; Thippraphan, P.; Semmarath, W.; Yodkeeree, S.; Chiewchanvit, S.; Piyamongkol, W.; Limtrakul, P. Photochemoprotective Effects of *Spirulina platensis* Extract against UVB Irradiated Human Skin Fibroblasts. *South. Afr. J. Bot.* **2020**, *130*, 198–207. [CrossRef]
219. Ragusa, I.; Nardone, G.N.; Zanatta, S.; Bertin, W.; Amadio, E. Spirulina for Skin Care: A Bright Blue Future. *Cosmetics* **2021**, *8*, 7. [CrossRef]

220. Reportlinker. Global Market Study on Spirulina: Powder Product form Segment Anticipated to Dominate the Global Market in Terms of both Value and Volume during 2016–2026. Available online: https://www.prnewswire.com/news-releases/global-market-study-on-spirulina-powder-product-form-segment-anticipated-to-dominate-the-global-market-in-terms-of-both-value-and-volume-during-2016---2026-300443004.html (accessed on 15 November 2023).
221. Qin, X. Bilirubin Would Be the Indispensable Component for Some of the Most Important Therapeutic Effects of Calculus Bovis (Niuhuang). *Chin. Med. J.* **2008**, *121*, 480. [CrossRef]
222. Yu, Z.-J.; Xu, Y.; Peng, W.; Liu, Y.-J.; Zhang, J.-M.; Li, J.-S.; Sun, T.; Wang, P. Calculus Bovis: A Review of the Traditional Usages, Origin, Chemistry, Pharmacological Activities and Toxicology. *J. Ethnopharmacol.* **2020**, *254*, 112649. [CrossRef]
223. Zhang, S.; Jiang, X.; Wang, Y.; Lin, K.; Zhang, Z.; Zhang, Z.; Zhu, P.; Ng, M.L.; Qu, S.; Sze, S.C.W.; et al. Protective Effect of An-Gong-Niu-Huang Wan Pre-Treatment Against Experimental Cerebral Ischemia Injury via Regulating GSK-3β/HO-1 Pathway. *Front. Pharmacol.* **2021**, *12*, 640297. [CrossRef]
224. Banjerdpongchai, R.; Wudtiwai, B.; Khawon, P. Induction of Human Hepatocellular Carcinoma HepG2 Cell Apoptosis by Naringin. *Asian Pac. J. Cancer Prev.* **2016**, *17*, 3289–3294.
225. Martins, T.; Barros, A.N.; Rosa, E.; Antunes, L. Enhancing Health Benefits through Chlorophylls and Chlorophyll-Rich Agro-Food: A Comprehensive Review. *Molecules* **2023**, *28*, 5344. [CrossRef] [PubMed]
226. Mishra, V.K.; Bachheti, R.; Husen, A. Medicinal Uses of Chlorophyll: A Critical Overview. In *Chlorophyll: Structure, Production and Medicinal Uses*; Nova Biomedical: Waltham, MA, USA, 2011; pp. 177–196.
227. Ferruzzi, M.G.; Blakeslee, J. Digestion, Absorption, and Cancer Preventative Activity of Dietary Chlorophyll Derivatives. *Nutr. Res.* **2007**, *27*, 1–12. [CrossRef]
228. Vaňková, K.; Marková, I.; Jašprová, J.; Dvořák, A.; Subhanová, I.; Zelenka, J.; Novosádová, I.; Rasl, J.; Vomastek, T.; Sobotka, R.; et al. Chlorophyll-Mediated Changes in the Redox Status of Pancreatic Cancer Cells Are Associated with Its Anticancer Effects. *Oxid. Med. Cell Longev.* **2018**, *2018*, 4069167. [CrossRef] [PubMed]
229. Pittala, V.; Vanella, L.; Salerno, L.; Di Giacomo, C.; Acquaviva, R.; Raffaele, M.; Romeo, G.; Modica, M.N.; Prezzavento, O.; Sorrenti, V. Novel Caffeic Acid Phenethyl Ester (Cape) Analogues as Inducers of Heme Oxygenase-1. *Curr. Pharm. Des.* **2017**, *23*, 2657–2664. [CrossRef]
230. Šmíd, V.; Šuk, J.; Kachamakova-Trojanowska, N.; Jašprová, J.; Valášková, P.; Józkowicz, A.; Dulak, J.; Šmíd, F.; Vítek, L.; Muchová, L. Heme Oxygenase-1 May Affect Cell Signalling via Modulation of Ganglioside Composition. *Oxid. Med. Cell Longev.* **2018**, *2018*, 3845027. [CrossRef]
231. Moon, S.; Kim, C.-H.; Park, J.; Kim, M.; Jeon, H.S.; Kim, Y.-M.; Choi, Y.K. Induction of BVR-A Expression by Korean Red Ginseng in Murine Hippocampal Astrocytes: Role of Bilirubin in Mitochondrial Function via the LKB1–SIRT1–ERRα Axis. *Antioxidants* **2022**, *11*, 1742. [CrossRef]
232. Wu, L.-X.; Guo, C.-X.; Qu, Q.; Yu, J.; Chen, W.-Q.; Wang, G.; Fan, L.; Li, Q.; Zhang, W.; Zhou, H.-H. Effects of Natural Products on the Function of Human Organic Anion Transporting Polypeptide 1B1. *Xenobiotica* **2012**, *42*, 339–348. [CrossRef]
233. Vítek, L.; Carey, M.C. Enterohepatic Cycling of Bilirubin as a Cause of "black" Pigment Gallstones in Adult Life. *Eur. J. Clin. Investig.* **2003**, *33*, 799–810. [CrossRef]
234. Vítek, L.; Zelenka, J.; Zadinová, M.; Malina, J. The Impact of Intestinal Microflora on Serum Bilirubin Levels. *J. Hepatol.* **2005**, *42*, 238–243. [CrossRef]
235. Hall, B.; Levy, S.; Dufault-Thompson, K.; Ndjite, G.M.; Weiss, A.; Braccia, D.; Jenkins, C.; Yang, Y.; Arp, G.; Abeysinghe, S.; et al. Discovery of the Gut Microbial Enzyme Responsible for Bilirubin Reduction to Urobilinogen. *bioRxiv* **2023**, Preprint. [CrossRef]

biomolecules

MDPI

*Review*

# Ellagic Acid and Cancer Hallmarks: Insights from Experimental Evidence

**Martina Čižmáriková [1,†], Radka Michalková [1,†], Ladislav Mirossay [1,*], Gabriela Mojžišová [2], Martina Zigová [1], Annamária Bardelčíková [1] and Ján Mojžiš [1,*]**

[1] Department of Pharmacology, Faculty of Medicine, Pavol Jozef Šafárik University, 040 01 Košice, Slovakia; martina.cizmarikova@upjs.sk (M.Č.); radka.michalkova@upjs.sk (R.M.); chripkova.martina@gmail.com (M.Z.); annamaria.bardelcikova@upjs.sk (A.B.)

[2] Center of Clinical and Preclinical Research MEDIPARK, Faculty of Medicine, Pavol Jozef Šafárik University, 040 01 Košice, Slovakia; gabriela.mojzisova@upjs.sk

* Correspondence: ladislav.mirossay@upjs.sk (L.M.); jan.mojzis@upjs.sk (J.M.)

† These authors contributed equally to this work.

**Abstract:** Cancer is a complex and multifaceted disease with a high global incidence and mortality rate. Although cancer therapy has evolved significantly over the years, numerous challenges persist on the path to effectively combating this multifaceted disease. Natural compounds derived from plants, fungi, or marine organisms have garnered considerable attention as potential therapeutic agents in the field of cancer research. Ellagic acid (EA), a natural polyphenolic compound found in various fruits and nuts, has emerged as a potential cancer prevention and treatment agent. This review summarizes the experimental evidence supporting the role of EA in targeting key hallmarks of cancer, including proliferation, angiogenesis, apoptosis evasion, immune evasion, inflammation, genomic instability, and more. We discuss the molecular mechanisms by which EA modulates signaling pathways and molecular targets involved in these cancer hallmarks, based on in vitro and in vivo studies. The multifaceted actions of EA make it a promising candidate for cancer prevention and therapy. Understanding its impact on cancer biology can pave the way for developing novel strategies to combat this complex disease.

**Keywords:** cancer; hallmarks; ellagic acid; chemoprevention; chemotherapy

## 1. Introduction

According to the WHO, cancer is the second leading cause of death globally, following cardiovascular diseases [1]. Furthermore, the Global Cancer Observatory (GLOBOCAN) reported that more than 19 million new cases and approximately 10 million deaths were associated with cancer worldwide in 2020. In addition, predictions for the future indicate that by the year 2040, the global population could experience a surge in cancer incidence reaching 28.4 million cases [2].

Cancer is a complex and multifaceted disease with many different types, each with its own distinct characteristics and behaviors, and is generally characterized by the uncontrolled and abnormal growth of cells within the body [3]. Cancer risk factors include genetics, exposure to carcinogens, lifestyle factors, and certain infections [4].

In contrast to normal cells, which adhere to stringent regulations governing their growth and replication, cancer cells circumvent these checks and possess the capability to not only infiltrate adjacent tissues but also disperse to distant locations within the body via the bloodstream or lymphatic system [5].

## 2. Cancer Hallmarks

Cancer cells display a series of unique characteristics that differentiate them from normal cells. These characteristics collectively play a role in the unrestrained proliferation and invasive nature that define cancer. These traits are commonly referred to as "cancer hallmarks".

More than two decades ago, Hanahan and Weinberg [6] published a seminal review known as "The Hallmarks of Cancer" to provide structure to the intricate landscape of cancer biology. In this publication, they sought to categorize the complex facets of cancer into six key hallmarks. A decade later, in a subsequent review [7], the original six hallmarks were expanded to eight, with two enabling characteristics (tumor-promoting inflammation and genome instability and mutation). However, twelve years later, it has become clear that these newer additions, similar to the original six hallmarks, can now be regarded as fundamental characteristics of cancer. They are now recognized and incorporated as such into the current representation of cancer hallmarks (Figure 1).

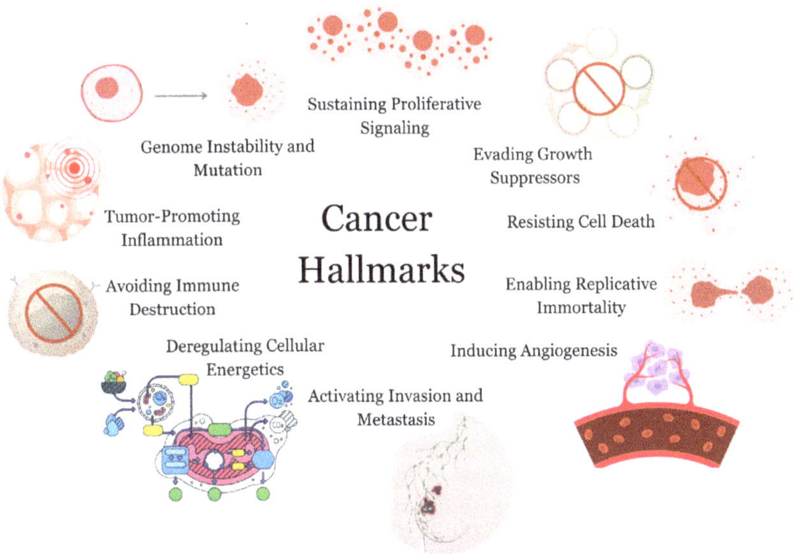


**Figure 1.** The Hallmarks of Cancer. This diagram illustrates the hallmark traits that are typically possessed by the majority of cancers.

*Sustaining Proliferative Signaling*: Cancer cells can receive and send signals that stimulate their own uncontrolled growth and division, even without external growth signals.

*Evading Growth Suppressors*: Cancer cells can bypass the mechanisms that normally inhibit their growth, such as checkpoints that prevent damaged cells from proliferating.

*Resisting Cell Death*: Cancer cells can evade programmed cell death (apoptosis), which is a mechanism that eliminates damaged or abnormal cells.

*Deregulating Cellular Energetics*: Cancer cells can rewire their metabolic pathways to support their high energy needs for rapid growth and division.

*Enabling Replicative Immortality*: Cancer cells can maintain their ability to divide indefinitely by avoiding the usual limitations on the number of divisions that normal cells experience.

*Inducing Angiogenesis*: Cancer cells can stimulate the growth of new blood vessels (angiogenesis) to provide them with the necessary nutrients and oxygen for continued growth and survival.

*Activating Invasion and Metastasis*: Cancer cells can invade nearby tissues and spread to distant sites in the body, a process known as metastasis, which is a major cause of cancer-related deaths.

*Avoiding Immune Destruction*: Cancer cells can evade recognition and attack by the immune system, allowing them to escape immune surveillance.

*Tumor-Promoting Inflammation*: Chronic inflammation in the tumor microenvironment can promote cancer development by creating a favorable environment for cancer cells to grow and evade the immune system.

*Genome Instability and Mutation*: Cancer cells often exhibit high levels of genetic instability and mutations, which contribute to the diversity of cancer cell populations and their ability to adapt to different conditions.

These hallmarks are not exclusive; they often interact with and reinforce each other to drive the complex process of cancer development. Understanding these fundamental traits has led to the development of targeted therapies that aim to interfere with specific hallmark features, offering new strategies for cancer treatment. However, it is important to note that each cancer type and even individual tumors can exhibit variations in the prominence of these hallmarks, making each case unique and challenging to treat effectively.

## 3. Ellagic Acid

Ellagic acid (EA) is a naturally derived polyphenol present in a variety of fruits, nuts, and vegetables. Notably concentrated in fruits such as pomegranates, strawberries, raspberries, blackberries, and nuts such as walnuts. This compound also shows higher levels in certain tree species' wood and bark, including *Quercus* spp., *Eucalyptus* spp., and *Castanea* spp. [8].

In addition to the naturally occurring unbound EA found in plant-based foods, a substantial portion of this compound is produced within the gastrointestinal tract of both humans and animals. This production is a result of the enzymatic or non-enzymatic breakdown of dietary polyphenolic molecules called ellagitannins [9]. Within the gastrointestinal tract, EA has restricted bioavailability, primarily due to its hydrophobic nature and very low water solubility. Consequently, only a fraction of EA is absorbed in the small intestine. The unabsorbed EA molecules undergo further metabolic transformations within the large intestine, facilitated by intestinal microorganisms. This process leads to the formation of a group of lipophilic metabolites known as urolithins, which are subsequently absorbed into the bloodstream [10]. Readers can find more about the fate and biological effects of EA in a recent article authored by Sharifi-Rad and colleagues [11].

EA has gained attention in the last decades due to its broad range of biological effects and potential health benefits. It is known for its strong antioxidant properties [12,13], anti-inflammatory effects [14,15], and antimicrobial [16–18] and antimutagenic effects [19]. Furthermore, many studies have reported cardioprotective [20], neuroprotective [21,22], gastroprotective [23], hepatoprotective [24], and nephroprotective [25] effects of EA.

Moreover, numerous studies indicate that EA possesses properties that can inhibit cell proliferation and potentially counteract cancer development. In light of this perspective, the aim of this review is to clarify the mechanisms by which EA impacts fundamental cancer characteristics. The chemical structure of EA is shown in Figure 2.

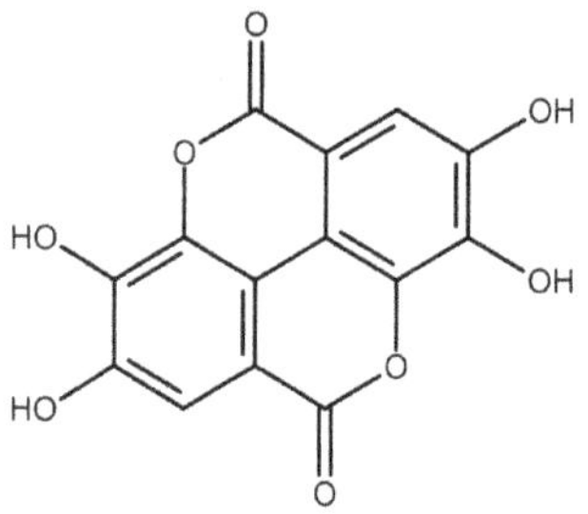


**Figure 2.** Chemical structure of ellagic acid.

### 3.1. Ellagic Acid and Sustaining Proliferative Signaling

One of the fundamental hallmarks of cancer is sustained (uncontrolled chronic cell proliferation) [7], often arising from deregulations in the normal cell cycle [26]. Generally, gap (G) phases represent the essential cell cycle checkpoints, in which the cell cycle is slowed down or stopped to ensure the accuracy and integrity of cell division. The importance of these checkpoints lies in their ability to maintain genomic stability and prevent the proliferation of damaged or mutated cells. Growing evidence supports the view that the main restriction checkpoint is G1-S transition [27], and the complex of cyclin D with two cyclin-dependent kinases (CDK4 and CDK6) seems to be its primary positive regulatory element [28]. The primary function of this complex is to sustain the phosphorylation of the retinoblastoma protein (pRB), ultimately leading to the progression of the cell cycle. On the other hand, the dephosphorylated state of this protein is widely recognized as a tumor growth suppressor, as it halts the cell cycle through the pRB pathway. Moreover, DNA damage signaling is the primary hindrance to entry into the S phase of the cell cycle.

Generally, CDK4/6 activity is modulated via the association with D-type cyclins (D1, D2, D3), which activates it, and via the binding of cyclin-dependent kinase inhibitors (CKIs) from the INK4 family (p16, p15, p18, p19), which inhibits it, ensuring proper control of the cell cycle [29]. However, additional kinases and CDKs can be involved in cell cycle control at other checkpoint sites [30].

Previous data indicate that the sustained proliferation in cancer cells may be attributed to various factors, including but not limited to altered expression or activity of molecules regulating the cell cycle checkpoints and the constitutive activation of various signal transduction pathways [26]. In brief, these mechanisms include increased generation of ligands (e.g., epidermal growth factor, EGF) for growth factor receptors (GFRs), increased receptor signaling as a result of upregulated GFRs at the surface of cancer cells, receptor activation without the need for ligands (ligand-independent firing), and constitutive activation of the elements within signaling pathways. However, mitogenic signaling in cancer cells is generally better comprehended than the action of mitogenic factors themselves. Furthermore, the signaling molecules (e.g., protein-kinases) of the pathways regulating the cell cycle progression have pivotal roles in various other processes that are important in oncogenesis (e.g., apoptosis, cell differentiation, angiogenesis, cell migration, and adhesion or stress response) [31]. Consequently, studying these pathways in cancer development presents considerable challenges.

Interestingly, signaling pathways that influence the G1 phase progression of the cell cycle typically do not control CDK levels directly. Instead, they primarily impact CDK activity by regulating the expression of their associated cyclin partners and a set of CKIs [27]. In the control of the cell cycle, cyclin D fulfills its role in a complex with specific CDKs (CDK4 and CDK6), acting as their allosteric modulator. The newly established complex can interact with CDK-activating kinases (CAKs), initiating phosphorylation and subsequent activation of CDKs [32]. Referring to the prior text in the article, the active cyclin D-CDK4/CDK6 complex is a pivotal regulator of pRB phosphorylation [28]. The literature on cyclin D also reported its additional functions, such as the sequestration of two important CKIs p21 and p27 [33], phosphorylation of other transcription factors involved in the regulation of the cell cycle Smad3 and forkhead box M1 (FOXM1) [34], effects on mitochondrial metabolism [35], and the effects in the cellular response to genotoxicity [36]. The available evidence seems to suggest that transcription and activity of cyclin D are significantly reliant on the mitogenic signals from three key signaling pathways: the MAPK pathway (Ras/Raf/MEK-ERK for cyclin D synthesis), the PI3K/Akt pathway (for cyclin D stability), and the Wnt/ß-catenin pathway (ß-catenin/TCF-LEF) [37]. Moreover, the transcription of cyclin D can also be regulated by other signals, such as the Notch pathway, extracellular matrix, cytokines (through STAT3), and immune signals (through nuclear factor kappa B, NF-κB) [27]. Based on the literature, there are three distinct types of D-type cyclins (cyclin D1–3), each of which is encoded by the genes *CCND1*, *CCND2*, and *CCND3*, respectively [38]. All three members of this family can interact with CDK4 and CDK6, playing important roles in the regulation of the cell cycle. In addition to their presence in various healthy cells, they have also been identified in tumor cells. However, their levels may differ depending on the particular tissue and type of cancer. Next, it has been reported that their oncogenic potential is typically linked to genetic deregulation, often stemming from gene amplification, mutations, or other genetic alterations [37]. Interestingly, the amplification of *CCND1* was found to be the most frequently observed genetic event in cancer [39].

Another key player in the narrative of sustaining proliferative signaling is the Myc protein, encoded by the c-Myc gene. Abnormalities in Myc expression or function are closely linked to various cancers [40]. Myc operates as a transcription factor, wielding control over a multitude of target genes involved in cell cycle progression, cell proliferation, differentiation, and apoptosis [41]. Its pivotal role lies in promoting the transition from G1 to S phase by upregulating cyclin D2 expression and inhibiting the expression of genes encoding CKIs such as p21 and p15. Furthermore, Myc can influence other crucial cell cycle regulators, including CDK4, E2F1, and CDC25A, while downregulating p27 [40]. The regulation of the Myc protein is a complex process involving multiple critical signaling pathways, such as Notch1 [42,43], MAPK [44], PI3K/Akt/GSK-3β [45], Wnt-ß-catenin [46], Janus kinase 2, and signal transducer and activator of transcription 3 (JAK-STAT3) [47] or Hedgehog [48].

Beyond cyclins and Myc, emerging research has shed light on the role of mini-chromosome maintenance (MCM) proteins in cell cycle checkpoints [49]. MCM3, in particular, has been implicated in regulating the cell cycle, with its knockdown leading to reduced expression of key cell cycle progression elements and G1/S transition suppression [50]. Elevated expression of MCM2, MCM5, and MCM7 has also been identified as potential prognostic biomarkers associated with unfavorable outcomes in several cancer types [51].

Classical isoenzymes of the protein kinase C (PKC) family can also be involved in cell cycle regulation and cell growth by influencing its downstream targets such as Myc, cyclins, CDKs, CKIs [52,53], or NF-κB [54].

In addition, COX-2 may contribute to the modulation of the cell cycle control, e.g., via inhibition of cyclin D generation [55].

Based on the above, the inhibition of sustaining proliferative signaling pathways represents an important strategy in the quest for effective cancer therapy.

Accumulated evidence has confirmed that EA can attenuate several specified signal molecules or signal pathways involved in cell cycle control and proliferation in different cancer cells (Figure 3).

For instance, EA effectively suppressed cell proliferation of two human ovarian cancer cells (ES-2 and PA-1). It achieved this by causing cell cycle arrest at the G1 phase in both cell lines. This arrest was attributed to the reduction of cyclin D1 and E levels, as well as to the upregulation of p53 and p21 proteins [56].

In another study, the analysis of murine xenografts of human pancreatic cancer cells (PANC-1) highlighted that the antitumor effect of EA may manifest via the inhibition of various signaling pathways and molecules [57]. Some of the signaling pathways (Notch, PI3K/Akt, or Hedgehog) and molecules (cyclin D1, CDK2, CDK6, COX-2, IL-6, and IL-8) investigated in this study, based on the knowledge from previous research, could have an impact on cell cycle regulation to some extent. Subsequent studies have substantiated the capacity of EA to impede the proliferation of pancreatic cancer cells in laboratory settings (in vitro using PANC-1 cells) and within living organisms (in vivo, employing mice bearing PANC-1 tumors) [58]. Furthermore, in vitro analyses conducted on pancreatic cancer cells have revealed a connection between EA treatment and the arrest of the cell cycle at the G1 phase. Additionally, there has been a notable reduction in the protein expression of COX-2 and NF-κB as a result of this treatment.

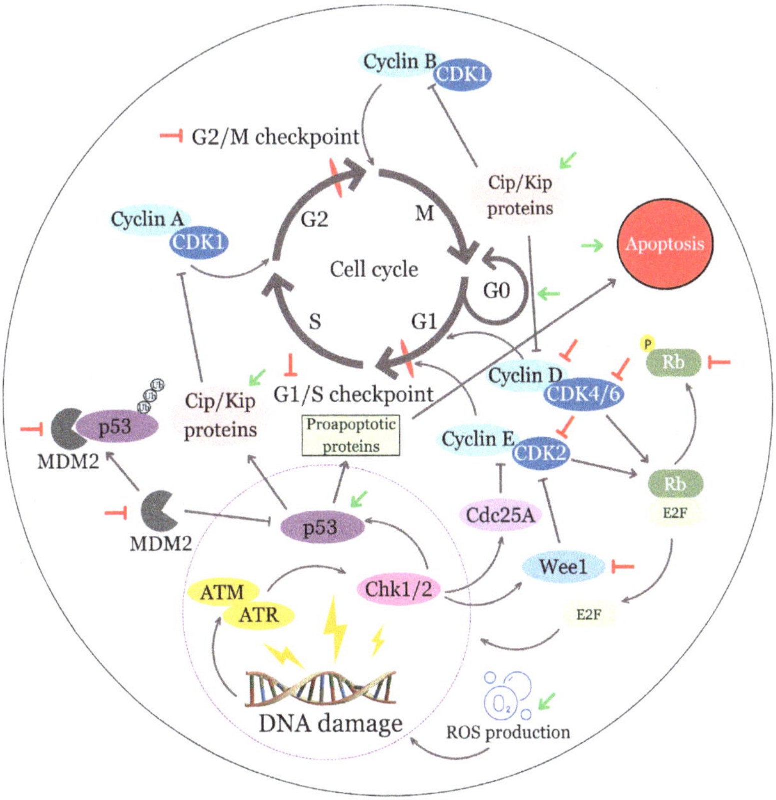


**Figure 3.** Ellagic acid eliminates self-sufficiency in growth signals and restores responsiveness to growth-inhibitory signals. Upward-pointing green arrows signify heightened expression/activation, while red T bars indicate lowered expression or reduced activity.

Next, in vivo, experiments on U87 glioblastoma mouse xenografts showed that EA can mitigate cell proliferation by suppressing Akt and Notch signaling pathways, as well as reducing the levels of several signaling molecules, including cyclin D1, CDK2, and CDK6 [59]. In human bladder cancer cells (TSGH8301), EA demonstrated the capacity to hinder cell proliferation by concurrently suppressing the expression of two proteins governing the cell cycle, namely CDK2 and WEE1 kinase [60]. Additionally, it prompted a state of cell cycle arrest specifically at the G1 phase. Other researchers have documented a decrease in cell proliferation, a halt in the cell cycle at the G1 phase, and the induction of apoptosis in human cervical carcinoma cells (HeLa) treated with EA. This was observed concurrently with the suppression of the JAK/STAT3 pathway [61]. Furthermore, there is a suggestion that the reduced cell proliferation in HeLa cells induced via EA treatment might be associated with the decreased expression of epidermal growth factor receptor (EGFR) protein [62]. In the identical cell line, EA also demonstrated this effect when used in combination with curcumin. Moreover, the authors observed cell cycle arrest either in the S phase when EA was administered alone or in the G2/M phase when EA was combined with curcumin. As reported in the literature, hormones are other regulators of cell cycle progression via upregulation of factors promoting cell cycle progression (e.g., cyclin D1 or Cdk-activating phosphatase, Cdc25A) and downregulation of CKIs (p21, p27) expression [63,64]. Interestingly, EA acted as a natural estrogen antagonist in HeLa cells, primarily acting through ERbeta [65]. In addition, in endometrial cancer cells (KLE and AN3CA), EA has also demonstrated a reduction in cell proliferation as a result of inducing cell cycle arrest in the G1 phase, ultimately leading to apoptosis [66]. Furthermore, both bioinformatics analyses and in vitro experiments have underscored the significance of two key signaling molecules, phosphatidylinositol-4,5-bisphosphate 3-kinase catalytic subunit alpha (PIK3CA) and phosphatidylinositol-4,5-bisphosphate 3-kinase regulatory subunit 1 (PIK3R1), in the anticancer effects of EA. It has been documented that EA downregulates their gene and protein expression. While this study has examined the downregulation of these two signaling molecules in the context of inhibiting cell migration, it is important to acknowledge that these actions may potentially contribute to various other anti-tumor effects of EA as well through involvement in several signaling pathways.

EA was also found to inhibit signaling pathways that regulate cell proliferation in prostate cancer cells (PC3), even in the presence of upregulated gene expression of IL-6 [67]. It specifically downregulated the expression of phosphorylated ERK1/2, AKT, and STAT3, effectively suppressing cell proliferation.

Earlier studies have identified the downregulation of the TGF-β/Smad3 signaling pathway as a critical factor contributing to the inhibition of cell proliferation and the induction of cell cycle arrest in EA-treated MCF-7 breast cancer cells [68,69]. Recent research on human breast cancer cells (MCF-7 and MDA-MB-231) documented EA as a potent inhibitor of CDK6 [70]. The compound not only impeded tumor growth but also suppressed the gene and protein expression of CDK6 within MCF-7 and MDA-MB-231 cancer cells. Additionally, EA demonstrated the capacity to bind to CDK6 and effectively inhibit its activity.

In metastatic human melanoma cell lines (namely 1205Lu, WM852c, and A375), EA exhibited several noteworthy in vitro effects. Notably, it proficiently inhibited cell proliferation, induced cell cycle arrest at the G1 phase suppressed NF-κB activity, and significantly diminished gene expression of pro-inflammatory cytokines such as IL-1β and IL-8 [71]. Furthermore, apoptosis was also observed in the study. Later experiments performed on melanoma cancer cells (WM115 and A375) elucidated that EA can hinder cell proliferation by inhibiting the EGFR signaling pathway [72]. Specifically, significantly reduced protein expression of the phosphorylated form of EGFR was one of the obtained results. A subsequent in vivo analysis using a nude mouse A375 tumor xenograft model confirmed the ability of EA to inhibit phosphorylation of EGFR and reduce tumor size and weight.

In human colon cancer cells, specifically HT29 and HT116, EA administration was observed to notably inhibit cell proliferation and downregulate the Wnt/β-catenin signaling pathway [73]. Inhibition of this signaling cascade was linked to a reduction in the protein expression of β-catenin. Additionally, there was an elevation in the levels of phosphorylated β-catenin, axin 1, and axin 2, suggesting enhanced degradation of β-catenin. Moreover, the downstream components of this cascade, specifically Myc, cyclin D1, and survivin, exhibited decreased levels. Furthermore, a diminished CDK8 level was observed. In the next study, EA was also found to impede the proliferation of HCT-15 colon adenocarcinoma cells [74]. Subsequent analyses showed prompt cell cycle arrest (G2/M) and apoptosis induction via modulation of the PI3K/Akt signaling pathway in EA-treated HCT-15 cancer cells. In other human colorectal cancer cell lines (HCT-116 and CaCo-2), EA exhibited a significant capacity to impede cell proliferation. It also demonstrated the ability to suppress the expression of K-ras protein, along with the significant suppression of Akt phosphorylation at both Thr308 and Ser473. Moreover, EA caused cell cycle arrest at either the G1 or S phase [75]. Later, the experimental data conducted on HCT-116 cells pointed to the ability of EA to retard cell proliferation and elicit G0/G1 cell cycle arrest and apoptosis, all likely as a result of stimulation of the TGF-ß1/Smad3 signaling pathway [76]. Triggering of this pathway resulted in an upregulation of p15, a well-established CKI responsible for dampening the activity of the CDK4/CDK6 complex.

In hepatocellular carcinoma cells (HepG2), EA treatment resulted in a notable decrease in both gene and protein expression among several members of the MCM family (MCM2-7) and CDKs (CDK2 and CDK4), concomitantly with upregulated p21 and decreased phosphorylation of pRB [77]. Furthermore, cell cycle arrest in the G1 phase was observed.

The treatment with EA was also observed to be associated with the suppression of PKC signaling in Dalton's lymphoma-bearing (DL) mice, resulting in the inhibition of cell proliferation [78]. Specifically, EA has demonstrated a notable downregulatory impact on the expression of classical isoenzymes of the PKC family (PKC$\alpha$, PKC$\beta$, and PKC$\gamma$). Moreover, EA also exhibited the capacity to diminish the overall activity of PKC and the expression of downstream molecules, including NF-kB and c-Myc. On the other hand, the expression of TGF-$\beta$1 was upregulated. According to the DrugBank database (Edmonton, AB, Canada), EA serves as a competitive inhibitor for both protein kinase C alpha and beta types [66].

### *3.2. Ellagic Acid and Evading Growth Suppressors*

Insensitivity to anti-growth signals is another typical characteristic of many malignancies. It occurs when the function of growth suppressors encoded by tumor suppressor genes is deregulated. Under normal conditions, the role of tumor suppressors is to control the cell cycle and ensure the repair of DNA damage, limit inappropriate cell growth, and reduce invasive and metastatic potential. Their normal function thus prevents the development and progression of cancer or the failure of cancer treatment. However, genetic and epigenetic alterations, as well as other factors, can alter their function and subsequently they may even gain oncogenic potential [79,80]. Therefore, the complexity of molecular interactions requires further research.

Anti-growth signaling includes several important molecules [80]. However, mainly p53, pRB, p16, and p21 proteins were reported as common tumor suppressors controlling the cell cycle [81]. Based on evidence from the literature, EA may modulate specifically three of them (p53, p21, and pRB). In addition, the effects of EA on p53 and p21 have often been evaluated together in research.

Among a variety of known growth suppressors, p53, dubbed "guardian of the genome" [82], is the most studied. It is a transcription factor induced by many endogenous and exogenous stressors [83]. Its function is to regulate several types of programmed cell death [84] and pathways for the generation of reactive oxygen species (ROS) [85], but it is also involved in controlling cell cycle and DNA repair [81] besides many other functions associated with cancer progression and cancer treatment failure [79]. The protein is frequently genetically altered or absent in various cancer cells. Genetic alterations occur in more than 50% of all cancers with the highest genetic alterations frequency in small cell lung cancer, non-melanoma cancer, and ovarian cancer [84]. In this sense, downstream target genes of p53 are not activated, and the protein loses its tumor-suppressive activity. Moreover, mutated forms of p53 independently activate several signaling pathways or molecules, e.g., NF-Y/p300 [86], Y/NF-κB–MAP2K3 [87], EGFR/ERK due to suppression miR-27a [88], YAP–cyclin A/cyclin B/CDK1 [89], YAP/TEAD–circPVT1 [90], STAT3 [91] and CDK4–cyclin D1– c-MYC [79] to encourage cell proliferation. Interestingly, the crucial negative regulator of p53 is a nuclear protein known as a murine double minute clone 2 (MDM2). It functions mostly as an E3 ubiquitin ligase that targets p53 [92] and some other proteins involved in cell cycle regulation and apoptosis [93] for proteasomal degradation. However, it also blocks the activity of intact p53 by inhibiting the p53 transactivation [94]. In short, cancer development and progression can occur despite wild-type p53 as a consequence of MDM2 upregulation. The prevalence of MDM2 amplification has been reported, in various types of cancer, from 3.5% [95] to 5.5% [96], mostly in sarcomas [97]. Moreover, MDM2 amplification may be associated with drug resistance [98]. Recently, several MDM2 inhibitors have even been discovered, which are currently under investigation in clinical studies for their possible capacity to treat cancer, e.g., glioblastoma [99].

Protein p21 (CIP1) or cyclin-dependent kinase inhibitor 1A (*CDKN1A*) is another important tumor suppressor controlling cell cycle checkpoint mechanisms that inhibits cyclins, CDKs, or their complexes (cyclin-CDK2, cyclin-CDK1, and cyclin-CDK4,6) (Al Bitar and Gali-Muhtasib, 2019). Due to its effects, cell cycle progression is ultimately inhibited in several phases [100,101]. However, the pro-oncogenic function of p21 was also reported [102]. According to the other authors, the opposite role of p21 in carcinogenesis depends on cytoplasmic or nuclear p21 expression [103]. Recently, it has been suggested that p21 inducers might be valuable, especially in the treatment of cancers with downregulated p21 [104]. On the other hand, patients suffering from cancers with increased levels of p21 might be candidates for the treatment of p21 inhibitors. The expression of p21 is primarily regulated by p53, although p53-independent regulation exists [105].

Next, pRB is the protein associated with the development of retinoblastoma and many other cancers. This protein is encoded by the RB transcription corepressor gene 1 (*Rb1*), which was the first tumor suppressor gene discovered [106]. According to the literature, pRB can control the cell cycle via both canonical (E2F-dependent) and non-canonical (E2F-independent) mechanisms [107]. Canonical mechanisms are more common and depend on the phosphorylation status of the protein. The non-phosphorylated form of pRB binds and inhibits the transcription function of several members of the E2F family (E2F1, E2F2, E2F3), which results in suppression of the cell cycle progression and finally in the arrest of cell proliferation. Conversely, phosphorylation of pRB induced by cyclin-dependent kinases (mainly by CDK4/6) during G1/S phases disturbs pRB/E2F interaction allowing the transcription factor to fulfill its function to promote the cell cycle [108]. The non-canonical mechanism of cell cycle suppression involves increasing the levels of p27 protein, which is thought to be a CDK inhibitor. The expression and function of pRB are usually diminished by genetic [109] and epigenetic [110] inhibition, but also other regulating factors of the pRB pathway were determined. For example, oncoprotein from human papillomavirus HPV E7 can suppress the pRB-E2F interaction [111]. Additionally, repression of pRB occurs after several post-translational modifications [112]; for example, MDM2 targets pRB for degradation by the proteasome [113]. Furthermore, upstream regulators can be modified [107]. For example, genes for cyclins and CDKs can be amplificated or genes for cyclin-dependent kinase inhibitors can be deleted. As a result, phosphorylation and functional repression of pRB occur.

EA-induced upregulation of p53, p21, or both together has been demonstrated in several studies in various cancer cells. However, the molecular mechanisms underlying their overexpression and implications of this phenomenon are not yet fully elucidated. For instance, increased mRNA levels of p53 and p21 accompanied by reduced cell viability, G0/G1-phase arrest of the cell cycle, and apoptosis induction were found in case of exposure to EA in human bladder cancer cells (T24) [114]. Later investigations confirmed increased protein levels of p53 and p21 in other human bladder cancer cells (TSGH8301) treated with EA, which the authors attributed to the subsequent arrest of the cell cycle at the G0/G1 phase [60]. The next study has also documented the suppressive impact of EA on cell viability, in addition to its capacity to arrest the cell cycle at the G1 phase and induce the upregulation of protein expression for both p53 and p21 in two distinct ovarian carcinoma cells (ES-2 and PA-1) [56]. Other authors evaluated the antiproliferative effects of EA in human prostate cancer cell lines [115]. In this study, EA inhibited cell proliferation in the androgen-dependent cell line (LNCaP), whereas it had no impact on proliferation in the androgen-independent cells (PC-3 and DU145). Furthermore, the treatment of LNCaP cells with EA resulted in an upregulation of protein expression levels of several cell cycle regulatory proteins, including p21 and p27, which play a crucial role in regulating the transition of cells from the G1 phase to the S phase in the cell cycle. Thus, the authors have suggested that EA hindered the proliferation of prostate cancer cells by initiating both cell cycle arrest and apoptosis. In HepG2 cells, EA decreased cell viability and enhanced the expression of p53 and p21 [116]. In addition, these effects were further intensified when EA was combined with gamma irradiation. Later, it was observed that EA-treated HepG2 cancer cells exhibited elevated protein levels of p53, possibly contributing to apoptosis, which could be linked to an increase in ROS [117]. The recent research performed on EA-treated HepG2 cells and aimed at differentially expressed gene analysis declared the upregulation of *CDKN1A*, a gene encoding p21, as an essential mechanism responsible for cell growth inhibition via induction of cell cycle arrest at the G1 phase [77]. The significance of this discovery was affirmed via subsequent *CDKN1A* knockdown analysis. EA also had notable antiproliferative and apoptotic effects in rat C6 glioma cells, as well as the ability to increase protein expression of p53 [118,119]. The cell cycle arrest at the G0/G1 phase was another consequence of the action of EA in this cell line. In human breast cancer cells (MCF-7), EA inhibited cell proliferation and halted the cell cycle at the G0/G1 phase via an increase in p21 induced via overexpression of a signal molecule called mothers against decapentaplegic homolog 3 (Smad3) [68]. The antiproliferative and pro-apoptotic properties of EA were also demonstrated in human gastric adenocarcinoma cells (AGS), while one of the observed changes was a significant increase in mRNA expression of the p53 protein [120]. In another study using human cervical carcinoma cells (HeLa), EA significantly enhanced p53 and p21 only in combination with curcumin [121]. Nonetheless, a prior investigation involving various cervical carcinoma cells (CaSki) reported that EA induced an upregulation of p21 protein expression without a substantial elevation in p53 levels [122]. These findings suggest that EA could potentially induce G1 arrest and apoptosis via the upregulation of p21, employing a p53-independent mechanism. EA additionally potentiated the impact of quercetin in promoting the phosphorylation of p53 at Ser 15 and increasing p21 protein levels in the human leukemia cell line (MOLT-4), ultimately leading to apoptosis [123].

Additionally, in the study conducted by Mohammed Saleem and Selim [124] using different models of prostate cancer (LNCaP, $p53^{+/+}$; 22RV1, $p53^{-/+}$, and PC3, $p53^{-/-}$), EA was reported as effective in reducing MDM2 at the gene and protein levels in all investigated cancer cell lines. These effects were also associated with an enhancement of protein expression of its downstream target p53 in LNCaP and 22RV1, but not in PC3 cells. Moreover, this change was not accompanied by a subsequent increase in MDM2 (a p53 target molecule) within an autoregulatory feedback loop. On the other hand, increased levels of p21 have been found in all prostate cancer cell lines, independent of the status of p53. This finding points to the ability of reduced MDM2 levels to exert p53-independent anticancer effects, which has been demonstrated by the downregulation of X-linked inhibitor of apoptosis protein (XIAP). However, most effects of EA observed in this study were primarily assessed in the context of apoptosis induction. Next, the authors revealed enhanced phosphorylation of p53 at ser15 and ser20 induced by EA in 22RV1 cells. It is noteworthy to mention that earlier research had elucidated that phosphorylation of p53 at ser20 diminishes the interaction between p53 and MDM2 [125]. Interestingly, urolithin A, a metabolite of EA, has also been shown to stabilize p53 [126] and attenuate MDM2-mediated p53 ubiquitination despite increased levels of MDM2 as a result of autoregulatory feedback loop [127]. In colorectal cancer, upregulation of p21 induced via specific microRNA is thought to be an important mechanism involved in anticancer properties (inhibition of cell growth, cell cycle arrest, and apoptosis) of ellagitannins [128]. However, some urolithins generated by gut microbiota were much more potent than EA in these effects.

Less is known about the effects of EA on pRB. As it was reported, EA exerted a decrease in phosphorylation of pRB and G0/G1 cell cycle arrest in MCF-7 breast cancer cells [68]. In addition, the results from the MTT assay confirmed decreased cancer cell proliferation. Finally, the authors highlighted that decreased pRB phosphorylation and effects on cell proliferation might be mediated by the activation of transforming growth factor β (TGF-β)/Smad3 signaling and subsequent overexpression of p21.

Equally important, several tumor suppressors may be non-functional at the same time in cancer cells. For example, the combined loss of *TP53* and *RB1* led to increased cell proliferation and diminished therapeutic outcomes for several androgen receptor antagonists [129]. Another essential point is the overlapping of both p53 and pRB signaling pathways via the function of p21, which can inhibit CD/CDKN complexes responsible for pRB phosphorylation [130]. Also, ubiquitination by MDM2 is common for both p53 and pRB proteins [112]. To sum up, EA may represent a promising potential drug, as it regulates the activity of all three major components of the p53–p21–pRB signaling pathway.

Furthermore, EA can also affect less important tumor suppressor proteins, such as PTEN (phosphatase and tensin homolog deleted on chromosome 10) that contribute to several aspects of cancer development and progression, including cell proliferation. Evidence has been provided in the literature that EA enhances the protein phosphatase activity of PTEN in melanoma cells (B16F10), resulting in reduced phosphorylation of focal adhesion kinase (FAK) [131]. Active phosphorylated FAK can stimulate cell cycle progression and proliferation signals, while its dephosphorylated form does not contribute to these processes. In this study, increased activity of PTEN also resulted in the accumulation of p53 as an indirect effect of EA. Figure 3 shows the findings described in this topic.

### 3.3. Ellagic Acid and Resisting Cell Death

The accompanying phenomenon of most malignant diseases is the deregulated response of tumor cells to proapoptotic stimuli. Therefore, the induction of cell death via apoptosis has been considered an essential mechanism of the action of cytotoxic drugs from the beginning of targeted research on antitumor chemotherapeutics, either directly or indirectly [132]. Under physiological conditions, apoptosis ensures normal tissue homeostasis regulation and a balance between cell proliferation and cell death [133,134]. This process is a response to cell damage caused by endogenous and exogenous stress factors such as DNA damage, endoplasmic reticulum stress, irradiation, UV radiation, heat shock, growth factor deficiency, viruses, Fas ligand (FasL), cytokines from the TNF family, and others [135,136].

In response to stress, the death machinery is activated via the intrinsic mitochondrial pathway of apoptosis, which is regulated by antiapoptotic and proapoptotic proteins from the Bcl-2 family. After mitochondrial membrane damage and permeabilization, cytochrome *c*, AIF, and other proapoptotic molecules are released, leading to the formation of the apoptosome and the activation of effector caspases. The extrinsic receptor-mediated pathway mediates signals from the extracellular space via ligands (TRAIL, TNF$\alpha$, FasL) and death receptors (TNFR1, DR4, DR5), followed by the cleavage of initiator caspase 8 and effector caspases 3 and 7 [137]. Apoptosis can also be mediated by the significant tumor suppressor p53 after DNA damage or endoplasmic reticulum stress in a response known as the unfolded protein response (UPR) [138,139]. Morphologically, these complex molecular mechanisms result in chromatin condensation, nuclear fragmentation, cell shrinkage, cytoskeletal degradation, plasma membrane blebbing, and the formation of apoptotic bodies without eliciting an inflammatory response [140,141].

The disruption of some of these key regulatory mechanisms plays a major role in suppressing the apoptosis of tumor cells. Sublethal cell damage such as oxidative stress, hypoxia, DNA damage, starvation, and others, along with acquired genomic changes with oncogenic potential, lead to the development of drug-tolerant persistent or resistant cells. Therefore, it is necessary to explore additional forms of programmed cell death, such as necroptosis, ferroptosis, pyroptosis, and others, to complement apoptosis-inducing anti-tumor therapy [142,143]. Malignant cells with a reduced response to apoptotic stimuli or resistance to apoptosis are characterized by an imbalance in members of the Bcl-2 family, specifically overexpression of antiapoptotic proteins such as Bcl-2, Mcl-1, Bcl-xL, and/or underexpression of proapoptotic proteins such as Bax, Bad, Bim, and others. Reduced expression of death receptors and ligands, or expression of death receptors without death domains incapable of transmitting the stimulus into the cell, can disrupt the induction of the extrinsic pathway. In addition to increased expression of apoptosis inhibitors known as IAPs, decreased caspase expression, defects, and mutations in the tumor suppressor p53, which upregulates proapoptotic proteins and significantly influences the metabolism of tumor cells, are also involved in cell resistance to apoptosis [144,145]. Changes in the expression itself and the associated transcription of apoptosis-associated proteins, along with post-translational regulation changes, significantly affect not only the levels but also the function of expressed proteins. These processes not only hinder programmed cell death and contribute to the survival of malignant cells but can also accelerate tumorigenesis [146].

Currently, ongoing work is focused on new drugs and therapeutic approaches, such as cyclotherapy, which would be effective against tumor cells while sparing healthy non-tumor cells [147]. New drugs directly inducing apoptosis include Bcl-2 inhibitors, such as the FDA-approved venetoclax, Bcl-xL and Mcl-1 inhibitors, IAP inhibitors, Smac mimetics, PARP inhibitors, or other molecules in various stages of clinical trials, whose main mechanism of action is the induction of the extrinsic pathway of apoptosis [148–152].

Studies from recent years have shown that EA is capable of modulating several molecular targets, including those essential for the inhibition and induction of apoptosis (Figure 4). Along with influencing the activity of proteins regulating the cell cycle, such as p53, p21, cyclins, cyclin-dependent kinases, and their regulators, EA also affects members of the Bcl-2 family under in vitro conditions in human bladder cancer T24 cells and in vivo in glioblastoma U87 cells in xenografted Balb/c nude mice [59,114]. Reducing the ratio of levels between the antiapoptotic protein Bcl-2 and the proapoptotic protein Bax signifies a stimulus for cells that reduces their resistance to apoptosis and leads to cell death via the induction of the mitochondrial pathway of apoptosis [153]. This is due to changes in interactions between proapoptotic BH3-only proteins that inhibit Bcl-2, Mcl-1, Bcl-xL, and others. This activation leads to the oligomerization of Bax and Bak, allowing them to permeabilize the mitochondrial outer membrane [154]. Several studies confirm that EA significantly increases the Bax/Bcl-2 ratio with increasing concentration and exposure time and downregulates Bcl-xL and Mcl-1. EA also increases the expression of Bcl-2 inhibitory proapoptotic proteins PUMA and Noxa in prostate cancer cells [124]. This effect has been demonstrated in prostate cancer cell lines LNCaP (androgen-responsive prostate cancer), ES-2 and PA-1 (human ovarian carcinoma), cervical cancer HeLa cells, and non-small cell lung cancer A549 cells [56,61,115,155].

After the permeabilization of the mitochondrial outer membrane (MOMP), depolarization and loss of the transmembrane potential ($\Delta\psi m$) generated by proton pumps (Complexes I, III, and IV) occur. These are essential for oxidative phosphorylation and ATP production [156]. EA, via the upregulation of Bax and Bak, forms pores, significantly reducing the mitochondrial membrane potential (MMP) in the lung (HOP62 and H1975) and hepatocellular (HepG2) cancer cells. In these cells, ATP levels were significantly reduced [157–159]. A similar effect was observed in MCF-7 cells (human breast adenocarcinoma) after exposure to an extract from the plant *Terminalia chebula* from the *Combretaceae* family, which also contains EA as a significant component [160].

Following the induction of the intrinsic pathway of apoptosis by EA, in addition to the reduction in MMP, the release of cytochrome *c* into the cytosol occurs in pancreatic cancer cells MIA PaCa-2 and PANC-1 and colon adenocarcinoma cells HCT-15. Although EA does not directly affect mitochondria, it induces apoptosis via dose-dependent inhibition of the binding ability of the transcription factor NF-κB [74,158,161]. In addition to cytochrome *c*, proteins Smac/Diablo and Omi/HtrA2, and endonucleases apoptosis-inducing factor (AIF) and endonuclease G (EndoG) are released from the mitochondrial intermembrane space. Their role is oligonucleosomal DNA fragmentation [162,163]. Cytochrome *c* in a healthy cell shuttles electrons during oxidative phosphorylation from Complex III to Complex IV, but in the case of mitochondrial damage, it is one of the main inducers of cell death [164]. The mechanism of EA-induced caspase-dependent apoptosis involves the interaction of cytochrome *c* with APAF1 (Apoptotic protease activating factor 1), which is essential for the formation of the procaspase-9-activating heptameric protein complex known as the apoptosome. Ho et al. [60] studied the complex anti-proliferative and pro-apoptotic effect of EA on TSGH8301 (human bladder cancer) cells. They observed increased levels of apoptosis-inducing Bak, Bax, and Bid, along with mitochondrial membrane depolarization, induction of ROS production, and changes in intracellular calcium concentration, leading to increased levels of EndoG, Smac/DIABLO, AIF, cytochrome *c*, and APAF1 in the cytosol. In a study by Cheshomi et al. (2022) on gastric cancer cells AGS, in addition to upregulation of APAF1, significant increases in the expression of Bax, the tumor suppressor p53, and decreased expression of Bcl-2, NF-κB, and iNOS were observed after exposure to EA at concentrations of 15 and 30 μg/mL [120].

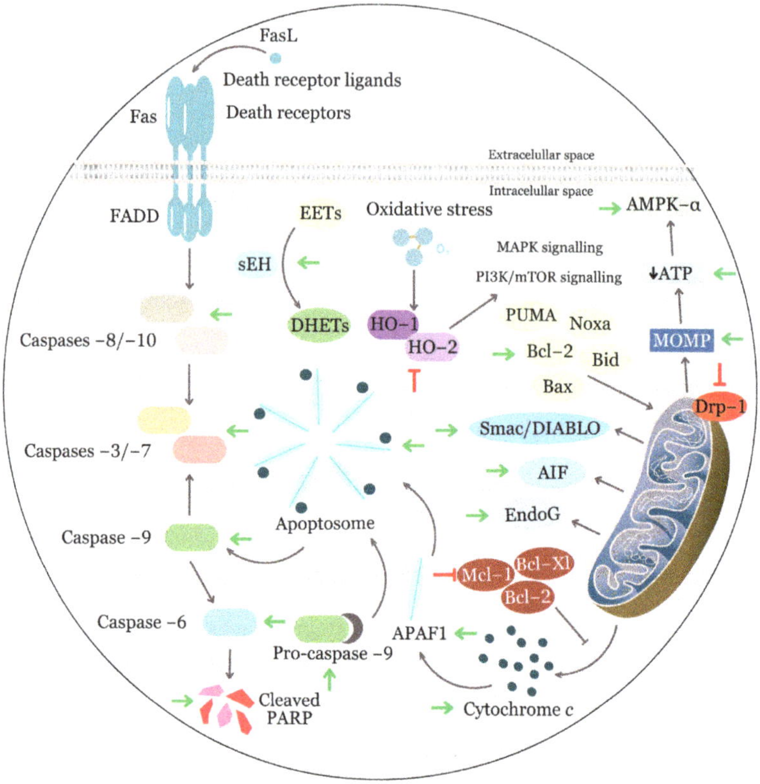


**Figure 4.** Effect of ellagic acid on intrinsic and extrinsic pathways of apoptosis. Upward-pointing green arrows signify heightened expression/activation, while red T bars indicate lowered expression or reduced activity.

After proteolytic cleavage and activation of pro-caspase 9 through the apoptosome, active initiator caspase 9 is formed. It is capable of further activating effector caspases 3, 6, and 7 via intrachain activation cleavage [165]. Interestingly, in the presence of a mutant procaspase 9 that has lost its function, the apoptosome can directly activate caspase 3 [166]. The increase in caspase 9 activity in EA-treated pancreatic cancer cells PANC-1 and AsPC-1 subsequently led to increased activity and expression of one of the effector caspases, caspase 3 [167]. This effect is closely related to the modulation of some signaling pathways, such as PI3K/Akt, which was observed in colorectal cancer cells HCT-15 and lung cancer A549 cells [74,155]. Similar results were also reported by Edderkaoui et al., who, using a pan-caspase inhibitor, demonstrated that cell death induced by EA is caspase-dependent [161]. Suppression of cell proliferation, irreversible DNA damage, and activation of caspase 3 were also demonstrated in an animal model of prostate adenocarcinoma [115].

Unlike caspase activation cleavage, the cleavage of DNA repair enzymes and damage to DNA repair mechanisms stimulate the progression of cell death. Poly(ADP-ribose) polymerases (PARPs) are DNA-dependent nuclear enzymes that significantly impact cellular energetics and determine the fate of the cell. PARP1 stabilizes p53 and enhances its function, but under excessive damage, it can paradoxically facilitate the translocation of AIF into the nucleus, promote DNA fragmentation, and induce caspase-independent cell death known as parthanatos. To preserve the required levels of NAD+ and ATP and allow the cell to undergo programmed cell death, caspases cleave PARP into fragments, thereby inhibiting its activity [168,169]. In various p53-expressing ovarian cancer cells ES-2 and PA-1 and prostate cancer cells LNCaP, 22RV1, and PC3, a direct correlation between EA-induced caspase 3 activation and PARP cleavage in a dose-dependent manner (10–100 μmol/L) was demonstrated, culminating in a significant increase in the number of apoptotic cells [56,124]. Furthermore, EA sensitizes and reduces the resistance of breast cancer MCF-7 cells to apoptosis induced by $\gamma$-radiation [170], Ba/F3-insH cells (a murine interleukin-3 dependent pro-B cell line) treated with erlotinib [171], and HT-29 cells treated with 5-fluorouracil [172].

The fact that mitochondria are one of the main targets of EA's anti-tumor action is also supported by the detection of additional mechanisms that contribute to mitochondrial dysfunction. Mitochondrial dynamin-related protein 1 (Drp-1) is a cytosolic guanosine triphosphatase (GTPase) responsible for the balance between mitochondrial division and fusion [173]. EA inhibits Drp-1-mediated mitochondrial fusion, contributing to the induction of the mitochondrial pathway of apoptosis in colorectal cancer (HCT116) and breast adenocarcinoma (MCF7) cells. This leads to a decrease in the respiratory capacity, loss of MMP, and cell cycle arrest [174]. Increased ATP production and intensified mitochondrial respiration are considered mechanisms that lead to the inhibition of cell death and increased resistance to anti-tumor treatment [175]. Results from a recent study show that EA reduced the expression of HIF-1$\alpha$ (Hypoxia-inducible factor-1$\alpha$), a regulator of mitochondrial oxygen metabolism often overexpressed in cancer cells, while increasing phosphorylation and total levels of AMPK-$\alpha$ and ACC protein in HOP62 and H1975 lung cancer cells [157,176]. Changes in ATP-sensitive kinase AMPK regulation suggest that EA may also induce other mechanisms, such as autophagy and autophagy-associated cell death [177,178].

In addition to inducing the intrinsic mitochondrial apoptotic pathway mediated by Bcl-2 family proteins and the activation of effector caspases 3 and 6, and initiator caspase 9, Malik et al. also observed EA-mediated activation of caspase 8 in PC3 prostate cancer cells, which is considered essential for initiating the extrinsic, receptor-mediated pathway of apoptosis [179,180].

The machinery involving both the extrinsic and intrinsic apoptotic cascades further continues with morphological changes to cells and the formation of apoptotic bodies. One of these changes is the damage to the cell membrane and the flippase-mediated externalization of the phospholipid phosphatidylserine from the inner side of the plasma membrane to the outer side, which serves as an "eat me" signal for activated phagocytes. Such damaged membranes are permeable to other molecules such as propidium iodide (PI), 7-amino-actinomycin (7-AAD), and others [181,182]. Dual staining using a combination of the aforementioned components allows for quantifying the proapoptotic effect of various substances on cells. It was found that EA dose-dependently (10–40 μmol/L) induces apoptosis in cervical cancer cells (HeLa), gastric cancer cells (AGS), and colorectal cancer cells (HCT-116), increasing the population of Annexin V-positive, PI-positive, and double-positive cells compared to control cells [61,76,120]. This effect was observed in combination with EA and erlotinib [171]. Similar staining with acridine orange/propidium iodide (AO/PI) can be used to identify EA-induced apoptotic cells, which, in the presence of cell death, nuclear penetration, and nucleic acid intercalation, emit orange and red fluorescence instead of green fluorescence of live cells [56,160].

Another mechanism by which cells resist apoptosis is the induction of cytoprotective and anti-inflammatory proteins. Heme oxygenase-1 (HO-1) is an inducible enzyme capable of responding to oxidative stress, inhibiting the activity of proapoptotic K+ channels, and activating signaling pathways such as p38 MAPK and PI3K/Akt to protect cells from apoptosis [183]. It has been shown that EA significantly reduced the levels of two isoforms of this enzyme, HO-1, and HO-2, and increased the levels of sEH (Soluble epoxide hydrolase) in LnCap (prostate cancer cell line) cells [184]. sEH is the main metabolizer of epoxyeicosatrienoic acids (EETs), which can also contribute to the escape of cancer cells from apoptosis [185]. Other antiapoptotic proteins, such as RNA-binding protein HuR or sirtuins, with dual functions in carcinogenesis are often strongly expressed in cancer cells. Sirt1 can have antiapoptotic activity by deacetylating proapoptotic proteins and helping cells survive under genotoxic and oxidative stresses [186,187]. Vanella et al. [188] found that EA-induced apoptosis was associated with reduced expression of HuR and Sirt1, as well as decreased levels of transcription factors mTOR and β-catenin, with subsequent increased activity of caspase 3 and AIF.

A significant advantage of EA as a potential chemopreventive, anti-tumor, or adjuvant therapeutic agent in cancer treatment is its relative selectivity. In a study conducted by Salimi et al. (2015) [158] on CLL (chronic lymphocytic leukemia) B-lymphocytes and healthy B-lymphocytes, it was shown that EA significantly reduced the viability of cancer cells at a concentration of 10 μmol/L, while in healthy cells, this effect was observed only at a concentration of 200 μmol/L. Similarly, selective production of ROS, a decrease of MMP, mitochondrial swelling, cytochrome *c* release, and increased caspase 3 activity occurred in cancer cells [158].

### 3.4. *Ellagic Acid and Deregulating Cellular Energetics*

Cancer cells exhibit distinct energy requirements in comparison to normal cells, and these deviations are a hallmark of cancer that significantly contribute to tumor initiation, progression, and survival. One of the most widely recognized metabolic changes in cancer cells is known as the Warburg effect, wherein cancer cells predominantly rely on glycolysis, even when oxygen is available (aerobic glycolysis). This metabolic shift leads to increased glucose uptake and the production of lactate [189,190]. Additionally, cancer cells frequently undergo substantial modifications in their protein and lipid metabolism to support their rapid proliferation and growth. Apart from various metabolic and biochemical pathways, cancer cells hijack numerous signaling pathways to enhance metabolic flux through intermediary metabolism. These pathways, such as the PI3K/Akt/mTOR, AMPK, Wnt/β-Catenin, HIF, NF-κB, or MYC pathway, are often dysregulated, resulting in metabolic alterations that facilitate the growth and survival of cancer cells [191,192].

Despite the many unresolved questions in the realm of cancer metabolism, the pursuit of targeting it remains a highly promising avenue for creating more efficient cancer treatments while minimizing harm to healthy cells.

EA has demonstrated the potential to impact the metabolism of cancer cells via various mechanisms, showing promise in this regard (Figure 5).

The mammalian target of rapamycin (mTOR), also known as the mechanistic target of rapamycin, is a central regulator of cancer cell metabolism and plays a pivotal role in coordinating various cellular processes related to growth, proliferation, and energy utilization. mTOR exists in two main complexes, mTOR Complex 1 (mTORC1) and mTOR Complex 2 (mTORC2). Phosphorylation events are key for activating these complexes [193]. Furthermore, AMP-activated protein kinase (AMPK) is another important regulator that plays a central role in maintaining energy homeostasis and metabolic balance within cells. The relationship between AMPK and mTOR is intimately connected, with AMPK playing a pivotal role as a crucial inhibitory regulator of the mTOR pathway, primarily via mTORC1 [194]. Targeting the AMPK/mTOR pathway is an active area of research for developing cancer therapies that disrupt cancer cell metabolism and growth.

Ni and co-workers recently documented the pro-apoptotic effect of EA-treated colon cancer cells. They found that EA reduced mTOR phosphorylation together with AMP-activated protein kinase (AMPK) phosphorylation resulting in autophagy induction [177]. Another study showed that EA inhibited mTOR signaling in vitro as well as in vivo. Xia et al. (2020) [195] studied the antiproliferative and anticancer effect of EA on HeLa cells and on a cervical tumor-bearing BALB/c mouse model. In vitro, EA significantly decreased cancer cell proliferation, migration, and invasion, and this effect has been associated with decreased phosphorylation of the members of PI3K/AKT/mTOR/STAT3 signaling pathway. In vivo, EA suppressed cervical cancer growth and among other parameters, the authors documented decreased expression of phosphorylated form of mTOR detected via immunohistochemistry. Furthermore, a study conducted by Elsaid et al. [196] reported the antiproliferative and anti-invasive effects of EA associated with increased levels of phosphorylated AMPK in the SKOV-3 ovarian cancer cells. Moreover, EA prevented the phosphorylation of S6K1 and 4EBP, downstream targets of mTORC1 as well as Akt, indicating that inhibition of the Akt/mTOR pathway can also play a role in the antiproliferative effect of EA in ovarian cancer cells. Activation of AMPK in EA-treated lung cancer cells has also been documented by Duan et al. [157]. They reported activation of AMPK and, on the other hand, suppressed levels of HIF-1α. Given that HIF-1 activation in cancer cells induces substantial metabolic alterations that facilitate tumor growth and advancement (Infantino 2021), the potential for targeting the AMPK/HIF-1 pathway as a therapeutic approach in cancer treatment appears promising.

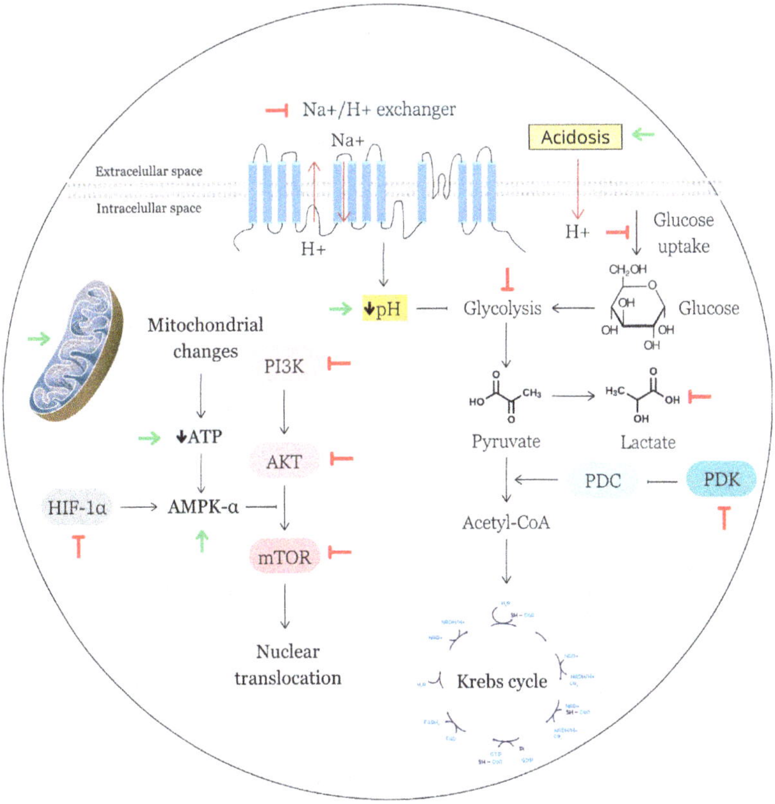


**Figure 5.** The effect of ellagic acid on selected metabolic pathways in cancer cells. Upward-pointing green arrows signify heightened expression/activation, while red T bars indicate lowered expression or reduced activity.

As it was mentioned above, aerobic glycolysis is the main source of energy for cancer cells. In a study carried out by Abdelazeem and colleagues in 2017, it was observed that EA had the capacity to regulate cytosolic pH by downregulating the expression of the $Na^+/H^+$

exchanger (NHE1) [197]. This led to intracellular acidification with subsequent impairment of glycolysis in endometrial carcinoma cells. In addition, this effect was also associated with a decrease in the cellular uptake of glucose and a notable reduction in lactate levels in supernatant. Another study showed the potential of EA to inhibit pyruvate dehydrogenase kinase (PDK), an enzyme that plays a key role in regulating the conversion of pyruvate to acetyl-CoA, a critical step in mitochondrial energy production. PDKs reduce the conversion of pyruvate to acetyl-CoA, favoring the metabolic pathway of glycolysis over oxidative phosphorylation. EA has been found to bind and inhibit PDK3, one of the four known isoforms of PDK [198].

As it was mentioned above, cancer cells undergo direct alterations in metabolic pathways, such as increased glycolysis or altered nutrient utilization. However, several indirect regulatory mechanisms also contribute to cancer cell-associated metabolic changes.

It is well known that the tumor suppressor gene TP53, which encodes the protein p53, plays a central role in maintaining genomic stability and preventing the development of cancer [199]. In addition to these activities, TP53 also regulates metabolic pathways involved in glycolysis, or oxidative phosphorylation [200]. However, frequent TP53 mutations result in divergent effects on metabolic reprogramming between wild-type (wt) p53 and gain-of-function (GOF) mutant p53. While wt p53 generally inhibits glycolysis, GOF mutant p53 has been noted to promote glycolysis in cancer cells [201], and the restoration of p53's metabolic functions appears to be one of the potential options for anticancer therapy. Several experimental studies have showed a close relationship between the antiproliferative effect of EA and the activation of p53 [121,124,131]. While the predominant effects of p53 induction via EA have been reported to involve cell cycle arrest and the triggering of apoptosis, it remains a possibility that p53-induced alterations in cancer cell metabolism could also play a role in the antiproliferative effects of EA.

Mutations or amplifications in oncogenes, such as Myc, can directly or indirectly promote changes in metabolism. Myc, for example, regulates the expression of genes involved in glycolysis, glutamine metabolism, and nucleotide biosynthesis [202]. In a study conducted by Lim et al. [203], EA has been shown to influence extracellular acidosis, a metabolic characteristic of cancer tissue. This modulation of extracellular acidosis by EA has been linked to alterations in cancer cell behaviors, particularly reduced invasiveness. Furthermore, this effect has been associated with the downregulation of cancer-related genes, including COX1, COX2, snail, twist1, and c-Myc.

### 3.5. Ellagic Acid and Enabling Replicative Immortality

Continuous cell proliferation or tumor cell immortality of malignant cells is promoted by the reactivation of a ribonucleoprotein enzyme called telomerase (terminal transferase), which prevents the shortening of telomeres during repeated replications [204]. It is well known that telomeres, repetitive (TTAGGG) DNA-protein complexes at the $3'$ ends of linear chromosomes, prevent genome instability, and conversely, their shortening leads to the cell growth arrest, replicative cellular senescence, and aging of the organism [205]. On the other hand, the maintenance of their length by telomerase is crucial for carcinogenesis [206]. Increased expression of this enzyme is characteristic of the majority of cancer cells, while most normal human somatic cells lack it [207]. In humans, telomerase is composed of a catalytic protein subunit known as human telomerase reverse transcriptase (hTERT), encoded by the *hTERT* gene, and an RNA component named human telomerase RNA (hTR) or human telomerase RNA component (hTERC), encoded by the *hTERC* gene [206]. It has been suggested that reactivation of telomerase occurs primarily via increased expression of its key catalytic subunit hTERT [208].

In addition to regulating telomere length (canonical function), telomerase and hTERT perform many other functions such as impact on proliferation (due to Wnt/β-catenin pathway or NF-kB and p65 pathway), cell survival, apoptotic resistance, adhesion, migration, stem cell renewal, inflammation, chromatin configuration and DNA repair, mitochondrial DNA protection and resistance to ionizing radiation (non-canonical functions) [209,210].

Thus, telomerase and hTERT were identified as potential therapeutic targets for cancer treatment [211].

Few studies have focused on EA and factors influencing replicative immortality. An older study showed that *Phyllanthus urinaria* extract containing several bioactive phytochemicals, including EA, might dose-dependently inhibit telomerase activity, as well as the mRNA expression level of hTERT and human telomerase-associated protein 1 in nasopharyngeal cancer cells [212]. On the other hand, it has been demonstrated that EA might have a dual effect on hTERT. When used alone, it upregulated hTERT alpha + beta + mRNA levels in estrogen receptor-positive MCF-7 breast cancer cells but significantly decreased upregulated level of hTERT induced by hormonal growth stimulator (17beta-estradiol) [213]. Finally, a huge study aimed at molecular modeling analyses of several phytochemicals from *Acacia nilotica* identified EA as a potential inhibitor of hTERT [214].

Interestingly, hTERT activity may also be downregulated by the increased level of p53 (Chen et al., 2017). Other authors reported that p53 can downregulate various genes needed for telomere metabolism [215]. The ability of EA to upregulate p53 has been already described in the previous section of the article.

### 3.6. Ellagic Acid and Inducing Angiogenesis

In 1971, Dr. Judah Folkman published a basic paper that laid the foundation for the field of angiogenesis research [216]. As tumors grow, the existing blood supply may become insufficient to meet their demands. Angiogenesis enables tumors to establish new blood vessels, ensuring a constant supply of nutrients, oxygen, and other essential factors to cells, which supports their expansion. This phenomenon plays a crucial role in the growth, progression, and metastasis of cancer [217]. Cancer cells and other cells within the tumor microenvironment (TME) release signaling molecules that stimulate the growth of new blood vessels. The TME influences angiogenesis via various mechanisms including secretion of angiogenic factors, hypoxia, interaction with immune cells, or extracellular matrix remodeling [218,219]. Therefore, regulating angiogenesis could potentially serve as an effective approach to treating malignant tumors. Nowadays, there are numerous antiangiogenic drugs in clinical practice including drugs targeted to vascular endothelial growth factor (VEGF) and its receptors (e.g., bevacizumab, ramucirumab, aflibercept), several receptor tyrosine kinases (e.g., apatinib, sunitinib, sorafenib, levantinib), hypoxia-inducible factor (e.g., belzutifan), receptor fusion proteins (e.g., Ziv-afibercept), or immunomodulatory agents with anti-angiogenic effects (e.g., lenalidomide, thalidomide) [220–222].

Notably, the VEGF and its corresponding receptor (VEGFR-2) hold significant importance as therapeutic targets for addressing diverse cancer types [223].

Moreover, several natural compounds have been reported to modulate angiogenesis [224–226].

EA has been studied for its potential anti-cancer properties, including its ability to inhibit angiogenesis. The mechanisms via which EA may inhibit angiogenesis involve its interactions with various molecular pathways and factors involved in blood vessel formation (Figure 6).

In a study by Losso et al. [227], it was observed that EA hindered the proliferation of human umbilical vein endothelial cells (HUVECs) and impeded the formation of tube-like structures by these cells. Furthermore, cancer cells treated with EA demonstrated a notable reduction in the secretion of matrix metalloproteinase (MMP)-2 and MMP-9. Additionally, the release of vascular endothelial growth factor (VEGF) was suppressed in cancer cells treated with EA. The ability of EA to inhibit MMP-2 activity and secretion has been confirmed by Huang et al. [228] in EA-treated HUVECs. Moreover, they also documented in vivo antiangiogenic effect of EA using CAM assay. At concentrations of 5 and 10 μM, EA led to a reduction in the angiogenic index by 62% and 83%, respectively. Another study showed that EA effectively suppresses the phosphorylation of VEGFR-2 induced by VEGF in endothelial cells (ECs). Additionally, EA inhibits the phosphorylation of platelet-derived growth factor receptors (PDGFR) in smooth muscle cells, which inhibits the downstream signaling pathways of these receptors. Moreover, EA inhibits the migration and tube-like structure formation of VEGF-treated ECs [229]. Later, in the study conducted by Vanella et al. [184], EA has been reported to decline activity/expression of different angiogenic factors in prostate cancer cells including heme oxygenase (HO), a potent modulator of cell growth and angiogenesis, VEGF, fibroblast growth factor (FGF), granulocyte colony-stimulating factor (G-CSF), and hepatocyte growth factor (HGF) expression. Recently, Kisacam [230] documented the dose-dependent effect of EA on some angiogenic factors in the human intestinal cancer cell line. At lower concentrations (10 and 20 μM), EA led to a substantial increase in VEGF levels. However, at higher doses (40 and 100 μM), a notable reduction in VEGF levels was observed in EA-treated cancer cells. Conversely, with the exception of exposure to 40 μM EA, there was a significant decrease in HO protein levels in Caco-2 cells as compared to the control group. The interference of EA with angiogenesis was also clearly demonstrated in a study by Wang and co-workers [231]. The antiangiogenic potential of EA was thoroughly examined via investigations conducted in vitro, ex vivo, and in vivo settings. In vitro experiments revealed that EA markedly decreased VEGF-induced EC proliferation, migration, and the formation of tube-like structures. These effects were linked to the inhibition of VEGFR-2 tyrosine kinase activity, along with the suppression of downstream signaling pathways in EC like MAPK and PI3K/Akt. Ex vivo assessments demonstrated that EA notably diminished the length and density of EC sprouts around chick aortic rings after stimulation with VEGF. Furthermore, the antiangiogenic effect of EA has also been confirmed in vivo using CAM assay and a breast cancer xenograft model. A detailed study showed that EA has the capability to establish hydrogen bonds and aromatic interactions within the ATP-binding domain of the VEGFR-2 kinase unit and thus markedly modulate several VEGF-dependent angiogenesis processes. Furthermore, in a study conducted by Ceci et al. [232], EA exhibited a reduction in VEGFR-2 expression in human bladder cancer cells, with varying degrees of inhibition ranging from 39% to 72% depending on the cell line analyzed. Surprisingly, EA did not affect the VEGF-A-induced phosphorylation of VEGFR-2. In another study, EA was highlighted as a potent inhibitor of angiogenesis in a pancreatic cancer cell xenograft model conducted on Balb C nude mice [57]. Administration of EA to the experimental animals resulted in a decrease in the expression of various angiogenesis markers, including COX-2, HIF1$\alpha$, VEGF, and VEGFR, alongside other pro-angiogenic molecules such as IL-6, IL-8, MMP-2, and MMP-9. Immunohistochemical analysis of the tumor tissue demonstrated a notable reduction in the formation of new blood vessels in EA-treated animals. On the other hand, the findings reported by Pitchacarn et al. [233] indicated that EA treatment did not induce significant alterations in the secretions of MMP-2 and MMP-9 from prostate cancer cells. Interestingly, they found that EA significantly suppressed the activity of these gelatinases. Based on this observation, the authors hypothesized that EA has the potential to diminish the invasive attributes of prostate cancer cells by targeting protease activity. Regarding matrix metalloproteinase, studies showed that MMPs are known to require zinc ions for their enzymatic activity, and this zinc-dependent activation is a key aspect of their function [234]. Huang et al. [235] conducted a study to explore whether the antiangiogenic effect of EA is associated with MMP-2 inhibition and if this inhibition could be counteracted by zinc supplementation. They found that EA-induced suppression of the MMP-2 activity and secretion was reversed by the zinc addition. Based on these results, the authors suggest that the zinc-chelation properties of EA play an important role in the antiangiogenic effect via MMP inhibition.

In summary, EA has emerged as a promising natural compound with the potential to target cancer angiogenesis, a critical process that fuels tumor growth, invasion, and metastasis.

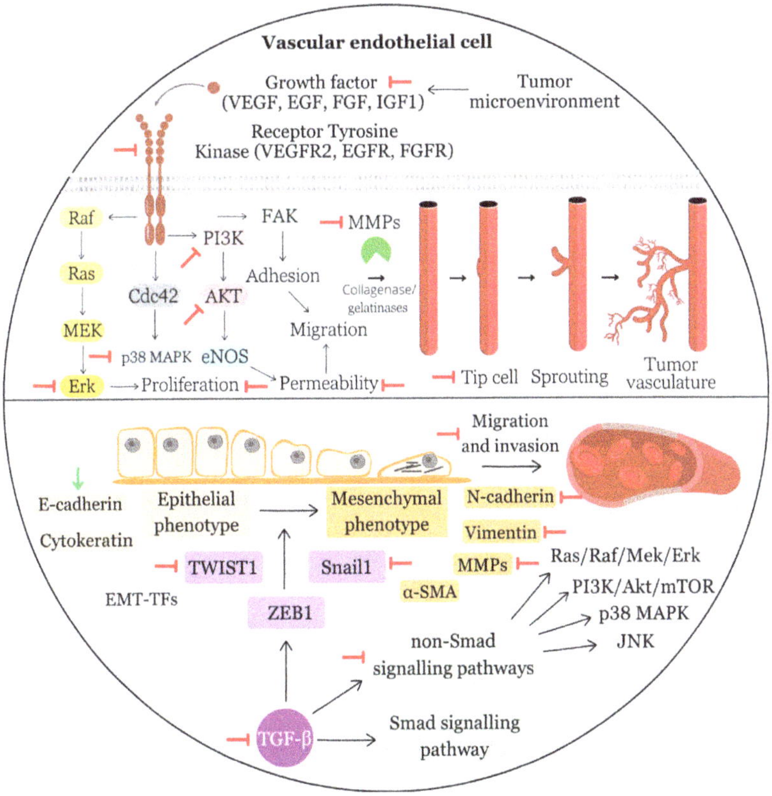


**Figure 6.** Ellagic acid's impact on angiogenesis, cancer cell migration, and invasiveness. Upward-pointing green arrows signify heightened expression/activation, while red T bars indicate lowered expression or reduced activity.

### *3.7. Ellagic Acid and Activating Invasion and Metastasis*

Tumor infiltration and metastasis remain the most challenging hallmarks of cancer and are responsible for the majority of cancer-related fatalities [236]. Despite being the best predictor of poor prognosis, it is still inadequately understood, with a substantial portion of its molecular mechanisms primarily elucidated from experiments using cancer cell lines [237]. The metastasis of tumor cells, often described as the invasion–metastasis cascade, encompasses a series of stages. These stages include initial local invasion into the neighboring stroma, followed by intravasation into the bloodstream, withstanding shear stress, extravasation to distant locations, and finally, adaptation to new microenvironments. This progression culminates in the emergence of micro-metastases, which subsequently evolve into discernible secondary tumors [238]. Numerous signaling pathways contribute to the invasive and metastatic properties of cancer cells. The epithelial–mesenchymal transition (EMT) process, for instance, facilitates the conversion of stationary epithelial cells into motile mesenchymal cells, promoting invasion and dissemination [239]. Several key molecular players, including transcription factors such as Snail, Slug, and Twist, orchestrate this transition [240]. Moreover, growth factors such as TGFβ and EGF stimulate the invasive behavior of cancer cells and are implicated in various stages of the metastatic process [241]. Tumor invasion requires intricate remodeling of the extracellular matrix (ECM) to enable cancer cells to breach physical barriers. Matrix metalloproteinases (MMPs) represent a family of $Ca^{2+}$-dependent $Zn^{2+}$-containing endopeptidases capable of breaking down proteins in the extracellular matrix, thereby facilitating cancer cell migration, invasion, and metastasis [242]. Understanding the molecular basis of invasion and metastasis opens up avenues for targeted therapies. Emerging treatments that target EMT, ECM remodeling, and specific signaling pathways show promise in limiting the spread of metastasis.

Recent studies have suggested that EA may play a multifaceted role in inhibiting invasion and metastasis (Figure 6). EA has been shown to influence the expression of various genes and proteins involved in cellular adhesion, migration, and invasion. It downregulates factors that promote invasion while upregulating those that inhibit it, thus creating an unfavorable environment for cancer cells seeking to metastasize. Several studies have demonstrated that EA may exert inhibitory effects on MMPs, specifically on MMP-2 and MMP-9. Among the numerous MMPs identified (over 20 in total), MMP2 and MMP9 have been linked to the aggressive behavior of cancer [243]. By downregulating these MMPs, EA may play a crucial role in impeding the degradation of the ECM and suppressing cancer cell migration and invasion [203,244–246]. Furthermore, a study conducted by Pitchakarn et al. (2013) revealed that EA had a significant impact on reducing the enzymatic activity of collagenase/gelatinase produced by the rat prostate cancer cell line (PLS-10). Additionally, EA exhibited concentration-dependent inhibition of collagenase IV activity. These findings suggest that EA possesses the capacity to hinder the metastasis and invasion of prostate cancer cells by targeting protease activity [233].

Another study showed that EA exhibited a dose-dependent inhibition of the growth, migration, and invasion of pancreatic cancer cells (PANC-1). The effect was potentially mediated by upregulating E-cadherin and downregulating vimentin, consequently impeding the progression of EMT [58]. During EMT, the pivotal protein E-cadherin in epithelial cells is downregulated, while vimentin, a significant mesenchymal protein in interstitial cells, undergoes upregulation [247]. In the study conducted by Kim et al. [167], it was demonstrated that EA inhibited both of these processes in pancreatic cancer cells (PANC-1, AsPC-1). The inhibition of proliferation was explained by an increase in apoptosis through the caspase-3 and caspase-9 pathways, a consequence of EA's action. The suppression of migration was associated with an upregulation of E-cadherin and decreased levels of TGF-β, MMP-2, and MMP-9. Notably, these effects were only observed in PANC-1 and AsPC-1 cells, not in MIA PaCA-2 cells. This implied that the impact of EA on pancreatic cancer was influenced by cell characteristics. Additionally, a study by Wang et al. (2017) revealed that EA suppressed the metastasis of glioblastoma cells (U87 and U118) by controlling the expression of EMT-related transcription factors and MMPs. EA inhibited glioblastoma invasion in the U87 xenograft model by upregulating E-cadherin and downregulating the expression of transcription factors, such as Snail, MMP-2, and MMP-9 [59]. The results of a study by Lim et al. (2019) suggested that EA suppressed acidity-enhanced migration and invasion of gastric cancer cells by inhibiting the expression of multiple factors (COX1, COX2, snail, twist1, and c-myc). Due to this factor, it may be a promising candidate for treating cancer in the presence of acidosis. Extracellular acidosis, a key metabolic factor in tumor tissue, is a significant metastatic factor in many types of cancer cells [248–250]. It alters the expression of oncogenes and the aforementioned genes linked to EMT (*SNAIL1*, *TWIST1*, *MYC*). Under acidic conditions, gastric cancer cells become more invasive, with elevated expression of MMP-7 and MMP-9. In concentrations below 10 μM, EA effectively suppressed MMP-7 and MMP-9 expression triggered by acidity, restraining the migratory and matrigel-infiltrating capacity of gastric cancer cells [203].

By specifically targeting and reducing the expression of TGF-β1, FSCN1 (Fascin Actin-Bundling Protein 1), and vimentin, EA exhibited the potential to obstruct pivotal processes closely associated with the development and progression of hepatocellular carcinoma (HCC). TGF-β1 assumes a critical role in HCC development by instigating the EMT process [245]. Meanwhile, FSCN1 plays a pivotal role in cell invasion and migration, both essential aspects of cancer metastasis. Upon activation of TGF-β1, FSCN1 experiences overexpression in HCC, thereby further exacerbating the progression of EMT. Moreover, vimentin acts as a significant marker during EMT, with heightened levels observed in cells undergoing this transformation, thereby contributing to the aggressive nature of cancer cells [251,252]. Furthermore, intervention with EA inhibited the migration, invasion, and proliferation of melanoma cells (WM115 and A375) by reducing the activation of EGFR [72]. EGFR is typically strongly activated in various tumorigenic processes. It also significantly contributes to the progression and metastasis of melanoma [253]. Elevated EGFR expression in tumor cells enhances cell adhesion to the extracellular matrix and facilitates distant metastasis [254]. EGFR has emerged as a prognostic marker in various cancer types, indicating a poor prognosis. Consequently, the administration of EA presents a promising adjunctive therapeutic approach for melanoma patients [72].

In another study, it was reported that EA effectively inhibited the proliferation, migration, and invasion of anaplastic thyroid cancer cells (ATC) by suppressing the Wnt/β-catenin and PI3K/Akt pathways [255]. Abnormal activation of these signaling pathways collectively correlates with cancer progression and contributes to the promotion of EMT via epigenetic control [256]. Regarding the PI3K pathway, the concurrent administration of EA and the PI3K inhibitor GDC-0941 to various breast cancer cell lines resulted in a significant reduction in cell proliferation, migration, and in vitro invasion. This combination also led to a decreased tumor initiation rate and a reduction in metastasis formation in vivo [257]. Other authors demonstrated that EA could effectively inhibit the invasion and migration of endometrial carcinoma cell lines KLE and AN3CA by suppressing PI3K phosphorylation and downregulating the expression of MMP9. Additionally, EA was observed to inhibit lung metastasis in BALB/c nude mice, which were intravenously injected with KLE and AN3CA cells. Collectively, these findings strongly suggest that EA possesses potent antitumor properties against endometrial cancer, effectively suppressing cell invasion and migration by targeting the PI3K signaling pathway both in vitro and in vivo [66].

Moreover, EA effectively impeded the growth and metastasis of breast cancer by directly targeting ACTN4 (Alpha-actinin-4). This intervention was concomitant with a reduction in the population of cancer stem cells. Cancer stem cells expressing ACTN4 displayed heightened capabilities, including increased mammosphere formation and enhanced tumorigenesis in vivo [258]. Increased ACTN4 expression is directly correlated with more advanced cancer stages, elevated incidence of metastasis, poor patient prognosis, and tumor chemoresistance [259]. Silencing of ACTN4 by EA led to the inhibition of malignant cell proliferation, formation of cell colonies, and improved suppression of metastatic potential. Taken together, these findings suggest that ACTN4 could potentially serve as a critical target for EA treatment, particularly in the context of breast cancer stem cell-related metastasis [258].

### 3.8. Ellagic Acid and Avoiding Immune Destruction

The immune system plays an important role in carcinogenesis. Several suppressive mechanisms are involved in the control and elimination of mutated malignant cells. However, cancer cells influenced by humoral and cellular components of TME can escape from the control and promote cancer development. To eliminate these negative inhibitory mechanisms and reintroduce an effective malignant cell-clearing activity of immunocompetent cells, several immunotherapeutic approaches were proposed [260].

The acute inflammatory process is a part of the defense mechanisms. However, chronic inflammation, which affects many cellular pathways, can lead to the initiation and development of different forms of cancer [261]. Inflammatory cells produce a number of growth factors, cytokines, ROS, proteases, matrix metalloproteinases, and a large scale of other enzymes. They can activate transcriptional factors and bring about cellular proliferation, genomic instability, angiogenesis, resistance to apoptosis, invasion, and metastasis [262]. Cells and their products form part of the TME and ensure dynamic interactions between tumor cells and the TME and modulate response to immunotherapy. Some of those inflammatory molecules allow cancer cells to proliferate, survive, and metastasize [263]. On the other hand, there exist molecules that can influence inflammatory parameters and exert an anti-carcinogenic potential by different mechanisms. For example, some cytokines have shown antitumor activity by influencing tumor antigen presentation, T-cell priming and activation, T-cell infiltration, and cancer cell death via stimulation of the adaptive and innate cell immunity [264,265]. All these data indicate that inflammation and related factors can play a dual role in the initiation and progression of the malignant process, modulating immune system processes and their interaction with tumor cells in the TME.

The discovery of checkpoint inhibitory proteins, e.g., CTLA-4, PD-1, PD-L1, and the development of antibodies that block their reducing immunologic responses resulted in one of the most promising therapeutic modalities in anticancer immunotherapy. However, despite considerable clinical progress, a number of interacting factors within the TME can influence the therapeutic response to immunotherapies [266]. Moreover, a limited number of patients can benefit from checkpoint inhibition treatment due to the lack of appropriate initial indication markers, development of resistance, or occurrence of severe immunity overstimulating adverse reactions. These negative effects result in intensive need and reasonable search for subsequent therapeutic combinatorial modalities, which include the use of naturally occurring substances.

EA exerts potent anticarcinogenic effects by several mechanisms, including inhibition of tumor cell proliferation, induction of tumor cell apoptosis and autophagy, break of DNA binding to carcinogens, blocking virus infection, disturbing inflammation, angiogenesis, inhibition of tumor cell metastasis and invasion, affection of tumor metabolic reprogramming and drug-resistance processes [261,267]. It is also characterized by its antioxidant properties and ability to influence the production of pro-inflammatory mediators (e.g., TNF-$\alpha$, IL-1$\beta$, IL-6) resulting in anti-inflammatory effects (see below) tightly related to carcinogenesis. Please note that the role of EA in the modulation of anticancer immunity reactions seems to be also important. All those properties together with above mentioned malignant process suppressing mechanisms make EA a highly interesting potential anticancer compound [268].

The effects of EA on immunity parameters have been studied in several in vivo experiments and on different immune cells in in vitro conditions. An early study investigating the effect of EA on common immunological surveillance was performed in B6C3F1 mice (a hybrid strain extensively used in carcinogenicity bioassays) exposed continuously to EA in drinking water at 0.5, 1.0, or 2.0 mg/kg/day for 28 days. Among several different parameters, responses of natural killer (NK) cell activity, cytotoxic T lymphocyte activity, and IgM antibody plaque-forming cells were investigated. Obtained data have shown that subchronic exposure to EA (2.0 mg/kg) caused significant suppression of specific IgM antibody responses in the treated group and suppressed cytotoxic T-cell function in the 0.5 and 1.0 mg/kg EA-treated groups. All other immunological parameters were within normal ranges [269].

Additional immunomodulating effects of EA in different subpopulations of immunocompetent cells were discovered and described in the following experimental studies. The results of these experiments were summarized in a review by Rahimi et al. [270]. It has been found that the influence of EA on immune cell subpopulations results in strong anti-inflammatory effects. For example, EA in phytohemagglutinin stimulated human peripheral blood mononuclear cells (PBMC) resulting in decreased cell proliferation, reduced levels of IL-13 and TNF-$\alpha$, and increased IL-2 production [271]. In the same cell population, EA antagonized the stimulatory effect of macrophage migration inhibitory factor and inhibited NF-$\kappa$B nuclear translocation and chemotaxis [272]. Similar immunomodulatory effects with inflammation-suppressing activity were observed in PE regularly containing EA as an active constituent. In this case, the extract was used to influence neutrophil activity stimulated by FN-formyl-methionyl-Leucyl-phenylalanine and phorbol myristate acetate, inhibition of myeloperoxidase activity and decreased ROS level were noted [273]. The functions of human basophilic cell line KU812 (chronic myelogenous leukemia) stimulated by phorbol-12-myristate 13-acetate plus calcium ionophore A23187 were also affected by PE. The effect consisted of the inhibition of pro-inflammatory cytokines IL-6 and IL-8 production and NF-$\kappa$B gene expression [274].

In another study, the authors investigated some additional immunological responses to EA. The experiments studying the effects of EA were performed on the lipopolysaccharide (LPS)-induced bone marrow-derived dendritic cells. The results have shown that EA suppressed LPS-induced expression of co-stimulatory molecules (CD80 and CD86). However, the expression of major histocompatibility complex (MHC) class I and class II were not suppressed. These reactions were mediated via the EA-induced block of LPS-induced c-Jun N-terminal kinase (JNK) activation. At the same time, LPS-mediated expression of proinflammatory cytokines (IL-12 and IFN-$\gamma$) was diminished by EA. This was confirmed by the concurrent use of a JNK inhibitor, which antagonized EA effects on the levels of IL-12 and IFN-$\gamma$. It has been suggested that EA demonstrated promising immunomodulatory effects in this preclinical model through the regulation of dendritic cell maturation and suppression of humoral immunity [275].

Taken together, these studies show that EA is capable of downmodulating pro-inflammatory mediators and stimulating the production of anti-inflammatory cytokines [268]. Decreasing inflammatory and aforementioned interfering immunity reactions EA may interfere with carcinogenic processes and suppress initiation and development of malignant cell transformation and progression.

On the other hand, immune regulatory mechanisms could have a dual effect on carcinogenesis. Progression in tumoral growth, invasion, and metastasis formation is among the other factors attributed to tumor cell escape from immunity system control. This control is effectively diminished via overexpression of immune checkpoint proteins and their receptors, mainly PD-1, PD-L1, and CTLA-4, among others. A restart of immunocompetent cell activities in recognition and subsequent elimination of cancer cells are believed to participate in the effective inhibition of tumoral growth and the overall spread of invasive malignant cells.

The following experimental study advanced toward the evaluation of EA effects on PD-L1 expression in human urinary bladder cell lines. This research was performed because of FDA approval of atezolizumab (humanized monoclonal antibody against the immune checkpoint PD-L1) for platinum-treated advanced urothelial cancer. Analysis of UM-UC-3 (hypertriploid epithelial human bladder cells) and T24 (transitional cell human bladder carcinoma) cells showed that EA downregulated the expression of the immune checkpoint PD-L1 in tumor cells. It was suggested that the effect of EA might potentially enhance the efficacy of anti-PD-L1 treatment and contribute to a decrease in immunosuppressive mechanisms that favor disease progression [232].

Another more recent study evaluated the effects of *Rubus coreanus* extract containing a large amount of EA on cellular PD-1/PD-L1 blocking activity. *Rubus coreanus* extract and EA dose-dependently inhibited PD-1/PD-L1 binding in PD-1/NFAT Jurkat (Jurkat recombinant cell line containing a firefly luciferase gene under the control of the nuclear factor of activated T cells response element) and PD-L1/aAPC CHO-K1 (cells stably expressing human PD-L1 and a cell surface protein designed to activate cognate T-cell receptors in an antigen-independent manner) cell co-cultures. These results were confirmed in in vivo experiments in humanized PD-1 mice bearing MC38 colorectal tumor. The efficacy of oral application of *Rubus coreanus* extract was comparable to PD-1 antibody activity [276].

The involvement of EA and its beneficial effect in anti-cancer immunomodulation was demonstrated as well in the evolution of anti-PD-1/PD-L1 resistance in murine and human lung tumors. The mechanism of resistance resulted from increased collagen levels in the tumor extracellular matrix which induced T cell exhaustion via the leukocyte-associated immunoglobulin-like receptor 1 (LAIR1). EA was found to decrease tumor collagen deposition through lysyl oxidase-like protein-2 (LOXL2) suppression. Additionally, a reduction in tumor growth was observed when EA was combined with anti-PD-L1 treatment, showing its effectiveness [277]. Figure 7 shows the findings described in this topic.

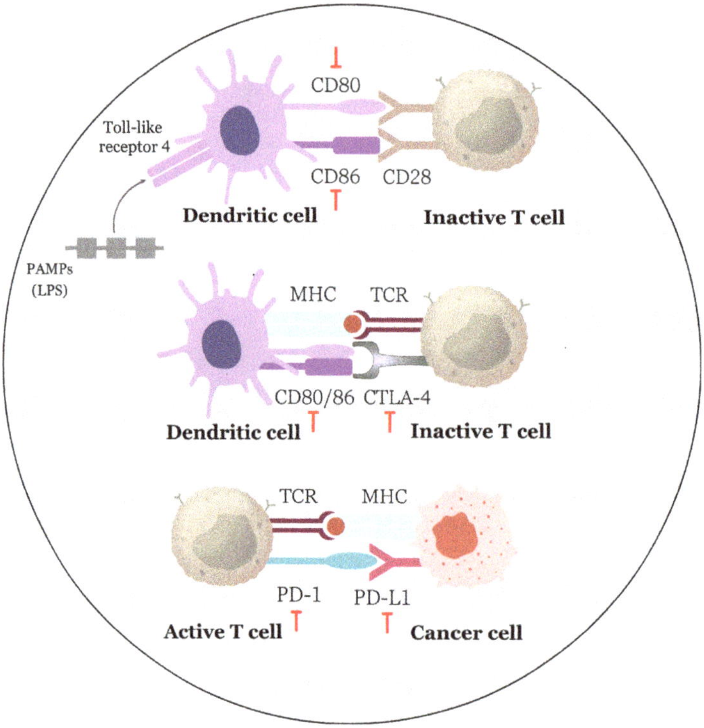


**Figure 7.** Effect of ellagic acid on specific aspects of the immune system. Red T bars indicate lowered expression or reduced activity.

### 3.9. Ellagic Acid and Tumor-Promoting Inflammation

Oxidative stress, inflammation, and the associated constant production of pro-inflammatory cytokines, inflammatory mediators, growth factors, and prostaglandins produced by activated macrophages lead to the development of chronic inflammation. Inflammation is the body's response to cell damage accompanied by massive production of inflammatory cytokines (interleukins and tumor necrosis factor (TNF-$\alpha$), transcription factors (IFN-$\gamma$), and nuclear factor (NF-$\kappa$B) [278]. Prolonged chronic inflammation, subsequent genetic mutations, and cellular differentiation can instigate a precancerous state that ultimately culminates in the development of cancer. Statistically, it was evaluated that approximately 15% of cancer diseases are associated with chronic inflammation [279]. Factors from the external environment as well as infection and autoimmune diseases influence the development of chronic inflammation. The most well-known examples of chronic inflammation that can lead to cancer are inflammatory bowel disease, chronic hepatitis, Helicobacter pylori-induced gastritis, pneumonia caused by inhaling tobacco smoke, inflammation associated with obesity, hyperglycemia, excessive accumulation of lipids, or bacterial products [280]. Due to the interplay between chronic inflammation and cancer, targeting chronic inflammation represents a promising avenue for cancer prevention.

As it was mentioned above, EA is an excellent antioxidant with the ability to capture free radicals and ROS, thereby protecting cells from oxidative stress, inhibiting the mutagenicity of carcinogens, and at the same time stimulating the immune system to produce anticancer molecules [281]. EA has also been shown to have very strong anti-inflammatory effects. Experimental works inspired by the results of the antioxidant and anti-inflammatory effects of EA currently support the idea of its chemopreventive and antitumor effects [11,282]. The anti-inflammatory effects of EA in relation to the inhibition of carcinogenesis have been demonstrated by numerous in vivo and in vitro studies.

Several results show that natural substances suppressing NF-κB or products regulated by NF-κB should have potential in the prevention and treatment of cancer [283]. From the results of the work by Khan et al. [284], it is evident that EA interacts with the Rel homologous domain of NF-κB. This interaction affects the expression of genes encoding pro-inflammatory molecules, resulting in a significant reduction in the levels of pro-inflammatory cytokines and chemokines. Another study focused on the effect of EA on inflammatory responses in pancreatic cancer cells, showed suppression of the NF-κB activity in EA-treated cancer cells. Furthermore, it was reported that EA triggered the loss of mitochondrial membrane potential and the activation of mitochondrial apoptotic pathways in these cells [161]. Additionally, in a study conducted by Cheng et al. [58], EA has been reported to suppress the proliferation of human pancreatic carcinoma PANC-1 cells both in vitro and in vivo. In vitro, EA inhibited the proliferation, migration, and invasion of PANC-1 cells. In animal experiments, EA significantly reduced tumor growth and prolonged the survival of experimental animals. Western blot analyses showed that these effects have been associated with decreased expression of pro-inflammatory markers COX-2 and NF-κB. Another group of researchers conducted a study examining the impact of EA on colon cancer induced by 1,2-dimethylhydrazine in rats. They observed heightened expressions of pro-inflammatory proteins and cytokines in 1,2-dimethylhydrazine-treated rats. However, oral administration of EA led to a significant reduction in NF-κB levels, subsequently resulting in a decrease in inflammatory markers, including iNOS, COX-2, TNF-α, and IL-6 [285]. In addition, treatment of the human gastric adenocarcinoma cell line (AGS) with EA reduced cell proliferation and migration. Moreover, EA treatment led to alterations in the expression of genes involved in inflammation including, iNOS, NF-κB, IL-8, and TNF-α [120]. In another study, EA exhibited a substantial increase in both the expression and functionality of antioxidant enzymes, such as catalase, superoxide dismutase, and glutathione reductase, in the livers of mice bearing Dalton lymphoma. Furthermore, EA downregulated protein kinase C expression and prevented the activation of NF-κB. This outcome can be attributed to the fact that PKC is activated by ROS. Consequently, the heightened antioxidant defense induced by EA inhibited PKC activation, subsequently resulting in the suppression of NF-κB activation [286]. A significant study of the effect of EA on HepG2 hepatocellular carcinoma cells after irradiation demonstrated that EA stimulated the formation of ROS, and increased the expression of p53 and p21 proteins as well as markers of apoptosis (Bax and caspase-3). At the same time, EA reduced the levels of survival markers as well as the inflammatory markers TNF-α and IL-6 [116].

Additionally, numerous studies have centered on pomegranate as a significant source of EA. For instance, a study conducted by Bishayee et al. in 2013 explored the impact of PE on inflammation in diethylnitrosamine (DENA)-induced liver carcinogenesis in rats [287]. Notably, DENA alone led to a significant increase in the expression of various inflammatory markers, including iNOS, 3-nitrotyrosine, COX-2, and NF-κB. However, when PE was administered preventively, it exhibited a remarkable decrease in the expression of all these inflammatory markers. According to the authors' suggestions, the modulation of the NF-κB signaling pathway could play a pivotal role in the anticancer effects observed with the use of PE. In a separate investigation, the ingestion of PE was found to inhibit chemically induced lung carcinogenesis. Among the multitude of identified proteins, PE notably repressed inflammatory markers such as CD31, iNOS, and NF-κB [288].

The analysis of the level of secreted pro-inflammatory cytokines/chemokines by prostate tumor cells DU145 and PC3 cells was devoted to the work of Wang et al. [289]. Luminex Multiplex analysis revealed that pro-inflammatory cytokines/chemokines such as IL-6, IL-12p40, IL-1β, and RANTES were reduced by PE with high content of EA. Therefore, they believe that EA has the potential to reduce inflammation and influence cancer progression. Similar results were also achieved by Rettig et al. [290] and Wang et al. [289] when PE reduced the levels of pro-inflammatory markers TNF-α and NF-κB. Subsequently,

Mandal et al. (2017) [291] investigated the anti-inflammatory mechanism of PE in a rat mammary carcinogenesis model induced by dimethylbenz(a)anthracene. Their findings indicated that PE exhibited a significant anti-inflammatory effect, as evidenced by the reduced expression/activity of several proteins involved in inflammatory processes, including COX-2, HSP90, and NF-κB. Conversely, PE increased the expression and nuclear localization of Nrf2, a transcription factor that plays a pivotal role in safeguarding cells against inflammation and oxidative stress [291]. In summary, Figure 8 shows the actions of EA targeting inflammation.

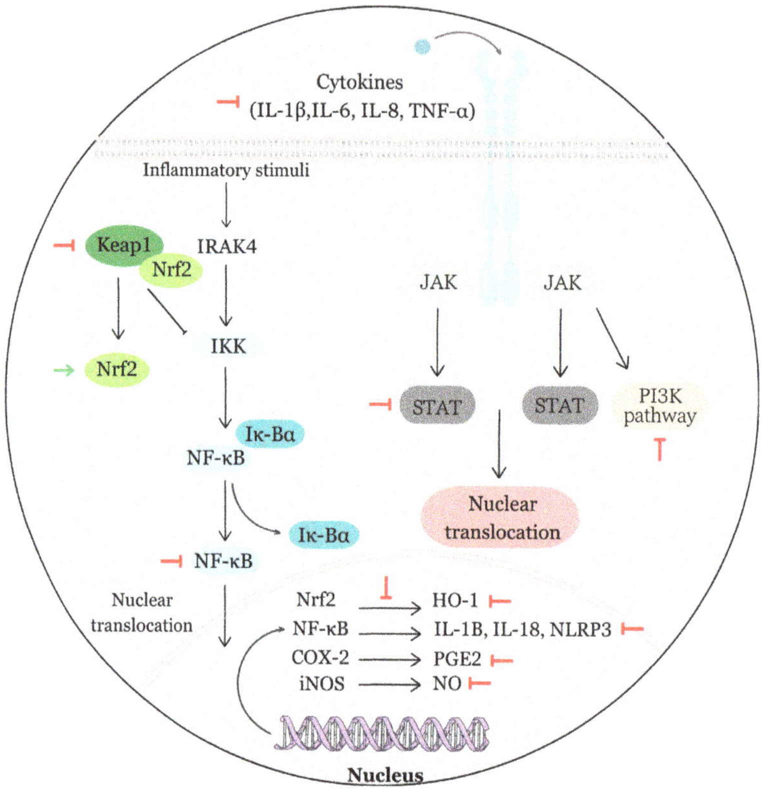


**Figure 8.** Anti-inflammatory effect of ellagic acid. Green arrows indicate increased expression/activation, while red T bars indicate lowered expression or reduced activity.

### 3.10. Ellagic Acid and Genome Instability and Mutation

Cells undergo a continuous cycle of damage and repair. DNA damage arises from various endogenous and exogenous sources, such as such as radiation [292], chemical mutagens [293], DNA repair deficiencies [294], oxidative stress-inducing high ROS levels [295], viral infections and chronic inflammation [296], cell-cell fusion [297] and others. Moreover, telomere shortening can lead to next-genomic instability and promote cancer development [298]. DNA repair pathways aim to correct these errors and damages, ensuring the preservation of the genetic code and preventing the accumulation of mutations that could lead to diseases such as cancer. Key players in this intricate mechanism are the Ataxia Telangiectasia Mutated (ATM) and RAD3-related (ATR) kinases [299,300].

A few years ago, five priority targets against genomic instability were identified: prevention of DNA damage, promotion of DNA repair mechanisms, restoration of deficient DNA repair; impairing centrosome clustering, and inhibition of telomerase activity [301].

In line with these, an attenuation of genome instability is relevant in the context of chemoprevention used to prevent the initiation or progression of cancer. The main goal of chemopreventive agents is to reduce the genetic alterations that can lead to carcinogenesis.

In a study conducted by Aiyer et al. [302], EA was reported to prevent oxidative stress-induced DNA damage. Co-incubation of 4-hydroxy-17ß-estradiol E2 (4E2) and $CuCl_2$ with EA significantly decreased levels of 8-oxodeoxyguosine (8-oxodG), a marker of oxidative DNA damage compared to 4E2/$CuCl_2$ only. Moreover, EA significantly decreased levels of oxidative adducts in vivo. In addition, EA has been found to upregulate DNA repair genes, including xeroderma pigmentosum group A complementing protein (XPA), DNA excision repair protein (ERCC5), and DNA ligase III (DNL3). Later, Berni and co-workers [303] demonstrated a protective effect of EA in an N-methyl-N′-nitro-N-nitrosoguanidine (MNNG) model of genotoxicity. MNNG is a well-known carcinogen targeting primarily guanine resulting in the formation of $O^6$-methylguanine adducts [304]. The outcomes of their study indicated that EA effectively decreased the micronuclei formation both in vitro and in vivo. However, this reduction was observed only when EA was administered before exposure to MNNG. Other authors documented the protective effect of EA against benzo[a]pyrene (BaP)-induced DNA adducts. BaP is widely recognized as a potent carcinogen and mutagen. When metabolized into benzo[a]pyrene-7,8-diol-9,10-epoxide, it can bind to DNA, leading to the creation of DNA adducts. These adducts can introduce mistakes during DNA replication and repair processes, ultimately elevating the risk of cancer [305]. In a study conducted by Zahin et al. [306], EA significantly decreased number of BaP-induced DNA adducts. Moreover, using different salmonella strains in the Ames test, EA significantly decreased the number of mutations induced by several mutagens including sodium azide, methyl methanesulfonate, BaP, and 2-aminoflourine. Additionally, at the tested concentrations ranging from 50 to 500 μM, EA displayed no mutagenic effects. A similar protective effect of EA in BaP-induced genotoxicity has also been confirmed by Varshney and co-workers [307] in both in vitro (Ames test) and in vivo (mouse bone marrow micronucleus) assays. Administration of EA led to a decrease in the frequency of BaP-induced revertants in the *S. typhimurium* strain TA100. Furthermore, administration of EA resulted in a reduction in the average number of BaP-induced polychromatic erythrocytes in the micronucleus test in mice. Additionally, Western blotting results indicated a downregulation of the cytochrome P450 (1A1) isoform, which in turn prevented the activation of benzo[α]pyrene. In another study, the protective effect of EA against acrylamide-induced genotoxicity in human lymphocytes has been studied [308]. This study found that exposure to acrylamide at a concentration of 50 μM resulted in increased production of reactive oxygen species (ROS), elevated lipid peroxidation, disruption of the mitochondrial membrane potential, lysosomal damage, and DNA damage. Conversely, co-administration of EA at concentrations of 25 and 50 μM, among others, suppressed acrylamide-induced DNA damage as evidenced by a reduction in the levels of 8-hydroxy-2′-deoxyguanosine, a marker of oxidative stress-induced DNA damage.

Furthermore, a recent research analysis performed on rat embryonic fibroblasts [309] demonstrated that EA acid can restore telomerase activity inhibited by exposure to phosalone, a common agricultural organophosphate pesticide capable of ROS induction [310] and genotoxicity [311].

Interestingly, EA exhibited statistically significant upregulation of sirtuin 6 at the protein level in Caco2 cells [312]. This member of the sirtuin family is a multifaced protein that contributes to the maintenance of telomere function, DNA integrity, and genome stability [313].

Furthermore, multiple studies have established the antimutagenic and antigenotoxic properties of EA when exposed to various mutagenic or genotoxic substances. These include benzene [314], cyclophosphamide [315], nicotine [316], and aflatoxin B1 [317], as well as sodium azide and N-methyl-N′-nitro-N-nitrosoguanidine [318].

Additionally, in a study conducted by Matić et al. [319], the results demonstrated that EA did not exhibit genotoxic effects when tested on somatic and germ cells of *Drosophila melanogaster*.

However, it is important to mention that several in vitro studies also reported an increase in DNA damage or increased ROS formation under EA exposure in cancer cells. As an example, EA promoted DNA damage and ROS generation in the TSGH8301 human bladder cancer cell line [60]. In addition, significantly increased EA-induced DNA damage was confirmed in human glioblastoma cell lines U87 and U118 by determining the presence of the phosphorylated form of H2AX histone variant (γ-H2AX), which serves as a molecular marker of DNA damage [59]. In some studies, the administration of EA led to the induction of PARP-cleavage in pancreatic cancer cells (PANC-1) [57] or prostate cancer cell lines [124]. PARP cleavage serves as a marker of apoptosis, but the cleavage of PARP has also functional implications [168]. The cleaved PARP fragments lose their ability to perform their roles in DNA repair, further promoting the apoptotic process.

In summary, the dual impact of EA on DNA damage and ROS production underscores the need for comprehensive research to elucidate the precise mechanisms governing these processes. Understanding the context in which ellagic acid promotes DNA protection or induces damage is vital for harnessing its potential benefits effectively, particularly in cancer therapy and prevention.

Molecular and cellular mechanisms of EA in relation to cancer hallmarks are presented in Table 1.

**Table 1.** Molecular mechanisms of ellagic acid in influencing cancer hallmarks. Arrows indicate an increase (↑) or decrease (↓) in the levels/activity of the molecules.

| Cancer Hallmark | Mechanism of Action | References |
|---|---|---|
| *Sustaining Proliferative Signaling and Evading Growth Suppressor* | Suppression of cell proliferation, S, G1, and G2/M phase cell cycle arrest, inhibition of signaling pathways, estrogen antagonism,<br>↑ p53, p21, p15, PTEN<br>↓ cyclin D1, CDK2, CDK4, CDK6, COX-2, IL-1β, IL-6, IL-8, NF-κB, Notch pathway, PI3K/Akt pathway, Hedgehog pathway, Akt pathway, Wee1, JAK/STAT3 pathway, EGFR, PIK3CA, PIK3R1, TGF-β/Smad3 signaling, β-catenin, axin 1, axin 2, Myc, survivin, K-ras, MCM2-7, phospho-pRb, PKCα, PKCβ, and PKCγ, MDM2, XIAP, FAK | [56–62,65,66,68,69,71,73–78,124,125,131] |
| *Resisting Cell Death* | Induction of mitochondrial and extrinsic pathway of apoptosis, permeabilization of the mitochondrial outer membrane, DNA damage and fragmentation, inhibition of mitochondrial respiration, and mitochondrial damage<br>↑ Bax, Bak, Bid, PUMA, Noxa, cytochrome *c*, EndoG, Smac/DIABLO, AIF and APAF1 release, the activity of caspases-3, -8, -9, phosphorylation and expression of AMPK-α and ACC, PS externalization<br>↓ Bcl-2, Bcl-xL, Mcl-1, MMP, Drp-1, HIF-1α, HO-1, HO-2, sEH, HuR, Sirt | [56,59,60,74,114,120,124,153,157–159,161,167,174,176,179,180,184,188] |
| *Deregulating Cellular Energetics* | Inhibition of aerobic glycolysis, inhibition of Akt/mTOR signaling pathway<br>↑ phospho-AMPK, intracellular acidification<br>↓ phospho-mTOR, phospho-S6K1, 4EBP, HNE1, PDK, acetyl-CoA | [177,196–198] |
| *Enabling Replicative Immortality* | Inhibition of telomerase activity, modulation of hTERT activity | [212–214] |

**Table 1.** *Cont.*

| Cancer Hallmark | Mechanism of Action | References |
|---|---|---|
| *Inducing Angiogenesis* | Inhibition of angiogenesis, inhibition of formation tube-like structures of endothelial cells and migration, reduction of angiogenic index, inhibition of VEGF signaling pathways<br>↓ MMP-2, MMP-9, VEGF secretion, phospho-VEGFR-2, phospho-PDGFR, HO activity, MAPK signaling, PI3K/Akt signaling | [184,227–229,231–233] |
| *Activating Invasion and Metastasis* | Impeding the degradation of the ECM, suppression of cell migration and invasion, inhibition of epithelial-mesenchymal transition<br>↑ E-cadherin<br>↓ enzymatic activity of collagenase/gelatinase, vimentin, TGF-β, MMP-2, MMP-9, Snail, COX1, COX2, c-Myc, Twist1, FSCN1, Wnt/β-catenin pathway, PI3K/Akt pathway, ACTN4 | [59,167,203,233,244–246,255,258] |
| *Avoiding Immune Destruction* | Antioxidant activity, modulation of the production of pro-inflammatory mediators, inhibition of PD-1/PD-L1 binding<br>↑ NK cells response, cytotoxic T lymphocyte activity, IL-2, effectiveness of anti-PD-L1 treatment<br>↓ IL-13, TNF-α, macrophage migration inhibitory factor, NF-κB nuclear translocation, ROS production, myeloperoxidase activity, expression of CD80 and CD86, JNK activation, IL-12, IFN-γ, PD-L1, LOXL2, collagen deposition | [232,268,269,271–273,275,277] |
| *Tumor-Promoting Inflammation* | ROS scavenging, oxidative stress protection, inhibition of mutagenicity of carcinogens, anti-inflammatory effect, reduction in the levels of pro-inflammatory cytokines and chemokines<br>↑ catalase, superoxide dismutase, glutathione reductase<br>↓ NF-κB, COX-2, iNOS, TNF-α, IL-6, IL-8, PKC, 3-nitrotyrosine, CD31, IL-12p40, IL-1β, RANTES, HSP90, Nrf2 | [58,120,281,284–288,291] |
| *Genome Instability and Mutation* | Prevention of oxidative stress-induced DNA damage, upregulation of DNA repair genes, micronuclei formation<br>↑ XPA, ERCC5, DNL3, restore telomerase activity, Sirt6<br>↓ 8-oxodG, cytochrome P450 (1A1) | [302,304,306–309,312] |

## 4. Conclusions

The experimental evidence presented in this review underscores the promising role of EA as a potential therapeutic agent in addressing the fundamental hallmarks of cancer. Our analysis reveals that EA exhibits multifaceted actions, impacting key aspects of cancer biology making it a compelling subject for future investigations and the development of innovative strategies to enhance cancer prevention and therapy. While the experimental evidence presented here is promising, it is crucial to acknowledge the complexity of cancer as a disease and the need for further research, including clinical trials, to validate the efficacy and safety of EA-based therapies. Additionally, considerations such as bioavailability and optimal dosing regimens should be explored in greater depth.

**Author Contributions:** Conceptualization, J.M.; investigation, M.Č., R.M., L.M., G.M., M.Z., A.B. and J.M.; writing—original draft preparation, M.Č., R.M., L.M., G.M., M.Z., A.B. and J.M.; writing—review and editing, L.M. and J.M.; visualization, R.M.; supervision, J.M.; funding acquisition, M.Č., L.M. and J.M. All authors have read and agreed to the published version of the manuscript.

**Funding:** This research was funded in part by the Grant Agency of the Ministry of the Education, Science, Research and Sport of the Slovak Republic (VEGA 1/0539/21, VEGA 1/0513/21, VEGA 1/0446/22) and the Slovak Research and Development Agency under contract No. APVV-16–0446. Moreover, this publication is the result of the project implementation: "Open scientific community for modern interdisciplinary research in medicine (OPENMED)", ITMS2014+: 313011V455, supported by the Operational Programme Integrated Infrastructure, funded by the ERDF. This publication was also supported by the Operational program Integrated Infra-structure within the project: Demand-driven research for the sustainable and innovative food, Drive4SIFood 313011V336, co-financed by the European Regional Development Fund.

**Conflicts of Interest:** The authors declare no conflict of interest.

## References

1. WHO. Noncommunicable Diseases. 2023. Available online: https://www.who.int/news-room/fact-sheets/detail/noncommunicable-diseases (accessed on 16 September 2023).
2. Sung, H.; Ferlay, J.; Siegel, R.L.; Laversanne, M.; Soerjomataram, I.; Jemal, A.; Bray, F. Global Cancer Statistics 2020: GLOBOCAN Estimates of Incidence and Mortality Worldwide for 36 Cancers in 185 Countries. *CA Cancer J. Clin.* **2021**, *71*, 209–249. [CrossRef] [PubMed]
3. Bidram, E.; Esmaeili, Y.; Ranji-Burachaloo, H.; Al-Zaubai, N.; Zarrabi, A.; Stewart, A.; Dunstan, D.E. A concise review on cancer treatment methods and delivery systems. *J. Drug Deliv. Sci. Technol.* **2019**, *54*, 101350. [CrossRef]
4. Budreviciute, A.; Damiati, S.; Sabir, D.K.; Onder, K.; Schuller-Goetzburg, P.; Plakys, G.; Katileviciute, A.; Khoja, S.; Kodzius, R. Management and Prevention Strategies for Non-communicable Diseases (NCDs) and Their Risk Factors. *Front. Public Health* **2020**, *8*, 574111. [CrossRef] [PubMed]
5. Krakhmal, N.V.; Zavyalova, M.V.; Denisov, E.V.; Vtorushin, S.V.; Perelmuter, V.M. Cancer Invasion: Patterns and Mechanisms. *Acta Nature* **2015**, *7*, 17–28. [CrossRef]
6. Hanahan, D.; Weinberg, R.A. The hallmarks of cancer. *Cell* **2000**, *100*, 57–70. [CrossRef]
7. Hanahan, D.; Weinberg, R.A. Hallmarks of cancer: The next generation. *Cell* **2011**, *144*, 646–674. [CrossRef]
8. Evtyugin, D.D.; Magina, S.; Evtuguin, D.V. Recent Advances in the Production and Applications of Ellagic Acid and Its Derivatives. A Review. *Molecules* **2020**, *25*, 2745. [CrossRef]
9. Landete, J.M. Ellagitannins, ellagic acid and their derived metabolites: A review about source, metabolism, functions and health. *Food Res. Int.* **2011**, *44*, 1150–1160. [CrossRef]
10. Espin, J.C.; Larrosa, M.; Garcia-Conesa, M.T.; Tomas-Barberan, F. Biological significance of urolithins, the gut microbial ellagic Acid-derived metabolites: The evidence so far. Evid. Based Complement. *Altern. Med.* **2013**, *2013*, 270418. [CrossRef]
11. Sharifi-Rad, J.; Quispe, C.; Castillo, C.M.S.; Caroca, R.; Lazo-Velez, M.A.; Antonyak, H.; Polishchuk, A.; Lysiuk, R.; Oliinyk, P.; De Masi, L.; et al. Ellagic Acid: A Review on Its Natural Sources, Chemical Stability, and Therapeutic Potential. *Oxidative Med. Cell. Longev.* **2022**, *2022*, 3848084. [CrossRef]
12. Zeb, A. Ellagic acid in suppressing in vivo and in vitro oxidative stresses. *Mol. Cell. Biochem.* **2018**, *448*, 27–41. [CrossRef] [PubMed]
13. Tosovic, J.; Bren, U. Antioxidative Action of Ellagic Acid-A Kinetic DFT Study. *Antioxidants* **2020**, *9*, 587. [CrossRef]
14. Mannino, F.; Imbesi, C.; Bitto, A.; Minutoli, L.; Squadrito, F.; D'Angelo, T.; Booz, C.; Pallio, G.; Irrera, N. Anti-oxidant and anti-inflammatory effects of ellagic and punicic acid in an in vitro model of cardiac fibrosis. *Biomed. Pharmacother.* **2023**, *162*, 114666. [CrossRef] [PubMed]
15. Bains, M.; Kaur, J.; Akhtar, A.; Kuhad, A.; Sah, S.P. Anti-inflammatory effects of ellagic acid and vanillic acid against quinolinic acid-induced rat model of Huntington's disease by targeting IKK-NF-kappaB pathway. *Eur. J. Pharmacol.* **2022**, *934*, 175316. [CrossRef] [PubMed]
16. Cota, D.; Patil, D. Antibacterial potential of ellagic acid and gallic acid against IBD bacterial isolates and cytotoxicity against colorectal cancer. *Nat. Prod. Res.* **2023**, *37*, 1998–2002. [CrossRef] [PubMed]
17. De, R.; Sarkar, A.; Ghosh, P.; Ganguly, M.; Karmakar, B.C.; Saha, D.R.; Halder, A.; Chowdhury, A.; Mukhopadhyay, A.K. Antimicrobial activity of ellagic acid against Helicobacter pylori isolates from India and during infections in mice. *J. Antimicrob. Chemother.* **2018**, *73*, 1595–1603. [CrossRef] [PubMed]
18. Cui, Q.; Du, R.; Anantpadma, M.; Schafer, A.; Hou, L.; Tian, J.; Davey, R.A.; Cheng, H.; Rong, L. Identification of Ellagic Acid from Plant *Rhodiola rosea* L. as an Anti-Ebola Virus Entry Inhibitor. *Viruses* **2018**, *10*, 152. [CrossRef]
19. Mandal, S.; Ahuja, A.; Shivapurkar, N.M.; Cheng, S.J.; Groopman, J.D.; Stoner, G.D. Inhibition of aflatoxin B1 mutagenesis in Salmonella typhimurium and DNA damage in cultured rat and human tracheobronchial tissues by ellagic acid. *Carcinogenesis* **1987**, *8*, 1651–1656. [CrossRef]
20. Salinger-Martinovic, S.; Cosic, V.; Stojiljkovic, N.; Ilic, S.; Stojanovic, N.; Dencic, T. Impact of ellagic acid application on doxorubicin-induced cardiovascular toxicity model. *Can. J. Physiol. Pharmacol.* **2021**, *99*, 185–191. [CrossRef]
21. Zhu, H.; Yan, Y.; Jiang, Y.; Meng, X. Ellagic Acid and Its Anti-Aging Effects on Central Nervous System. *Int. J. Mol. Sci.* **2022**, *23*, 10937. [CrossRef]

22. Gupta, A.; Singh, A.K.; Kumar, R.; Jamieson, S.; Pandey, A.K.; Bishayee, A. Neuroprotective Potential of Ellagic Acid: A Critical Review. *Adv. Nutr.* **2021**, *12*, 1211–1238. [CrossRef] [PubMed]
23. Beserra, A.M.; Calegari, P.I.; Souza Mdo, C.; Dos Santos, R.A.; Lima, J.C.; Silva, R.M.; Balogun, S.O.; Martins, D.T. Gastroprotective and ulcer-healing mechanisms of ellagic acid in experimental rats. *J. Agric. Food Chem.* **2011**, *59*, 6957–6965. [CrossRef] [PubMed]
24. Aishwarya, V.; Solaipriya, S.; Sivaramakrishnan, V. Role of ellagic acid for the prevention and treatment of liver diseases. *Phytother. Res.* **2021**, *35*, 2925–2944. [CrossRef] [PubMed]
25. Garcia-Nino, W.R.; Ibarra-Lara, L.; Cuevas-Magana, M.Y.; Sanchez-Mendoza, A.; Armada, E. Protective activities of ellagic acid and urolithins against kidney toxicity of environmental pollutants: A review. *Environ. Toxicol. Pharmacol.* **2022**, *95*, 103960. [CrossRef] [PubMed]
26. Feitelson, M.A.; Arzumanyan, A.; Kulathinal, R.J.; Blain, S.W.; Holcombe, R.F.; Mahajna, J.; Marino, M.; Martinez-Chantar, M.L.; Nawroth, R.; Sanchez-Garcia, I.; et al. Sustained proliferation in cancer: Mechanisms and novel therapeutic targets. *Semin. Cancer Biol.* **2015**, *35*, S25–S54. [CrossRef]
27. Duronio, R.J.; Xiong, Y. Signaling pathways that control cell proliferation. *Cold Spring Harb. Perspect. Biol.* **2013**, *5*, a008904. [CrossRef]
28. Hume, S.; Dianov, G.L.; Ramadan, K. A unified model for the G1/S cell cycle transition. *Nucleic Acids Res.* **2020**, *48*, 12483–12501. [CrossRef]
29. Malumbres, M.; Barbacid, M. To cycle or not to cycle: A critical decision in cancer. *Nat. Rev. Cancer* **2001**, *1*, 222–231. [CrossRef]
30. Ding, L.; Cao, J.; Lin, W.; Chen, H.; Xiong, X.; Ao, H.; Yu, M.; Lin, J.; Cui, Q. The Roles of Cyclin-Dependent Kinases in Cell-Cycle Progression and Therapeutic Strategies in Human Breast Cancer. *Int. J. Mol. Sci.* **2020**, *21*, 1960. [CrossRef]
31. Baier, A.; Szyszka, R. Compounds from Natural Sources as Protein Kinase Inhibitors. *Biomolecules* **2020**, *10*, 1546. [CrossRef]
32. Lolli, G.; Johnson, L.N. CAK-Cyclin-dependent Activating Kinase: A key kinase in cell cycle control and a target for drugs? *Cell Cycle* **2005**, *4*, 572–577. [CrossRef] [PubMed]
33. Sherr, C.J.; Roberts, J.M. CDK inhibitors: Positive and negative regulators of G1-phase progression. *Genes Dev.* **1999**, *13*, 1501–1512. [CrossRef] [PubMed]
34. Matsuura, I.; Denissova, N.G.; Wang, G.; He, D.; Long, J.; Liu, F. Cyclin-dependent kinases regulate the antiproliferative function of Smads. *Nature* **2004**, *430*, 226–231. [CrossRef] [PubMed]
35. Sakamaki, T.; Casimiro, M.C.; Ju, X.; Quong, A.A.; Katiyar, S.; Liu, M.; Jiao, X.; Li, A.; Zhang, X.; Lu, Y.; et al. Cyclin D1 determines mitochondrial function in vivo. *Mol. Cell. Biol.* **2006**, *26*, 5449–5469. [CrossRef]
36. Agami, R.; Bernards, R. Distinct initiation and maintenance mechanisms cooperate to induce G1 cell cycle arrest in response to DNA damage. *Cell* **2000**, *102*, 55–66. [CrossRef]
37. Saleban, M.; Harris, E.L.; Poulter, J.A. D-Type Cyclins in Development and Disease. *Genes* **2023**, *14*, 1445. [CrossRef]
38. Inaba, T.; Matsushime, H.; Valentine, M.; Roussel, M.F.; Sherr, C.J.; Look, A.T. Genomic organization, chromosomal localization, and independent expression of human cyclin D genes. *Genomics* **1992**, *13*, 565–574. [CrossRef]
39. The ICGC/TCGA Pan-Cancer Analysis of Whole Genomes Consortium; Campbell, P.J.; Getz, G. Pan-cancer analysis of whole genomes. *Nature* **2020**, *578*, 82–93. [CrossRef]
40. Huang, H.; Weng, H.; Zhou, H.; Qu, L. Attacking c-Myc: Targeted and combined therapies for cancer. *Curr. Pharm. Des.* **2014**, *20*, 6543–6554. [CrossRef]
41. Pelengaris, S.; Khan, M.; Evan, G. c-MYC: More than just a matter of life and death. *Nat. Rev. Cancer* **2002**, *2*, 764–776. [CrossRef]
42. Palomero, T.; Lim, W.K.; Odom, D.T.; Sulis, M.L.; Real, P.J.; Margolin, A.; Barnes, K.C.; O'Neil, J.; Neuberg, D.; Weng, A.P.; et al. NOTCH1 directly regulates c-MYC and activates a feed-forward-loop transcriptional network promoting leukemic cell growth. *Proc. Natl. Acad. Sci. USA* **2006**, *103*, 18261–18266. [CrossRef]
43. Feng, L.; Wang, K.; Tang, P.; Chen, S.; Liu, T.; Lei, J.; Yuan, R.; Hu, Z.; Li, W.; Yu, X. Deubiquitinase USP18 promotes the progression of pancreatic cancer via enhancing the Notch1-c-Myc axis. *Aging* **2020**, *12*, 19273–19292. [CrossRef] [PubMed]
44. Lepore Signorile, M.; Grossi, V.; Fasano, C.; Forte, G.; Disciglio, V.; Sanese, P.; De Marco, K.; La Rocca, F.; Armentano, R.; Valentini, A.M.; et al. c-MYC Protein Stability Is Sustained by MAPKs in Colorectal Cancer. *Cancers* **2022**, *14*, 4840. [CrossRef]
45. Thongsom, S.; Racha, S.; Ei, Z.Z.; Petsri, K.; Aksorn, N.; Chamni, S.; Panpuang, V.; Zou, H.; Chanvorachote, P. N,N′-Diarylurea Derivatives (CTPPU) Inhibited NSCLC Cell Growth and Induced Cell Cycle Arrest through Akt/GSK-3beta/c-Myc Signaling Pathway. *Int. J. Mol. Sci.* **2023**, *24*, 1357. [CrossRef] [PubMed]
46. Zhang, S.; Li, Y.; Wu, Y.; Shi, K.; Bing, L.; Hao, J. Wnt/beta-catenin signaling pathway upregulates c-Myc expression to promote cell proliferation of P19 teratocarcinoma cells. *Anat. Rec.* **2012**, *295*, 2104–2113. [CrossRef] [PubMed]
47. Subramaniam, K.S.; Omar, I.S.; Kwong, S.C.; Mohamed, Z.; Woo, Y.L.; Mat Adenan, N.A.; Chung, I. Cancer-associated fibroblasts promote endometrial cancer growth via activation of interleukin-6/STAT-3/c-Myc pathway. *Am. J. Cancer Res.* **2016**, *6*, 200–213.
48. Usui, T.; Sakurai, M.; Umata, K.; Elbadawy, M.; Ohama, T.; Yamawaki, H.; Hazama, S.; Takenouchi, H.; Nakajima, M.; Tsunedomi, R.; et al. Hedgehog Signals Mediate Anti-Cancer Drug Resistance in Three-Dimensional Primary Colorectal Cancer Organoid Culture. *Int. J. Mol. Sci.* **2018**, *19*, 1098. [CrossRef] [PubMed]
49. Wei, Q.; Li, J.; Liu, T.; Tong, X.; Ye, X. Phosphorylation of minichromosome maintenance protein 7 (MCM7) by cyclin/cyclin-dependent kinase affects its function in cell cycle regulation. *J. Biol. Chem.* **2013**, *288*, 19715–19725. [CrossRef] [PubMed]

50. Zhou, H.; Xiong, Y.; Zhang, G.; Liu, Z.; Li, L.; Hou, S.; Zhou, T. Elevated expression of minichromosome maintenance 3 indicates poor outcomes and promotes G1/S cell cycle progression, proliferation, migration and invasion in colorectal cancer. *Biosci. Rep.* **2020**, *40*, BSR20201503. [CrossRef]
51. Gou, K.; Liu, J.; Feng, X.; Li, H.; Yuan, Y.; Xing, C. Expression of Minichromosome Maintenance Proteins (MCM) and Cancer Prognosis: A meta-analysis. *J. Cancer* **2018**, *9*, 1518–1526. [CrossRef]
52. Bernstein, S.H.; Kharbanda, S.M.; Sherman, M.L.; Stone, R.M.; Kufe, D.W. Inhibition of protein kinase C is associated with a decrease in c-myc expression in human myeloid leukemia cells. *FEBS Lett.* **1991**, *294*, 73–76. [CrossRef] [PubMed]
53. Black, A.R.; Black, J.D. Protein kinase C signaling and cell cycle regulation. *Front. Immunol.* **2012**, *3*, 423. [CrossRef] [PubMed]
54. Saijo, K.; Mecklenbrauker, I.; Santana, A.; Leitger, M.; Schmedt, C.; Tarakhovsky, A. Protein kinase C beta controls nuclear factor kappaB activation in B cells through selective regulation of the IkappaB kinase alpha. *J. Exp. Med.* **2002**, *195*, 1647–1652. [CrossRef] [PubMed]
55. Lv, X.; Chen, Z.; Li, S.; Xie, H. Knockdown of cyclooxygenase-2 leads to growth inhibition and cell cycle arrest in hepatocellular carcinoma cells. *Onco Targets Ther.* **2019**, *12*, 4341–4349. [CrossRef]
56. Chung, Y.C.; Lu, L.C.; Tsai, M.H.; Chen, Y.J.; Chen, Y.Y.; Yao, S.P.; Hsu, C.P. The inhibitory effect of ellagic Acid on cell growth of ovarian carcinoma cells. *Evid.-Based Complement. Altern. Med.* **2013**, *2013*, 306705. [CrossRef]
57. Zhao, M.; Tang, S.N.; Marsh, J.L.; Shankar, S.; Srivastava, R.K. Ellagic acid inhibits human pancreatic cancer growth in Balb c nude mice. *Cancer Lett.* **2013**, *337*, 210–217. [CrossRef]
58. Cheng, H.; Lu, C.; Tang, R.; Pan, Y.; Bao, S.; Qiu, Y.; Xie, M. Ellagic acid inhibits the proliferation of human pancreatic carcinoma PANC-1 cells in vitro and in vivo. *Oncotarget* **2017**, *8*, 12301–12310. [CrossRef]
59. Wang, D.; Chen, Q.; Tan, Y.; Liu, B.; Liu, C. Ellagic acid inhibits human glioblastoma growth in vitro and in vivo. *Oncol. Rep.* **2017**, *37*, 1084–1092. [CrossRef]
60. Ho, C.C.; Huang, A.C.; Yu, C.S.; Lien, J.C.; Wu, S.H.; Huang, Y.P.; Huang, H.Y.; Kuo, J.H.; Liao, W.Y.; Yang, J.S.; et al. Ellagic acid induces apoptosis in TSGH8301 human bladder cancer cells through the endoplasmic reticulum stress- and mitochondria-dependent signaling pathways. *Environ. Toxicol.* **2014**, *29*, 1262–1274. [CrossRef]
61. Li, L.W.; Na, C.; Tian, S.Y.; Chen, J.; Ma, R.; Gao, Y.; Lou, G. Ellagic acid induces HeLa cell apoptosis via regulating signal transducer and activator of transcription 3 signaling. *Exp. Ther. Med.* **2018**, *16*, 29–36. [CrossRef]
62. Pani, S.; Mohapatra, S.; Sahoo, A.; Baral, B.; Debata, P.R. Shifting of cell cycle arrest from the S-phase to G2/M phase and downregulation of EGFR expression by phytochemical combinations in HeLa cervical cancer cells. *J. Biochem. Mol. Toxicol.* **2022**, *36*, e22947. [CrossRef] [PubMed]
63. Watts, C.K.; Sweeney, K.J.; Warlters, A.; Musgrove, E.A.; Sutherland, R.L. Antiestrogen regulation of cell cycle progression and cyclin D1 gene expression in MCF-7 human breast cancer cells. *Breast Cancer Res. Treat.* **1994**, *31*, 95–105. [CrossRef] [PubMed]
64. Foster, J.S.; Henley, D.C.; Bukovsky, A.; Seth, P.; Wimalasena, J. Multifaceted regulation of cell cycle progression by estrogen: Regulation of Cdk inhibitors and Cdc25A independent of cyclin D1-Cdk4 function. *Mol. Cell. Biol.* **2001**, *21*, 794–810. [CrossRef] [PubMed]
65. Papoutsi, Z.; Kassi, E.; Tsiapara, A.; Fokialakis, N.; Chrousos, G.P.; Moutsatsou, P. Evaluation of estrogenic/antiestrogenic activity of ellagic acid via the estrogen receptor subtypes ERalpha and ERbeta. *J. Agric. Food Chem.* **2005**, *53*, 7715–7720. [CrossRef]
66. Wang, Y.; Ren, F.; Li, B.; Song, Z.; Chen, P.; Ouyang, L. Ellagic acid exerts antitumor effects via the PI3K signaling pathway in endometrial cancer. *J. Cancer* **2019**, *10*, 3303–3314. [CrossRef] [PubMed]
67. Eskandari, E.; Heidarian, E.; Amini, S.A.; Saffari-Chaleshtori, J. Evaluating the effects of ellagic acid on pSTAT3, pAKT, and pERK1/2 signaling pathways in prostate cancer PC3 cells. *J. Cancer Res. Ther.* **2016**, *12*, 1266–1271. [CrossRef]
68. Zhang, T.; Chen, H.S.; Wang, L.F.; Bai, M.H.; Wang, Y.C.; Jiang, X.F.; Liu, M. Ellagic acid exerts anti-proliferation effects via modulation of Tgf-beta/Smad3 signaling in MCF-7 breast cancer cells. *Asian Pac. J. Cancer Prev.* **2014**, *15*, 273–276. [CrossRef]
69. Chen, H.S.; Bai, M.H.; Zhang, T.; Li, G.D.; Liu, M. Ellagic acid induces cell cycle arrest and apoptosis through TGF-beta/Smad3 signaling pathway in human breast cancer MCF-7 cells. *Int. J. Oncol.* **2015**, *46*, 1730–1738. [CrossRef]
70. Yousuf, M.; Shamsi, A.; Khan, P.; Shahbaaz, M.; AlAjmi, M.F.; Hussain, A.; Hassan, G.M.; Islam, A.; Haque, Q.M.R.; Hassan, M.I. Ellagic Acid Controls Cell Proliferation and Induces Apoptosis in Breast Cancer Cells via Inhibition of Cyclin-Dependent Kinase 6. *Int. J. Mol. Sci.* **2020**, *21*, 3526. [CrossRef]
71. Jensen, J.D.; Dunn, J.H.; Luo, Y.; Liu, W.; Fujita, M.; Dellavalle, R.P. Ellagic acid inhibits melanoma growth in vitro. *Dermatol. Rep.* **2011**, *3*, e36. [CrossRef]
72. Wang, F.; Chen, J.; Xiang, D.; Lian, X.; Wu, C.; Quan, J. Ellagic acid inhibits cell proliferation, migration, and invasion in melanoma via EGFR pathway. *Am. J. Transl. Res.* **2020**, *12*, 2295–2304.
73. Fang, Y.; Zhou, H.; Xia, J.-F.; Lin, J.-J.; Li, R.-Z.; Yang, D.-Q.; Xu, M.-Y.; Li, X.-Y. Ellagic acid regulates Wnt/β-catenin signaling pathway and CDK8 in HCT 116 and HT 29 colon cancer cells. *Bangladesh J. Pharmacol.* **2015**, *10*, 47–56. [CrossRef]
74. Umesalma, S.; Nagendraprabhu, P.; Sudhandiran, G. Ellagic acid inhibits proliferation and induced apoptosis via the Akt signaling pathway in HCT-15 colon adenocarcinoma cells. *Mol. Cell. Biochem.* **2015**, *399*, 303–313. [CrossRef]
75. Yousef, A.I.; El-Masry, O.S.; Mohsen, M.A.A. Impact of Cellular Genetic Make-up on Colorectal Cancer Cell Lines Response to Ellagic Acid: Implications of Small Interfering RNA. *Asian Pac. J. Cancer Prev.* **2016**, *17*, 743–748. [CrossRef]

76. Zhao, J.; Li, G.; Wei, J.; Dang, S.; Yu, X.; Ding, L.; Shang, C.; Zhang, H.; Zhang, Z.; Chen, H.; et al. Ellagic acid induces cell cycle arrest and apoptosis via the TGF-beta1/Smad3 signaling pathway in human colon cancer HCT-116 cells. *Oncol. Rep.* **2020**, *44*, 768–776. [CrossRef]
77. Qiu, S.; Zhong, C.; Zhao, B.; Li, G.; Wang, J.; Jehan, S.; Li, J.; Zhao, X.; Li, D.; Sui, G. Transcriptome analysis of signaling pathways targeted by Ellagic acid in hepatocellular carcinoma cells. *Biochim. Biophys. Acta Gen. Subj.* **2021**, *1865*, 129911. [CrossRef]
78. Mishra, S.; Vinayak, M. Ellagic acid checks lymphoma promotion via regulation of PKC signaling pathway. *Mol. Biol. Rep.* **2013**, *40*, 1417–1428. [CrossRef]
79. Zhang, C.; Liu, J.; Xu, D.; Zhang, T.; Hu, W.; Feng, Z. Gain-of-function mutant p53 in cancer progression and therapy. *J. Mol. Cell Biol.* **2020**, *12*, 674–687. [CrossRef]
80. Datta, N.; Chakraborty, S.; Basu, M.; Ghosh, M.K. Tumor Suppressors Having Oncogenic Functions: The Double Agents. *Cells* **2020**, *10*, 46. [CrossRef]
81. Gordon, E.M.; Ravicz, J.R.; Liu, S.; Chawla, S.P.; Hall, F.L. Cell cycle checkpoint control: The cyclin G1/Mdm2/p53 axis emerges as a strategic target for broad-spectrum cancer gene therapy—A review of molecular mechanisms for oncologists. *Mol. Clin. Oncol.* **2018**, *9*, 115–134. [CrossRef]
82. Lane, D.P. Cancer. p53, guardian of the genome. *Nature* **1992**, *358*, 15–16. [CrossRef] [PubMed]
83. Hu, J.; Cao, J.; Topatana, W.; Juengpanich, S.; Li, S.; Zhang, B.; Shen, J.; Cai, L.; Cai, X.; Chen, M. Targeting mutant p53 for cancer therapy: Direct and indirect strategies. *J. Hematol. Oncol.* **2021**, *14*, 157. [CrossRef] [PubMed]
84. Su, Y.; Sai, Y.; Zhou, L.; Liu, Z.; Du, P.; Wu, J.; Zhang, J. Current insights into the regulation of programmed cell death by TP53 mutation in cancer. *Front. Oncol.* **2022**, *12*, 1023427. [CrossRef] [PubMed]
85. Cordani, M.; Butera, G.; Pacchiana, R.; Masetto, F.; Mullappilly, N.; Riganti, C.; Donadelli, M. Mutant p53-Associated Molecular Mechanisms of ROS Regulation in Cancer Cells. *Biomolecules* **2020**, *10*, 361. [CrossRef] [PubMed]
86. Di Agostino, S.; Strano, S.; Emiliozzi, V.; Zerbini, V.; Mottolese, M.; Sacchi, A.; Blandino, G.; Piaggio, G. Gain of function of mutant p53: The mutant p53/NF-Y protein complex reveals an aberrant transcriptional mechanism of cell cycle regulation. *Cancer Cell* **2006**, *10*, 191–202. [CrossRef] [PubMed]
87. Gurtner, A.; Starace, G.; Norelli, G.; Piaggio, G.; Sacchi, A.; Bossi, G. Mutant p53-induced up-regulation of mitogen-activated protein kinase kinase 3 contributes to gain of function. *J. Biol. Chem.* **2010**, *285*, 14160–14169. [CrossRef]
88. Wang, W.; Cheng, B.; Miao, L.; Mei, Y.; Wu, M. Mutant p53-R273H gains new function in sustained activation of EGFR signaling via suppressing miR-27a expression. *Cell Death Dis.* **2013**, *4*, e574. [CrossRef]
89. Di Agostino, S.; Sorrentino, G.; Ingallina, E.; Valenti, F.; Ferraiuolo, M.; Bicciato, S.; Piazza, S.; Strano, S.; Del Sal, G.; Blandino, G. YAP enhances the pro-proliferative transcriptional activity of mutant p53 proteins. *EMBO Rep.* **2016**, *17*, 188–201. [CrossRef]
90. Verduci, L.; Ferraiuolo, M.; Sacconi, A.; Ganci, F.; Vitale, J.; Colombo, T.; Paci, P.; Strano, S.; Macino, G.; Rajewsky, N.; et al. The oncogenic role of circPVT1 in head and neck squamous cell carcinoma is mediated through the mutant p53/YAP/TEAD transcription-competent complex. *Genome Biol.* **2017**, *18*, 237. [CrossRef]
91. Schulz-Heddergott, R.; Stark, N.; Edmunds, S.J.; Li, J.; Conradi, L.C.; Bohnenberger, H.; Ceteci, F.; Greten, F.R.; Dobbelstein, M.; Moll, U.M. Therapeutic Ablation of Gain-of-Function Mutant p53 in Colorectal Cancer Inhibits Stat3-Mediated Tumor Growth and Invasion. *Cancer Cell* **2018**, *34*, 298–314. [CrossRef]
92. Zhao, K.; Yang, Y.; Zhang, G.; Wang, C.; Wang, D.; Wu, M.; Mei, Y. Regulation of the Mdm2-p53 pathway by the ubiquitin E3 ligase MARCH7. *EMBO Rep.* **2018**, *19*, 305–319. [CrossRef] [PubMed]
93. Bohlman, S.; Manfredi, J.J. p53-independent effects of Mdm2. *Subcell. Biochem.* **2014**, *85*, 235–246. [CrossRef]
94. Momand, J.; Zambetti, G.P.; Olson, D.C.; George, D.; Levine, A.J. The mdm-2 oncogene product forms a complex with the p53 protein and inhibits p53-mediated transactivation. *Cell* **1992**, *69*, 1237–1245. [CrossRef] [PubMed]
95. Kato, S.; Ross, J.S.; Gay, L.; Dayyani, F.; Roszik, J.; Subbiah, V.; Kurzrock, R. Analysis of MDM2 Amplification: Next-Generation Sequencing of Patients with Diverse Malignancies. *JCO Precis. Oncol.* **2018**, *2018*, PO.17.00235. [CrossRef] [PubMed]
96. Consortium, A.P.G. AACR Project GENIE: Powering Precision Medicine through an International Consortium. *Cancer Discov.* **2017**, *7*, 818–831. [CrossRef]
97. Sciot, R. MDM2 Amplified Sarcomas: A Literature Review. *Diagnostics* **2021**, *11*, 496. [CrossRef]
98. Hou, H.; Sun, D.; Zhang, X. The role of MDM2 amplification and overexpression in therapeutic resistance of malignant tumors. *Cancer Cell Int.* **2019**, *19*, 216. [CrossRef]
99. Pellot Ortiz, K.I.; Rechberger, J.S.; Nonnenbroich, L.F.; Daniels, D.J.; Sarkaria, J.N. MDM2 Inhibition in the Treatment of Glioblastoma: From Concept to Clinical Investigation. *Biomedicines* **2023**, *11*, 1879. [CrossRef]
100. Niculescu, A.B., 3rd; Chen, X.; Smeets, M.; Hengst, L.; Prives, C.; Reed, S.I. Effects of p21 (Cip1/Waf1) at both the G1/S and the G2/M cell cycle transitions: pRb is a critical determinant in blocking DNA replication and in preventing endoreduplication. *Mol. Cell. Biol.* **1998**, *18*, 629–643. [CrossRef]
101. Bertoli, C.; Skotheim, J.M.; de Bruin, R.A. Control of cell cycle transcription during G1 and S phases. *Nat. Rev. Mol. Cell. Biol.* **2013**, *14*, 518–528. [CrossRef]
102. Marhenke, S.; Buitrago-Molina, L.E.; Endig, J.; Orlik, J.; Schweitzer, N.; Klett, S.; Longerich, T.; Geffers, R.; Sanchez Munoz, A.; Dorrell, C.; et al. p21 promotes sustained liver regeneration and hepatocarcinogenesis in chronic cholestatic liver injury. *Gut* **2014**, *63*, 1501–1512. [CrossRef] [PubMed]

103. Huang, Y.; Wang, W.; Chen, Y.; Huang, Y.; Zhang, J.; He, S.; Tan, Y.; Qiang, F.; Li, A.; Roe, O.D.; et al. The opposite prognostic significance of nuclear and cytoplasmic p21 expression in resectable gastric cancer patients. *J. Gastroenterol.* **2014**, *49*, 1441–1452. [CrossRef]
104. Al Bitar, S.; Gali-Muhtasib, H. The Role of the Cyclin Dependent Kinase Inhibitor p21(cip1/waf1) in Targeting Cancer: Molecular Mechanisms and Novel Therapeutics. *Cancers* **2019**, *11*, 1475. [CrossRef] [PubMed]
105. Macleod, K.F.; Sherry, N.; Hannon, G.; Beach, D.; Tokino, T.; Kinzler, K.; Vogelstein, B.; Jacks, T. p53-dependent and independent expression of p21 during cell growth, differentiation, and DNA damage. *Genes Dev.* **1995**, *9*, 935–944. [CrossRef] [PubMed]
106. Friend, S.H.; Bernards, R.; Rogelj, S.; Weinberg, R.A.; Rapaport, J.M.; Albert, D.M.; Dryja, T.P. A human DNA segment with properties of the gene that predisposes to retinoblastoma and osteosarcoma. *Nature* **1986**, *323*, 643–646. [CrossRef]
107. Janostiak, R.; Torres-Sanchez, A.; Posas, F.; de Nadal, E. Understanding Retinoblastoma Post-Translational Regulation for the Design of Targeted Cancer Therapies. *Cancers* **2022**, *14*, 1265. [CrossRef]
108. Kim, S.; Leong, A.; Kim, M.; Yang, H.W. CDK4/6 initiates Rb inactivation and CDK2 activity coordinates cell-cycle commitment and G1/S transition. *Sci. Rep.* **2022**, *12*, 16810. [CrossRef]
109. Yao, Y.; Gu, X.; Xu, X.; Ge, S.; Jia, R. Novel insights into RB1 mutation. *Cancer Lett.* **2022**, *547*, 215870. [CrossRef]
110. Price, E.A.; Kolkiewicz, K.; Patel, R.; Hashim, S.; Karaa, E.; Scheimberg, I.; Sagoo, M.S.; Reddy, M.A.; Onadim, Z. Detection and reporting of RB1 promoter hypermethylation in diagnostic screening. *Ophthalmic Genet.* **2018**, *39*, 526–531. [CrossRef]
111. Felsani, A.; Mileo, A.M.; Paggi, M.G. Retinoblastoma family proteins as key targets of the small DNA virus oncoproteins. *Oncogene* **2006**, *25*, 5277–5285. [CrossRef]
112. Zhou, L.; Ng, D.S.; Yam, J.C.; Chen, L.J.; Tham, C.C.; Pang, C.P.; Chu, W.K. Post-translational modifications on the retinoblastoma protein. *J. Biomed. Sci.* **2022**, *29*, 33. [CrossRef] [PubMed]
113. Uchida, C.; Miwa, S.; Kitagawa, K.; Hattori, T.; Isobe, T.; Otani, S.; Oda, T.; Sugimura, H.; Kamijo, T.; Ookawa, K.; et al. Enhanced Mdm2 activity inhibits pRB function via ubiquitin-dependent degradation. *EMBO J.* **2005**, *24*, 160–169. [CrossRef] [PubMed]
114. Li, T.M.; Chen, G.W.; Su, C.C.; Lin, J.G.; Yeh, C.C.; Cheng, K.C.; Chung, J.G. Ellagic acid induced p53/p21 expression, G1 arrest and apoptosis in human bladder cancer T24 cells. *Anticancer Res.* **2005**, *25*, 971–979. [PubMed]
115. Naiki-Ito, A.; Chewonarin, T.; Tang, M.; Pitchakarn, P.; Kuno, T.; Ogawa, K.; Asamoto, M.; Shirai, T.; Takahashi, S. Ellagic acid, a component of pomegranate fruit juice, suppresses androgen-dependent prostate carcinogenesis via induction of apoptosis. *Prostate* **2015**, *75*, 151–160. [CrossRef]
116. Das, U.; Biswas, S.; Chattopadhyay, S.; Chakraborty, A.; Dey Sharma, R.; Banerji, A.; Dey, S. Radiosensitizing effect of ellagic acid on growth of Hepatocellular carcinoma cells: An in vitro study. *Sci. Rep.* **2017**, *7*, 14043. [CrossRef]
117. Li, J.; Wang, G.; Hou, C.; Li, J.; Luo, Y.; Li, B. Punicalagin and ellagic acid from pomegranate peel induce apoptosis and inhibits proliferation in human HepG2 hepatoma cells through targeting mitochondria. *Food Agric. Immunol.* **2019**, *30*, 897–912. [CrossRef]
118. Cetin, A.; Biltekin, B.; Degirmencioglu, S. Ellagic Acid Enhances the Antitumor Efficacy of Bevacizumab in an In Vitro Glioblastoma Model. *World Neurosurg.* **2019**, *132*, e59–e65. [CrossRef]
119. Cetin, A.; Biltekin, B. Ellagic Acid Enhances Antitumor Efficacy of Temozolomide in an in vitro Glioblastoma Model. *Turk. Neurosurg.* **2020**, *30*, 813–821. [CrossRef]
120. Cheshomi, H.; Bahrami, A.R.; Rafatpanah, H.; Matin, M.M. The effects of ellagic acid and other pomegranate (*Punica granatum* L.) derivatives on human gastric cancer AGS cells. *Hum. Exp. Toxicol.* **2022**, *41*, 9603271211064534. [CrossRef]
121. Kumar, D.; Basu, S.; Parija, L.; Rout, D.; Manna, S.; Dandapat, J.; Debata, P.R. Curcumin and Ellagic acid synergistically induce ROS generation, DNA damage, p53 accumulation and apoptosis in HeLa cervical carcinoma cells. *Biomed. Pharmacother.* **2016**, *81*, 31–37. [CrossRef]
122. Narayanan, B.A.; Geoffroy, O.; Willingham, M.C.; Re, G.G.; Nixon, D.W. p53/p21(WAF1/CIP1) expression and its possible role in G1 arrest and apoptosis in ellagic acid treated cancer cells. *Cancer Lett.* **1999**, *136*, 215–221. [CrossRef]
123. Mertens-Talcott, S.U.; Bomser, J.A.; Romero, C.; Talcott, S.T.; Percival, S.S. Ellagic acid potentiates the effect of quercetin on p21waf1/cip1, p53, and MAP-kinases without affecting intracellular generation of reactive oxygen species in vitro. *J. Nutr.* **2005**, *135*, 609–614. [CrossRef] [PubMed]
124. Saleem, Y.I.M.; Selim, M.I. MDM2 as a target for ellagic acid-mediated suppression of prostate cancer cells in vitro. *Oncol. Rep.* **2020**, *44*, 1255–1265. [CrossRef] [PubMed]
125. Moll, U.M.; Petrenko, O. The MDM2-p53 interaction. *Mol. Cancer Res.* **2003**, *1*, 1001–1008. [PubMed]
126. Norden, E.; Heiss, E.H. Urolithin A gains in antiproliferative capacity by reducing the glycolytic potential via the p53/TIGAR axis in colon cancer cells. *Carcinogenesis* **2019**, *40*, 93–101. [CrossRef]
127. Mohammed Saleem, Y.I.; Albassam, H.; Selim, M. Urolithin A induces prostate cancer cell death in p53-dependent and in p53-independent manner. *Eur. J. Nutr.* **2020**, *59*, 1607–1618. [CrossRef]
128. Gonzalez-Sarrias, A.; Nunez-Sanchez, M.A.; Tome-Carneiro, J.; Tomas-Barberan, F.A.; Garcia-Conesa, M.T.; Espin, J.C. Comprehensive characterization of the effects of ellagic acid and urolithins on colorectal cancer and key-associated molecular hallmarks: MicroRNA cell specific induction of CDKN1A (p21) as a common mechanism involved. *Mol. Nutr. Food Res.* **2016**, *60*, 701–716. [CrossRef]
129. Nyquist, M.D.; Corella, A.; Coleman, I.; De Sarkar, N.; Kaipainen, A.; Ha, G.; Gulati, R.; Ang, L.; Chatterjee, P.; Lucas, J.; et al. Combined TP53 and RB1 Loss Promotes Prostate Cancer Resistance to a Spectrum of Therapeutics and Confers Vulnerability to Replication Stress. *Cell Rep.* **2020**, *31*, 107669. [CrossRef]

130. Engeland, K. Cell cycle regulation: p53-p21-RB signaling. *Cell Death Differ.* **2022**, *29*, 946–960. [CrossRef]
131. Tan, Y.H.; Shudo, T.; Yoshida, T.; Sugiyama, Y.; Si, J.Y.; Tsukano, C.; Takemoto, Y.; Kakizuka, A. Ellagic acid, extracted from Sanguisorba officinalis, induces G1 arrest by modulating PTEN activity in B16F10 melanoma cells. *Genes Cells* **2019**, *24*, 688–704. [CrossRef]
132. Falzone, L.; Salomone, S.; Libra, M. Evolution of Cancer Pharmacological Treatments at the Turn of the Third Millennium. *Front. Pharmacol.* **2018**, *9*, 1300. [CrossRef] [PubMed]
133. Ha, K.T. Recent advances in cell death. *Exp. Mol. Med.* **2023**, *55*, 1571–1572. [CrossRef] [PubMed]
134. Chowdhury, I.; Tharakan, B.; Bhat, G.K. Current concepts in apoptosis: The physiological suicide program revisited. *Cell. Mol. Biol. Lett.* **2006**, *11*, 506–525. [CrossRef] [PubMed]
135. Bedoui, S.; Herold, M.J.; Strasser, A. Emerging connectivity of programmed cell death pathways and its physiological implications. *Nat. Rev. Mol. Cell. Biol.* **2020**, *21*, 678–695. [CrossRef]
136. Gulbins, E.; Jekle, A.; Ferlinz, K.; Grassme, H.; Lang, F. Physiology of apoptosis. *Am. J. Physiol. Ren. Physiol.* **2000**, *279*, F605–F615. [CrossRef]
137. Pfeffer, C.M.; Singh, A.T.K. Apoptosis: A Target for Anticancer Therapy. *Int. J. Mol. Sci.* **2018**, *19*, 448. [CrossRef]
138. Aubrey, B.J.; Kelly, G.L.; Janic, A.; Herold, M.J.; Strasser, A. How does p53 induce apoptosis and how does this relate to p53-mediated tumour suppression? *Cell Death Differ.* **2018**, *25*, 104–113. [CrossRef] [PubMed]
139. Almanza, A.; Carlesso, A.; Chintha, C.; Creedican, S.; Doultsinos, D.; Leuzzi, B.; Luis, A.; McCarthy, N.; Montibeller, L.; More, S.; et al. Endoplasmic reticulum stress signalling—From basic mechanisms to clinical applications. *FEBS J.* **2019**, *286*, 241–278. [CrossRef]
140. Guo, M.; Lu, B.; Gan, J.; Wang, S.; Jiang, X.; Li, H. Apoptosis detection: A purpose-dependent approach selection. *Cell Cycle* **2021**, *20*, 1033–1040. [CrossRef]
141. Jan, R.; Chaudhry, G.E. Understanding Apoptosis and Apoptotic Pathways Targeted Cancer Therapeutics. *Adv. Pharm. Bull.* **2019**, *9*, 205–218. [CrossRef]
142. Hawkins, C.J.; Miles, M.A. Mutagenic Consequences of Sublethal Cell Death Signaling. *Int. J. Mol. Sci.* **2021**, *22*, 6144. [CrossRef] [PubMed]
143. Kalkavan, H.; Chen, M.J.; Crawford, J.C.; Quarato, G.; Fitzgerald, P.; Tait, S.W.G.; Goding, C.R.; Green, D.R. Sublethal cytochrome c release generates drug-tolerant persister cells. *Cell* **2022**, *185*, 3356–3374.e3322. [CrossRef] [PubMed]
144. Wong, R.S. Apoptosis in cancer: From pathogenesis to treatment. *J. Exp. Clin. Cancer Res.* **2011**, *30*, 87. [CrossRef] [PubMed]
145. Sharma, A.; Boise, L.H.; Shanmugam, M. Cancer Metabolism and the Evasion of Apoptotic Cell Death. *Cancers* **2019**, *11*, 1144. [CrossRef] [PubMed]
146. Fernald, K.; Kurokawa, M. Evading apoptosis in cancer. *Trends Cell Biol.* **2013**, *23*, 620–633. [CrossRef] [PubMed]
147. Blagosklonny, M.V. Selective protection of normal cells from chemotherapy, while killing drug-resistant cancer cells. *Oncotarget* **2023**, *14*, 193–206. [CrossRef]
148. Carneiro, B.A.; El-Deiry, W.S. Targeting apoptosis in cancer therapy. *Nat. Rev. Clin. Oncol.* **2020**, *17*, 395–417. [CrossRef]
149. Cang, S.; Iragavarapu, C.; Savooji, J.; Song, Y.; Liu, D. ABT-199 (venetoclax) and BCL-2 inhibitors in clinical development. *J. Hematol. Oncol.* **2015**, *8*, 129. [CrossRef]
150. Mihalyova, J.; Jelinek, T.; Growkova, K.; Hrdinka, M.; Simicek, M.; Hajek, R. Venetoclax: A new wave in hematooncology. *Exp. Hematol.* **2018**, *61*, 10–25. [CrossRef]
151. Alicke, B.; Varfolomeev, E.; Lee, S.H.K.; Frommlet, A.; Ubhayakar, S.; Quinn, J.G.; Fairbrother, W.J.; Jones, R.; Gould, S.E.; Vucic, D. IAP antagonist GDC-0917 is more potent than Debio1143 in promoting cell death, c-IAP1 degradation and tumor growth inhibition. *Cell Death Dis.* **2022**, *13*, 831. [CrossRef]
152. Lim, B.; Greer, Y.; Lipkowitz, S.; Takebe, N. Novel Apoptosis-Inducing Agents for the Treatment of Cancer, a New Arsenal in the Toolbox. *Cancers* **2019**, *11*, 1087. [CrossRef] [PubMed]
153. Gao, C.; Wang, A.Y. Significance of increased apoptosis and Bax expression in human small intestinal adenocarcinoma. *J. Histochem. Cytochem.* **2009**, *57*, 1139–1148. [CrossRef] [PubMed]
154. Westphal, D.; Kluck, R.M.; Dewson, G. Building blocks of the apoptotic pore: How Bax and Bak are activated and oligomerize during apoptosis. *Cell Death Differ.* **2014**, *21*, 196–205. [CrossRef] [PubMed]
155. Liu, Q.; Liang, X.; Niu, C.; Wang, X. Ellagic acid promotes A549 cell apoptosis via regulating the phosphoinositide 3-kinase/protein kinase B pathway. *Exp. Ther. Med.* **2018**, *16*, 347–352. [CrossRef]
156. Zorova, L.D.; Popkov, V.A.; Plotnikov, E.Y.; Silachev, D.N.; Pevzner, I.B.; Jankauskas, S.S.; Babenko, V.A.; Zorov, S.D.; Balakireva, A.V.; Juhaszova, M.; et al. Mitochondrial membrane potential. *Anal. Biochem.* **2018**, *552*, 50–59. [CrossRef]
157. Duan, J.; Li, Y.; Gao, H.; Yang, D.; He, X.; Fang, Y.; Zhou, G. Phenolic compound ellagic acid inhibits mitochondrial respiration and tumor growth in lung cancer. *Food Funct.* **2020**, *11*, 6332–6339. [CrossRef]
158. Salimi, A.; Roudkenar, M.H.; Sadeghi, L.; Mohseni, A.; Seydi, E.; Pirahmadi, N.; Pourahmad, J. Ellagic acid, a polyphenolic compound, selectively induces ROS-mediated apoptosis in cancerous B-lymphocytes of CLL patients by directly targeting mitochondria. *Redox Biol.* **2015**, *6*, 461–471. [CrossRef]
159. Boehning, A.L.; Essien, S.A.; Underwood, E.L.; Dash, P.K.; Boehning, D. Cell type-dependent effects of ellagic acid on cellular metabolism. *Biomed. Pharmacother.* **2018**, *106*, 411–418. [CrossRef]

160. Mathiyazhagan, J.; Kodiveri Muthukaliannan, G. Combined Zingiber officinale and Terminalia chebula Induces Apoptosis and Modulates mTOR and hTERT Gene Expressions in MCF-7 Cell Line. *Nutr. Cancer* **2021**, *73*, 1207–1216. [CrossRef]
161. Edderkaoui, M.; Odinokova, I.; Ohno, I.; Gukovsky, I.; Go, V.L.; Pandol, S.J.; Gukovskaya, A.S. Ellagic acid induces apoptosis through inhibition of nuclear factor kappa B in pancreatic cancer cells. *World J. Gastroenterol.* **2008**, *14*, 3672–3680. [CrossRef]
162. Uren, R.T.; Dewson, G.; Bonzon, C.; Lithgow, T.; Newmeyer, D.D.; Kluck, R.M. Mitochondrial release of pro-apoptotic proteins: Electrostatic interactions can hold cytochrome c but not Smac/DIABLO to mitochondrial membranes. *J. Biol. Chem.* **2005**, *280*, 2266–2274. [CrossRef] [PubMed]
163. Arnoult, D.; Gaume, B.; Karbowski, M.; Sharpe, J.C.; Cecconi, F.; Youle, R.J. Mitochondrial release of AIF and EndoG requires caspase activation downstream of Bax/Bak-mediated permeabilization. *EMBO J.* **2003**, *22*, 4385–4399. [CrossRef] [PubMed]
164. Vempati, U.D.; Diaz, F.; Barrientos, A.; Narisawa, S.; Mian, A.M.; Millan, J.L.; Boise, L.H.; Moraes, C.T. Role of cytochrome C in apoptosis: Increased sensitivity to tumor necrosis factor alpha is associated with respiratory defects but not with lack of cytochrome C release. *Mol. Cell. Biol.* **2007**, *27*, 1771–1783. [CrossRef] [PubMed]
165. Shi, Y. Caspase activation, inhibition, and reactivation: A mechanistic view. *Protein Sci.* **2004**, *13*, 1979–1987. [CrossRef]
166. Acehan, D.; Jiang, X.; Morgan, D.G.; Heuser, J.E.; Wang, X.; Akey, C.W. Three-dimensional structure of the apoptosome: Implications for assembly, procaspase-9 binding, and activation. *Mol. Cell* **2002**, *9*, 423–432. [CrossRef]
167. Kim, J.Y.; Choi, Y.J.; Kim, H.J. Determining the effect of ellagic acid on the proliferation and migration of pancreatic cancer cell lines. *Transl. Cancer Res.* **2021**, *10*, 424–433. [CrossRef]
168. Mashimo, M.; Onishi, M.; Uno, A.; Tanimichi, A.; Nobeyama, A.; Mori, M.; Yamada, S.; Negi, S.; Bu, X.; Kato, J.; et al. The 89-kDa PARP1 cleavage fragment serves as a cytoplasmic PAR carrier to induce AIF-mediated apoptosis. *J. Biol. Chem.* **2021**, *296*, 100046. [CrossRef]
169. Weaver, A.N.; Yang, E.S. Beyond DNA Repair: Additional Functions of PARP-1 in Cancer. *Front. Oncol.* **2013**, *3*, 290. [CrossRef]
170. Ahire, V.; Kumar, A.; Mishra, K.P.; Kulkarni, G. Ellagic Acid Enhances Apoptotic Sensitivity of Breast Cancer Cells to gamma-Radiation. *Nutr. Cancer* **2017**, *69*, 904–910. [CrossRef]
171. Xie, C.; Kong, J.; Miao, F.; Wang, X.; Sheng, J. Combination effects of ellagic acid with erlotinib in a Ba/F3 cell line expressing EGFR H773_V774 insH mutation. *Thorac. Cancer* **2020**, *11*, 2101–2111. [CrossRef]
172. Kao, T.Y.; Chung, Y.C.; Hou, Y.C.; Tsai, Y.W.; Chen, C.H.; Chang, H.P.; Chou, J.L.; Hsu, C.P. Effects of ellagic acid on chemosensitivity to 5-fluorouracil in colorectal carcinoma cells. *Anticancer Res.* **2012**, *32*, 4413–4418. [PubMed]
173. Genovese, I.; Carinci, M.; Modesti, L.; Aguiari, G.; Pinton, P.; Giorgi, C. Mitochondria: Insights into Crucial Features to Overcome Cancer Chemoresistance. *Int. J. Mol. Sci.* **2021**, *22*, 4770. [CrossRef] [PubMed]
174. Yakobov, S.; Dhingra, R.; Margulets, V.; Dhingra, A.; Crandall, M.; Kirshenbaum, L.A. Ellagic acid inhibits mitochondrial fission protein Drp-1 and cell proliferation in cancer. *Mol. Cell. Biochem.* **2023**, *478*, 2029–2040. [CrossRef] [PubMed]
175. Ghosh, P.; Vidal, C.; Dey, S.; Zhang, L. Mitochondria Targeting as an Effective Strategy for Cancer Therapy. *Int. J. Mol. Sci.* **2020**, *21*, 3363. [CrossRef]
176. Singh, D.; Arora, R.; Kaur, P.; Singh, B.; Mannan, R.; Arora, S. Overexpression of hypoxia-inducible factor and metabolic pathways: Possible targets of cancer. *Cell Biosci.* **2017**, *7*, 62. [CrossRef] [PubMed]
177. Ni, X.; Shang, F.S.; Wang, T.F.; Wu, D.J.; Chen, D.G.; Zhuang, B. Ellagic acid induces apoptosis and autophagy in colon cancer through the AMPK/mTOR pathway. *Tissue Cell* **2023**, *81*, 102032. [CrossRef]
178. Duan, J.; Zhan, J.C.; Wang, G.Z.; Zhao, X.C.; Huang, W.D.; Zhou, G.B. The red wine component ellagic acid induces autophagy and exhibits anti-lung cancer activity in vitro and in vivo. *J. Cell. Mol. Med.* **2019**, *23*, 143–154. [CrossRef]
179. Malik, A.; Afaq, S.; Shahid, M.; Akhtar, K.; Assiri, A. Influence of ellagic acid on prostate cancer cell proliferation: A caspase-dependent pathway. *Asian Pac. J. Trop. Med.* **2011**, *4*, 550–555. [CrossRef]
180. Fritsch, M.; Gunther, S.D.; Schwarzer, R.; Albert, M.C.; Schorn, F.; Werthenbach, J.P.; Schiffmann, L.M.; Stair, N.; Stocks, H.; Seeger, J.M.; et al. Caspase-8 is the molecular switch for apoptosis, necroptosis and pyroptosis. *Nature* **2019**, *575*, 683–687. [CrossRef]
181. Nagata, S.; Suzuki, J.; Segawa, K.; Fujii, T. Exposure of phosphatidylserine on the cell surface. *Cell Death Differ.* **2016**, *23*, 952–961. [CrossRef]
182. Kari, S.; Subramanian, K.; Altomonte, I.A.; Murugesan, A.; Yli-Harja, O.; Kandhavelu, M. Programmed cell death detection methods: A systematic review and a categorical comparison. *Apoptosis* **2022**, *27*, 482–508. [CrossRef] [PubMed]
183. Luu Hoang, K.N.; Anstee, J.E.; Arnold, J.N. The Diverse Roles of Heme Oxygenase-1 in Tumor Progression. *Front. Immunol.* **2021**, *12*, 658315. [CrossRef] [PubMed]
184. Vanella, L.; Di Giacomo, C.; Acquaviva, R.; Barbagallo, I.; Li Volti, G.; Cardile, V.; Abraham, N.G.; Sorrenti, V. Effects of ellagic Acid on angiogenic factors in prostate cancer cells. *Cancers* **2013**, *5*, 726–738. [CrossRef]
185. Li, X.; Wu, X. Soluble epoxide hydrolase (Ephx2) silencing attenuates the hydrogen peroxide-induced oxidative damage in IEC-6 cells. *Arch. Med. Sci.* **2021**, *17*, 1075–1086. [CrossRef] [PubMed]
186. Lal, A.; Kawai, T.; Yang, X.; Mazan-Mamczarz, K.; Gorospe, M. Antiapoptotic function of RNA-binding protein HuR effected through prothymosin alpha. *EMBO J.* **2005**, *24*, 1852–1862. [CrossRef] [PubMed]
187. Zhao, E.; Hou, J.; Ke, X.; Abbas, M.N.; Kausar, S.; Zhang, L.; Cui, H. The Roles of Sirtuin Family Proteins in Cancer Progression. *Cancers* **2019**, *11*, 1949. [CrossRef]
188. Vanella, L.; Di Giacomo, C.; Acquaviva, R.; Barbagallo, I.; Cardile, V.; Kim, D.H.; Abraham, N.G.; Sorrenti, V. Apoptotic markers in a prostate cancer cell line: Effect of ellagic acid. *Oncol. Rep.* **2013**, *30*, 2804–2810. [CrossRef]

189. Jang, M.; Kim, S.S.; Lee, J. Cancer cell metabolism: Implications for therapeutic targets. *Exp. Mol. Med.* **2013**, *45*, e45. [CrossRef]
190. Navarro, C.; Ortega, A.; Santeliz, R.; Garrido, B.; Chacin, M.; Galban, N.; Vera, I.; De Sanctis, J.B.; Bermudez, V. Metabolic Reprogramming in Cancer Cells: Emerging Molecular Mechanisms and Novel Therapeutic Approaches. *Pharmaceutics* **2022**, *14*, 1303. [CrossRef]
191. Cairns, R.A.; Harris, I.S.; Mak, T.W. Regulation of cancer cell metabolism. *Nat. Rev. Cancer* **2011**, *11*, 85–95. [CrossRef]
192. Schmidt, D.R.; Patel, R.; Kirsch, D.G.; Lewis, C.A.; Vander Heiden, M.G.; Locasale, J.W. Metabolomics in cancer research and emerging applications in clinical oncology. *CA Cancer J. Clin.* **2021**, *71*, 333–358. [CrossRef]
193. Bai, X.; Jiang, Y. Key factors in mTOR regulation. *Cell. Mol. Life Sci.* **2010**, *67*, 239–253. [CrossRef]
194. Xu, J.; Ji, J.; Yan, X.H. Cross-talk between AMPK and mTOR in regulating energy balance. *Crit. Rev. Food Sci. Nutr.* **2012**, *52*, 373–381. [CrossRef] [PubMed]
195. Xia, J.; Xue, C.; Yu, J. Ellagic acid inhibited cervical cancer growth via blocking the AKT/mTOR/STAT3 pathway. *Arch. Med. Sci.* **2020**, *16*, 1–15. [CrossRef]
196. Elsaid, F.G.; Alshehri, M.A.; Shati, A.A.; Al-Kahtani, M.A.; Alsheri, A.S.; Massoud, E.E.; El-Kott, A.F.; El-Mekkawy, H.I.; Al-Ramlawy, A.M.; Abdraboh, M.E. The anti-tumourigenic effect of ellagic acid in SKOV-3 ovarian cancer cells entails activation of autophagy mediated by inhibiting Akt and activating AMPK. *Clin. Exp. Pharmacol. Physiol.* **2020**, *47*, 1611–1621. [CrossRef] [PubMed]
197. Abdelazeem, K.N.M.; Singh, Y.; Lang, F.; Salker, M.S. Negative Effect of Ellagic Acid on Cytosolic pH Regulation and Glycolytic Flux in Human Endometrial Cancer Cells. *Cell Physiol. Biochem.* **2017**, *41*, 2374–2382. [CrossRef]
198. Dahiya, R.; Mohammad, T.; Gupta, P.; Haque, A.; Alajmi, M.F.; Hussain, A.; Hassan, M.I. Molecular interaction studies on ellagic acid for its anticancer potential targeting pyruvate dehydrogenase kinase 3. *Rsc Adv.* **2019**, *9*, 23302–23315. [CrossRef]
199. Aubrey, B.J.; Strasser, A.; Kelly, G.L. Tumor-Suppressor Functions of the TP53 Pathway. *Cold Spring Harb. Perspect. Med.* **2016**, *6*, a026062. [CrossRef]
200. Kamp, W.M.; Wang, P.Y.; Hwang, P.M. TP53 mutation, mitochondria and cancer. *Curr. Opin. Genet. Dev.* **2016**, *38*, 16–22. [CrossRef]
201. Zhang, C.; Liu, J.; Liang, Y.; Wu, R.; Zhao, Y.; Hong, X.; Lin, M.; Yu, H.; Liu, L.; Levine, A.J.; et al. Tumour-associated mutant p53 drives the Warburg effect. *Nat. Commun.* **2013**, *4*, 2935. [CrossRef]
202. Dejure, F.R.; Eilers, M. MYC and tumor metabolism: Chicken and egg. *EMBO J.* **2017**, *36*, 3409–3420. [CrossRef] [PubMed]
203. Lim, S.C.; Hwang, H.; Han, S.I. Ellagic Acid Inhibits Extracellular Acidity-Induced Invasiveness and Expression of COX1, COX2, Snail, Twist 1, and c-myc in Gastric Carcinoma Cells. *Nutrients* **2019**, *11*, 3023. [CrossRef] [PubMed]
204. Robinson, N.J.; Schiemann, W.P. Telomerase in Cancer: Function, Regulation, and Clinical Translation. *Cancers* **2022**, *14*, 808. [CrossRef]
205. Slawinska, N.; Krupa, R. Molecular Aspects of Senescence and Organismal Ageing-DNA Damage Response, Telomeres, Inflammation and Chromatin. *Int. J. Mol. Sci.* **2021**, *22*, 590. [CrossRef] [PubMed]
206. Jafri, M.A.; Ansari, S.A.; Alqahtani, M.H.; Shay, J.W. Roles of telomeres and telomerase in cancer, and advances in telomerase-targeted therapies. *Genome Med.* **2016**, *8*, 69. [CrossRef]
207. Sharma, S.; Chowdhury, S. Emerging mechanisms of telomerase reactivation in cancer. *Trends Cancer* **2022**, *8*, 632–641. [CrossRef]
208. Dratwa, M.; Wysoczanska, B.; Lacina, P.; Kubik, T.; Bogunia-Kubik, K. TERT-Regulation and Roles in Cancer Formation. *Front. Immunol.* **2020**, *11*, 589929. [CrossRef]
209. Romaniuk, A.; Paszel-Jaworska, A.; Totoń, E.; Lisiak, N.; Hołysz, H.; Królak, A.; Grodecka-Gazdecka, S.; Rubiś, B. The non-canonical functions of telomerase: To turn off or not to turn off. *Mol. Biol. Rep.* **2019**, *46*, 1401–1411. [CrossRef]
210. Judasz, E.; Lisiak, N.; Kopczynski, P.; Taube, M.; Rubis, B. The Role of Telomerase in Breast Cancer's Response to Therapy. *Int. J. Mol. Sci.* **2022**, *23*, 12844. [CrossRef]
211. Saretzki, G. Role of Telomeres and Telomerase in Cancer and Aging. *Int. J. Mol. Sci.* **2023**, *24*, 584–593. [CrossRef]
212. Huang, S.T.; Wang, C.Y.; Yang, R.C.; Chu, C.J.; Wu, H.T.; Pang, J.H. Phyllanthus urinaria increases apoptosis and reduces telomerase activity in human nasopharyngeal carcinoma cells. *Complement. Med. Res.* **2009**, *16*, 34–40. [CrossRef] [PubMed]
213. Strati, A.; Papoutsi, Z.; Lianidou, E.; Moutsatsou, P. Effect of ellagic acid on the expression of human telomerase reverse transcriptase (hTERT) alpha+beta+ transcript in estrogen receptor-positive MCF-7 breast cancer cells. *Clin. Biochem.* **2009**, *42*, 1358–1362. [CrossRef] [PubMed]
214. Al-Nour, M.Y.; Ibrahim, M.M.; Elsaman, T. Ellagic Acid, Kaempferol, and Quercetin from Acacia nilotica: Promising Combined Drug with Multiple Mechanisms of Action. *Curr. Pharmacol. Rep.* **2019**, *5*, 255–280. [CrossRef] [PubMed]
215. Toufektchan, E.; Toledo, F. The Guardian of the Genome Revisited: p53 Downregulates Genes Required for Telomere Maintenance, DNA Repair, and Centromere Structure. *Cancers* **2018**, *10*, 135. [CrossRef] [PubMed]
216. Folkman, J. Tumor angiogenesis: Therapeutic implications. *N. Engl. J. Med.* **1971**, *285*, 1182–1186. [CrossRef]
217. Carmeliet, P.; Jain, R.K. Angiogenesis in cancer and other diseases. *Nature* **2000**, *407*, 249–257. [CrossRef]
218. Mittal, K.; Ebos, J.; Rini, B. Angiogenesis and the tumor microenvironment: Vascular endothelial growth factor and beyond. *Semin. Oncol.* **2014**, *41*, 235–251. [CrossRef]
219. Nyberg, P.; Salo, T.; Kalluri, R. Tumor microenvironment and angiogenesis. *Front. Biosci.* **2008**, *13*, 6537–6553. [CrossRef]
220. Patel, S.A.; Nilsson, M.B.; Le, X.; Cascone, T.; Jain, R.K.; Heymach, J.V. Molecular Mechanisms and Future Implications of VEGF/VEGFR in Cancer Therapy. *Clin. Cancer Res.* **2023**, *29*, 30–39. [CrossRef]

221. Cao, Y.; Langer, R.; Ferrara, N. Targeting angiogenesis in oncology, ophthalmology and beyond. *Nat. Rev. Drug Discov.* **2023**, *22*, 476–495. [CrossRef]
222. Lugano, R.; Ramachandran, M.; Dimberg, A. Tumor angiogenesis: Causes, consequences, challenges and opportunities. *Cell. Mol. Life Sci.* **2020**, *77*, 1745–1770. [CrossRef] [PubMed]
223. Petrovic, N. Targeting Angiogenesis in Cancer Treatments: Where do we Stand? *J. Pharm. Pharm. Sci.* **2016**, *19*, 226–238. [CrossRef]
224. Mirossay, L.; Varinska, L.; Mojzis, J. Antiangiogenic Effect of Flavonoids and Chalcones: An Update. *Int. J. Mol. Sci.* **2018**, *19*, 27. [CrossRef] [PubMed]
225. Turrini, E.; Maffei, F.; Fimognari, C. Ten Years of Research on Fucoidan and Cancer: Focus on Its Antiangiogenic and Antimetastatic Effects. *Mar. Drugs* **2023**, *21*, 307. [CrossRef]
226. Marrero, A.D.; Quesada, A.R.; Martinez-Poveda, B.; Medina, M.A. Antiangiogenic Phytochemicals Constituent of Diet as Promising Candidates for Chemoprevention of Cancer. *Antioxidants* **2022**, *11*, 302. [CrossRef] [PubMed]
227. Losso, J.N.; Bansode, R.R.; Trappey, A., 2nd; Bawadi, H.A.; Truax, R. In vitro anti-proliferative activities of ellagic acid. *J. Nutr. Biochem.* **2004**, *15*, 672–678. [CrossRef]
228. Huang, S.T.; Wang, C.Y.; Yang, R.C.; Wu, H.T.; Yang, S.H.; Cheng, Y.C.; Pang, J.H. Ellagic Acid, the Active Compound of Phyllanthus urinaria, Exerts In Vivo Anti-Angiogenic Effect and Inhibits MMP-2 Activity. *Evid. Based Complement. Altern. Med.* **2011**, *2011*, 215035. [CrossRef]
229. Labrecque, L.; Lamy, S.; Chapus, A.; Mihoubi, S.; Durocher, Y.; Cass, B.; Bojanowski, M.W.; Gingras, D.; Beliveau, R. Combined inhibition of PDGF and VEGF receptors by ellagic acid, a dietary-derived phenolic compound. *Carcinogenesis* **2005**, *26*, 821–826. [CrossRef]
230. Kisacam, M.A.; Ali, M. Ellagic Acid Exerts Apoptotic and Antiangiogenic Activity on Human Intestinal Cancer (Caco-2) Cell Line. *Int. J. Acad. Med. Pharm.* **2022**, *4*, 59–63.
231. Wang, N.; Wang, Z.Y.; Mo, S.L.; Loo, T.Y.; Wang, D.M.; Luo, H.B.; Yang, D.P.; Chen, Y.L.; Shen, J.G.; Chen, J.P. Ellagic acid, a phenolic compound, exerts anti-angiogenesis effects via VEGFR-2 signaling pathway in breast cancer. *Breast Cancer Res. Treat.* **2012**, *134*, 943–955. [CrossRef]
232. Ceci, C.; Tentori, L.; Atzori, M.G.; Lacal, P.M.; Bonanno, E.; Scimeca, M.; Cicconi, R.; Mattei, M.; de Martino, M.G.; Vespasiani, G.; et al. Ellagic Acid Inhibits Bladder Cancer Invasiveness and In Vivo Tumor Growth. *Nutrients* **2016**, *8*, 744. [CrossRef]
233. Pitchakarn, P.; Chewonarin, T.; Ogawa, K.; Suzuki, S.; Asamoto, M.; Takahashi, S.; Shirai, T.; Limtrakul, P. Ellagic acid inhibits migration and invasion by prostate cancer cell lines. *Asian Pac. J. Cancer Prev.* **2013**, *14*, 2859–2863. [CrossRef]
234. Klein, T.; Geurink, P.P.; Overkleeft, H.S.; Kauffman, H.K.; Bischoff, R. Functional proteomics on zinc-dependent metalloproteinases using inhibitor probes. *ChemMedChem* **2009**, *4*, 164–170. [CrossRef]
235. Huang, S.T.; Yang, R.C.; Wu, H.T.; Wang, C.N.; Pang, J.H. Zinc-chelation contributes to the anti-angiogenic effect of ellagic acid on inhibiting MMP-2 activity, cell migration and tube formation. *PLoS ONE* **2011**, *6*, e18986. [CrossRef]
236. Meirson, T.; Gil-Henn, H.; Samson, A.O. Invasion and metastasis: The elusive hallmark of cancer. *Oncogene* **2020**, *39*, 2024–2026. [CrossRef]
237. Liu, K.; Newbury, P.A.; Glicksberg, B.S.; Zeng, W.Z.D.; Paithankar, S.; Andrechek, E.R.; Chen, B. Evaluating cell lines as models for metastatic breast cancer through integrative analysis of genomic data. *Nat. Commun.* **2019**, *10*, 2138. [CrossRef]
238. Saha, T.; Solomon, J.; Samson, A.O.; Gil-Henn, H. Invasion and Metastasis as a Central Hallmark of Breast Cancer. *J. Clin. Med.* **2021**, *10*, 3498. [CrossRef]
239. Kim, D.H.; Xing, T.; Yang, Z.; Dudek, R.; Lu, Q.; Chen, Y.H. Epithelial Mesenchymal Transition in Embryonic Development, Tissue Repair and Cancer: A Comprehensive Overview. *J. Clin. Med.* **2017**, *7*, 1. [CrossRef]
240. Wang, Y.; Zhou, B.P. Epithelial-mesenchymal Transition—A Hallmark of Breast Cancer Metastasis. *Cancer Hallm.* **2013**, *1*, 38–49. [CrossRef]
241. Schelch, K.; Vogel, L.; Schneller, A.; Brankovic, J.; Mohr, T.; Mayer, R.L.; Slany, A.; Gerner, C.; Grusch, M. EGF Induces Migration Independent of EMT or Invasion in A549 Lung Adenocarcinoma Cells. *Front. Cell Dev. Biol.* **2021**, *9*, 634371. [CrossRef]
242. Verma, R.P.; Hansch, C. Matrix metalloproteinases (MMPs): Chemical-biological functions and (Q)SARs. *Bioorg Med. Chem.* **2007**, *15*, 2223–2268. [CrossRef] [PubMed]
243. Jiang, H.; Li, H. Prognostic values of tumoral MMP2 and MMP9 overexpression in breast cancer: A systematic review and meta-analysis. *BMC Cancer* **2021**, *21*, 149. [CrossRef] [PubMed]
244. Liu, H.; Zeng, Z.; Wang, S.; Li, T.; Mastriani, E.; Li, Q.H.; Bao, H.X.; Zhou, Y.J.; Wang, X.; Liu, Y.; et al. Main components of pomegranate, ellagic acid and luteolin, inhibit metastasis of ovarian cancer by down-regulating MMP2 and MMP9. *Cancer Biol. Ther.* **2017**, *18*, 990–999. [CrossRef] [PubMed]
245. Abdallah, H.M.I.; El Awdan, S.A.; Abdel-Rahman, R.F.; Farrag, A.R.H.; Allam, R.M. 1,8 Cineole and Ellagic acid inhibit hepatocarcinogenesis via upregulation of MiR-122 and suppression of TGF-beta1, FSCN1, Vimentin, VEGF, and MMP-9. *PLoS ONE* **2022**, *17*, e0258998. [CrossRef]
246. Ceci, C.; Lacal, P.M.; Tentori, L.; De Martino, M.G.; Miano, R.; Graziani, G. Experimental Evidence of the Antitumor, Antimetastatic and Antiangiogenic Activity of Ellagic Acid. *Nutrients* **2018**, *10*, 1756. [CrossRef]
247. Akhtar, K.; Ara, A.; Siddiqui, S.A.; Sherwani, R.K. Transition of Immunohistochemical Expression of E-Cadherin and Vimentin from Premalignant to Malignant Lesions of Oral Cavity and Oropharynx. *Oman Med. J.* **2016**, *31*, 165–169. [CrossRef]

248. Thews, O.; Gassner, B.; Kelleher, D.K.; Schwerdt, G.; Gekle, M. Impact of hypoxic and acidic extracellular conditions on cytotoxicity of chemotherapeutic drugs. *Adv. Exp. Med. Biol.* **2007**, *599*, 155–161. [CrossRef]
249. Reichert, M.; Steinbach, J.P.; Supra, P.; Weller, M. Modulation of growth and radiochemosensitivity of human malignant glioma cells by acidosis. *Cancer* **2002**, *95*, 1113–1119. [CrossRef]
250. Huang, S.; Tang, Y.; Peng, X.; Cai, X.; Wa, Q.; Ren, D.; Li, Q.; Luo, J.; Li, L.; Zou, X.; et al. Acidic extracellular pH promotes prostate cancer bone metastasis by enhancing PC-3 stem cell characteristics, cell invasiveness and VEGF-induced vasculogenesis of BM-EPCs. *Oncol. Rep.* **2016**, *36*, 2025–2032. [CrossRef]
251. Zhang, Y.; Lu, Y.; Zhang, C.; Huang, D.; Wu, W.; Zhang, Y.; Shen, J.; Cai, Y.; Chen, W.; Yao, W. FSCN-1 increases doxorubicin resistance in hepatocellular carcinoma through promotion of epithelial-mesenchymal transition. *Int. J. Oncol.* **2018**, *52*, 1455–1464. [CrossRef]
252. Strouhalova, K.; Prechova, M.; Gandalovicova, A.; Brabek, J.; Gregor, M.; Rosel, D. Vimentin Intermediate Filaments as Potential Target for Cancer Treatment. *Cancers* **2020**, *12*, 184. [CrossRef] [PubMed]
253. Boone, B.; Jacobs, K.; Ferdinande, L.; Taildeman, J.; Lambert, J.; Peeters, M.; Bracke, M.; Pauwels, P.; Brochez, L. EGFR in melanoma: Clinical significance and potential therapeutic target. *J. Cutan. Pathol.* **2011**, *38*, 492–502. [CrossRef] [PubMed]
254. Hassanein, S.S.; Abdel-Mawgood, A.L.; Ibrahim, S.A. EGFR-Dependent Extracellular Matrix Protein Interactions Might Light a Candle in Cell Behavior of Non-Small Cell Lung Cancer. *Front. Oncol.* **2021**, *11*, 766659. [CrossRef] [PubMed]
255. Meng, X.; Cui, Z.; Shi, H.; Ma, X.; Li, W.; Liu, X.; Jiang, Y. Ellagic acid inhibits cell proliferation, migration, and invasion of anaplastic thyroid cancer cells via the Wnt/beta-catenin and PI3K/Akt pathways. *Acta Biochim. Pol.* **2023**, *70*, 109–115. [CrossRef] [PubMed]
256. Song, Y.; Li, Z.X.; Liu, X.; Wang, R.; Li, L.W.; Zhang, Q. The Wnt/beta-catenin and PI3K/Akt signaling pathways promote EMT in gastric cancer by epigenetic regulation via H3 lysine 27 acetylation. *Tumour Biol.* **2017**, *39*, 1010428317712617. [CrossRef] [PubMed]
257. Shi, L.; Gao, X.; Li, X.; Jiang, N.; Luo, F.; Gu, C.; Chen, M.; Cheng, H.; Liu, P. Ellagic Acid Enhances the Efficacy of PI3K Inhibitor GDC-0941 in Breast Cancer Cells. *Curr. Mol. Med.* **2015**, *15*, 478–486. [CrossRef] [PubMed]
258. Wang, N.; Wang, Q.; Tang, H.; Zhang, F.; Zheng, Y.; Wang, S.; Zhang, J.; Wang, Z.; Xie, X. Direct inhibition of ACTN4 by ellagic acid limits breast cancer metastasis via regulation of beta-catenin stabilization in cancer stem cells. *J. Exp. Clin. Cancer Res.* **2017**, *36*, 172. [CrossRef]
259. Yamamoto, S.; Tsuda, H.; Honda, K.; Onozato, K.; Takano, M.; Tamai, S.; Imoto, I.; Inazawa, J.; Yamada, T.; Matsubara, O. Actinin-4 gene amplification in ovarian cancer: A candidate oncogene associated with poor patient prognosis and tumor chemoresistance. *Mod. Pathol.* **2009**, *22*, 499–507. [CrossRef]
260. Galluzzi, L.; Vacchelli, E.; Bravo-San Pedro, J.M.; Buqué, A.; Senovilla, L.; Baracco, E.E.; Bloy, N.; Castoldi, F.; Abastado, J.P.; Agostinis, P.; et al. Classification of current anticancer immunotherapies. *Oncotarget* **2014**, *5*, 12472–12508. [CrossRef]
261. Zhang, H.M.; Zhao, L.; Li, H.; Xu, H.; Chen, W.W.; Tao, L. Research progress on the anticarcinogenic actions and mechanisms of ellagic acid. *Cancer Biol. Med.* **2014**, *11*, 92–100. [CrossRef]
262. Shrihari, T.G. Dual role of inflammatory mediators in cancer. *Ecancermedicalscience* **2017**, *11*, 721. [CrossRef] [PubMed]
263. Cadona, F.C.; Dantas, R.F.; de Mello, G.H.; Silva-Jr, F.P. Natural products targeting into cancer hallmarks: An update on caffeine, theobromine, and (+)-catechin. *Crit. Rev. Food Sci. Nutr.* **2022**, *62*, 7222–7241. [CrossRef] [PubMed]
264. Qiu, Y.; Su, M.; Liu, L.; Tang, Y.; Pan, Y.; Sun, J. Clinical Application of Cytokines in Cancer Immunotherapy. *Drug Des. Dev. Ther.* **2021**, *15*, 2269–2287. [CrossRef]
265. Atallah-Yunes, S.A.; Robertson, M.J. Cytokine Based Immunotherapy for Cancer and Lymphoma: Biology, Challenges and Future Perspectives. *Front. Immunol.* **2022**, *13*, 872010. [CrossRef]
266. Rupp, T.; Genest, L.; Babin, D.; Legrand, C.; Hunault, M.; Froget, G.; Castagné, V. Anti-CTLA-4 and anti-PD-1 immunotherapies repress tumor progression in preclinical breast and colon model with independent regulatory T cells response. *Transl. Oncol.* **2022**, *20*, 101405. [CrossRef]
267. Lu, G.; Wang, X.; Cheng, M.; Wang, S.; Ma, K. The multifaceted mechanisms of ellagic acid in the treatment of tumors: State-of-the-art. *Biomed. Pharmacother.* **2023**, *165*, 115132. [CrossRef] [PubMed]
268. Ríos, J.L.; Giner, R.M.; Marín, M.; Recio, M.C. A Pharmacological Update of Ellagic Acid. *Planta Med.* **2018**, *84*, 1068–1093. [CrossRef]
269. Allen, C.T.; Peden-Adams, M.M.; EuDaly, J.; Keil, D.E. Subchronic exposure to ellagic acid impairs cytotoxic T-cell function and suppresses humoral immunity in mice. *Immunopharmacol. Immunotoxicol.* **2003**, *25*, 409–422. [CrossRef]
270. Baradaran Rahimi, V.; Ghadiri, M.; Ramezani, M.; Askari, V.R. Antiinflammatory and anti-cancer activities of pomegranate and its constituent, ellagic acid: Evidence from cellular, animal, and clinical studies. *Phytother. Res.* **2020**, *34*, 685–720. [CrossRef]
271. Anderson, K.C.; Teuber, S.S. Ellagic acid and polyphenolics present in walnut kernels inhibit in vitro human peripheral blood mononuclear cell proliferation and alter cytokine production. *Ann. N. Y. Acad. Sci.* **2010**, *1190*, 86–96. [CrossRef]
272. Sarkar, S.; Siddiqui, A.A.; Mazumder, S.; De, R.; Saha, S.J.; Banerjee, C.; Iqbal, M.S.; Adhikari, S.; Alam, A.; Roy, S.; et al. Ellagic Acid, a Dietary Polyphenol, Inhibits Tautomerase Activity of Human Macrophage Migration Inhibitory Factor and Its Pro-inflammatory Responses in Human Peripheral Blood Mononuclear Cells. *J. Agric. Food Chem.* **2015**, *63*, 4988–4998. [CrossRef] [PubMed]

273. Bachoual, R.; Talmoudi, W.; Boussetta, T.; Braut, F.; El-Benna, J. An aqueous pomegranate peel extract inhibits neutrophil myeloperoxidase in vitro and attenuates lung inflammation in mice. *Food Chem. Toxicol.* **2011**, *49*, 1224–1228. [CrossRef] [PubMed]
274. Rasheed, Z.; Akhtar, N.; Anbazhagan, A.N.; Ramamurthy, S.; Shukla, M.; Haqqi, T.M. Polyphenol-rich pomegranate fruit extract (POMx) suppresses PMACI-induced expression of pro-inflammatory cytokines by inhibiting the activation of MAP Kinases and NF-kappaB in human KU812 cells. *J. Inflamm.* **2009**, *6*, 1. [CrossRef] [PubMed]
275. Noh, K.T.; Cha, G.S.; Kim, H.C.; Lee, J.H.; Ahn, S.C.; Kim, D.K.; Park, Y.M. Ellagic acid modulates LPS-induced maturation of dendritic cells through the regulation of JNK activity. *J. Med. Food* **2014**, *17*, 996–1002. [CrossRef] [PubMed]
276. Kim, J.H.; Kim, Y.S.; Kim, T.I.; Li, W.; Mun, J.G.; Jeon, H.D.; Kee, J.Y.; Choi, J.G.; Chung, H.S. Unripe Black Raspberry (Rubus coreanus Miquel) Extract and Its Constitute, Ellagic Acid Induces T Cell Activation and Antitumor Immunity by Blocking PD-1/PD-L1 Interaction. *Foods* **2020**, *9*, 1590. [CrossRef]
277. Peng, D.H.; Rodriguez, B.L.; Diao, L.; Chen, L.; Wang, J.; Byers, L.A.; Wei, Y.; Chapman, H.A.; Yamauchi, M.; Behrens, C.; et al. Collagen promotes anti-PD-1/PD-L1 resistance in cancer through LAIR1-dependent CD8(+) T cell exhaustion. *Nat. Commun.* **2020**, *11*, 4520. [CrossRef]
278. Trinchieri, G. Cancer and inflammation: An old intuition with rapidly evolving new concepts. *Annu. Rev. Immunol.* **2012**, *30*, 677–706. [CrossRef]
279. Pantuck, A.J.; Leppert, J.T.; Zomorodian, N.; Aronson, W.; Hong, J.; Barnard, R.J.; Seeram, N.; Liker, H.; Wang, H.; Elashoff, R.; et al. Phase II study of pomegranate juice for men with rising prostate-specific antigen following surgery or radiation for prostate cancer. *Clin. Cancer Res.* **2006**, *12*, 4018–4026. [CrossRef]
280. Singh, N.; Baby, D.; Rajguru, J.P.; Patil, P.B.; Thakkannavar, S.S.; Pujari, V.B. Inflammation and cancer. *Ann. Afr. Med.* **2019**, *18*, 121–126. [CrossRef]
281. Seeram, N.P.; Adams, L.S.; Henning, S.M.; Niu, Y.; Zhang, Y.; Nair, M.G.; Heber, D. In vitro antiproliferative, apoptotic and antioxidant activities of punicalagin, ellagic acid and a total pomegranate tannin extract are enhanced in combination with other polyphenols as found in pomegranate juice. *J. Nutr. Biochem.* **2005**, *16*, 360–367. [CrossRef]
282. Kilic, I.; Yesiloglu, Y.; Bayrak, Y. Spectroscopic studies on the antioxidant activity of ellagic acid. *Spectrochim. Acta A Mol. Biomol. Spectrosc.* **2014**, *130*, 447–452. [CrossRef] [PubMed]
283. Aggarwal, B.B.; Shishodia, S.; Sandur, S.K.; Pandey, M.K.; Sethi, G. Inflammation and cancer: How hot is the link? *Biochem. Pharmacol.* **2006**, *72*, 1605–1621. [CrossRef] [PubMed]
284. Khan, M.K.; Ansari, I.A.; Khan, M.S. Dietary phytochemicals as potent chemotherapeutic agents against breast cancer: Inhibition of NF-kappaB pathway via molecular interactions in rel homology domain of its precursor protein p105. *Pharmacogn. Mag.* **2013**, *9*, 51–57. [CrossRef] [PubMed]
285. Umesalma, S.; Sudhandiran, G. Differential inhibitory effects of the polyphenol ellagic acid on inflammatory mediators NF-kappaB, iNOS, COX-2, TNF-alpha, and IL-6 in 1,2-dimethylhydrazine-induced rat colon carcinogenesis. *Basic Clin. Pharmacol. Toxicol.* **2010**, *107*, 650–655. [CrossRef]
286. Mishra, S.; Vinayak, M. Ellagic acid inhibits PKC signaling by improving antioxidant defense system in murine T cell lymphoma. *Mol. Biol. Rep.* **2014**, *41*, 4187–4197. [CrossRef]
287. Bishayee, A.; Thoppil, R.J.; Darvesh, A.S.; Ohanyan, V.; Meszaros, J.G.; Bhatia, D. Pomegranate phytoconstituents blunt the inflammatory cascade in a chemically induced rodent model of hepatocellular carcinogenesis. *J. Nutr. Biochem.* **2013**, *24*, 178–187. [CrossRef]
288. Khan, N.; Afaq, F.; Kweon, M.H.; Kim, K.; Mukhtar, H. Oral consumption of pomegranate fruit extract inhibits growth and progression of primary lung tumors in mice. *Cancer Res.* **2007**, *67*, 3475–3482. [CrossRef]
289. Wang, L.; Alcon, A.; Yuan, H.; Ho, J.; Li, Q.J.; Martins-Green, M. Cellular and molecular mechanisms of pomegranate juice-induced anti-metastatic effect on prostate cancer cells. *Integr. Biol.* **2011**, *3*, 742–754. [CrossRef]
290. Rettig, M.B.; Heber, D.; An, J.; Seeram, N.P.; Rao, J.Y.; Liu, H.; Klatte, T.; Belldegrun, A.; Moro, A.; Henning, S.M.; et al. Pomegranate extract inhibits androgen-independent prostate cancer growth through a nuclear factor-kappaB-dependent mechanism. *Mol. Cancer Ther.* **2008**, *7*, 2662–2671. [CrossRef]
291. Mandal, A.; Bhatia, D.; Bishayee, A. Anti-Inflammatory Mechanism Involved in Pomegranate-Mediated Prevention of Breast Cancer: The Role of NF-kappaB and Nrf2 Signaling Pathways. *Nutrients* **2017**, *9*, 436. [CrossRef]
292. Morgan, W.F.; Day, J.P.; Kaplan, M.I.; McGhee, E.M.; Limoli, C.L. Genomic Instability Induced by Ionizing Radiation. *Radiat. Res.* **1996**, *146*, 247–258. [CrossRef] [PubMed]
293. Langie, S.A.; Koppen, G.; Desaulniers, D.; Al-Mulla, F.; Al-Temaimi, R.; Amedei, A.; Azqueta, A.; Bisson, W.H.; Brown, D.G.; Brunborg, G.; et al. Causes of genome instability: The effect of low dose chemical exposures in modern society. *Carcinogenesis* **2015**, *36* (Suppl. S1), S61–S88. [CrossRef] [PubMed]
294. Kalimutho, M.; Nones, K.; Srihari, S.; Duijf, P.H.G.; Waddell, N.; Khanna, K.K. Patterns of Genomic Instability in Breast Cancer. *Trends Pharmacol. Sci.* **2019**, *40*, 198–211. [CrossRef] [PubMed]
295. Nakamura, H.; Takada, K. Reactive oxygen species in cancer: Current findings and future directions. *Cancer Sci.* **2021**, *112*, 3945–3952. [CrossRef]
296. Porter, V.L.; Marra, M.A. The Drivers, Mechanisms, and Consequences of Genome Instability in HPV-Driven Cancers. *Cancers* **2022**, *14*, 4623. [CrossRef]
297. Dittmar, T.; Hass, R. Intrinsic signalling factors associated with cancer cell-cell fusion. *Cell Commun. Signal.* **2023**, *21*, 68. [CrossRef]

298. Blasco, M.A.; Lee, H.W.; Hande, M.P.; Samper, E.; Lansdorp, P.M.; DePinho, R.A.; Greider, C.W. Telomere shortening and tumor formation by mouse cells lacking telomerase RNA. *Cell* **1997**, *91*, 25–34. [CrossRef]
299. Srinivas, U.S.; Tan, B.W.Q.; Vellayappan, B.A.; Jeyasekharan, A.D. ROS and the DNA damage response in cancer. *Redox Biol.* **2019**, *25*, 101084. [CrossRef]
300. Chatterjee, N.; Walker, G.C. Mechanisms of DNA damage, repair, and mutagenesis. *Environ. Mol. Mutagen.* **2017**, *58*, 235–263. [CrossRef]
301. Ferguson, L.R.; Chen, H.; Collins, A.R.; Connell, M.; Damia, G.; Dasgupta, S.; Malhotra, M.; Meeker, A.K.; Amedei, A.; Amin, A.; et al. Genomic instability in human cancer: Molecular insights and opportunities for therapeutic attack and prevention through diet and nutrition. *Semin. Cancer Biol.* **2015**, *35*, S5–S24. [CrossRef]
302. Aiyer, H.S.; Vadhanam, M.V.; Stoyanova, R.; Caprio, G.D.; Clapper, M.L.; Gupta, R.C. Dietary berries and ellagic acid prevent oxidative DNA damage and modulate expression of DNA repair genes. *Int. J. Mol. Sci.* **2008**, *9*, 327–341. [CrossRef] [PubMed]
303. Berni, A.; Grossi, M.R.; Pepe, G.; Filippi, S.; Muthukumar, S.; Papeschi, C.; Natarajan, A.T.; Palitti, F. Protective effect of ellagic acid (EA) on micronucleus formation induced by N-methyl-N′-nitro-N-nitrosoguanidine (MNNG) in mammalian cells, in in vitro assays and in vivo. *Mutat. Res.* **2012**, *746*, 60–65. [CrossRef]
304. Wyatt, M.D.; Pittman, D.L. Methylating agents and DNA repair responses: Methylated bases and sources of strand breaks. *Chem. Res. Toxicol.* **2006**, *19*, 1580–1594. [CrossRef]
305. Gerhards, M.; Bohme, A.; Schubert, K.; Kodritsch, B.; Ulrich, N. DNA adducts as link between in vitro and in vivo carcinogenicity—A case study with benzo[a]pyrene. *Curr. Res. Toxicol.* **2023**, *4*, 100097. [CrossRef] [PubMed]
306. Zahin, M.; Ahmad, I.; Gupta, R.C.; Aqil, F. Punicalagin and ellagic acid demonstrate antimutagenic activity and inhibition of benzo[a]pyrene induced DNA adducts. *BioMed Res. Int.* **2014**, *2014*, 467465. [CrossRef] [PubMed]
307. Varshney, M.; Vijayan, V.; Meshram, G.P. Chemopreventive effects of ellagic acid against genotoxicity induced by benzo(a)pyrene. *Toxicol. Environ. Chem.* **2015**, *97*, 786–798. [CrossRef]
308. Salimi, A.; Baghal, E.; Ghobadi, H.; Hashemidanesh, N.; Khodaparast, F.; Seydi, E. Mitochondrial, lysosomal and DNA damages induced by acrylamide attenuate by ellagic acid in human lymphocyte. *PLoS ONE* **2021**, *16*, e0247776. [CrossRef]
309. Baeeri, M.; Momtaz, S.; Navaei-Nigjeh, M.; Niaz, K.; Rahimifard, M.; Ghasemi-Niri, S.F.; Sanadgol, N.; Hodjat, M.; Sharifzadeh, M.; Abdollahi, M. Molecular evidence on the protective effect of ellagic acid on phosalone-induced senescence in rat embryonic fibroblast cells. *Food Chem. Toxicol.* **2017**, *100*, 8–23. [CrossRef]
310. Ghasemi-Niri, S.F.; Maqbool, F.; Baeeri, M.; Gholami, M.; Abdollahi, M. Phosalone-induced inflammation and oxidative stress in the colon: Evaluation and treatment. *World J. Gastroenterol.* **2016**, *22*, 4999–5011. [CrossRef]
311. Khodabandeh, Z.; Etebari, M.; Aliomrani, M. Study of the probable genotoxic effects of Zolone (Phosalone) exposure in mice bone marrow derived cells. *Genes Environ.* **2021**, *43*, 18. [CrossRef]
312. Rahnasto-Rilla, M.; Jarvenpaa, J.; Huovinen, M.; Schroderus, A.M.; Ihantola, E.L.; Kublbeck, J.; Khadeer, M.; Moaddel, R.; Lahtela-Kakkonen, M. Effects of galloflavin and ellagic acid on sirtuin 6 and its anti-tumorigenic activities. *Biomed. Pharmacother.* **2020**, *131*, 110701. [CrossRef]
313. Amano, H.; Sahin, E. Telomeres and sirtuins: At the end we meet again. *Mol. Cell. Oncol.* **2019**, *6*, e1632613. [CrossRef] [PubMed]
314. Mottola, F.; Finelli, R.; Iovine, C.; Carannante, M.; Santonastaso, M.; Rocco, L. Anti-Genotoxicity Evaluation of Ellagic Acid and Curcumin—An In Vitro Study on Zebrafish Blood Cells. *Appl. Sci.* **2021**, *11*, 8142. [CrossRef]
315. Rehman, M.U.; Tahir, M.; Ali, F.; Qamar, W.; Lateef, A.; Khan, R.; Quaiyoom, A.; Oday, O.H.; Sultana, S. Cyclophosphamide-induced nephrotoxicity, genotoxicity, and damage in kidney genomic DNA of Swiss albino mice: The protective effect of Ellagic acid. *Mol. Cell. Biochem.* **2012**, *365*, 119–127. [CrossRef] [PubMed]
316. Sudheer, A.R.; Muthukumaran, S.; Devipriya, N.; Menon, V.P. Ellagic acid, a natural polyphenol protects rat peripheral blood lymphocytes against nicotine-induced cellular and DNA damage in vitro: With the comparison of N-acetylcysteine. *Toxicology* **2007**, *230*, 11–21. [CrossRef]
317. Loarca-Pina, G.; Kuzmicky, P.A.; de Mejia, E.G.; Kado, N.Y. Inhibitory effects of ellagic acid on the direct-acting mutagenicity of aflatoxin B1 in the Salmonella microsuspension assay. *Mutat. Res.* **1998**, *398*, 183–187. [CrossRef]
318. Vattem, D.A.; Jang, H.D.; Levin, R.; Shetty, K. Synergism of cranberry phenolics with ellagic acid and rosmarinic acid for antimutagenic and DNA protection functions. *J. Food Biochem.* **2006**, *30*, 98–116. [CrossRef]
319. Matić, S.; Stanić, S.; Kanjevac, M. Genotoxic effect of gallic and ellagic acids in somatic and germ cells of Drosophila melanogaster. *Kragujev. J. Sci.* **2019**, *41*, 69–76. [CrossRef]

biomolecules

MDPI

*Review*

# Synergistically Anti-Multiple Myeloma Effects: Flavonoid, Non-Flavonoid Polyphenols, and Bortezomib

**Kaixi Ding †, Wei Jiang †, Huanan Jia and Ming Lei ***

Hospital of Chengdu University of Traditional Chinese Medicine, Chengdu 610075, China
* Correspondence: minglei2828cdutcm@gmail.com
† These authors contributed equally to this work.

**Abstract:** Multiple myeloma (MM) is a clonal plasma cell tumor originating from a post-mitotic lymphoid B-cell lineage. Bortezomib (BTZ), a first-generation protease inhibitor, has increased overall survival, progression-free survival, and remission rates in patients with MM since its clinical approval in 2003. However, the use of BTZ is challenged by the malignant features of MM and drug resistance. Polyphenols, classified into flavonoid and non-flavonoid polyphenols, have potential health-promoting activities, including anti-cancer. Previous preclinical studies have demonstrated the anti-MM potential of some dietary polyphenols. Therefore, these dietary polyphenols have the potential to be alternative therapies in anti-MM treatment regimens. This systematic review examines the synergistic effects of flavonoids and non-flavonoid polyphenols on the anti-MM impacts of BTZ. Preclinical studies on flavonoids and non-flavonoid polyphenols-BTZ synergism in MM were collected from PubMed, Web of Science, and Embase published between 2008 and 2020. 19 valid preclinical studies (Published from 2008 to 2020) were included in this systematic review. These studies demonstrated that eight flavonoids (icariin, icariside II, (-)-epigallocatechin-3-gallate, scutellarein, wogonin, morin, formononetin, daidzin), one plant extract rich in flavonoids (Punica granatum juice) and four non-flavonoid polyphenols (silibinin, resveratrol, curcumin, caffeic acid) synergistically enhanced the anti-MM effect of BTZ. These synergistic effects are mediated through the regulation of cellular signaling pathways associated with proliferation, apoptosis, and drug resistance. Given the above, flavonoids and non-flavonoid polyphenols can benefit MM patients by overcoming the challenges faced in BTZ treatment. Despite the positive nature of this preclinical evidence, some additional investigations are still needed before proceeding with clinical studies. For this purpose, we conclude by providing some suggestions for future research directions.

**Keywords:** multiple myeloma; bortezomib; flavonoids; non-flavonoid polyphenols; synergy

## 1. Introduction

Multiple myeloma (MM) is a clonal plasma cell tumor originating from the post-germinal lymphoid B cell line. It has an incidence of approximately 2.1 out of 100,000 people worldwide [1]. Its incidence varies notably among different races, with older individuals at higher risk [2]. Moreover, it often leads to poor outcomes in low and middle-income countries due to the lack of specialized health care, diagnosis, and advanced treatments. The 5-year survival rate for MM patients treated in Nigeria is only 7.6% [3].

At present, internal medicine intervention predominates in MM therapy. Standard treatment options include (1) induction therapy with a combination of injectable proteasome inhibitors (i.e., bortezomib (BTZ)), oral immunomodulators (i.e., lenalidomide), and dexamethasone; and (2) autologous stem cell transplantation (ASCT) with subsequent maintenance lenalidomide [4]. According to the clinical evidence, Velcade, Revlimid, and Dexamethasone (VRD)-based pre-ASCT induction therapy enhances the prognosis of MM patients [5,6]. Nevertheless, these interventions remain challenged by MM's malignant qualities. For example, ASCT is not recommended for MM patients who are over the age of 65 or who have severe comorbid conditions like renal or pulmonary impairment, cardiac disease, or hepatic disease [7]. Most significantly, intrinsic clonal heterogeneity and genomic instability of plasma cells influence both inherent and acquired drug resistance [8]. For most MM patients, after several remissions and relapses, drug resistance develops into almost all available therapies [9]. Additionally, these MM medications could also have adverse effects. BTZ, for instance, may have adverse effects on the cardiovascular system, bone marrow suppression, and peripheral neuropathy [10]. Therefore, in order to overcome these difficulties, alternative and supportive therapies are required.

In recent years, natural dietary compounds have attracted the attention of researchers. They have a wide range of biological activities, such as anti-inflammatory, antioxidant, anti-aging, immune enhancement, as well as anti-tumor [11–15]. Among them, flavonoids, a class of polyphenolic plant secondary metabolites, are of great interest. Through inhibition of cell proliferation, stimulation of apoptosis, targeting of cancer stem cells, inhibition of angiogenesis and metastasis, and chemotherapy sensitization, flavonoids have anti-cancer effects [16]. In addition, the cancer prevention and treatment potential of some non-flavonoid polyphenols are equally noteworthy [17,18]. Jöhrer and Pojero reviewed the tumor activity inhibition and targeting cancer effects of some natural compounds (including polyphenols) in MM in preclinical trials, respectively [19,20]. As mentioned earlier, the main challenge MM drug interventions face is drug resistance. Therefore, exploring possible alternative treatments to overcome it is clinically valuable. This systematic review focuses on the synergistic anti-MM effects of flavonoids, other non-flavonoid polyphenols, and BTZ.

## 2. Research Methodology

The systematic review adheres to the guidance of the Preferred Reporting Items for Reference Systems Evaluation and Meta-Analysis (PRISMA) statement [21]. Preclinical studies on flavonoids or non-flavonoid polyphenols -bortezomib synergistic potentiation of anti-MM effects were collected by searching PubMed, Web of Science(WOS), and Embase for the keywords "flavonoid" or "polyphenol," "bortezomib" or "Velcade," and "Multiple myeloma" or "Myeloma, Plasma-Cell."

Searching with the search formula "Multiple myeloma" or "Myeloma, plasma cell" (subject) and "Bortezomib or Velcade" (all fields) and "flavonoids or polyphenols" (all fields) were put into WOS, 33 publications were found. Performing a search at PubMed with the search formula: ((Multiple myeloma or "Myeloma, Plasma-Cell") AND (Bortezomib or Velcade)) AND (flavonoids or polyphenols), 38 publications were retrieved. Searching in Embase with the search terms: ((Multiple myeloma or "Myeloma, Plasma-Cell") AND (Bortezomib or Velcade)) AND (flavonoids or polyphenols), 144 publications were retrieved. The filters had not been applied in any of the above three databases. Endnote 20 was used to check for duplicates automatically and by hand for 215 publications, and 41 duplicates were taken out. The titles and abstracts of 174 publications were reviewed by two independent reviewers (Kaixi Ding and Wei Jiang). In case no consensus was reached, a third independent reviewer made the final decision. 121 publications that were not relevant to the topic of this review were removed.

The remaining 53 publications were then read in full by two independent reviewers. According to the inclusion and exclusion criteria: (1) treatment with flavonoid compounds or non-flavonoid polyphenolic compounds or plant extracts containing polyphenols; (2) animal studies or cell cultures; (3) interventions in animals or cell models of multiple myeloma; (4) studies included flavonoid compounds or non-flavonoid polyphenolic compounds or plant extracts containing polyphenols and BTZ synergistic anti-MM effects or mechanisms (5) Excluding review literature, meta-analyses, case reports, editorials, abstracts and conference proceedings and other types of literature. A total of 38 publications were discharged after a detailed review by the reviewers. In addition, two reviewers agreed that four publications with relevant citation searches from the research and the review literature should be added as secondary sources.

Eventually, 19 valid articles (Published from 2008 to 2020) were included in this review (Figure 1) (Table 1).

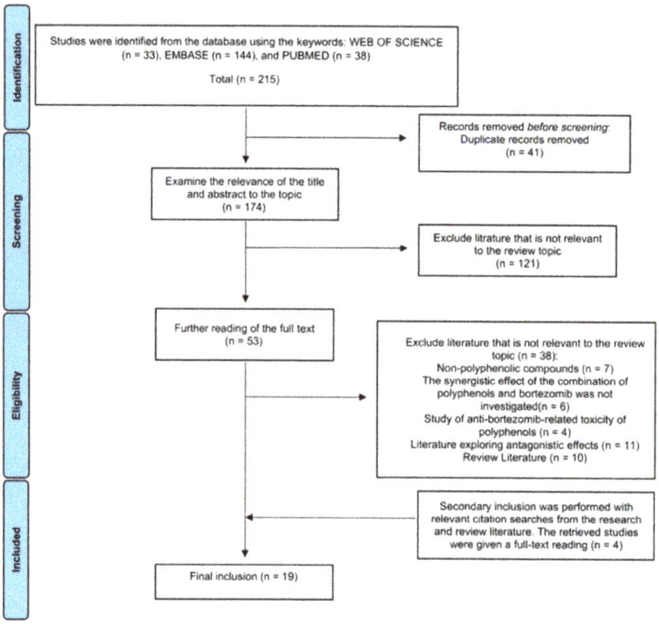


**Figure 1.** Diagram of the process for including and excluding the literature for this review.

**Table 1.** General information on the 19 publications included in this systematic review.

| Authors | Year | Title | Polyphenols, or Plant Extracts Rich in Polyphenols | Journal Title, Volume and Page Number | Doi |
|---|---|---|---|---|---|
| Li, Z. Y., et al. [22] | 2017 | Icaritin Reverses Multidrug Resistance of Multiple Myeloma Cell Line KM3/BTZ. | Icariin | Zhongguo shi yan xue ye xue za zhi 25(6): 1690–1695. | 10.19746/J.CNKI.ISSN.1009-2137.2019.02.025 |
| Kim, S. H., et al. [23] | 2011 | Janus activated kinase 2/signal transducer and activator of transcription 3 pathway mediates icariside II-induced apoptosis in U266 multiple myeloma cells. | Icariside II | Eur J Pharmacol 654(1): 10–16. | 10.1016/J.EJPHAR.2010.11.032 |
| Li, L., et al. [24] | 2020 | Scutellarin circumvents chemoresistance, promotes apoptosis, and represses tumor growth by HDAC/miR-34a-mediated down-modulation of Akt/mTOR and NF-κB-orchestrated signaling pathways in multiple myeloma. | Scutellarein | Int J Clin Exp Pathol 13(2): 212–219. | Not available |
| Shi, L., et al. [25] | 2019 | Scutellarein selectively targets multiple myeloma cells by increasing mitochondrial superoxide production and activating intrinsic apoptosis pathway. | Scutellarein | Biomed Pharmacother 109: 2109–2118. | 10.1016/j.biopha.2018.09.024 |

**Table 1.** *Cont.*

| Authors | Year | Title | Polyphenols, or Plant Extracts Rich in Polyphenols | Journal Title, Volume and Page Number | Doi |
|---|---|---|---|---|---|
| Fu, R., et al. [26] | 2016 | Wogonin inhibits multiple myeloma-stimulated angiogenesis via c-Myc/VHL/HIF-1α signaling axis. | Wogonin | Oncotarget 7(5): 5715–5727. | 10.18632/ONCOTARGET.6796 |
| Gupta, S. C., et al. [27] | 2013 | Morin inhibits STAT3 tyrosine 705 phosphorylation in tumor cells through activation of protein tyrosine phosphatase SHP1. | Morin | Biochem Pharmacol 85(7): 898–912. | 10.1016/J.BCP.2012.12.018 |
| Kim, C., et al. [28] | 2018 | Formononetin-induced oxidative stress abrogates the activation of STAT3/5 signaling axis and suppresses the tumor growth in multiple myeloma preclinical model. | Formononetin | Cancer Lett 431: 123–141. | 10.1016/j.canlet.2018.05.038 |
| Kim, C., et al. [29] | 2019 | Formononetin Regulates Multiple Oncogenic Signaling Cascades and Enhances Sensitivity to Bortezomib in a Multiple Myeloma Mouse Model. | Formononetin | Biomolecules 9(7): 262. | 10.3390/biom9070262 |
| Yang, M. H., et al. [30] | 2019 | Attenuation of STAT3 Signaling Cascade by Daidzin Can Enhance the Apoptotic Potential of Bortezomib against Multiple Myeloma. | Daidzin | Biomolecules 10(1): 23 | 10.3390/biom10010023 |
| Tibullo, D., et al. [31] | 2016 | Antiproliferative and Antiangiogenic Effects of Punica granatum Juice (PGJ) in Multiple Myeloma (MM). | PGJ | Nutrients 8(10):611. | 10.3390/nu8100611 |
| Geraldes, C., et al. [32] | 2015 | New targeted drugs in multiple myeloma therapy-in vitro studies. | Silibinin | Clinical Lymphoma, and Leukemia 15: e254. | 10.1016/j.clml.2015.07.539 |
| Ge, X. P., et al. [33] | 2019 | Curcumin Increases the Chemosensitivity of Multiple Myeloma to Bortezomib by Inhibiting the Notch1 Signaling Pathway. | Curcumin | Zhongguo shi yan xue ye xue za zhi 27(2): 464–471. | 10.19746/j.cnki.issn.1009-2137.2019.02.025 |
| Altayli, E., et al. [34] | 2015 | An in vitro and in vivo investigation of the cytotoxic effects of caffeic acid (3,4-dihydroxycinnamic acid) phenethyl ester and bortezomib in multiple myeloma cells. | Caffeic acid | Turk J Med Sci 45(1): 38–46. | 10.3906/SAG-1401-127 |
| Jung, Y. Y., et al. [35] | 2018 | Anti-myeloma Effects of Icariin Are Mediated Through the Attenuation of JAK/STAT3-Dependent Signaling Cascade. | Icariin | Front Pharmacol 9: 531. | 10.3389/fphar.2018.00531 |

**Table 1.** *Cont.*

| Authors | Year | Title | Polyphenols, or Plant Extracts Rich in Polyphenols | Journal Title, Volume and Page Number | Doi |
|---|---|---|---|---|---|
| Wang, Q., et al. [36] | 2009 | Potentiation of (-)-epigallocatechin-3-gallate-induced apoptosis by bortezomib in multiple myeloma cells. | EGCG | Acta Biochim Biophys Sin (Shanghai) 41(12): 1018–1026. | 10.1093/ABBS/GMP094 |
| Reis-Sobreiro, M., et al. [37] | 2009 | Involvement of mitochondria and recruitment of Fas/CD95 signaling in lipid rafts in resveratrol-mediated antimyeloma and antileukemia actions. | Resveratrol | ONCOGENE 28(36): 3221–3234. | 10.1038/onc.2009.183 |
| Park, J., et al. [38] | 2008 | Curcumin in combination with bortezomib synergistically induced apoptosis in human multiple myeloma U266 cells. | Curcumin | Mol Oncol 2(4): 317–326. | 10.1016/J.MOLONC.2008.09.006 |
| Sung, B., et al. [39] | 2009 | Curcumin circumvents chemoresistance in vitro and potentiates the effect of thalidomide and bortezomib against human multiple myeloma in nude mice model. | Curcumin | Mol Cancer Ther 8(4): 959–970. | 10.1158/1535-7163.MCT-08-0905 |
| Mujtaba, T., et al. [40] | 2012 | Sensitizing human multiple myeloma cells to the proteasome inhibitor bortezomib by novel curcumin analogs. | Novel curcumin analogs | Int J Mol Med 29(1): 102–106. | 10.3892/ijmm.2011.814 |

## 3. Mechanisms of MM

MM's onset, progression, metastasis, and osteolytic destruction are associated with the regulation of multiple signaling pathways (Figure 2). Throughout the development of MM, rapid cell proliferation was mainly regulated by PI3K/Akt, Ras/Raf/MEK/Erk, JAK/STAT, Wnt/β-catenin, and RANK/RANKL/OPG signal pathways [41]. Uncontrolled cell proliferation was attributed to the inhibition of normal cell cycle regulation (e.g., FOXO1 and Cyclin D1), and the upregulation of downstream key anti-apoptotic factors (e.g., Bcl2, Bcl-xl, Mcl-1, and Caspase) [42]. Angiogenesis, mainly promoted by VEGF, provides oxygen and nutrients for tumor proliferation. Additionally, the activation of the Wnt/-catenin and RANK/RANKL/OPG signal pathways is primarily responsible for osteolytic bone disease in MM, which is brought on by an increase in osteoclast activity [41].

Finally, MM cells can further develop drug resistance, and this may be related to mechanisms such as the upregulation of permeability-glycoprotein (P-gp) (leading to reduced drug aggregation in vivo). IL-6 and insulin-like growth factor (IGF), two cytokines produced by bone marrow mesenchymal stem cells, also encourage drug resistance in MM cells(activating specific signaling pathways that lead to drug resistance) [43].

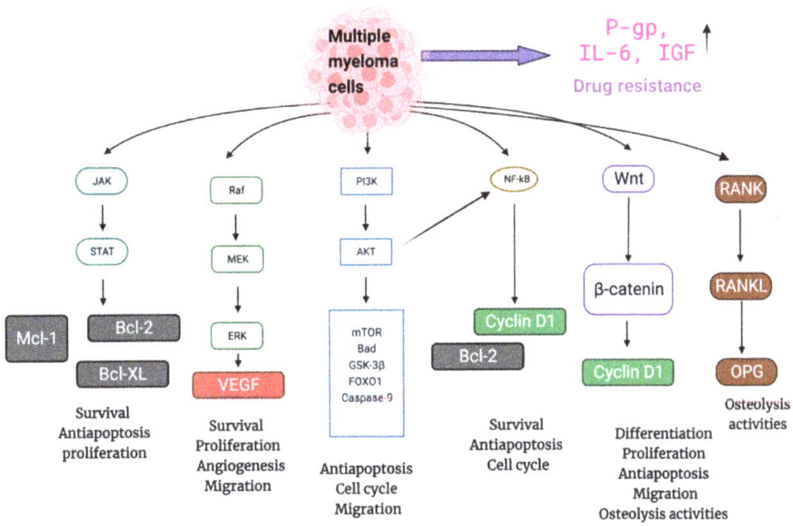


**Figure 2.** Multiple signaling pathways are involved in the tumor formation and progression of MM. Mechanisms promoting anti-apoptosis, proliferation, angiogenesis, migration, and drug resistance were upregulated, while mechanisms related to apoptosis and normal cell cycle regulation were suppressed. Abbreviations: Bad: BCL2 associated agonist of cell death; Bcl2: B-cell lymphoma 2; Bcl-xl: B-cell lymphoma-extra-large; FOXO1: Forkhead box protein O1; GSK-3β: Glycogen synthase kinase-3 beta; JAK/STAT: Janus kinases/signal transducer and activator of transcription proteins; Mcl-1: Myeloid cell leukemia 1; MEK/Erk: MAP kinase–ERK kinase/extracellular-signal-regulated kinases; TOR: mammalian target of rapamycin; PI3K/Akt: Phosphoinositide 3-kinases/Protein Kinase B; RANK/RANKL/OPG: Receptor activator of nuclear factor κB/Receptor activator of nuclear factor kappa-B ligand/Osteoprotegerin; Wnt: Wingless and Int-1.

## 4. Bortezomib in MM Therapy

An important class of medications for the treatment of MM is protease inhibitors. Based on targeted inhibition of the 26S proteasome, protease inhibitors are crucial in the pathogenesis and proliferation of MM [44]. BTZ, the first protease inhibitor, approved for clinical use in 2003, has resulted in gains in overall survival, progression-free survival, and remission rates in patients with MM (Figure 3) [45]. Its main anti-cancer mechanism is the inhibition of the chymotrypsin-like site of the 20S protein hydrolysis core within the 26S proteasome, which induces cell cycle arrest and apoptosis [46]. The main signaling mechanisms of BTZ-induced apoptosis in MM cells are NF-κB blockade and JNK activation. In addition, BTZ stabilizes various tumor suppressor proteins, such as P53, inhibiting MM cell cycle progression [47]. Currently, the drug is commonly used in MM patients in first-line, relapsed, and/or refractory settings [48]. However, drug resistance in MM therapy is the main challenge currently faced when using BTZ [43].

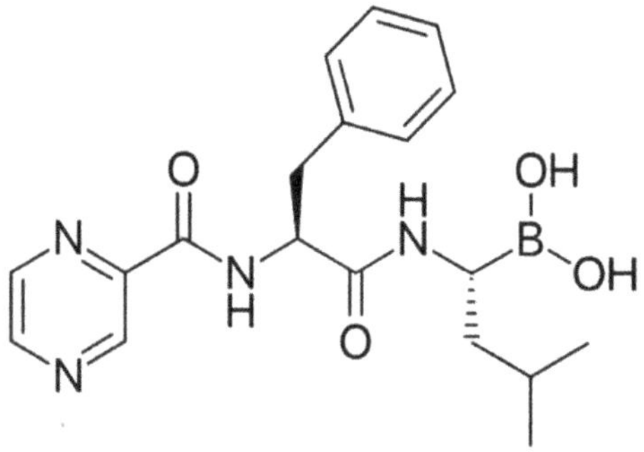


**Figure 3.** Chemical structure of bortezomib, an anti-MM Protease inhibitor.

## 5. Exploration of Synergistic Drug Combinations in MM Therapy

When two or more medications are combined to improve their therapeutic effects, this is referred to as drug synergism [49]. For various diseases, including MM, appropriate drug combinations can reduce drug resistance or maximize efficacy [50–53]. For instance, the previously mentioned VRD is currently an effective and well-tolerated pre-ASCT induction protocol, benefiting newly diagnosed MM patients [6]. BTZ, as a clinically used protease inhibitor, is a cornerstone of VRD. However, significant obstacles to using BTZ to treat MM are drug resistance and the malignancy of MM. To cross these obstacles, researchers have investigated the BTZ resistance-reducing effects of various drug combinations and their synergistic anti-MM effects, such as daratumumab (anti-CD38 antibody), BTZ, and dexamethasone (CASTOR trial), TAK-243 (novel and specific UAE inhibitor) and BTZ, decitabine (epigenetic modulator) and BTZ [54–56]. Additionally, recent studies have demonstrated that some naturally occurring polyphenolic compounds have anti-MM activity. These polyphenolic compounds also have the potential to reduce BTZ resistance. Given the low toxicity of natural polyphenolic compounds, their synergistic combination with BTZ may offer MM patients extra therapeutic options [19,20].

## 6. Flavonoids and Non-Flavonoid Polyphenols in MM Therapy

Polyphenols are a class of phytochemicals divided into flavonoids and non-flavonoids. A group of compounds known as flavonoids consists of two benzene rings connected by phenolic hydroxyl groups by a central three-carbon atom. They are secondary metabolites in fruits, vegetables, and other herbal plants. The main flavonoid subgroup includes flavones, flavonols, flavan-3-ols, anthocyanins, isoflavones, and chalcones (Figure 4) [57]. The main non-flavonoids contained phenolic, hydroxycinnamic, lignans, stilbenes, and tannins [58]. Based on the results of this systematic review, various polyphenols, including flavonols(icariin, icariside II), flavan-3-ols((-)-epigallocatechin-3-gallate), flavone(scutellarein, wogonin, morin), isoflavone(formononetin, daidzin), plant extracts rich in flavonoids(punica granatum juice) and non-flavonoid polyphenols(silibinin, resveratrol, curcumin, caffeic acid) exerted anti-MM synergistic effects in combination with BTZ in vivo and/or in vitro (Table 2) (Figure 5).

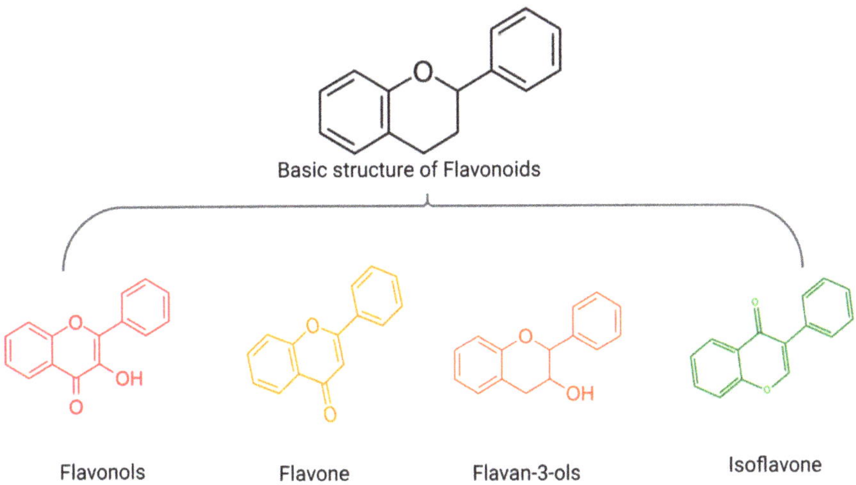


**Figure 4.** The basic structure of flavonoids(black) and the general structure of flavonols (red), flavone (yellow), flavan-3-ols (orange), isoflavone (green).

**Table 2.** Classes and sources of 13 polyphenols (9 flavonoids and 4 non-flavonoids) that synergize with BTZ against MM.

| Polyphenol | Class | Source | Reference |
|---|---|---|---|
| Icariin | Flavonol | Dried stems and leaves of arrow leaf Epimedium, pilose Epimedium, Wushan Epimedium, Korean Epimedium. | [22] |
| icariside II | Flavonol | Stems and leaves of Epimedium koreanum Nakai (Berberidaceae). | [23] |
| (-)-epigallocatechin-3-gallate | Flavan-3-ol | Green tea | [59] |
| Scutellarein | Flavone | *S. baicalensis* or *S. barbata* | [24,25] |
| Wogonin | Flavone | Scutellaria baicalensis | [26] |
| Morin | Flavone | Mulberry figs and old fustic (Chlorophora tinctoria) | [27] |
| Formononetin | Isoflavone | Roots of Astragalus membranaceus, Trifolium pratense, Glycyrrhiza glabra, and Pueraria lobate | [28,29] |
| Daidzin | Isoflavone | Soy and soy products | [30] |
| Punica granatum Juice | Plant extract rich in polyphenols | Punica granatum | [31] |
| Silibinin | Non-flavonoid polyphenol | Silybum marianum | [32,60] |
| Resveratrol | Non-flavonoid polyphenol | Grape, blueberry, raspberry, and mulberries peel and peanut | [61] |
| Curcumin | Non-flavonoid polyphenol | Turmeric rhizome | [33] |
| Caffeic acid | Non-flavonoid polyphenol | Honey Bee Propolis | [34] |

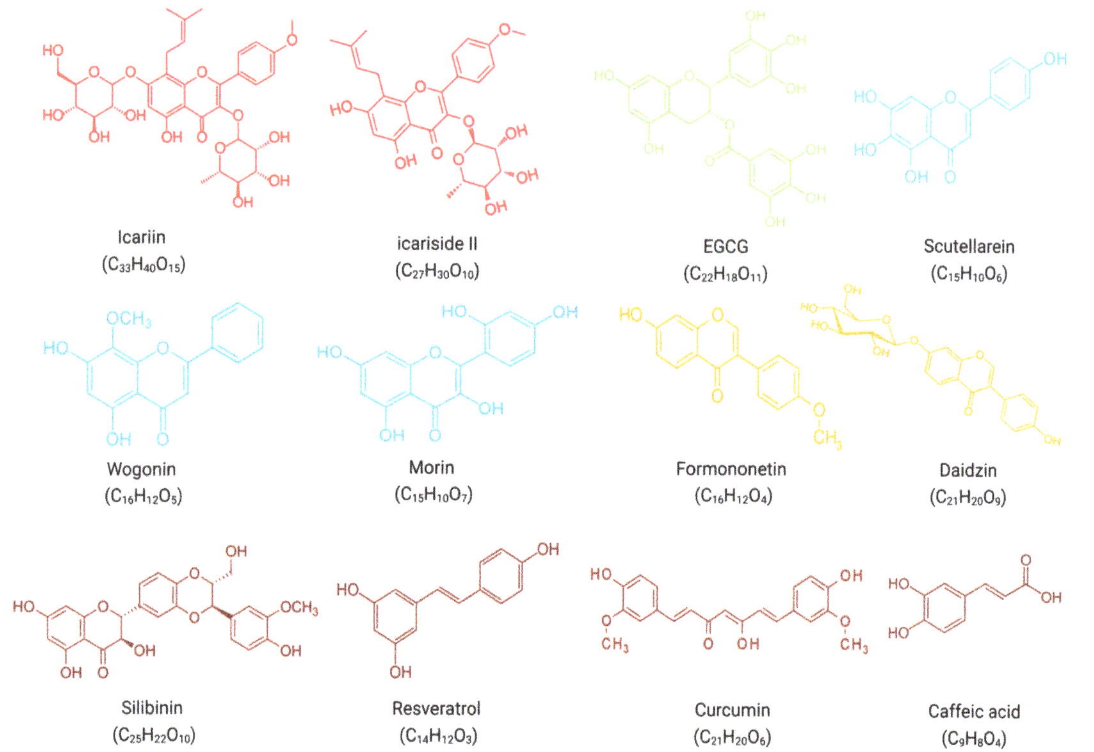


**Figure 5.** Molecular formulae and chemical structures of flavonoid compounds and non-flavonoid polyphenolic compounds(excluding PGJ, a plant extract rich in flavonoids). Flavonols(icariin, icariside II) have a red chemical formula. Flavan-3-ol (EGCG) is green in color. The formula of flavones (scutellarein, wogonin, morin) is blue. The chemical formula of isoflavones (formononetin, daidzin) is yellow. Non-flavonoid polyphenols (silibinin, resveratrol, curcumin, caffeic acid) have a brown chemical formula.

Flavonols are a class of flavonoids with a 3-hydroxyflavonoid backbone. Two flavonols are of interest in synergistic anti-MM therapy. Icariin, an active ingredient in the stem and leaves of Epimedium, exerts significant anti-tumor effects on a variety of human tumor cells [22]. Similarly, icariside II, isolated from E. koreanum, inhibits the proliferation and induces apoptosis of many human cancer cells, such as MM, breast cancer, and prostate cancer [23].

Flavan-3-ols are flavan derivatives with a 2-phenyl-3,4-dihydro-2H-chroman-3-ol backbone. One flavan-3-ol is of interest in synergistic anti-MM therapy. (-)-epigallocatechin-3-gallate(EGCG) belongs to the category of catechins, commonly found in green tea. Catechins have anti-mutagenic, tumor progression-inhibiting, apoptosis-promoting, and antioxidant effects in various solid tumors [59,62].

Flavones are a class of flavonoids with a 2-phenylchromen-4-one backbone. Three flavones are of interest in synergistic anti-MM therapy. Scutellarein, obtained from *S. baicalensis* or *S. barbata*, exerts anticancer activity against various tumors, including breast, lymphoma, colorectal, lung, and liver cancers [24,25]. Similarly, wogonin, an active monoflavone from Scutellaria baicalensis with anti-angiogenic activity, is a potential anti-cancer drug with low toxicity [26]. Morin, isolated from members of the mulberry family, such as mulberry figs and old figs, has anti-proliferative activity against some tumors, such as oral squamous cell carcinoma, leukemia, and colorectal cancer [27].

In contrast to flavones, isoflavones have a chemical structure based on a 3-phenylchromium-4-one backbone. Two isoflavones are of interest in synergistic anti-MM therapy. Formononetin, mainly isolated from the roots of Astragalus membranaceus, Trifolium pratense, Glycyrrhiza glabra, and Pueraria lobate, has anti-inflammatory, antioxidant, antiviral, neuroprotective, wound healing, and antitumor biological activities [28,29]. Daidzin can be isolated from Pueraria lobate. It has demonstrated anticancer activity in preventing and treating breast and prostate cancers [30].

In addition, Punica granatum juice(PGJ), one plant extract rich in flavonoids, is also of attention. The drug activity of PGJ is related to many flavonoid phytoactive components, including the anthocyanins catechin, quercetin, kaempferol, apigenin, and lignan. The anticancer effects of PGJ are widely used to prevent and treat colon, lung, skin, and prostate cancers [31].

Finally, polyphenols other than flavonoids belong to non-flavonoid polyphenols. Four non-flavonoid polyphenols are of interest in synergistic anti-MM therapy. Silibinin, an extract of Milk Thistle, exerts very high antioxidant and antitumor properties [32,60]. Resveratrol is widely found in grapes, berries, and peanuts and has anticancer activity against most human cancers [61]. Another non-flavonoid polyphenol, curcumin, is present in turmeric root. It exerts anti-inflammatory, anti-atherosclerotic, and anti-tumor pharmacological effects [33]. In addition, caffeic acid, an active component of honeybee propolis, has cytotoxic, apoptosis, and anti-proliferation effects [34].

## 7. Synergistic Effects of Flavonoids and Bortezomib in Anti-MM

Nine flavonoids (icariin, icariside II, EGCG, scutellarein, wogonin, morin, formononetin, daidzin) plus a flavonoid-rich plant extract (*PGJ*) synergize with BTZ in anti-MM by regulating proliferation, apoptosis, and drug resistance-related signaling pathways (Figure 6) (Table 3).

**Table 3.** Mechanistic anti-MM effects of flavonoid-BTZ combinations, as demonstrated in vitro and in vivo.

| Flavonoid | Cell Line | Flavonoid Concentration or Dose | Bortezomib Concentration or Dose | Effects | Reference |
|---|---|---|---|---|---|
| Icariin | Drug-resistant cell line KM3/BTZ | 0, 5, 10, 20, 40 mol/L | 0.08, 0.16, 0. 32, 0.64, 1.28 μg/mL | Increased drug sensitivity and reversed drug resistance | [22] |
| Icariin | U266 | 10 μM | 1 nM | Enhanced cytotoxic effect, G0/G1 phase cell cycle arrest, and apoptosis | [35] |
| Icariside II | U266 | 0, 25, 50, 100 μM | 20, 40 nM | Enhance cell apoptosis | [23] |
| EGCG | KM3 cell line | 0, 25, 50, 100 μM | 20 nM | Synergistically inhibit cell growth and induce cell apoptosis | [36] |
| Scutellarein | MM.1R, IM-9, U266B1, RPMI 8226<br>IM-9 cells injected subcutaneously in BALB/C nude mice | 200 μg/mL<br>0, 30, 60, 120 mg/kg | 30, 60 mg/kg | Inhibition of cell viability and selective induction of cell apoptosis.<br>The tumor burden of xenograft in nude mice was significantly reduced, and body weight was not affected. | [25] |
| Scutellarein | MM.1S cells injected subcutaneously in the BALB/C nude mice | 30, 60 mg/kg | 30, 60 mg/kg | The tumor weight was significantly reduced and Pharmacochemical resistance was reduced | [24] |
| Wogonin | RPMI 8226, U266 | 20, 40, 80 μM | 10 nM | Synergistically inhibit tumor-stimulated angiogenesis. | [26] |
| Morin | U266, RPMI 8226, MM.1S | 5, 10, 25 μM | 20 nmol/L | Enhanced drug-induced apoptosis | [27] |

**Table 3.** *Cont.*

| Flavonoid | Cell Line | Flavonoid Concentration or Dose | Bortezomib Concentration or Dose | Effects | Reference |
|---|---|---|---|---|---|
| Formononetin | U266 | 50 μM | 10 nM | Enhanced drug-induced apoptosis | [28] |
| Formononetin | U266, RPMI 8226<br>U266 cells injected subcutaneously in the athymic nu/nu female mice | 50 μM<br>20 mg/kg | 10 nM<br>0.25 mg/kg | Synergistically enhanced anti-tumor and apoptotic effects | [29] |
| Daidzin | U266, MM1.S | 10, 20, 30 μM | 1, 2.5, 5 nM | Decreased cell growth, induced higher cell death, increased SubG1 arrest and increased cell apoptosis. | [30] |
| Punica granatum juice | KMS26, MM1S, U266 | 3, 6, 12% | 15 nM | Improved the cytotoxic effect | [31] |

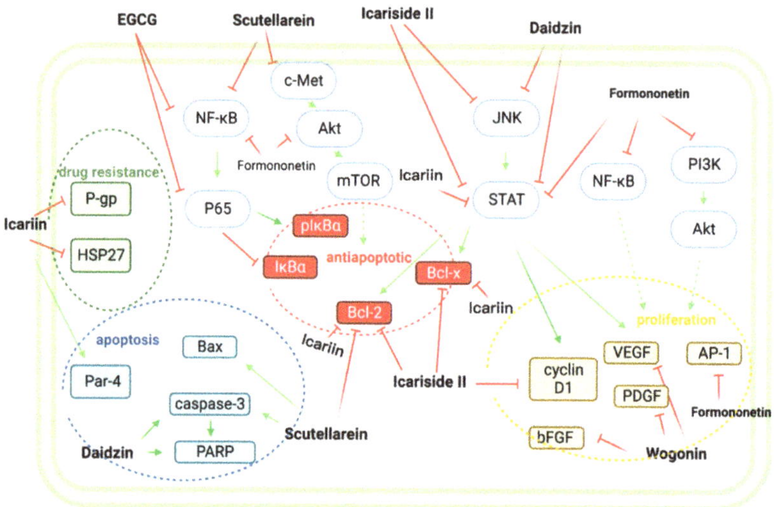


**Figure 6.** Nine flavonoids, icariin, icariside II, EGCG, scutellarein, wogonin, morin, formononetin, daidzin, and isoginkgetin, together exerted multiple synergistic BTZ anti-MM effects. Icariin synergistically improved the drug sensitivity of MM to BTZ by reducing the expression of drug resistance proteins HSP27 and P-gp. EGCG, scutellarein, icariin, icariside II, and formononetin down-regulated the downstream anti-apoptotic proteins Bcl2, BclXl, pIκBα by inhibiting NF-κB, c-Met/Akt/mTOR, and JAK/STAT pathways, thereby synergistically reversing anti-apoptosis. Icariside II, formononetin, and Wogonin, with BTZ synergistically, exerted antiproliferative effects by downregulating VEGF, PDGF, bFGF, cyclin D1, and AP-1. Finally, scutellarein, Daidzin, and Icariin enhanced BTZ-related apoptosis by upregulating the expression of Par-4, Bax, caspase-3, and PARP. Abbreviations: AP-1: Activator protein 1; Bax: bcl-2-like protein 4; Bcl2: B-cell lymphoma 2; Bcl-xl: B-cell lymphoma-extra-large; bFGF: basic fibroblast growth factor; c-Met: tyrosine-protein kinase Met; NF-κB/P56: Nuclear factor-kappa-B/p65; HSP27: Heat shock protein 27; JAK/STAT: Janus kinases/signal transducer and activator of transcription proteins; IκBα: nuclear factor of kappa light polypeptide gene enhancer in B-cells inhibitor, alpha; TOR: mammalian target of rapamycin; PARP: Poly (ADP-ribose) polymerase; Par-4: Prostate apoptosis response-4; PDGF: Platelet-derived growth factor; P-gp: P-glycoprotein; PI3K/Akt: Phosphoinositide 3-kinases/Protein Kinase B.

### 7.1. Icariin and Bortezomib

In KM3/BTZ-resistant cells, icariin increased the sensitivity of KM3/BTZ cells to BTZ and partially reversed the drug resistance. Its effect on changing drug resistance may be mediated by upregulating the expression of pro-apoptotic cytokine Par-4, decreasing the expression of drug resistance proteins HSP27 and P-gp [22]. Another study found that the combination of icariin and BTZ promoted apoptosis in U266 cells by blocking the JAK/STAT pathway and lowering the expression of anti-apoptotic proteins like Bcl-2, Bcl-xl, and Survivin. This combination also delayed the cell cycle progression, as shown by the result that more U266 cells were in the G0/G1 phase [23].

### 7.2. Icariside II and Bortezomib

While icariside II enhanced the apoptotic effect of BTZ in U266 cells, it also inhibited the JAK/STAT pathway and down-regulated the expression of STAT3 target genes Bcl-2, Bcl-xl, Survivin, cyclin D1, COX-2, and VEGF, beneficially inhibiting proliferation and promoting apoptosis [35].

### 7.3. EGCG and Bortezomib

EGCG, a catechin in green tea, and BTZ synergistically inhibited KM3 cell growth and induced apoptosis by inhibiting NF-κB/P65 expression, down-regulating pIκBα, and up-regulating IκBα expression [36].

### 7.4. Scutellarein and Bortezomib

In vitro, scutellarein alone time-dependently reduced cell viability and significantly induced apoptosis in MM.1R and IM-9 cells. In vivo, dual intervention with scutellarein and BTZ significantly reduced xenograft tumor burden in nude mice with no significant effect on mouse body weight. In parallel, protein expression levels of some apoptotic markers were altered, such as active caspase-3 (upregulated), Bax (upregulated), and Bcl-2 (downregulated) [25]. In another in vivo study, scutellarein eliminated MM cell resistance to BTZ through multiple mechanistic pathways. This result was caused by the HDAC/miR-34a-mediated epigenetic regulation of the c-Met/Akt/mTOR pathway and the NF-κB-mediated activation of the apoptosis cascade [24].

### 7.5. Wogonin and Bortezomib

Wogonin and BTZ synergistically inhibited the secretion levels of pro-angiogenic factors VEGF, PDGF, and bFGF in RPMI 8226 cells. The angiogenesis-inhibitory effect of wogonin may be related to its inhibition of the c-Myc/HIF-1α signaling axis [26].

### 7.6. Morin and Bortezomib

Morin potentiates BTZ-induced apoptosis in U266 cells, from 18.7% to 51.2%, by inhibiting the STAT3 pathway, leading to downregulation of STAT3-dependent gene expressions, such as XIAP, cFLIP, Mcl-1, Survivin, Bcl2, BclXl, and c-IAP-2. And morin's inhibition of STAT3 phosphorylation was more pronounced than other flavonols (galangin, kaempferol, quercetin, and myricetin) with different positions and numbers of its hydroxyl groups on the B-loop [27].

### 7.7. Formononetin and Bortezomib

In vitro, formononetin and BTZ enhanced STAT3 inhibition and promoted the death of U266 cells [28]. Subsequently, a follow-up study by the same research group found that formononetin and BTZ exert synergistic enhancement of anti-proliferation and pro-apoptosis by blocking the activation of NF-κB, PI3K/AKT, and AP-1 [29].

### *7.8. Daidzin and Bortezomib*

Daidzin synergistically increased the apoptotic and cytotoxic effects of BTZ by inhibiting the activation of STAT3 and its upstream kinases (JAK1, JAK2, and c-Src). In addition, this drug combination increased caspase-3 activation and PARP cleavage, leading to the downregulation of the expression of various oncogenic apoptotic proteins [30].

### *7.9. Plant Extracts and Bortezomib*

In vitro, PGJ inhibited angiogenesis, microvascular growth outside of aortic rings, cell migration, and invasion in MM cells. In addition, After BTZ exposure, PGJ intervention increased the cytotoxic effects on U266 cells [31].

## 8. Synergistic Effects of Non-Flavonoid Polyphenols and Bortezomib in Anti-MM

Four flavonoids (silibinin, resveratrol, curcumin, caffeic acid) synergize with BTZ in anti-MM effect by regulating proliferation, apoptosis, and drug resistance-related signaling pathways (Table 4).

**Table 4.** Mechanistic anti-MM effects of non-flavonoid polyphenols-BTZ combinations, as demonstrated in vitro and in vivo.

| Non-Flavonoid Polyphenol | Cell Line | Non-Flavonoid Polyphenol Concentration or Dose | Bortezomib Concentration or Dose | Effects | Reference |
|---|---|---|---|---|---|
| Silibinin | NCI H929 | Not available | Not available | Increased cytotoxic effects | [32] |
| resveratrol | MM1S, MM144, U266 | 10 mM | 10 nM | Enhanced cell apoptosis | [37] |
| Curcumin | U266 | 4, 8 μM | 0.5, 4 nM | Synergistically inhibit cell growth and promote apoptosis | [38] |
| Curcumin | U266, RPMI-8226, RPMI-8226-Dox-6, RPMI-8226-LR-5, MM.1S, MM.1R<br>U266 cells injected subcutaneously into mice | 5,10 μmol/L<br>1 g/kg | 0.25 mg/kg, 100 μL | Enhanced anti-tumor effect | [39] |
| Curcumin | RPMI-8266, U266 | 0, 20, 40 μM | 5 nmol/L | Enhanced cell apoptosis | [33] |
| novel curcumin analogs | RPMI-1640 | 2.5, 5, 10 μM | 10 nM | Enhanced cell proliferation inhibition, apoptosis induction and proteasome inhibition | [40] |
| caffeic acid | ARH-77 | 5, 10, 20, 40, 80, 160 g/mL | 1, 10, 20, 30, 50, 100 nM | Synergistically enhance Cytotoxic effects and anti-proliferation | [34] |

### *8.1. Silibinin and Bortezomib*

In combination with low concentrations of BTZ, silibinin increased the cytotoxic effect of BTZ by increasing the expression of activated caspases, which promoted apoptosis [32].

## 8.2. Resveratrol and Bortezomib

In vitro, resveratrol induced apoptosis in MM144 cells by upregulating the Fas/CD95 signaling pathway and caspase-8 and caspase-10. Moreover, when used in combination, resveratrol, and BTZ highly enhanced U266 cell apoptosis [37].

## 8.3. Curcumin and Bortezomib

Curcumin enhances the apoptotic effect of BTZ on MM cells by regulating multiple signaling pathways (Figure 7). The bone marrow microenvironment, in which bone marrow stromal cells (BMSCs) interact with MM, influences the survival and growth of MM cells. In vitro, curcumin inhibited the activation of JAK/STAT and MAPK pathways in U266 cells after treatment with BMSCs cell supernatant. Additionally, curcumin and BTZ co-treatment efficiently prevented IL-6-induced STAT3 and Erk phosphorylation, increased PARP cleavage, and decreased pro-caspase-3 levels. Through these mechanisms mentioned above, this combination inhibited the growth of U266 cells and promoted apoptosis [38]. After that, in another trial, curcumin enhanced the inhibitory effect of BTZ on NF-κB activation in U266 cells, leading to increased apoptosis. In vivo, the combination group was more potent than the BTZ group in reducing tumor volume [39]. Moreover, curcumin increased the expression of cleaved caspase-3 protein by inhibiting the Notch1 signaling pathway, which increased the drug sensitivity of RPMI-8226 cells and U266 cells to BTZ [33].

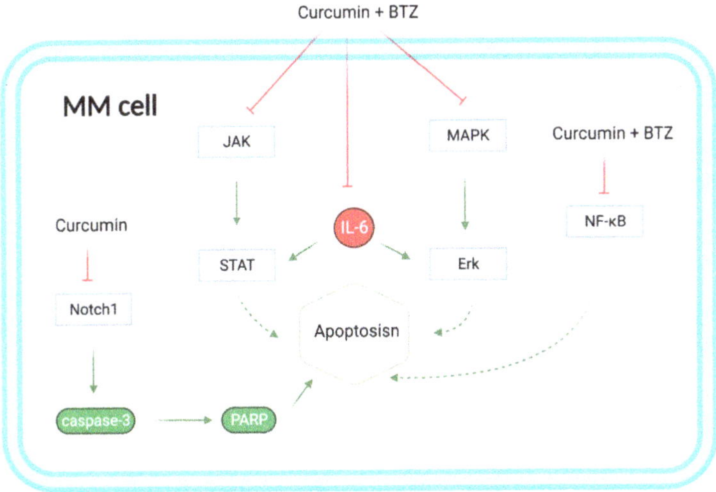


**Figure 7.** The combination of curcumin and BTZ synergistically enhances the effect on MM apoptosis by regulating multiple signaling pathways. Curcumin+BTZ synergistically enhanced apoptosis through potent inhibition of IL-6-induced JNK/STAT and MAPK/ERK, as well as NF-κB. Additional inhibition of the Notch1 signaling pathway by curcumin increased the expression of caspase3 protein and PARP and therefore enhanced apoptosis. Abbreviations: BTZ: bortezomib; NF-κB: Nuclear factor-kappa-B; MAPK/ERK: mitogen-activated protein kinases/extracellular signal-regulated kinases; MM: Multiple myeloma; Notch1: Neurogenic locus notch homolog protein 1; JAK/STAT: Janus kinases/signal transducer and activator of transcription proteins; PARP: Poly (ADP-ribose) polymerase.

A class of novel curcumin analogs known as amino acid adducts of curcumin improved the proteasomal inhibitory effect of BTZ on MM cells and they enhanced the proliferation inhibition and apoptosis induction of BTZ. Notably, similar to curcumin, the twelfth curcumin analog increased PARP and caspase-3 cleavage, enhancing BTZ-induced apoptosis. The water-soluble, highly bioavailable twelfth curcumin analog has the potential to replace curcumin in anti-MM therapy in the future [40].

### *8.4. Caffeic acid and Bortezomib*

In vitro, caffeic acid alone inhibits NF-κB-binding activity and IL-6 levels, which are closely linked to apoptosis and growth of tumor cells. Moreover, the combination of caffeic acid and BTZ synergistically increased cytotoxic and anti-proliferative effects on ARH-77 cells [34].

## 9. Conclusions and Future Directions

Although treatment options for MM continue to be optimized, the prognosis remains unsatisfactory. The aggressiveness and drug resistance of malignant tumors hinder the current treatment with protease inhibitors—especially BTZ. In this regard, flavonoids and non-flavonoid polyphenols are potential supportive therapies to address some of the challenges faced in BTZ treatment. Flavonoids(icariin, icariside II, EGCG, scutellarein, wogonin, morin, formononetin, daidzin), plant extract rich in flavonoids(PGJ), and non-flavonoid polyphenols(silibinin, resveratrol, curcumin, caffeic acid) combined with BTZ demonstrate the synergistic anti-MM effect. These synergistic anti-MM effects were achieved by anti-proliferative, pro-apoptotic, and anti-drug resistance. Based on those pieces of evidence, flavonoids, plant extract rich in flavonoids, and non-flavonoid polyphenols may benefit patients with MM, especially by overcoming the challenges faced in BTZ therapy.

According to some other relevant preclinical evidence, some polyphenols alone have anti-MM activity, such as isoginkgetin (a biflavone) and virola oleifera (a polyphenol-rich plant extract) [63,64]. Their synergistic effects, in combination with BTZ for the treatment of MM, need to be further explored. In addition, most of the synergistic anti-MM impacts of polyphenols + BTZ are only verified at the cellular level, such as icariin, icariside, EGCG, wogonin, morin, daidzin, PGJ, silibinin, and resveratrol. Their anti-MM synergistic effects in combination with BTZ need further clinically relevant animal models for validation. These polyphenols + BTZ exert synergistic effects through modulation of MM-related signaling pathways such as NF-κB, PI3K/Akt, Ras/Raf/MEK/Erk, JAK/STAT, and Wnt/β-catenin. However, the signaling pathways of each combination of anti-MM are still rudimentary, and more research is needed to refine the signaling pathway network. Although these results show promise for a synergistic MM treatment using these polyphenolic compounds and BTZ, the evidence at this time is only preclinical. Therefore, further clinical studies must demonstrate these synergistic effects. Moreover, preclinical studies on the anti-MM synergistic effects of these polyphenols with two other clinically available protease inhibitors, carfilzomib, and estazomib, need to be conducted to provide more options for clinical combination use. Adverse effects of protease inhibitor use are another challenge in the treatment of MM, although not within the scope of the review. Some polyphenols have the ameliorative potential for the non-tumor toxicity of protease inhibitors. For instance, rutin mitigates carfilzomib-induced cardiotoxicity by inhibiting NF-κB, mast gene expression, and reducing oxidative stress [65]. Resveratrol, by activating SIRT1, improves BTZ mechanical nociceptive sensitization [66]. For the comprehensive management of MM, additional studies on the antagonism of polyphenols against protease inhibitor-associated toxicity are also of clinical value. Finally, BTZ could alleviate MM-associated bone disease by regulating the RANK/RANKL/OPG signaling pathway [67]. By concentrating on the modulation of the RANK/RANKL/OPG signaling pathway, it is possible to investigate whether combining polyphenols and BTZ has beneficial effects on MM-associated bone disease.

Some polyphenols contain catechol moieties, such as EGCG, quercetin, and myricetin. Notably, they reduce the protease inhibitory activity of BTZ by forming stable cyclic boronic esters, thereby decreasing the anti-MM effect of BTZ [68,69]. A representative potential BTZ antagonistic polyphenol, EGCG, demonstrated dose-dependent inhibition of BTZ antitumor activity [62]. An in vitro study found that, by activating Wnt/β-catenin, EGCG antagonized the antitumor effect of BTZ [70]. Therefore, clinical staff members should be cautious about BTZ drug interactions with these polyphenols in clinical use. Last but not least, the problem of bioavailability is an important reason why polyphenols do not work as well as they could in the body. In the future, Low bioavailability polyphenols will eventually need to be combined with new drug delivery systems or other derivatives to utilize their anticancer capabilities fully.

**Author Contributions:** Conceptualization, K.D. and H.J.; methodology, K.D. and W.J.; software, W.J.; validation, K.D.; writing—original draft preparation, K.D.; writing—review and editing, K.D. and W.J.; visualization, W.J.; supervision, M.L. and H.J.; project administration, M.L. All authors have read and agreed to the published version of the manuscript.

**Funding:** This research received no external funding.

**Institutional Review Board Statement:** Not applicable.

**Informed Consent Statement:** Agreed to publish.

**Data Availability Statement:** Not applicable.

**Acknowledgments:** The authors would like to thank Xueping Li of the department of geriatrics, Hospital of Chengdu University of Traditional Chinese Medicine for critically reviewing the manuscript.

**Conflicts of Interest:** The authors declare no competing interest.

## References


1. Ludwig, H.; Novis Durie, S.; Meckl, A.; Hinke, A.; Durie, B. Multiple Myeloma Incidence and Mortality Around the Globe; Interrelations Between Health Access and Quality, Economic Resources, and Patient Empowerment. *Oncologist* **2020**, *25*, e1406–e1413. [CrossRef] [PubMed]
2. Kazandjian, D. Multiple myeloma epidemiology and survival: A unique malignancy. *Semin. Oncol.* **2016**, *43*, 676–681. [CrossRef] [PubMed]
3. Cowan, A.J.; Allen, C.; Barac, A.; Basaleem, H.; Bensenor, I.; Curado, M.P.; Foreman, K.; Gupta, R.; Harvey, J.; Hosgood, H.D.; et al. Global Burden of Multiple Myeloma: A Systematic Analysis for the Global Burden of Disease Study 2016. *JAMA Oncol.* **2018**, *4*, 1221–1227. [CrossRef] [PubMed]
4. Cowan, A.J.; Green, D.J.; Kwok, M.; Lee, S.; Coffey, D.G.; Holmberg, L.A.; Tuazon, S.; Gopal, A.K.; Libby, E.N. Diagnosis and Management of Multiple Myeloma. *JAMA: J. Am. Med. Assoc.* **2022**, *327*, 464–477. [CrossRef] [PubMed]
5. Kumar, L.; Gundu, N.; Kancharia, H.; Sahoo, R.K.; Malik, P.S.; Sharma, A.; Gupta, R.; Sharma, O.; Biswas, A.; Kumar, R.; et al. Multiple Myeloma—Effect of Induction Therapy on Transplant Outcomes. *Clin. Lymphoma Myeloma Leuk.* **2020**, *21*, 80–90.e5. [CrossRef]
6. Rosiñol, L.; Oriol, A.; Rios, R.; Sureda, A.; Blanchard, M.J.; Hernández, M.T.; Martínez-Martínez, R.; Moraleda, J.M.; Jarque, I.; Bargay, J.; et al. Bortezomib, lenalidomide, and dexamethasone as induction therapy prior to autologous transplant in multiple myeloma. *Blood* **2019**, *134*, 1337–1345. [CrossRef]
7. Facon, T.; Venner, C.P.; Bahlis, N.J.; Offner, F.; White, D.J.; Karlin, L.; Benboubker, L.; Rigaudeau, S.; Rodon, P.; Voog, E.; et al. Oral ixazomib, lenalidomide, and dexamethasone for transplant-ineligible patients with newly diagnosed multiple myeloma. *Blood* **2021**, *137*, 3616–3628. [CrossRef]
8. Al Hadidi, S.; Yellapragada, S. Treatment Options for Relapsed and Refractory Multiple Myeloma: A Luxury or a Challenge? *JAMA Oncol.* **2021**, *7*, 1449. [CrossRef]
9. Kastritis, E.; Terpos, E.; Dimopoulos, M.A. How I treat relapsed multiple myeloma. *Blood* **2022**, *139*, 2904–2917. [CrossRef]
10. Pancheri, E.; Guglielmi, V.; Wilczynski, G.M.; Malatesta, M.; Tonin, P.; Tomelleri, G.; Nowis, D.; Vattemi, G. Non-Hematologic Toxicity of Bortezomib in Multiple Myeloma: The Neuromuscular and Cardiovascular Adverse Effects. *Cancers* **2020**, *12*, 2540. [CrossRef]
11. Argyropoulou, A.; Aligiannis, N.; Trougakos, I.P.; Skaltsounis, A.-L. Natural compounds with anti-ageing activity. *Nat. Prod. Rep.* **2013**, *30*, 1412–1437. [CrossRef] [PubMed]
12. Haque, A.; Brazeau, D.; Amin, A.R. Perspectives on natural compounds in chemoprevention and treatment of cancer: An update with new promising compounds. *Eur. J. Cancer* **2021**, *149*, 165–183. [CrossRef] [PubMed]
13. Mrityunjaya, M.; Pavithra, V.; Neelam, R.; Janhavi, P.; Halami, P.M.; Ravindra, P.V. Immune-Boosting, Antioxidant and Anti-inflammatory Food Supplements Targeting Pathogenesis of COVID-19. *Front. Immunol.* **2020**, *11*, 570122. [CrossRef] [PubMed]
14. Neha, K.; Haider, R.; Pathak, A.; Yar, M.S. Medicinal prospects of antioxidants: A review. *Eur. J. Med. Chem.* **2019**, *178*, 687–704. [CrossRef] [PubMed]
15. Rondanelli, M.; Faliva, M.A.; Miccono, A.; Naso, M.; Nichetti, M.; Riva, A.; Guerriero, F.; De Gregori, M.; Peroni, G.; Perna, S. Food pyramid for subjects with chronic pain: Foods and dietary constituents as anti-inflammatory and antioxidant agents. *Nutr. Res. Rev.* **2018**, *31*, 131–151. [CrossRef] [PubMed]
16. Forni, C.; Rossi, M.; Borromeo, I.; Feriotto, G.; Platamone, G.; Tabolacci, C.; Mischiati, C.; Beninati, S. Flavonoids: A Myth or a Reality for Cancer Therapy? *Molecules* **2021**, *26*, 3583. [CrossRef]

17. Maiuolo, J.; Gliozzi, M.; Carresi, C.; Musolino, V.; Oppedisano, F.; Scarano, F.; Nucera, S.; Scicchitano, M.; Bosco, F.; Macri, R.; et al. Nutraceuticals and Cancer: Potential for Natural Polyphenols. *Nutrients* **2021**, *13*, 3834. [CrossRef]
18. Zhou, Y.; Zheng, J.; Li, Y.; Xu, D.-P.; Li, S.; Chen, Y.-M.; Li, H.-B. Natural Polyphenols for Prevention and Treatment of Cancer. *Nutrients* **2016**, *8*, 515. [CrossRef]
19. Jöhrer, K.; Çiçek, S. Multiple Myeloma Inhibitory Activity of Plant Natural Products. *Cancers* **2021**, *13*, 2678. [CrossRef]
20. Pojero, F.; Poma, P.; Spanò, V.; Montalbano, A.; Barraja, P.; Notarbartolo, M. Targeting multiple myeloma with natural polyphenols. *Eur. J. Med. Chem.* **2019**, *180*, 465–485. [CrossRef]
21. Moher, D.; Liberati, A.; Tetzlaff, J.; Altman, D.G.; the PRISMA Group. Preferred reporting items for systematic reviews and meta-analyses: The PRISMA statement. *BMJ* **2009**, *339*, b2535. [CrossRef] [PubMed]
22. Li, Z.Y.; Li, Z.J.; Chen, X.; Huang, X.R.; Fang, Z.Q.; Zhang, J.G.; Fang, J. Icaritin Reverses Multidrug Resistance of Multiple Myeloma Cell Line KM3/BTZ. *Zhongguo Shi Yan Xue Ye Xue Za Zhi* **2017**, *25*, 1690–1695. [CrossRef] [PubMed]
23. Kim, S.-H.; Ahn, K.S.; Jeong, S.-J.; Kwon, T.-R.; Jung, J.H.; Yun, S.-M.; Han, I.; Lee, S.-G.; Kim, D.K.; Kang, M.; et al. Janus activated kinase 2/signal transducer and activator of transcription 3 pathway mediates icariside II-induced apoptosis in U266 multiple myeloma cells. *Eur. J. Pharmacol.* **2011**, *654*, 10–16. [CrossRef] [PubMed]
24. Li, L.; Zheng, Y.; Zhang, W.; Hou, L.; Gao, Y. Scutellarin circumvents chemoresistance, promotes apoptosis, and represses tumor growth by HDAC/miR-34a-mediated down-modulation of Akt/mTOR and NF-κB-orchestrated signaling pathways in multiple myeloma. *Int. J. Clin. Exp. Pathol.* **2020**, *13*, 212–219. [PubMed]
25. Shi, L.; Wu, Y.; Lv, D.L.; Feng, L. Scutellarein selectively targets multiple myeloma cells by increasing mitochondrial superoxide production and activating intrinsic apoptosis pathway. *Biomed. Pharmacother.* **2018**, *109*, 2109–2118. [CrossRef]
26. Fu, R.; Chen, Y.; Wang, X.-P.; An, T.; Tao, L.; Zhou, Y.-X.; Huang, Y.-J.; Chen, B.-A.; Li, Z.-Y.; You, Q.-D.; et al. Wogonin inhibits multiple myeloma-stimulated angiogenesis via c-Myc/VHL/HIF-1α signaling axis. *Oncotarget* **2015**, *7*, 5715–5727. [CrossRef]
27. Gupta, S.C.; Phromnoi, K.; Aggarwal, B.B. Morin inhibits STAT3 tyrosine 705 phosphorylation in tumor cells through activation of protein tyrosine phosphatase SHP1. *Biochem. Pharmacol.* **2012**, *85*, 898–912. [CrossRef]
28. Kim, C.; Lee, S.-G.; Yang, W.M.; Arfuso, F.; Um, J.-Y.; Kumar, A.P.; Bian, J.; Sethi, G.; Ahn, K.S. Formononetin-induced oxidative stress abrogates the activation of STAT3/5 signaling axis and suppresses the tumor growth in multiple myeloma preclinical model. *Cancer Lett.* **2018**, *431*, 123–141. [CrossRef]
29. Kim, C.; Lee, J.H.; Ko, J.-H.; Chinnathambi, A.; Alharbi, S.A.; Shair, O.H.; Sethi, G.; Ahn, K.S. Formononetin Regulates Multiple Oncogenic Signaling Cascades and Enhances Sensitivity to Bortezomib in a Multiple Myeloma Mouse Model. *Biomolecules* **2019**, *9*, 262. [CrossRef]
30. Yang, M.H.; Jung, S.H.; Chinnathambi, A.; Alahmadi, T.A.; Alharbi, S.A.; Sethi, G.; Ahn, K.S. Attenuation of STAT3 Signaling Cascade by Daidzin Can Enhance the Apoptotic Potential of Bortezomib against Multiple Myeloma. *Biomolecules* **2019**, *10*, 23. [CrossRef]
31. Tibullo, D.; Caporarello, N.; Giallongo, C.; Anfuso, C.D.; Genovese, C.; Arlotta, C.; Puglisi, F.; Parrinello, N.L.; Bramanti, V.; Romano, A.; et al. Antiproliferative and Antiangiogenic Effects of Punica granatum Juice (PGJ) in Multiple Myeloma (MM). *Nutrients* **2016**, *8*, 611. [CrossRef] [PubMed]
32. Geraldes, C.; Goncalves, A.; Alves, R.; Roque, A.; Manuel, J.; Costa, N.; Sarmento-Ribeiro, A. New Targeted Drugs in Multiple Myeloma Therapy—in Vitro Studies. *Clin. Lymphoma Myeloma Leuk.* **2015**, *15*, e254. [CrossRef]
33. Ge, X.-P.; Hong, Y.-T.; Li, Y.-L.; Li, B.-Z. Curcumin Increases the Chemosensitivity of Multiple Myeloma to Bortezomib by Inhibiting the Notch1 Signaling Pathway. *Zhongguo Shi Yan Xue Ye Xue Za Zhi* **2019**, *27*, 464–471.
34. Altayli, E.; Koru, Ö.; ÖNGÜRÜ, Ö.; Ide, T.; Acikel, C.; Sarper, M.; Elci, M.P.; Sağkan, R.I.; Astarci, E.; Tok, D.; et al. An in vitro and in vivo investigation of the cytotoxic effects of caffeic acid (3,4-dihydroxycinnamic acid) phenethyl ester and bortezomib in multiple myeloma cells. *Turk. J. Med. Sci.* **2015**, *45*, 38–46. [CrossRef] [PubMed]
35. Jung, Y.Y.; Lee, J.H.; Nam, D.; Narula, A.S.; Namjoshi, O.A.; Blough, B.E.; Um, J.-Y.; Sethi, G.; Ahn, K.S. Anti-myeloma Effects of Icariin Are Mediated Through the Attenuation of JAK/STAT3-Dependent Signaling Cascade. *Front. Pharmacol.* **2018**, *9*, 531. [CrossRef]
36. Wang, Q.; Li, J.; Gu, J.; Huang, B.; Zhao, Y.; Zheng, D.; Ding, Y.; Zeng, L. Potentiation of (−)-epigallocatechin-3-gallate-induced apoptosis by bortezomib in multiple myeloma cells. *Acta Biochim. et Biophys. Sin.* **2009**, *41*, 1018–1026. [CrossRef] [PubMed]
37. Reis-Sobreiro, M.; Gajate, C.; Mollinedo, F. Involvement of mitochondria and recruitment of Fas/CD95 signaling in lipid rafts in resveratrol-mediated antimyeloma and antileukemia actions. *Oncogene* **2009**, *28*, 3221–3234. [CrossRef]
38. Park, J.; Ayyappan, V.; Bae, E.-K.; Lee, C.; Kim, B.-S.; Kim, B.K.; Lee, Y.-Y.; Ahn, K.-S.; Yoon, S.-S. Curcumin in combination with bortezomib synergistically induced apoptosis in human multiple myeloma U266 cells. *Mol. Oncol.* **2008**, *2*, 317–326. [CrossRef]
39. Sung, B.; Kunnumakkara, A.B.; Sethi, G.; Anand, P.; Guha, S.; Aggarwal, B.B. Curcumin circumvents chemoresistance in vitro and potentiates the effect of thalidomide and bortezomib against human multiple myeloma in nude mice model. *Mol. Cancer Ther.* **2009**, *8*, 959–970. [CrossRef]
40. Chan, T.H.; Mujtaba, T.; Kanwar, J.; Wan, S.B.; Dou, Q.P. Sensitizing human multiple myeloma cells to the proteasome inhibitor bortezomib by novel curcumin analogs. *Int. J. Mol. Med.* **2011**, *29*, 102–106. [CrossRef]
41. Hu, J.; Hu, W. X. Targeting signaling pathways in multiple myeloma: Pathogenesis and implication for treatments. *Cancer Lett.* **2018**, *414*, 214–221. [CrossRef] [PubMed]
42. Anderson, K.C.; Carrasco, R.D. Pathogenesis of Myeloma. *Annu. Rev. Pathol. Mech. Dis.* **2011**, *6*, 249–274. [CrossRef] [PubMed]

43. Robak, P.; Drozdz, I.; Szemraj, J.; Robak, T. Drug resistance in multiple myeloma. *Cancer Treat. Rev.* **2018**, *70*, 199–208. [CrossRef] [PubMed]
44. Gandolfi, S.; Laubach, J.P.; Hideshima, T.; Chauhan, D.; Anderson, K.C.; Richardson, P.G. The proteasome and proteasome inhibitors in multiple myeloma. *Cancer Metastasis Rev.* **2017**, *36*, 561–584. [CrossRef]
45. Scott, K.; Hayden, P.J.; Will, A.; Wheatley, K.; Coyne, I. Bortezomib for the treatment of multiple myeloma. *Cochrane Database Syst. Rev.* **2016**, *4*, CD010816. [CrossRef]
46. Tan, C.R.C.; Abdul-Majeed, S.; Cael, B.; Barta, S.K. Clinical Pharmacokinetics and Pharmacodynamics of Bortezomib. *Clin. Pharmacokinet.* **2018**, *58*, 157–168. [CrossRef]
47. Thibaudeau, T.A.; Smith, D.M. A Practical Review of Proteasome Pharmacology. *Pharmacol. Rev.* **2019**, *71*, 170–197. [CrossRef]
48. Manasanch, E.E.; Orlowski, R.Z. Proteasome inhibitors in cancer therapy. *Nat. Rev. Clin. Oncol.* **2017**, *14*, 417–433. [CrossRef]
49. Tallarida, R.J. Drug synergism: Its detection and applications. *J. Pharmacol. Exp. Ther.* **2001**, *298*, 865–872. Available online: https://pubmed.ncbi.nlm.nih.gov/11504778/ (accessed on 31 September 2001).
50. Coates, A.R.M.; Hu, Y.; Holt, J.; Yeh, P. Antibiotic combination therapy against resistant bacterial infections: Synergy, rejuvenation and resistance reduction. *Expert Rev. Anti-Infect. Ther.* **2020**, *18*, 5–15. [CrossRef]
51. Guieu, B.; Jourdan, J.-P.; Dreneau, A.; Willand, N.; Rochais, C.; Dallemagne, P. Desirable drug–drug interactions or when a matter of concern becomes a renewed therapeutic strategy. *Drug Discov. Today* **2020**, *26*, 315–328. [CrossRef] [PubMed]
52. Ianevski, A.; Giri, A.K.; Aittokallio, T. SynergyFinder 2.0: Visual analytics of multi-drug combination synergies. *Nucleic Acids Res.* **2020**, *48*, W488–W493. [CrossRef] [PubMed]
53. Rashid, M.B.M.A.; Toh, T.B.; Hooi, L.; Silva, A.; Zhang, Y.; Tan, P.F.; Teh, A.L.; Karnani, N.; Jha, S.; Ho, C.-M.; et al. Optimizing drug combinations against multiple myeloma using a quadratic phenotypic optimization platform (QPOP). *Sci. Transl. Med.* **2018**, *10*, eaan0941. [CrossRef]
54. Jin, Y.; Xu, L.; Wu, X.; Feng, J.; Shu, M.; Gu, H.; Gao, G.; Zhang, J.; Dong, B.; Chen, X. Synergistic Efficacy of the Demethylation Agent Decitabine in Combination With the Protease Inhibitor Bortezomib for Treating Multiple Myeloma Through the Wnt/β-Catenin Pathway. *Oncol. Res. Featur. Preclin. Clin. Cancer Ther.* **2019**, *27*, 729–737. [CrossRef]
55. Palumbo, A.; Chanan-Khan, A.; Weisel, K.; Nooka, A.K.; Masszi, T.; Beksac, M.; Spicka, I.; Hungria, V.; Munder, M.; Mateos, M.V.; et al. Daratumumab, Bortezomib, and Dexamethasone for Multiple Myeloma. *N. Engl. J. Med.* **2016**, *375*, 754–766. [CrossRef] [PubMed]
56. Zhuang, J.; Shirazi, F.; Singh, R.K.; Kuiatse, I.; Wang, H.; Lee, H.C.; Berkova, Z.; Berger, A.; Hyer, M.; Chattopadhyay, N.; et al. Ubiquitin-activating enzyme inhibition induces an unfolded protein response and overcomes drug resistance in myeloma. *Blood* **2019**, *133*, 1572–1584. [CrossRef] [PubMed]
57. Tzanova, M.; Atanasov, V.; Yaneva, Z.; Ivanova, D.; Dinev, T. Selectivity of Current Extraction Techniques for Flavonoids from Plant Materials. *Processes* **2020**, *8*, 1222. [CrossRef]
58. Di Lorenzo, C.; Colombo, F.; Biella, S.; Stockley, C.; Restani, P. Polyphenols and Human Health: The Role of Bioavailability. *Nutrients* **2021**, *13*, 273. [CrossRef]
59. Lu, X.; Saeed, M.E.M.; Hegazy, M.-E.F.; Kampf, C.J.; Efferth, T. Chemopreventive Property of Sencha Tea Extracts towards Sensitive and Multidrug-Resistant Leukemia and Multiple Myeloma Cells. *Biomolecules* **2020**, *10*, 1000. [CrossRef]
60. Tuli, H.S.; Mittal, S.; Aggarwal, D.; Parashar, G.; Parashar, N.C.; Upadhyay, S.K.; Barwal, T.S.; Jain, A.; Kaur, G.; Savla, R.; et al. Path of Silibinin from diet to medicine: A dietary polyphenolic flavonoid having potential anti-cancer therapeutic significance. *Semin. Cancer Biol.* **2020**, *73*, 196–218. [CrossRef]
61. Rauf, A.; Imran, M.; Butt, M.S.; Nadeem, M.; Peters, D.G.; Mubarak, M.S. Resveratrol as an anti-cancer agent: A review. *Crit. Rev. Food Sci. Nutr.* **2018**, *58*, 1428–1447. [CrossRef] [PubMed]
62. Bannerman, B.; Xu, L.; Jones, M.; Tsu, C.; Yu, J.; Hales, P.; Monbaliu, J.; Fleming, P.; Dick, L.; Manfredi, M.; et al. Preclinical evaluation of the antitumor activity of bortezomib in combination with vitamin C or with epigallocatechin gallate, a component of green tea. *Cancer Chemother. Pharmacol.* **2011**, *68*, 1145–1154. [CrossRef] [PubMed]
63. Francisco, V.; Ruiz-Fernández, C.; González-Rodríguez, M.; Cordero-Barreal, A.; Pino, J.; Viñuela, J.E.; Lago, F.; Conde, J.; Gómez, R.; Carvalho, G.R.; et al. Evaluation of Virola oleifera activity in musculoskeletal pathologies: Inhibition of human multiple myeloma cells proliferation and combination therapy with dexamethasone or bortezomib. *J. Ethnopharmacol.* **2021**, *272*, 113932. [CrossRef] [PubMed]
64. Tsalikis, J.; Abdel-Nour, M.; Farahvash, A.; Sorbara, M.T.; Poon, S.; Philpott, D.J.; Girardin, S.E. Isoginkgetin, a Natural Biflavonoid Proteasome Inhibitor, Sensitizes Cancer Cells to Apoptosis via Disruption of Lysosomal Homeostasis and Impaired Protein Clearance. *Mol. Cell. Biol.* **2019**, *39*, e00489-18. [CrossRef] [PubMed]
65. Imam, F.; Al-Harbi, N.O.; Al-Harbia, M.M.; Korashy, H.M.; Ansari, M.A.; Sayed-Ahmed, M.M.; Nagi, M.N.; Iqbal, M.; Anwer, K.; Kazmi, I.; et al. Rutin Attenuates Carfilzomib-Induced Cardiotoxicity Through Inhibition of NF-κB, Hypertrophic Gene Expression and Oxidative Stress. *Cardiovasc. Toxicol.* **2015**, *17*, 58–66. [CrossRef]
66. Chen, K.; Fan, J.; Luo, Z.-F.; Yang, Y.; Xin, W.-J.; Liu, C.-C. Reduction of SIRT1 epigenetically upregulates NALP1 expression and contributes to neuropathic pain induced by chemotherapeutic drug bortezomib. *J. Neuroinflamm.* **2018**, *15*, 292. [CrossRef]
67. Accardi, F.; Toscani, D.; Costa, F.; Aversa, F.; Giuliani, N. The Proteasome and Myeloma-Associated Bone Disease. *Calcif. Tissue Res.* **2017**, *102*, 210–226. [CrossRef]

68. Golden, E.B.; Lam, P.Y.; Kardosh, A.; Gaffney, K.; Cadenas, E.; Louie, S.G.; Petasis, N.; Chen, T.C.; Schönthal, A.H. Green tea polyphenols block the anticancer effects of bortezomib and other boronic acid–based proteasome inhibitors. *Blood* **2009**, *113*, 5927–5937. [CrossRef]
69. Sak, K. Dietary Flavonoids with Catechol Moiety Inhibit Anticancer Action of Bortezomib: What about the other Boronic Acid-Based Drugs? *Curr. Cancer Drug Targets* **2022**, *22*, 741–748. [CrossRef]
70. Qiu, X.; Wu, X.; He, W. (−)-Epigallocatechin-3-gallate plays an antagonistic role in the antitumor effect of bortezomib in myeloma cells via activating Wnt/β-catenin signaling pathway. *Adv. Clin. Exp. Med.* **2022**, *31*, 789–794. [CrossRef]

biomolecules

MDPI

Review

# Ethnobotany, Phytochemistry, and Biological Activities of the Genus Cordyline

**Romuald Tematio Fouedjou [1,†], Bienvenu Tsakem [1,†], Xavier Siwe-Noundou [2,*], Hervet P. Dongmo Fogang [3], Aphalaine Tiombou Donkia [1], Beaudelaire Kemvoufo Ponou [1], Madan Poka [2], Patrick H. Demana [2], Rémy B. Teponno [1,*] and Léon Azefack Tapondjou [1]**

[1] Research Unit of Environmental and Applied Chemistry, Department of Chemistry, Faculty of Science, University of Dschang, Dschang P.O. Box 67, Cameroon; tematioromy@yahoo.fr (R.T.F.); tsakemb@yahoo.fr (B.T.); atiombou@yahoo.com (A.T.D.); beaudelaireponou@yahoo.fr (B.K.P.); tapondjou2001@yahoo.fr (L.A.T.)

[2] Department of Pharmaceutical Sciences, School of Pharmacy, Sefako Makgatho Health Sciences University, P.O. Box 218, Pretoria 0208, South Africa; madan.poka@smu.ac.za (M.P.); patrick.demana@smu.ac.za (P.H.D.)

[3] Department of Physiological Sciences and Biochemistry, Faculty of Medicine and Biomedical, Sciences, University of Garoua, Garoua P.O. Box 317, Cameroon; hervpaulain@yahoo.fr

* Correspondence: xavier.siwenoundou@smu.ac.za (X.S.-N.); remyteponno@gmail.com (R.B.T.); Tel.: +27-12-521-5647 (X.S.-N.); +237-677312743 (R.B.T.)

† These authors contributed equally to this work.

**Abstract:** *Cordyline* species have a long history in traditional medicine as a basis of treatment for various ailments such as a bloody cough, dysentery, and a high fever. There are about 26 accepted species names in this genus distributed worldwide, including *C. fruticosa*, *C. autralis*, *C. stricta*, *C. cannifolia*, and *C. dracaenosides*. This work presents a comprehensive review of the traditional uses of plants of the genus *Cordylie* and their chemical constituents and biological activities. A bibliographic search was conducted to identify available information on ethnobotany, ethnopharmacology, chemical composition, and biological activities. A total of 98 isolated compounds potentially responsible for most of the traditional medicinal applications have been reported from eight species of *Cordyline* and are characterised as flavonoid, spirostane, furostane, and cholestane glycosides. Some of these pure compounds, as well as extracts from some species of *Cordyline*, have exhibited noteworthy anti-oxidant, antiproliferative, antimicrobial, and hypolipidemic activities. Although many of these species have not yet been investigated phytochemically or pharmacologically, they remain a potential source of new bioactive compounds.

**Keywords:** *Cordyline*; ethnopharmacology; phytochemistry; biological activities

## 1. Introduction

Traditional medicine frequently relies on medicinal plants for the treatment of various ailments and diseases. In developing countries, traditional medicine plays a significant role in delivering primary health care. Scientists have relied on the wealth of the floral kingdom for nearly half of all new drug discoveries to date [1,2]. Even ornamental plant species can contain a vast wealth of bioactive compounds valuable to medicine [3]. For example, drugs currently in clinical use for cancer treatment include paclitaxel, derived from *Taxus brevifolia* L., vincristine, derived from *Catharanthus roseus* (L.) G. Don., and epigallocatechin-3-gallate, a phenolic catechin from *Camellia sinensis* [4], all of which are better known as ornamental rather than medicinal plants.

Natural products and their derivatives are a significant source for novel drug development in modern medicine, and many of these derivatives form the active ingredient of drugs approved by the U.S. Food and Drug Administration (FDA) [5]. Dependency on direct medicinal plant derivatives is particularly marked in developing countries, where western medicine is often absent or simply too expensive. Plants of the genus *Cordyline*, well known as ornamental plants, have long been used as a source of traditional medicines and are of great interest in primary health care. As with *Taxus brevifolia*, *Catharanthus roseus*, and *Camelia sinensis*, these plants could be highly significant for drug discovery.

According to the literature, *Cordyline* species are characterised by the presence of flavonoids and saponins, particularly spirostane, furostanes, and cholestane glycosides, with various structural skeletons [6–8]. These compounds have been reported to show a wide range of biological activities, including anti-inflammatory, antiproliferative, antimicrobial, cytotoxic, and hypoglycaemic properties [7,9–11]. Some activities of these plants have been investigated, prompted by their anecdotal usage in traditional medicine, and various biological activities such as antidiabetic, anti-ulcer, antidiarrheal, wound healing, and anti-inflammatory properties have been reported in several scientific publications [7,12–14].

Despite literature surveys dating back to 1984, no reviews have summarised the ethnomedicinal uses, phytochemical constituents, and biological activities of plants of the genus *Cordyline*. Accordingly, this review provides an up-to-date synopsis of all bioactive compounds isolated from different parts of *Cordyline* species, as well as their biological activities and ethnopharmacological uses. This information is intended to summarise the therapeutic potential and to help guide future research. Databases such as Scopus, ScienceDirect, PubMed, and Google Scholar were used to source scientific articles. From ancient times up to 2022, eight species have been studied, and 56 saponins, 20 flavonoids, and 20 sapogenins were isolated and characterised from these plants. We have included in this paper the reported biological activities of some of these isolated compounds, as well as the activities of some extracts.

## 2. Botany and Taxonomy

The name *Cordyline* is derived from the Greek *kordyle*, which means *club*, which refers to the plants' thick underground stems or rhizomes. Royen was the first to report the name *Cordyline* in 1740 and to propose a classification, but due to the presence of *Yucca* and *Dracaena* in his classification, Adanson gave the plant a new generic classification in 1763 [15]. The genus *Cordyline*, one of the so-called 'tree lilies', was placed in the family Laxmanniaceae according to the Angiosperm Phylogeny Group (APG) in 2003, then in the Asparagaceae family in 2009. Also called Agavaceae, this family contains 128 genera, with more than 3000 species [16]. The genus *Cordyline* contains 26 species alone, of which *C. fruticosa*, *C. australis*, *C. stricta*, *C. annifolia*, *C. rubra*, and *C. dracaenoside* appear to have been the most researched. *Cordyline australis*, commonly called the cabbage tree and known as 'ti kouka' by the Maori people, is the most widespread of this genus and occurs in open places and at forest margins on the three main islands [17]. *Cordylines* form small trees or shrubs and have leathery leaves, which are often clustered and palm-like in appearance. *Cordyline* species have creamish white roots that often form root suckers starting from their rhizomes [18–20]. *Cordyline* belongs to the few genera of monocots in which a cambium is present in the stem. The most well-known ornamental and medicinal plant of this genus is *Cordyline fruticosa* L. (Chev.) (Figure 1), with its beautiful and attractive decorative foliage.

Plants of the genus *Cordyline* generally reach up to 3 m high and are remarkable for their colourful foliage and varied heights, colours, and the shape of their leaves [21–23]. Several hybrid combinations of the *Cordyline* species have been recorded. The species is believed to have originated in Southeast Asia and Papua New Guinea. The plants of the genus *Cordyline* are not always easy to distinguish from those of the *Dracaena*. However, the petiole of the former is generally larger than that of the latter (10 to 30 cm versus 1 to 8 cm).

**Figure 1.** *Cordyline fruticosa*.

## 3. Traditional Uses of the Genus *Cordyline*

Several plants of the genus *Cordyline*, and particularly their leaves, are used in traditional medicines throughout the tropical and subtropical regions for the treatment of various diseases such as dysentery, skin infection, rheumatism, inflammations, and periodic fever [24,25].

*Cordyline dracaenoides* is used as a traditional medicine in the south of Brazil as an anti-inflammatory preparation for the treatment of rheumatoid and related diseases [26]. *Cordyline stricta* is an evergreen shrub native to Australia, and an extract of its leaves is used as a hemostatic in traditional Chinese medicine [9]. *Cordyline fruticosa*, commonly known as the good luck plant, is grown in tropical and subtropical regions as an ornamental plant. In Polynesia, it is used for both cultural purposes and as a food crop [27]. The plant is widely utilised in both everyday life and as a source of traditional medicine. The leaves are used for costumes, decorations, clothing, sandals, packaging, and cooking. The rhizomes can be baked into a molasses-like food product and eaten [28]. In addition, the plant is used for the treatment of various diseases. The leaves are used to treat sore throat and neck pain [29], as a hemostatic [30,31], and to induce abortion [32], and the roots are used for toothache, laryngitis, and to treat infections of the mammary glands [31,33].

In Sabah, the Kadazan-Dusun people, who are the largest ethnic group in Malaysia, use the roots of *C. terminalis* L. Kunth (also known as *C. fruticosa*) (locally called *Pipisokalaganan*) as part of their local medicine for curing cough, bloody cough, dysentery, high fever, difficulties in urinating, bloody urine, kidney diseases, headaches, inflammation in the digestive tract, scurf, and joint pain. For example, the roots are pounded into a paste and applied to the stomach to relieve stomach pain [34]. In Hawaii, local people use the purple flowers of *C. fruticosa* to treat asthma and growths in the nose [35].

## 4. Other Important Uses of Plants of the Genus *Cordyline*

Ornamental plants are characterized by attractive foliage and/or flowers that can survive and grow indoors and/or outdoors. *Cordyline fruticosa*, with its attractive red decorative foliage, is one of India's most economically important ornamental houseplants [36]. The economic value of ornamental plants of the genus *Cordyline* has risen perceptibly worldwide and is still increasing to meet the steady demand of the floriculture industry. Reports that the roots of *Cordyline australis* may contain more than 60% fructofuranan suggest that the species could become a commercial source of fructose [37]. *Cordyline australis* has been used extensively by the Maori people in Australia, both as a source of fibre and as a food, and there is strong evidence to suggest that the plant was translocated extensively in pre-European times [38].

*Cordyline fruticosa* is often used ornamentally as a low-growing border shrub [39]. *C. fruticosa* is associated with cultural, material, and religious uses among the peoples of Oceania [18,40]. There is some evidence that the early Polynesians used *Cordyline fruticosa* for its importance in costume making, for wrapping food, for religious uses, and as a food source [41]. A hedge of *C. fruticosa* around the house was believed to ward off evil spirits and bring good luck [42]. The fleshy rhizome contains up to 20% sugar, mainly fructose, and is used as a natural sweetener in New Zealand and for the production of an alcoholic beverage in Hawaii [43]. In Hawaii, the fibre from the leaves of the 'Ti' plant (a common name for any plant of the genus in Hawaii) was used for making sandals, baskets, bird snares, sieves, and thatch for roofs, rope, and cord. The leaves of the plant (known as 'rau ti' in the Cook Islands) or the tender young shoots were eaten raw or roasted in the embers of a fire. Pith (commonly called Ti) was dried in the sun and cooked to make porridge [43]. Roots of Ti were used for making sweet drinks. Sugar was extracted by cooking the roots in an earth oven called an 'umu-ti'. Large pits were used to steam the roots. Long hedges of *Cordyline fruticosa* are planted by locals to guard the gardens against pigs. The handling of 'Ti' was often accompanied by ceremonial protocols [43].

The New Zealand cabbage tree, known as *Cordyline australis*, is internationally rated as one of their most famous indigenous plants. Its popularity is based not only on its characteristic habit, shape, and deliciously fragrant flowers but also on the relatively recent progress of coloured foliage forms [44]. Other species, also known as cabbage trees, include *C. kaspar*, *C. indivisa* (*Ti toi*), *C. banksii* (*Ti ngahere*), and *C. pumilio* (*Ti rauriki*) [45].

## 5. Chemical Constituents

A thorough review of the literature, based on the widespread uses of different parts of the plants of the *Cordyline* genus in traditional medicine, has resulted in considerable chemical analysis of their active compounds. A bibliographic investigation has revealed the isolation and identification of two principal classes of compounds with various structural skeletons. The process of isolating some of these compounds was carried out using simple column chromatography. In contrast, others used vacuum liquid chromatography on reversed-phase RP-18 silica gel medium-pressure liquid chromatography on silica gel 60. The structures of isolated compounds were established based on 1D and 2D NMR data, mass spectrometry, and chemical methods. Phytochemical studies have been mainly focused on the following eight *Cordyline* species: *C. australis*, *C. fruticosa*, *C. manners-suttoniae*, *C. stricta*, *C. rubra*, *C. dracaenoides*, *C. cannifolia*, and *C. indivisa*. The structures of the isolated secondary metabolites are grouped in Figures 2–5.

### 5.1. *Cordyline australis*

The first reports of the presence of steroidal sapogenins in *Cordyline* species originated in the pioneering work by Marker et al. in 1943 and Wall et al. in 1954 [46,47], who investigated natural plant sources of steroidal sapogenins such as smilagenin (**1**) and sarsasapogenin (**2**), as substrates for the commercial synthesis of steroidal hormones. Further studies concerning the distribution of steroidal sapogenins in the genus *Cordyline* were carried out [48–50]. Extracts of the leaves of *C. australis* obtained from plants grown in New Zealand and the United Kingdom gave a low yield of steroidal sapogenins [6]. The major compound was characterised as tigogenin or sarsasapogenin (**2**) based on TLC and IR data, while neotigogenin or smilagenin (**1**), diosgenin (**3**), yamogenin (**4**), and brisbagenin (**5**) were identified by TLC alone [51]. From the fruits of *Cordyline australis*, the new steroidal sapogenin, australigenin [5$\alpha$-spirost-25(27)-ene-1$\beta$,3$\beta$-diol] (**6**) was isolated, and it was also later reported from *C. manners-suttoniae* [8]. Other steroidal sapogenins have been isolated and characterised, including pompeygenin (**7**) and 1$\beta$-hydroxycrabbogenin (**8**) [12]. Korkashvili (2006) investigated the leaves and stems of *C. autralis* in partial fulfillment of a Ph.D. degree. He obtained seven steroidal glycosides (**9–15**) and an inseparable mixture of two flavonoid glycosides (**16** and **17**) (Figure 4) [10].

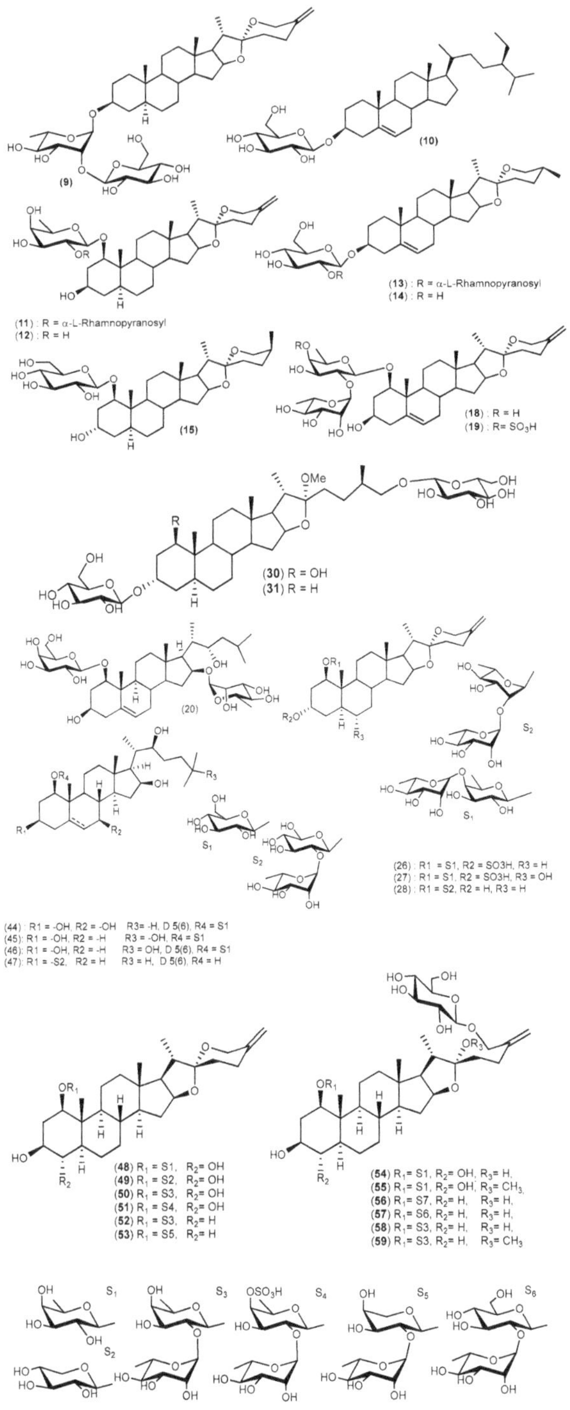


**Figure 2.** *Cont.*

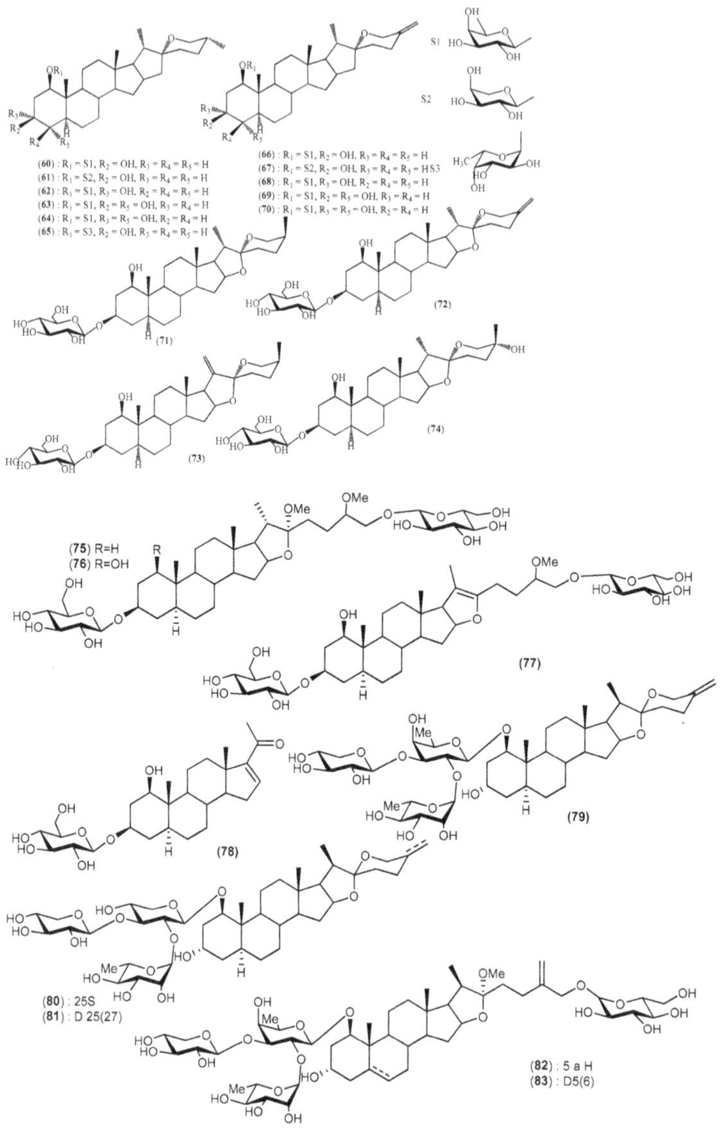

**Figure 2.** Structures of steroidal saponins isolated from some plants of the genus *Cordyline*.

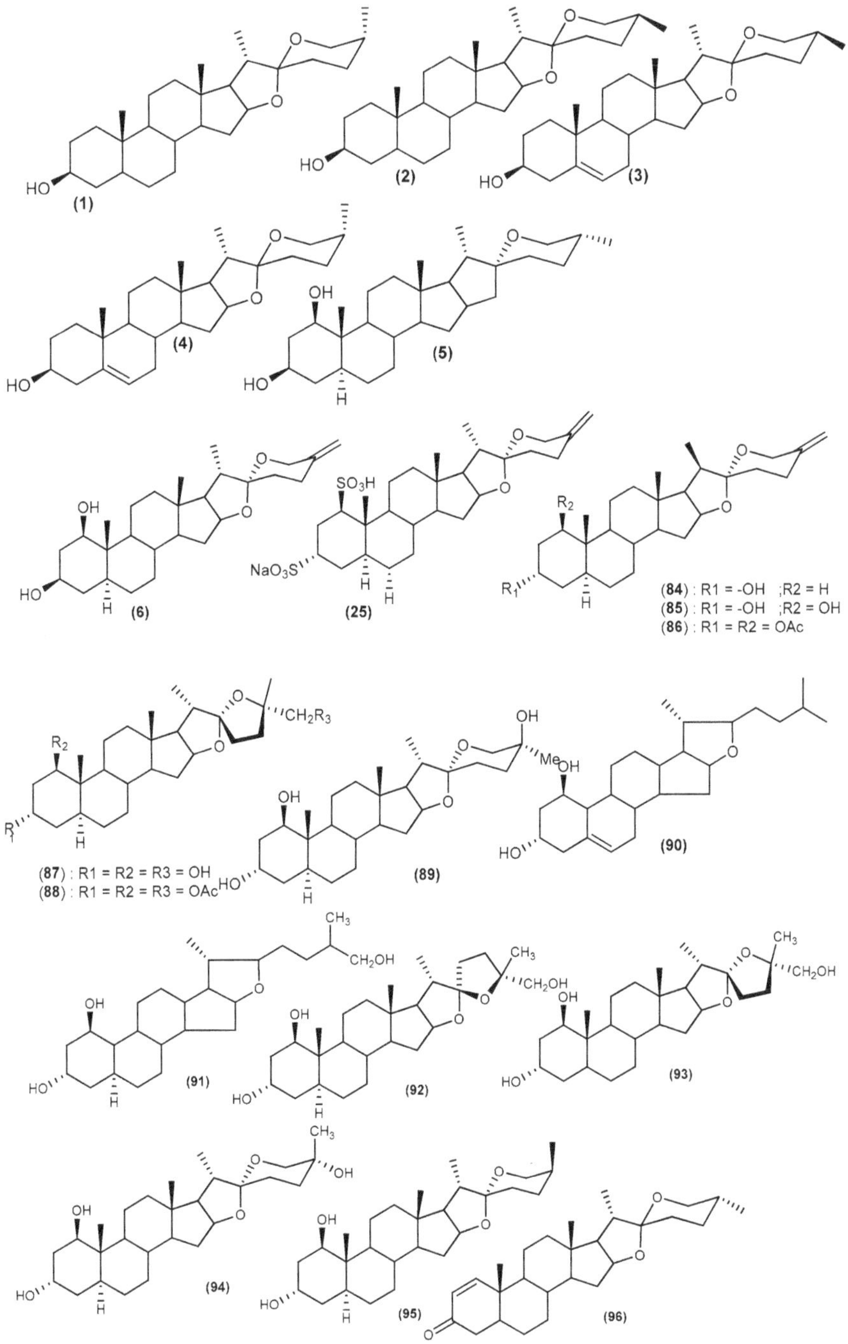


**Figure 3.** Main sapogenins isolated from plants of the genus *Cordyline*.

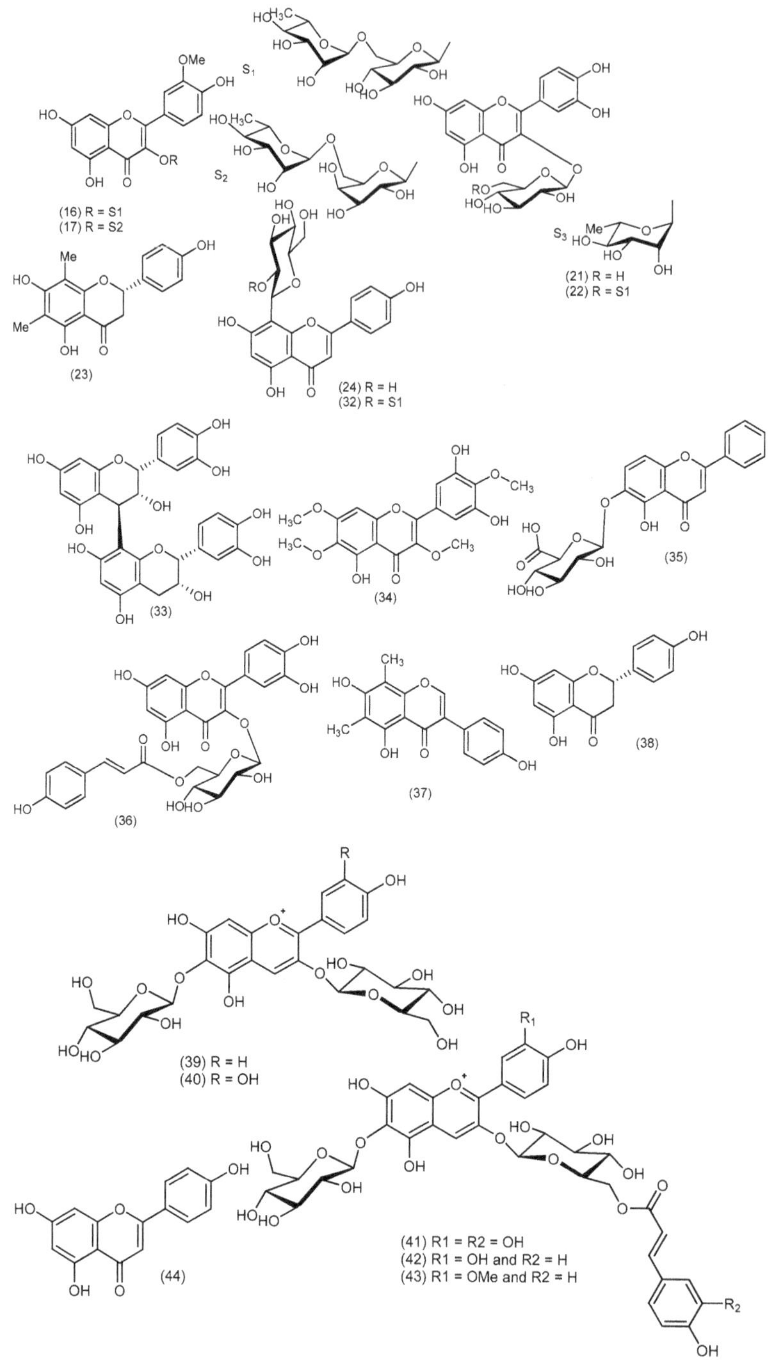


**Figure 4.** Flavonoids isolated from some plants of the genus *Cordyline*.

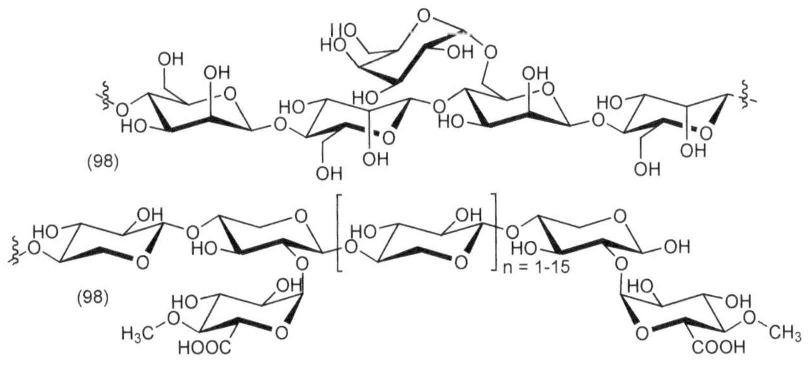

Figure 5. Polyosides from *Cordyline indivisa*.

## 5.2. *Cordyline fruticosa (Synonym: Cordyline terminalis)*

*Cordyline fruticosa* has been reported to contain cholestane glycosides, polyphenols, flavonoids, glucofructan, tannins, phytosterols, and steroidal alkaloid saponins. Three new steroidal saponins, spirosta-5,25(27)-diene-1$\beta$,3$\beta$-diol-1-*O*-$\alpha$-L-rhamnopyranosyl-(1→2)-$\beta$-D-fucopyranoside (fruticoside H) (**18**), 5$\alpha$-spirost-25(27)-ene-1$\beta$,3$\beta$-diol-1-*O*-$\alpha$-L-rhamnopyranosyl-(1→2)-(4-*O*-sulfo)-$\beta$-D-fucopyranoside (fruticoside I) (**19**), and (22S)-cholest-5-ene-1$\beta$,3$\beta$,16$\beta$,22-tetrol-1-*O*-$\beta$-D-galactopyranosyl-16-*O*-$\alpha$-L-rhamnopyranoside (fruticoside J) (**20**) were reported from the leaves of *Cordyline fruticosa* (Figure 2) in addition to the flavonoids quercetin 3-*O*-$\beta$-D-glucopyranoside (**21**), quercetin 3-rutinoside (**22**), Farrerol (**23**) and vitexin (**24**) (Figure 4) [7].

A new sulfated steroidal derivative fruticogenin A: 1-sulfo-australigenin-3-sodium sulphate, (**25**) (Figure 3) and three new steroidal saponins, namely, fruticoside K (3-sulfo-spirostan-25(27)-ene-1$\beta$,3$\beta$-diol-1-*O*-[$\alpha$-L-rhamnopyranosyl-(1→4)-$\beta$-D-fucopyranoside], (**26**), fruticoside L (3-sulfo-spirostan-25(27)-ene-1$\beta$,3$\beta$,6$\alpha$-triol-1-*O*-[$\alpha$-L-rhamnopyranosyl-(1→4)-$\beta$-D-fucopyranoside], (**27**) and fruticoside M (spirostan-25(27)-ene-1β,3α-diol-1-*O*-[α-L-rhamnopyranosyl-(1→2)-α-L-rhamnopyranoside], (**28**) were isolated from the aerial parts of *C. fruticosa* (Figure 2) [11]. Other reports have shown that these leaves contain apigenin (**29**) [52] as well as 12 other compounds, including four known saponins, namely, fruticoside M (**13**), fruticoside I (**15**), 26-*O*-$\beta$-D-glucopyranosyl-22-*O*-methyl-(25*S*)-5$\alpha$-furostane-1$\beta$,3$\alpha$,22$\varepsilon$,26-tetrol 3,26-*O*-$\beta$-D-glucopyranoside (**30**), and 26-*O*-$\beta$-D-glucopyranosyl-22-*O*-methyl-(22S)-5 afurostane-3$\alpha$,22$\varepsilon$-26-triol-3-*O*-$\beta$-D-glucopyranoside (**31**) and eight flavonoids, Farrerol (**23**), Vitexin 2″-*O*-rhamnoside or (5,7,4′-trihydroxyflavone 2″-*O*-$\alpha$-L-rhamnopyranosyl-8-*C*-$\beta$-D-glucopyranoside) (**32**), procyanidin B2 (**33**), 5,3′,5′-Trihydroxy-3,6,7,4′-tetramethoxyflavone (**34**), baicalein-6-*O*-$\beta$-glucuronopyranoside (**35**), quercetin-3-*O*-[6-trans-*p*-coumaryl]-$\beta$-D-glucoside (Helichrysoside) (**36**), 4′,5,7-trihydroxy-6,8-dimethylisoflavone (**37**), and naringenin (**38**) [53].

From the purple flowers of this plant, five anthocyanidins derivatives, namely, cyanidin 3,5-di-*O*-$\beta$-D-glucopyranoside (**39**), peonidin 3,5-di-*O*-$\beta$-D-glucopyranoside (**40**), cyanidin 3-*O*-$\beta$-D-(6″-*O*-*E*-p-caffeoylglucopyranoside)-5-*O*-$\beta$-D-glucopyranoside (**41**), cyanidin 3-*O*-$\beta$-D-(6″-*O*-*E*-p-coumaroylglucopyranoside)-5-*O*-$\beta$-D-glucopyranoside (**42**), and peonidin 3-*O*-$\beta$-D-(6″-*O*-*E*-p-coumaroylglucopyranoside)-5-*O*-$\beta$-D-glucopyranoside (**43**) have been isolated (Figure 4) [54]. Another study conducted on *Cordyline fruticosa* led to the isolation of four new cholestane glycosides, namely (22*S*)-3$\beta$,7$\beta$,16$\beta$,22-tetrahydroxycholest-5-en-1$\beta$-yl $\beta$-D-glucopyranoside (**44**), (22*S*)-3$\beta$,16$\beta$,22,25-tetrahydroxycholest-5-en-1$\beta$-yl $\beta$-D-glucopyranoside (**45**), (22S)-3$\beta$,16$\beta$,22,25-tetrahydroxy-5$\alpha$-cholestan-1$\beta$-yl $\beta$-D-glucopyranoside (**46**), and (22*S*)-16$\beta$,22,25-trihydroxycholest-5-en-3$\beta$-yl *O*-$\alpha$-L-rhamnopyranosyl-(1→2)-$\beta$-D-glucopyranoside (**47**) [55].

A recent study conducted on a C. *fruticose* cultivar called 'Fairchild red' from Vietnam, led to the isolation of twelve previously undescribed steroidal glycosides. Their structures were elucidated as spirostanol glycosides, 5$\alpha$-spirost-25(27)-ene-1$\beta$,3$\beta$,4$\alpha$-triol 1-*O*-$\beta$-

D-fucopyranoside (**48**), 5α-spirost-(25)27-ene-1β,3β,4α-triol 1-*O*-β-D-xylopyranoside (**49**), 5α-spirost-(25)27-ene-1β,3β,4α-triol 1-*O*-α-L-rhamnopyranosyl-(1→2)-β-D-fucopyranoside (**50**), 5α-spirost-(25)27-ene-1β,3β,4α-triol 1-*O*-α-L-rhamnopyranosyl-(1→2)-(4-O-sulfo)-β-D-fucopyranoside (**51**), 5α-spirost-25(27)-ene-1β,3β-diol 1-*O*-α-L-rhamnopyranosyl-(1→2)-β-D-fucopyranoside (**52**), and 5α-spirost-25(27)-ene-1β,3β-diol 1-*O*-α-L-rhamnopyranosyl-(1→2)-α-L-arabinopyranoside (**53**). Furostanol glycosides were also isolated as 26-*O*-β-D-glucopyranosyl-5α-furost-(25)27-ene-1β,3β,4α,22α,26-pentol 1-*O*-β-D-fucopyranoside (**54**), 26-*O*-β-D-glucopyranosyl-22α-methoxy-5α-furost-(25)27-ene-1β,3β,4α,26-tetrol 1-*O*-β-D-fucopyranoside (**55**), 26-*O*-β-D-glucopyranosyl-5α-furost-(25)27-ene-1β,3β,22α,26-tetrol 1-*O*-β-D-glucopyranoside (**56**), 26-*O*-β-D-glucopyranosyl-5α-furost-(25)27-ene-1β,3β,22α,26-tetrol 1-*O*-α-L-rhamnopyranosyl-(1→2)-β-D-glucopyranoside (**57**), 26-*O*-β-D-glucopyranosyl-5α-furost-(25)27-ene-1β,3β,22α,26-tetrol 1-*O*-α-L-rhamnopyranosyl-(1→2)-β-D-fucopyra noside (**58**), and 26-*O*-β-D-glucopyranosyl-22α-methoxy-5α-furost-(25)27-ene-1β,3β,26-triol 1-*O*-α-L-rhamnopyranosyl-(1→2)-β-D-fucopyranoside (**59**) (Figure 2) [56].

### 5.3. *Cordyline manners-suttoniae*

The chemical investigation of *C. manners-suttoniae* fruits led to the isolation and characterisation of a total of 12 secondary metabolites, 11 of which are the steroidal saponins (1*R*,3*R*,5*S*,22*S*,25*R*)-3-hydroxyspirostanyl-1-*O*-β-D-fucopyranoside (**60**), (1*R*,3*R*,5*S*,22*S*, 25*R*)-3-hydroxyspirostanyl-1-*O*-β-L-arabinopyranoside (**61**), (1*R*,3*S*,5*S*,22*S*,25*R*)-3-hydroxyspirostanyl-1-*O*-β-D-fucopyranoside (**62**), (1*R*,3*S*,4*S*,5*R*,22*S*,25*R*)-3,4-dihydroxyspirostanyl-1-*O*-β-D-fucopyranoside (**63**), (1*R*,3*R*,4*S*,5*R*,22*S*,25*R*)-3,4-dihydroxyspirostanyl-1-*O*-β-D-fucopyranoside (**64**), (1*R*,3*R*,5*S*,22*S*)-3-hydroxyspirost-25(27)-enyl-1-*O*-β-D-fucopyranoside (**65**), (1*R*,3*R*,5*S*,22*S*)-3-hydroxyspirost-25(27)-enyl-1-*O*-β-L-arabinopyranoside (**66**), (1*R*,3*S*,5*S*,22*S*)-3-hydroxyspirost-25(27)-enyl-1-*O*-β-D-fucopyranoside (**67**), (1*R*,3*S*,4*S*,5*R*,22*S*)-3,4-dihydroxyspirost-25(27)-enyl-1-*O*-β-D-fucopyranoside (**68**), and (1*R*,3*R*,4*S*,5*R*,22*S*)-3,4-dihydroxyspirost-25(27)-enyl-1-*O*-β-D-fucopyranoside (**69**) trivially named suttonigenins A-J, respectively, were reported for the first time, one reported brisbagenin 1-*O*-β-D-fucopyranoside (**70**) and one reported sapogenin (**6**) (Figures 2 and 4) [8].

### 5.4. *Cordyline stricta*

Phytochemical studies have been carried out on the leaves of *C. stricta* that resulted in the identification of four new spirostanol saponins, namely 25*S*-5α-spirostane-1β, 3α-diol 3-*O*-β-D-glucopyranoside (**71**), 5α-spirost-25(27) ène-1β,3α-diol 3-*O*-β-D-glucopyranoside (**72**), 25*S*- 5α-spirost-20- ene-1β,3α-diol 3-*O*-β-D-glucopyranoside (**73**), and 25*R*-5α-spirostane-1β, 3α,25-triol 3-*O*-β-D-glucopyranoside (**74**), three new furostanol saponins, 26-*O*-β-D-glucopyranosyl-22-*O*-methyl-25*S*-5α-furostane-3α-22γ-26 triols 3-*O*-β-D-glucopyranoside (**75**), 26-*O*-β-D-glucopyranosyl-22-*O*-methyl-25*S*-5α-furostane-1β-3α-22γ-26 tetrol 3-*O*-β-D-glucopyranoside (**76**), and 26-*O*-β-D-glucopyranosyl-5α-furost-20 (22)-ene1β-3α-26 triol 3-*O*-β-D-glucopyranoside (**77**), and a new pregnane glucoside 1β-3α-dihydryoxy-5α-pregn-16-en-20-one-3-*O*-β-D-glucopyranoside (**78**) (Figure 2) [9]. In another study, three new spirostanol saponins and two new furostanol saponins were isolated from the fresh leaves of *C. stricta*. Two of the isolated saponins contained a new branched triglycoside moiety assigned as *O*-α-L-rhamnopyranosyl-(1→2)-*O*-[β-D-xylopyranosyl-(1→3)]-β-D-xylopyranose with the formation of an *O*-glycosidic linkage to C-1 of the aglycone; namely 5α-spirost-25(27)-ene-1β, 3α- diol 1-*O*-{*O*-α-L-rhamnopyranosyl-(1, 2)-*O*-[β-*O*- xylopyranosyl-(1,3)]-β-D-fucopyranoside (**79**), (25*S*)-5α-spirostane- 1β, 3α-diol 1- *O*-{*O*-α-L-rhamnopyranosyl-(1, 2)-*O*-[β-D-xylo- pyranosyl-(1, 3)]-β-*O*-xylopyranoside (**80**), 5α-spirost-25(27)-ene-1β,3α-diol 1-*O*-{*O*-α-L-rhamnopyranosyl-(1→2)-*O*-[β-D-xylopyranosyl-(1→3)]-β-D-xylopyranoside} (**81**), 26-*O*-β-D-glucopyranosyl-22-*O*-methylfurosta- 5,25(27)-diene- 1 β,3β, 22α,26-tetrol 1-*O*-{*O*-α-L-rhamnopyranosyl-(1→2)-*O*-[β-D-xylopyranosyl-(1→3)]-β-*O*-fucopyranoside (**82**), and 26-*O*-β-D-glucopyranosyl-22-*O*-methylfurosta-5,25(27)-diene-1β,3β,22ξ,26-tetrol 1-*O*-{*O*-α-L-rhamnopyranosyl-(1→2)-*O*-[β-D-xylopyranosyl-(1→3)]-β-*O*-fucopyranoside} (**83**) (Figure 2) [4,57]. Further investigations on *Cordyline stricta* leaves have led to the isolation of four new steroidal sapogenins, crabbogenin {5α-spirost-25(27)-

en-3α-ol} (**84**), 1β-hydroxycrabbogenin (**85**), 1-β-hydroxycrabbogenin diacetate (**86**), strictagenin {(20*S*, 22*S*, 25*S*)-5a-furostan-22, 25-epoxy-lβ,3α-26-triol} (**87**), strictagenin triacetate (**88**) and pompeygenin ]{(25S)-5α-spirostane-1β,3α-25-triol} (**89**) (Figure 3) [6,51].

All the steroidal saponins presented here align with the results reported from studies on the Agavaceae family. In fact, this family is known to produce mainly steroidal saponins, as reviewed by Boyce and Tinto in 2006 [58]. In addition, one of the most widely studied genus of this family (the *Agave*) is reported to mostly contain steroidal saponins and sapogenins [59], and we noticed that the structures with $\Delta^{25(27)}$ are even more common in the genus *Cordyline* than in the genus *Agave*. The genus *Yucca* is also reported to produce such compounds [58]. This review presented several flavonoids isolated from plants of the genus *Cordyline*, while some research only highlighted the flavonoid profiles in the genus *Agave* [60].

### 5.5. *Cordyline rubra*

From the leaves of *C. rubra*, two new steroidal sapogenins 1β, 3α- dihydroxy-furost-5-ene (**90**) and 1β, 3α, 26- trihydroxy-5α-furostane (**91**) (Figure 3) [61] and three new steroidal sapogenin triols were isolated and identified as (22*S*,25*R*)-5α-furostan-22,25-epoxy-1β,3α,26-triol (rubragenin) (**92**), (22*R*,25*R*)-5α-furostan-22,25-epoxy-lβ,3α,26-triol (wallogenin) (**93**), and (22*R*,25*R*)-5α-spirostane-1β, 3α,25-triol (chenogenin) (**94**). Wallogenin is epimeric at C-22 to all the previously reported 'furanose' F-ring sapogenins. (Figure 3) [62].

### 5.6. *Cordyline cannifolia*

The first chemical study on *C. cannifolia* was undertaken in 1974 by Jewers et al. Cordylagenin (**95**) (1β,3α-dihydroxy-5α,22α,25β spirostane), a new steroidal sapogenin diol, was isolated from the leaves [48]. Other steroidal sapogeninins from the leaves of *C. cannifolia* and *C. stricta* were also reported. The sapogenin was obtained and identified as 1β,3α-dihydroxy-5α,25β-spirostane, brisbagenin (**5**), (Figure 3) [63].

### 5.7. *Cordyline dracaenoides*

Chemical analysis of *C. dracaenoides* revealed the presence of steroidal saponins and anthocyanidins in the stem, rhizome and root. Small amounts of tannins were also detected in its stem and rhizome. Other studies conducted on this plant have shown the existence of steroidal saponins, including brisbagenin (**5**) and brisbenone (**96**) [63,64].

### 5.8. *Cordyline indivisa*

The compound 4-*O*-methylglucuronoxylan (**97**) and a very small amount of a glucomannan (**98**) (Figure 5) have been isolated from the trunk of the tropical liliaceous tree *C. indivisa* (Steud.) [65].

The following figures summarise all reported compounds from the *Cordyline* genus.

## 6. Biological Activities

Phytochemical investigations have revealed the presence of several constituents belonging to different classes of compounds. Due to their antibacterial, antifungal, antiparasitic, and antioxidant activities, the genus *Cordyline* has mainly been investigated for its properties already observed in the natural environment. Bioactivities reported for *Cordyline* plant extracts and their pure compounds are summarised by activity in this section.

### 6.1. *Antimicrobial Activity*

The antimicrobial activity of fruticoside H (**18**), fruticoside I (**19**), and fruticoside J (**20**) was assessed against *Staphylococcus aureus, Escherichia coli, Pseudomonas aeruginosa, Enterococcus faecalis,* and *Candida albicans.* The three isolated saponins, even at the highest concentration used in the test (256 mg/L), were not able to inhibit growth except for compound **19** (Fructicoside I) against *E. faecalis.* In this case, the minimum inhibitory concentration (MIC) value was 128 mg/L, which, although still high, is indicative of a specific antibacterial activity. The minor structural difference between **18** and **19** suggested that the presence of the $SO_3H$ group at C-4′ and the absence of the double bond at C-5($\Delta^{5,6}$) can significantly decrease the MIC value (more than 256 μg/mL for **18** and 128 μg/mL for **19**) of the molecules against the enterococcal species [7]. The *n*-hexane, ethyl acetate, and methanol extracts of *C. fruticosa* were evaluated for their antibacterial capacity against Gram-negative *Escherichia coli* and *Salmonella typhi* as *Staphylococcus aureus* and Gram-positive *Bacillus subtilis* by the microdilution method. The ethyl acetate fraction was found to be the most active against *Salmonella typhi* (inhibition diameter ID 15.2 mm) and *S. aureus* (ID 14.2 mm) in comparison to penicillin (ID 83.5 and 73.2 mm), tetracycline (ID 76.8 and 72.1 mm) and streptomycin (ID 77.2 and 71.4 mm) used as standards [14]. The crude methanol extract of *C. fruticosa* showed a faint zone of inhibition against *E. fecalis* [13]. Firoj et al. (2003) reported antibacterial activity in methanol, chloroform, acetone, and hexane extracts of *Cordyline fruticosa*.

Moderate antibacterial activity was observed for methanol extracts against *Shigella boydii*, *Staphylococcus epidermis*, and *Streptococcus pyogenes*, with inhibition zone values of 14, 12, and 13 mm, respectively, in comparison to Kanamycin (38, 32, and 40, respectively). Hexane extracts have exhibited mild antibacterial activity against *Escherichia coli*, *Salmonella typhi*, *Shigella boydii*, and *Shigella dysenteriae* (with inhibition zone values of 8, 9, 8, and 9 mm, respectively) relative to 39, 30, 38, and 39 mm showed by Kanamycin. In contrast, acetone and chloroform extracts showed no significant activities [66].

Prihambodo et al. (2019) investigated the antibacterial potential of *C. fruticosa*. They found that the MeOH extract exhibited interesting activity against *Escherichia coli* and *Salmonella tyhpimurium*, with an inhibitory area of 6.72 ± 0.95 mm and 6.52 ± 1.21 mm, respectively, indicating that this extract possesses medium antibacterial activity (5–10 mm) [67]. *n*-Hexane, ethyl acetate, and methanol extracts were also investigated for their potential antibacterial activity by Elfita et al. (2019). The methanol extract displayed the highest antibacterial activity using the Kirby–Bauer method against Gram-negative (*Salmonella typhi*, ID 15.1 mm; *Escherichia coli*, ID 16.7 mm) and Gram-positive (*Staphylococcus aureus*, ID 17.0 mm; *Bacillus subtilis*, ID 12.5 mm) compared to 19.4, 19.5, 21.1, and 20.6 mm, respectively, for tetracycline [68].

The anticandidal activity of *C. fruticosa* leaf extracts was examined using agar well diffusion against *Candida albicans*. *N*-hexane, ethyl acetate, and ethanol extracts showed weak inhibition zones of 13, 8 and 12 mm, respectively, at a concentration of 400 μg/well, compared to 100 mm of the Chloramphenicol [69].

Antibacterial activities were assayed for fruticoside H (**18**), fruticoside I (**19**), and fruticoside J (**20**). Of these, only fruticoside H (**18**) displayed moderate antibacterial activity against *Enterococcus faecalis* (MIC = 128 μg/mL) [7]. The isolated compounds from the fruit of *Cordyline manners-suttoniae* (**6**, **60–69**) were assayed for their antibacterial activity by using a microdilution method against *Enterococcus faecium* (ATCC 35667 and C15), *Staphylococcus aureus* (ATCC 25923 and ATCC 29247), *Klebsiella pneumoniae* (ATCC 13883 and ATCC 12657), *Acinetobacter baumannii* (ATCC 19606 and ATCC 17978), *Pseudomonas aeruginosa* (ATCC 10145 and ATCC 49189), and *Enterobacter aerogenes* (ATCC 13048 and ATCC 13047). Suttonigenin A (**60**), suttonigenin C (**62**), suttonigenin D (**63**), suttonigenin E (**64**), suttonigenin G (**66**), suttonigenin H (**67**), suttonigenin I (**68**), and suttonigenin J (**69**) were found to inhibit the growth of the Gram-positive bacteria (*S. aureus* and *E. faecium*) with $MIC_{75}$ values ranging from 5.4 to 84.3 μM relative to 9.7 μM of Chloramphenicol (positive control). A dichloromethane extract also displayed antibacterial activity against these strains [8].

### 6.2. Antioxidant Activity

Antioxidant activity measured using the 2,2-diphenyl-1-picryl-hydrazyl (DPPH) method of crude extract of *C. fruticosa* purple flowers and isolated compounds showed the potential to scavenge free radicals. In fact, the methanol crude extract displayed an $IC_{50}$ of 13.1 μg/mL, while isolated compounds (**39–43**) showed $IC_{50}$ in the range of 13.8

to 16.4 μg/mL compared to that of quercetin (4.5 μg/mL) used as a positive control [54]. Antioxidant activity of encapsulated (with sodium caseinate beads) and unencapsulated phenolic content of *C. fruticosa* methanolic extract were assessed for their free radical scavenging capacity per gram of ascorbic acid equivalent (AAE). The encapsulated extracts showed good antioxidant activity after one year of conservation ($EC_{50}$ 27.38 mg/g AAE) in comparison to the unencapsulated extracts ($EC_{50}$ 13.4 mg/g AAE) [70]. A similar method was investigated on the methanol extract of *C. fruticosa* leaves; this extract was combined with copper oxide to form nanoparticles. Both crude extract and nanoparticles exhibited antioxidant activity following the DPPH method, with ascorbic acid used as a positive control [71]. As has been reported with several medicinal plants [72,73], this antioxidant activity is linked to the flavonoid content of *Cordyline fruticosa*.

### 6.3. Antiproliferative Activity

The antiproliferative effect of steroidal saponins is well reported and is known to affect cancer cell lines following several mechanisms such as apoptosis, differentiation, senescence, autophagy, oncosis, and other pathways [74–76]. The effect of fruticoside H-J (**18–20**) on the viability of the MDA-MB 231 human breast adenocarcinoma cell line, HCT 116 human colon carcinoma cell line and A375 human malignant melanoma cell line were evaluated in vitro using the MTT assay. Fruticoside H and I (**18** and **19**) showed moderate antiproliferative activity, with $IC_{50}$ values ranging from 37.83 to 69.68 μM for **18** on A375 and MDA-MB 231 cells, respectively, and from 46.59 to 59.97 μM for **19** on A375 and HCT116 cells, respectively. Fruticoside J (**20**) was inactive on the three tumour cell lines tested [7]. Using the same method against the same cell lines, investigators found that the steroidal saponins Fruticoside K and M (**26** and **28**) displayed weak cytotoxicity against melanoma (A375), breast adenocarcinoma (MDA-MB-231), and colon carcinoma (HCT116) human tumour cell lines [11].

Cytotoxicity of compounds **6**, **60**, and **69** was also evaluated against the cancer cell lines of human keratinocytes and human dermal fibroblasts. They all exhibited cytotoxicity with $IC_{50}$ values ranging from 18.2 to 74.9 μM for human keratinocyte cells and 4.2 to 16.5 μM for human dermal fibroblast cells [8].

Compounds **49–59** were evaluated for their cytotoxicity using the MTS method [3-(4,5-dimethylthiazol-2-yl)-5-(3-carboxymethoxyphenyl)-2-(4-sulfophenyl)-2H-tetrazolium] against the 4T1 cell line from a mouse mammary gland tissue, an animal model for stage IV human breast cancer. Compounds **49**, **54**, and **55** displayed good cytotoxic activity with the respective $IC_{50}$ values of 1.39, 1.52, and 3.24 μM, compared to 5-FU ($IC_{50}$ 6.70 Mm) [58]. The results obtained for **49**, **54**, **55**, and **69** seem to contradict those of Zhu et al. (2015), who reported that anticancer activity increased with the number of sugars. Several compounds tested with a high number of sugars exhibited low activity, and it is possible that the aglycone moieties could have contributed to this activity (likely the double bonds $\Delta^{5-6}$ and $\Delta^{25(27)}$). Additionally, the hydroxyl at C-3 could have had an effect [77]. This is in alignment with compound **20**, which had no activity, and the presence of the double bond $\Delta^{5-6}$ and the absence of the double bond $\Delta^{25(27)}$) could explain this inactivity.

### 6.4. Hypolipidemic Potential, Antiparasitic Activity, Nephrotoxicity and Hepatotoxicity Preventions

The methanol extract of the rhizomes of *C. fruticosa* has been investigated for its hypolipidemic potential in Wistar rats. Intake of 30 mg/Kg and 70 mg/Kg was found to lower lipid plasma levels in mice [78]. This extract showed the potential to lower blood fatty acid, glucose, and triglyceride levels in obese Wistar rats [79]. The *n*-butanol-soluble fraction of the methanol extract of senescent leaves of *C. australis* was assessed for its antiparasitic activity; it displayed a weak inhibition against *Plasmodium falciparum*, *Tripanosoma brucei*, and *Tripanosoma cruzi*, with an $IC_{50}$ of 12, 23 and 32 μM, respectively, relative to 0.018 μM of Chloroquine [10]. The aqueous extract of *C. fruticosa* also displayed antitrypanosomial activity by slowing the mitotic cycle in *Trypanosoma evansi*, with an $IC_{50}$ of 48.1 µg/mL [80].

The effect of *C. fruticosa* leaf extract against doxorubicin-cisplatin-induced nephrotoxicity and hepatotoxicity in rats was assessed. Administration of *C. fruticosa* extract induced a significant decrease in the aspartate aminotransferase, alanine amino transferase, alkaline phosphatase, urea and creatinine in doxorubicin-treated rats. It protected the renal tissues from necrosis induced by doxorubicin. The administration of the extract 10 days before cisplatin ingestion in the preventive process contributed to a pronounced protection against hepatic injury induced by cisplatin [55].

The effect of *C. fruticosa* leaves' methanol extract was studied on castor oil-induced diarrhoea and magnesium sulphate-induced diarrhoea. The extract significantly reduced the frequency of magnesium sulphate-induced diarrhoea at the dose of 800 mg/Kg of body weight [81].

Based on the number of studies that have reported the biological activities of steroidal saponins, we recommend that future biological assays address their effect on inflammatory cells, nervous cells, and fungi. The study of the action mechanism of the double bond $\Delta^{25(27)}$ could be carried out to shed light on their effect on cancer cells.

## 7. Toxicology

In a brine shrimp lethality test after 24 h, surviving brine shrimp nauplii were counted, and LC50 was assessed. The methanol extract of *C. fruticosa* exhibited considerable toxicity, with an LC50 value of 355.7 µg/mL, compared to a standard of vincristine sulfate (LC50 of 3.8 µg/mL). This result indicates that the methanol extract has a cytotoxicity effect and it could have antitumour molecules [81].

## 8. Conclusions

Due to their versatile traditional uses, *Cordyline* species have been extensively studied for their phytoconstituents as well as for their biological activities. Flavonoid, spirostane, furostane, and cholestane glycosides have been isolated from seven *Cordyline* species, while *C. indivisa* remains largely unstudied. In summary, the plant extracts and some of their pure compounds possess a broad spectrum of biological activities such as antibacterial, antifungal, antiproliferative, antiparasitic, and hypolipidemic effects. Saponins, sapogenins, and flavonoids are the most frequently isolated and identified classes of compounds in the species *C. australis*, *C. fruticosa*, *C. manners-suttoniae*, *C. stricta*, *C. rubra*, *C. cannifolia*, and *C. dracaenoides*, and currently constitute the only investigated and identified phytochemicals. *Cordyline* plants have been and continue to be used in traditional medicine against certain infections caused by bacteria, fungi, and parasites. This review of all available literature reveals that steroidal saponins, steroidal sapogenins, and flavonoids are the biomarkers of plants belonging to the *Cordyline* genus and present in one view the wide opportunities for the use of *Cordyline* plants in drug discovery with regard to bioactive molecules not yet characterised from this genus. The bioassay-guided isolation and identification of the bioactive components of plants of the genus *Cordyline* remain un-investigated, and more detailed in silico studies are required to propose detailed structure–activity relationships in order to determine the mechanisms of action of saponin and flavonoid constituents.

Although some compounds isolated from plants of the genus *Cordyline* have exhibited interesting biological activities, much more work is needed to exploit them in the pharmaceutical industry, as they were generally obtained in small amounts. Synthesis of some of these secondary metabolites from readily available reactants should be considered by pharmacochemists in order to obtain amounts sufficient for use in drug development.

**Author Contributions:** Writing—original draft preparation, conceptualisation, methodology, and supervision, R.T.F.; data collection, writing—original draft preparation, B.T.; data collection, A.T.D.; visualisation and original draft preparation, H.P.D.F.; conceptualisation, writing—original draft preparation, B.K.P. and R.B.T.; visualisation and validation, M.P., P.H.D., L.A.T. and X.S.-N. All authors have read and agreed to the published version of the manuscript.

**Funding:** X. Siwe-Noundou is grateful to the National Research Foundation / Competitive Support for Unrated Researchers (NRF/CSUR) Grant. Reference Number: SRUG2203291031.

**Institutional Review Board Statement:** Not applicable.

**Informed Consent Statement:** Not applicable.

**Data Availability Statement:** No new data were created or analysed in this study. Data sharing is not applicable in this article.

**Conflicts of Interest:** The authors declare no conflict of interest.

## References

1. Porter, R.J.; Medrum, B.S. Antiseizure drug. In *Basic & Clinical Pharmacology*, 12th ed.; Katzung, B.G., Masters, S.B., Trevor, A.J., Eds.; Prentice Hall International (UK) Ltd: London, UK, 1996; pp. 331–342.
2. Muazu, J.; Kaita, A.H. A review of traditional plants used in the treatment of epilepsy amongst the Hausa/Fulani tribes of northern Nigeria. *Afr. J. Tradit. Complement. Altern. Med.* **2008**, *5*, 387–390. [CrossRef] [PubMed]
3. Newman, D.J.; Cragg, G.M. Natural products as sources of new drugs over the 30 years from 1981 to 2010. *J. Nat. Prod.* **2012**, *75*, 311–355. [CrossRef] [PubMed]
4. Cragg, G.M.; Pezzuto, J.M. Natural products as a vital source for the discovery of cancer chemotherapeutic and chemopreventive agents. *Med. Princ. Pract.* **2016**, *25*, 41–59. [CrossRef] [PubMed]
5. Sofi, M.S.; Sateesh, M.K.; Bashir, M.; Ganie, M.A.; Nabi, S. Chemopreventive and anti-breast cancer activity of compounds isolated from leaves of *Abrus precatorius* L. *Biotech* **2018**, *8*, 371. [CrossRef] [PubMed]
6. Blunden, G.; Jaffer, A.J.; Jewers, J.K.; Griffin, W.J. Steroidal sapogenins from leaves of *Cordyline* species. *J. Nat. Prod.* **1981**, *44*, 441–447. [CrossRef]
7. Fouedjou, R.T.; Teponno, R.B.; Quassinti, L.; Bramucci, M.; Petrelli, D.; Vitali, L.A.; Fiorini, D.; Tapondjou, L.A. Steroidal saponins from the leaves of *Cordyline frutcosa* (L) A. Chev and their cytotoxic and antimicrobial activity. *Phytochem. Lett.* **2014**, *7*, 62–68. [CrossRef]
8. Tran, T.D.; Olsson, M.A.; Choudhury, M.A.; McMillan, D.; Cullen, J.; Parsons, P.J.; Bernhardt, V.P.; Reddell, P.W.; Ogbourne, M.S. Antibacterial 5$\alpha$-Spirostane saponins from the fruit of *Cordyline manners-suttoniae*. *J. Nat. Prod.* **2019**, *82*, 2809–2817. [CrossRef] [PubMed]
9. Mimaki, Y.; Kuroda, M.; Akaashi, Y.; Sashida, Y. Steroidal glucosides from leaves of *Cordyline stricta*. *Phytochemistry* **1997**, *45*, 1229–1234. [CrossRef]
10. Korkashvili, T. Steroidal glycosides of *Cordyline australis*. Master's Thesis, Philosophy in Chemistry, University of Waikato, Hamilton, New Zealand, 2006. Available online: http://hdl.handle.net/10289/2237 (accessed on 8 May 2022).
11. Ponou, K.B.; Teponno, B.R.; Tapondjou, A.L.; Lacaille-Dubois, M.; Quassinti, L.; Bramussi, M.; Barboni, L. Steroidal saponins from the aerial parts of *Cordyline fruticosa* L. Var strawberries. *Fitoterapia* **2019**, *134*, 454–458. [CrossRef]
12. Blunden, G.; Sitton, D.; Beach, S.; Turne, C.H. Australigenin, A new steroidal sapogenin from *Cordyline australis* fruits. *J. Nat. Prod.* **1984**, *47*, 266–269. [CrossRef]
13. Chunduri, J.R.; Shah, H.R. FTIR phytochemical fingerprinting and antioxidant analysis of selected indoor non-flowering indoor plants and their industrial importance. *Int. J. Curr. Pharm. Res.* **2016**, *8*, 37–43. [CrossRef]
14. Elfita, M.; Muharni; Mardiyanto; Fitrya. Chemical compounds from the antibacterial active fraction of *Cordyline fruticosa* (L). *Conf. Series: Earth Environ. Sci.* **2021**, *709*, 012048. [CrossRef]
15. Encyclopedia Britannica. Plant-Nomenclature-Taxonomy. Available online: https://www.britannica.com/biography/Michel-Adanson (accessed on 23 March 2022).
16. Mckenzie, E.H.C.; Buchanan, P.K.; Johnston, P.R. Checklist of fungi on cabbage trees (*Cordyline* spp.) and New Zealand flaxes (*Phormium* spp.) in New Zealand. *N. Z. J. Bot.* **2005**, *43*, 119–139. [CrossRef]
17. Harris, W.; Ross, E.; Beever, E.R.; Heenan, B.P. Phenotypic variation of leaves of and stems of wild stands of *Cordyline autralis* New *Zealand J. New Zealand J. Bot.* **1998**, *36*, 593–604. [CrossRef]
18. Ehrlich, C. The Ethnobotany of *Cordyline fruticosa* (L.) A. Chev.: The "Hawaiian Ti plant". Ph.D. Dissertation, State University at Buffalo, Buffalo, NY, USA, 1999.
19. Khan, S.; Naz, S.; Saeed, B. In vitro production of *Cordyline terminalis* for commercialization. *Pak. J. Bot.* **2004**, *36*, 757–761.
20. Kattoor, C.J. Micropropagation Studies in *Dracaena* and *Cordyline*. Master's Thesis, University of Agricultural Sciences, Dharwad, India, 2010.
21. Wilson, A. *Flora of Australia: Oceanic Islands, 49*; Australian Government Publishing Service: Canberra, Australia, 1994; ISBN -100644293853.
22. Conran, J.G. *Lomandraceae. The Families and Genera of Vascular Plants*; Kubitzki, K., Ed.; Volume 3: Flowering Plants Monocotyledons: Lilianae (Except Orchidaceae); Springer: Berlin/Heidelberg, Germany, 1998; pp. 354–365. ISBN 103642083773.
23. World Checklist of Selected Plant Families (WCSP). Available online: https://wcsp.science.kew.org/ (accessed on 1 November 2018).

24. Chifundera, K. Livestock diseases and the traditional medicine in the Bushi area. Kivu province, Democratic Republic of Congo. *Afr. Study Monogr.* **1998**, *19*, 13–33. [CrossRef]
25. Lense, O. The wild plants used as traditional medicines by indigenous people of Manokwari, West Papua. *Biodiversitas* **2012**, *13*, 98–106. [CrossRef]
26. Calixto, J.B.; Lima, T.C.M.; Morato, G.S.; Nicolau, M.; Takahashi, R.N.; Valle, R.M.R.; Schmidt, C.C.; Yunes, R.A. Chemical and pharmacological analysis of the crude aqueous/alcoholic extract from *Cordyline dracaenoides*. *Phytother. Res.* **1990**, *4*, 167. [CrossRef]
27. Melzer, M.; Sugano, J.; Uchida, J.; Kawate, M.; Borth, W.; Hu, J. Partial characterization of a novel emara-like virus from *Cordyline fruticosa* (L.) with ti ringspot disease. In Proceedings of the 2014 APS-CPS Joint Meeting, Minneapolis, MA, USA, 9–13 August 2014. [CrossRef]
28. Hinkle, A.E. Population structure of pacific *Cordyline fruticosa* (Laxmanniaceae) with implications for human settlement of Polynesia. *Am. J. Bot.* **2007**, *94*, 828–839. [CrossRef]
29. Whistler, W.A. Traditional and herbal medicine in the Cook Islands. *J. Ethnopharm.* **1985**, *13*, 239–280. [CrossRef]
30. Buttner, R. *Mansfeld's Encyclopedia of Agricultural and Horticultural Crops*; Hanelt, P., Ed.; Springer: Berlin/Heidelberg, Germany, 2001.
31. Dalimartha, S. *Atlas Tumbuhan Obat Indonesia*; Puspa Swara: Jakarta, Indonesia, 2007; Volume 4, ISBN 979323573X.
32. Nugent, J. *Permaculture Plants: Agaves and Cacti*, 2nd ed.; Sustainable Agriculture Research Institute: College Park, MD, USA, 2006.
33. Nombo, P.; Leach, J. *Reite Plants: An Ethnobotanical Study in Tok Pisin and English*; ANU E Press: Canberra, Australia, 2010.
34. Kulip, J. An ethnobotanical survey of medicinal and other useful plants of Muruts in Sabah, Malaysia. *Telopea* **2003**, *10*, 81–98. [CrossRef]
35. Kent, H.W. *Treasury of Hawaiian Words in One Hundred and One Categories. The Masonic Public Library of Hawaii*; University of Hawaii Press: Honolulu, HI, USA, 1995; 504p.
36. Ray, T.; Saha, P.; Roy, S.C. Commercial production of *Cordyline terminalis* (L) Kunth. From shoot apex meristem and assessment for genetic stability of somaclones by isozyme markers. *Sci. Hortic.* **2006**, *108*, 289–294. [CrossRef]
37. Brasch, D.J.; Fankhauser, B.L.; McDonald, A.G. A study of the glucofructofuranon from the New Zealand cabbage tree *Cordyline australis*. *Carbohydr. Res.* **1988**, *180*, 315–324. [CrossRef]
38. Simpson, P. *Dancing Leaves. The Story of New Zealand's Cabbage Tree, TıKouka*; Christchurch, Canterbury University Press: Canterbury, New Zealand, 2000.
39. Aufenanger, H. The *Cordyline* plant in the central highlands of new Guinea. *Anthropos* **1961**, *56*, 393–408.
40. Merlin, M. The traditional geographical range and ethnobotanical diversity of *Cordyline fruticosa* (L.) Chevalier. *Ethnobotany* **1989**, *1*, 25–39.
41. Whistler, W.A. *Polynesian Herbal Medicine. Lawa'i, Kaua'i, Hawai'i*; National Tropical Botanical Garden: Hong Kong, China, 1992.
42. Elbert, L.L.J.; Roger, G.S. *Common Forest Trees of Hawaii (Native and Introduced)*; Agriculture Hanbook No. 679; US Department of Agriculture, Forest Service: Washington, DC, USA, 1989.
43. Lim, T.K. *Edible Medicinal and Non-medicinal Plants. Volume 12: Modified Stems, Roots, Bulbs*; Springer: Berlin/Heidelberg, Germany, 2015. [CrossRef]
44. Duncalf, I. Conventional propagation of Cordyline australis. *Act. Hortic.* **2014**, 63–66. [CrossRef]
45. Simpson, P. Sudden decline in cabbage trees (*Cordyline australis*). *Sci. Res. Intern. Rep.* **1991**, *107*, 1–19.
46. Marker, R.E.; Wagner, R.B.; Ulshafer, P.R.; Wittbecker, E.L.; Goldsmith, D.P.J.; Ruof, C.H. Sterols. CLVII. Sapogenins. 69. Isolation and structures of thirteen new steroidal sapogenins. New sources for known sapogenins. *J. Am. Chem. Soc.* **1943**, *65*, 1199–1209. [CrossRef]
47. Wall, M.E.; Eddy, C.R.; Willaman, J.J.; Correll, D.S.; Schubert, B.G.; Gentry, H.S. Steroidal sapogenins. XII. Survey of plants for steroidal sapogenins and other constituents. *J. Am. Pharm. Assoc.* **1954**, *43*, 503–505. [CrossRef]
48. Jewers, A.H.K.; Manchanda, J.D.; Nagle, M.J. Cordylagenin, a new steroidal saponin diol from *Cordyline cannifolia* and *Cordyline stricta*. *Tetrahedron Lett.* **1974**, *16*, 1475–1478. [CrossRef]
49. Blunden, G.; Jaffer, J.A.; Jewers, K.; Robinson, J.M. Configurational assignments to spirostan-3-ols by mass spectrometry. *Steroids* **1980**, *36*, 299–303. [CrossRef] [PubMed]
50. Griffin, W.J.; Maunwongyathi, P. A comparison of four species of *Cordyline*. *Planta Med.* **1969**, *4*, 346–360. [CrossRef] [PubMed]
51. Blunden, G.; Jaffer, J.A.; Jewers, K.; Griffin, W.J. Crabbogenin 1$\beta$-hydroxycrabbogenin, strictagenin and pompeygenin, four new steroidal sapogenins from *Cordyline stricta* leaves. *Tetrahedron* **1981**, *37*, 2911–2915. [CrossRef]
52. Djohan, H.; Widiyantoro, A.; Shofiyani, A. Isolation of flavonoid from Andong leaves (*Cordyline fruticosa* (L.) A. Chev.) and its activity as complexor of $Fe^{2+}$. *Indones. J. Pure Appl. Chem.* **2019**, *2*, 39–45. [CrossRef]
53. Raslan, M.; Taher, R.F.; Al-Karmalawy, A.A.; El-Ebeedy, D.; Metwaly, A.G.; Elkateeb, N.M.; Ghanem, A.; Elghaish, R.A.; El Maksoud, A.I.A. *Cordyline fruticosa* (L.) A. Chev. leaves: Isolation, HPLC/MS profiling and evaluation of nephroprotective and hepatoprotective activities supported by molecular docking. *New J. Chem.* **2021**, *45*, 22216–22233. [CrossRef]
54. Adaku, C.; Skaar, I.; Byamukama, R.; Jordheim, M.; Andersen, Ø.M. Anthocyanin profile and antioxidant property of anti-asthma flowers of *Cordyline terminalis* (L.) Kunth (Agavaceae). *Nat. Prod. Commun.* **2020**, *15*, 1–7. [CrossRef]

55. Yokosuka, A.; Suzuki, T.; Mimaki, Y. New cholestane glycosides from the l eaves of *Cordyline terminalis*. *Chem. Pharm. Bull.* **2012**, *60*, 275–279. [CrossRef]
56. Nguyen, D.H.; Mitaine-Offer, A.C.; Miyamoto, T.; Tanaka, C.; Bellaye, P.S.; Collin, B.; Chambin, O.; Lacaille-Dubois, M.A. Steroidal glycosides from the Vietnamese cultivar *Cordyline fruticosa* "Fairchild red". *Phytochemistry* **2021**, *192*, 112966. [CrossRef]
57. Yang, M.; Blunden, G.; Patel, A.; Crabb, T.A.; Griffint, W.J. Two furostane sapogenins from *cordyline rubra*. *Phytochemistry* **1990**, *29*, 1332–1334. [CrossRef]
58. Simmons-Boycea, J.L.; and Tinto, W.F. Steroidal saponins and sapogenins from the Agavaceae family. *Nat. Prod. Commun.* **2006**, *2*, 99–114. [CrossRef]
59. Sidana, J.; Singh, B.; Sharma, O.P. Saponins of *Agave*: Chemistry and bioactivity. *Phytochemistry* **2016**, *130*, 22–46. [CrossRef] [PubMed]
60. Almaraz-Abarca, N.; Delgado-Alvarado, E.A.; Ávila-Reyes, J.A.; Uribe-Soto, J.N.; González-Valdez, L.S. The phenols of the genus *Agave* (Agavaceae). *J. Biomater. Nanobiotechnol.* **2013**, *4*, 9–16. [CrossRef]
61. Mimaki, Y.; Kuroda, M.; Akaashi, Y.; Sashida, Y. Steroidal saponins from the leaves of *Cordyline stricta*. *Phytochemistry* **1998**, *47*, 79–85. [CrossRef] [PubMed]
62. Jewers, K.; Burbage, M.B.; Blunden, G.; Griffin, W.J. Brisbagenin and brisbenone, two new spirostanes from *Cordyline* species. *Steroids* **1974**, *24*, 203–208. [CrossRef] [PubMed]
63. Griffin, W.; Blunden, G.; Jewers, K.; Burbage, B.M.; Nagler, M.J. Steroidal sapogenins from *Cordyline cannifolia* leaves. *Phytochemistry* **1976**, *15*, 1271–1274. [CrossRef]
64. Yang, M.; Blunden, G.; Patel, A.; Crabb, T.A.; Griffint, W.J. Rubragenin, chenogenin and wallogenin, steroidal sapogenins from *cordyline rubra*. *Phytochemistry* **1989**, *28*, 3171–3173. [CrossRef]
65. Sieber, R. Die hemicellulosen von *Cordyline indivisa*. *Phytochemistry* **1972**, *11*, 1433–1441. [CrossRef]
66. Firoj, A.; Prabir, K.D.; Islam, M.A.; Rahman, K.M.; Rahman, M.M.; Selim, M.S.T. Antibacterial activity of *Cordylline terminalis Kunth* leaves. *J. Med. Sci.* **2003**, *3*, 418–422. [CrossRef]
67. Prihambodo, T.R.; Nahrowi; Jayanegara, A. Antibacterial activity and phytochemical content of silage juice from tropical herbal leaves. *IOP Conf. Ser. Mater. Sci. Eng.* **2019**, *546*, 042032. [CrossRef]
68. Elfita, E.; Mardiyanto; Fitrya; Larasati, J.E.; Julinar; Widjajanti, H.; Muharni. Antibacterial activity of *Cordyline fruticosa* leaf extracts and its endophytic fungi extracts. *Biodiversitas* **2019**, *20*, 3804–3812. [CrossRef]
69. Kusuma, I.W.; Sari, N.M.; Murdiyanto; Kuspradini, H. Anticandidal activity of several plants used by Bentian tribe in East Kalimantan, Indonesia. *AIP Conf. Proc.* **2016**, *1755*, 040002. [CrossRef]
70. Reddy, C.; Noor, A.; Sarada, N.C.; Vijayalakshmi, M.A. Antioxidant properties of *Cordyline terminalis* (L.) Kunth and *Myristica fragrans* Houtt. Encapsulated separately into casein beads. *Curr. Sci.* **2011**, *101*, 416–420.
71. Sharmila, D.; Kumar, R.S. Antioxidant potential of copper oxide nanoparticles using leaves of *Cordyline fruticosa*. *Drug Invent. Today* **2019**, *12*, 2440–2443.
72. Banjarnahor, S.D.S.; Artanti, N. Antioxidant properties of flavonoids. *Med. J. Indones.* **2014**, *23*, 239–244. [CrossRef]
73. Tsakem, B.; Tchuenguem, R.T.; Siwe-Noundou, X.; Ponou, K.B.; Dzoyem, J.P.; Teponno, R.B.; Krause, R.W.M.; Tapondjou, A.L. New bioactive flavonoid glycosides with antioxidant activity from the stem bark of *Olax subscorpioidea* Oliv. *Nat. Prod. Res.* **2022**, *37*, 1641–1650. [CrossRef] [PubMed]
74. Yang, C.-R.; Zhang, Y.; Jacob, M.R.; Khan, S.I.; Zhang, Y.-J.; Li, X.-C. Antifungal activity of C-27 steroidal saponins. *Antimicrob. Agents Chemother.* **2006**, *50*, 1710–1714. [CrossRef] [PubMed]
75. Escobar-Sánchez, M.L.; Sánchez-Sánchez, L.; Sandoval-Ramírez, J. Steroidal Saponins and Cell Death in Cancer, Chapter 15. In *Cell Death—Autophagy, Apoptosis and Necrosis*; IntechOpen: London, UK, 2015; pp. 331–352. [CrossRef]
76. Sobolewska, D.; Galanty, A.; Grabowska, K.; Makowska-Was, J.; Wro'bel-Biedrawa, D.; Podolak, I. Saponins as cytotoxic agents: An update (2010–2018). Part I—Steroidal saponins. *Phytochem. Rev* **2020**, *19*, 139–189. [CrossRef]
77. Zhu, L.; Tan, J.; Wang, B.; Guan, L.; Liu, Y.; Zheng, C. In-vitro antitumor activity and antifungal activity of pennogenin steroidal saponins from *Paris Polyphylla* var. yunnanensis. *Iran. J. Pharm. Res.* **2011**, *10*, 279–286. [PubMed]
78. Bogoriani, N.W.; Ariati, N.K. The activity of Bali Andong rhizome extract of *Cordyline Terminalis* Kunth as Hypolipidemia agent in Wistar rats with high-cholesterol diet. *Int. J. Pharm. Phytopharmacol. Res.* **2018**, *8*, 75–80.
79. Bogoriani, N.W.; Suaniti, N.M.; Putra, A.A.B.; Lestari, K.D.P. The activity of *Cordyline terminalis*'s leaf extract as antidiabetic in obese wistar rats. *Int. J. Pharm. Res. Allied Sci.* **2019**, *8*, 206–213.
80. Dyary, H.O.; Arifah, A.K.; Sharma, R.S.K.; Rasedee, A. Antitrypanosomal and cytotoxic activities of selected medicinal plants and effect of *Cordyline terminalis* on trypanosomal nuclear and kinetoplast replication. *Pak. Vet. J.* **2014**, *34*, 444–448.
81. Naher, S.; Aziz, A.M.; Akter, I.; Rahman, S.M.; Sajon, S.R.; Mazumder, K. Anti-diarrheal activity and brine shrimp lethality bioassay of methanolic extract of *Cordyline fruticosa* (L.) A. Chev. Leaves. *Clin. Phytosci.* **2019**, *5*, 15. [CrossRef]

*biomolecules*

MDPI

*Review*

# Biochemical Composition and Biological Activities of Date Palm (*Phoenix dactylifera* L.) Seeds: A Review

**Maryam Abdulraheem Alkhoori [1], Amanda Shen-Yee Kong [2], Mariam Nasser Aljaafari [1], Aisha Abushelaibi [3], Swee-Hua Erin Lim [1], Wan-Hee Cheng [4], Chou-Min Chong [5] and Kok-Song Lai [1,*]**

1 Health Sciences Division, Abu Dhabi Women's College, Higher Colleges of Technology, Abu Dhabi P.O. Box 41012, United Arab Emirates
2 School of Biosciences, University of Nottingham Malaysia, Jalan Broga, Semenyih 43500, Malaysia
3 Office of the Executive Campus Director, Abu Dhabi Colleges, Higher Colleges of Technology, Abu Dhabi P.O. Box 41012, United Arab Emirates
4 Faculty of Health and Life Sciences, INTI International University, Persiaran Perdana BBN, Putra Nilai, Nilai 71800, Malaysia
5 Department of Aquaculture, Faculty of Agriculture, Universiti Putra Malaysia, Serdang 43400, Malaysia
* Correspondence: lkoksong@hct.ac.ae

**Abstract:** Date palm (*Phoenix dactylifera* L.) is an essential agricultural crop in most Middle Eastern countries, and its fruit, known as dates, is consumed by millions of people. Date seeds, a by-product of the date fruit processing industry, are a waste product used as food for domestic farm animals. Date seeds contain abundant sources of carbohydrates, oil, dietary fiber, and protein; they also contain bioactive phenolic compounds that may possess potential biological properties. In addition, its rich chemical composition makes date seeds suitable for use in food product formulation, cosmetics, and medicinal supplements. This review aims to provide a discourse on the nutritional value of date seeds. The latest data on the cytotoxicity of date seed compounds against cancer cell lines, its ability to combat diabetes, antioxidant potential, antimicrobial effect, and anti-inflammatory activity will be provided, considering its potential to be a nutritional therapeutic agent for chronic diseases. Application of date seeds in the form of powder and oil will also be discussed.

**Keywords:** *Phoenix dactylifera*; date seed; antimicrobial; antioxidant; anticancer; antidiabetic; anti-inflammatory

## 1. Introduction

Date palm (*Phoenix dactylifera* L.) is widely cultivated in tropical and subtropical regions, notably in the Middle East and North Africa [1]. Globally, the gross production value of date fruits has been rising considerably since the 20th century, with the highest value seen at more than US $14 billion in 2020 [2]. Date palm trees and fruits drive the economic and social aspects of date-producing countries [3,4]. During the holy month of Ramadan, since the era of Prophet Mohammed (Peace Be Upon Him), Muslims popularly break their fasts by eating dates due to the nutritional and medicinal values mentioned in the Holy Quran sections [5].

Date seeds, a by-product of date fruit production, are usually treated as waste, utilized as animal feed or just disposed of [1]. However, date seeds comprise 10–15% of date fruit fresh weight, and are rich in phenolic compounds [6]. There have been extensive studies carried out on the effects of date seeds in terms of pharmacological activities, such as antioxidant, anti-inflammation, antidiabetic, antibacterial, and antiviral properties [7,8]. Owing to its health benefits, date seeds possess high potential as a nutritional therapeutic agent for several chronic diseases [9]. This paper reviews the date seed nutritional components, biological activities, and its application in medicine and other non-conventional uses.

## 2. Nutritional Components

### 2.1. Amino Acids

The amino acid composition in date seeds is variable depending on the stage of date seed maturation. The major amino acids contained in the date seed are glutamic acid (16.44 g/100 g), phenylalanine (5.93 g/100 g), and leucine (6.10 g/100 g) [10]. Meanwhile, Shina and colleagues (2013) reported that aspartic acid (1.72 g/10 g), alanine (1.2 g/100 g), and tyrosine (1.2 g/100 g) are the major non-essential amino acids, while leucine (1.7 g/100 g), lysine (1.1 g/100 g), and phenylalanine (1.08 g/100 g) are the major essential amino acids, following implementation of an automatic amino acid analyzer [11]. Furthermore, non-proteogenic amino acids can bind with antibodies to produce T lymphocytes, eliminate toxic compounds in the liver, and reduce the creatinine levels in the body, including g-amino-n-butyric acid and 1-aminocyclopropane-1- carboxylic acid [12,13].

### 2.2. Dietary Fibers

Studies have found that dates contain more dietary fibers than cereals [14]. Functional dietary fibers can be found in date seeds, such as tannin, lignin, pectin, and hemicellulose components. Attia and colleagues (2021) reported a higher fiber content in date seeds than in yellow corn and barley [15]. Date seeds contain fiber fractions at a range of 39.6% to 57.5% for acid detergent fiber (cellulose + lignin), 51.6% to 75.0% for neutral detergent fiber (hemicellulose + cellulose + lignin), 12.0% to 17.5% for hemicellulose, 26.1% to 42.5% for cellulose, and 7.21% to 11.0% for lignin.

### 2.3. Minerals and Vitamins

Techniques such as gas chromatography and high-performance liquid chromatography (HPLC) have shown that date seeds contain vitamins consisting of vitamin A, vitamin B1 (thiamine), vitamin B2 (riboflavin), vitamin C, vitamin E, vitamin K, folate, and different types of minerals, with the most abundant being potassium [16,17]. Other minerals include boron, calcium, cobalt, copper, iron, magnesium, phosphorus, fluorine, selenium, sodium, and zinc [18]. Of these, potassium, phosphorus, and iron are present in higher concentrations among the minerals. Owing to their low sodium ratio level, consumption of date seeds is recommended for people with hypertension. Hinkaew and colleagues (2021) reported higher angiotensin-converting enzyme inhibitory activity in date seeds from cell culture origin compared to their flesh. Overall, date seeds exhibited 58.69–61.36% inhibition towards the key enzyme in regulating hypertension. In addition, date seeds also have high fluorine content and can help prevent tooth decay [19].

Date oil extracted from its seed is rich in amino acids and vitamins which can significantly promote hair growth [20]. Minerals extracted from date seed oils can generate energy and restore cellular function and growth of Riboflavin [21]. A study in the United Arab Emirates (UAE) has experimented on 18 types of date seed oil, including Khalas, Lulu, Fard, Raziz, and Sokkery, with several vitamins significantly observed through ultra-performance liquid chromatography analysis [22]. It was found that most date seed oils contain high $\alpha$-tocopherol, followed by $\alpha$-tocopheryl acetate, $\gamma$-tocopherol, and, lastly, Vitamin K1.

### 2.4. Sugars

The soluble sugars in date seeds contain 3.5 g/kg of glucose, 3.8 g/kg of fructose, 3.7 g/kg of stachyose, 3.5 g/kg of sucrose, and 2.2 g/kg of galactose, as reported through UV-visible spectrophotometry [23]. Date seed sugar can control blood sugar levels in diabetic patients with its beneficial powder of insulin-producing properties [24]. El Fouhil and colleagues (2013) reported the potential hypoglycemic efficacy of date seeds by observing an increased level of serum c-peptide in date seed extract (DSE)-insulin-treated diabetic rats in comparison to only insulin-treated diabetic rats [25]. This suggested an increase in endogenous insulin secretion, as confirmed by a compensatory beta-cell hypertrophy seen through immunohistochemistry analysis.

### 2.5. Fatty Acids

Oils extracted from date seeds contain five major fatty acids—oleic, lauric, myristic, linoleic, and palmitic [26]. Through gas chromatography, Biglar and colleagues (2012) reported oleic–lauric oil as the major fatty acid with 33.38% to 51.40% content, followed by lauric acid, with 18.78% to 31.61% content. Myristic acid was reported as the third major fatty acid, followed by palmitic and linoleic acid.

Attia and colleagues (2021) highlighted the importance of date seed oil as a natural source of medium-chain fatty acids and its ability to enhance poultry growth and immunity level [15]. Fatty acid fractions of date seeds were reported in a range of 12.20% to 23.06% for lauric, 9.70% to 11.30% for myristic, 10.11% to 12.70% for palmitic, 1.56% to 3.56% for stearic, 35.1% to 45.80% for oleic, 8.10% to 11.00% for linoleic, and 0.37% to 0.80% for linolenic acids. These values were similar to those from corn, except for myristic and oleic fatty acids which were higher than in corn.

### 2.6. Phenolic Compounds

Date seeds have been found to contain high levels of total polyphenols in comparison to other fruit, such as grapes, nut seeds, and even their flesh [27]. Seven different phenolic acids—including caffeic acid, chlorogenic acid, p-coumaric acid, ferulic acid, gallic acid, syringic acid, and vanillic acid—have been found to exist in date seeds, with the highest content ranging between 109.87 and 141.72 mg/100 g and seen in p-coumaric acid [28]. In addition, analysis of HPLC with a diode-array detector (DAD) data has suggested that rutin represents the most abundant flavonoid in date seeds, with a range of 71.74 to 86.32 mg/100 g, followed by quercetin (23.71–34.06 mg/100 g), and luteolin (9.17–13.24 mg/100 g). Table 1 summarizes the nutritional components of date seeds appraised in this review.

**Table 1.** A summary of the highly extractable and valuable components of data seeds.

| Component | Content (%) | References |
|---|---|---|
| Protein | 2.3–6.4 | [29] |
| Fat | 5.0–13.2 | [29] |
| Ash | 0.82–1.14 | [30] |
| Carbohydrate | 2.43–4.65 | [30] |
| Dietary fiber | 22.5–80.2 | [29] |
| Phenolics | 5.072 | [27] |
| Magnesium | 0.058–0.090 | [30] |
| Calcium | 0.014–0.034 | [30] |
| Phosphorus | 0.110–0.134 | [30] |
| Potassium | 0.175–0.240 | [30] |
| Sodium | 0.008–0.013 | [30] |

## 3. Biological Activities of Date Seeds

Based on the different nutritional values present in date seeds, numerous studies have looked into their biological activities which might be useful for human health. These activities include antimicrobial, antioxidant, anticancer, antidiabetic, and anti-inflammatory, which will be discussed in the following chapter.

### 3.1. Antimicrobial Activities

The date seed possesses an antimicrobial feature, being able to invade the cytoplasmic membrane of the bacteria [23]. Alrajhi and colleagues (2019) evaluated the antibacterial activity of date seed powder (DSP) extracted using ethyl acetate and hexane [31]. It was found that methicillin-resistant *Staphylococcus aureus* (MRSA) has the highest sensitivity, with an inhibition zone of 20 mm diameter, and was more effective than the common antibiotic gentamycin, which showed only a diameter of 5 mm inhibition. Among the tested gram-negative bacteria, only *Proteus mirabilis* showed significant inhibition towards DSEs.

Anwar and colleagues (2022) recently reported the antimicrobial activities of methanol Ajwa DSEs against gram-negative and gram-positive bacteria [32]. Notable inhibition zones by 100 mg/mL of Ajwa date seed against *S. aureus*, *Escherichia coli*, *Klebsiella pneumonia*, *Pseudomonas aeruginosa*, and *Enterococcus faecalis* were observed at 19 mm, 15 mm, 16 mm, 14 mm, and 16 mm, respectively.

Jassim and colleagues (2010) found that the antiviral activity of DSEs against lytic *Pseudomonas* phage ATCC 14209-B1 led to inhibition of the infectivity of *Pseudomonas* phage ATCC 14209-B1 to *P. aeruginosa*, reduction of the phage function, and disruption of the bacterial lysis [33].

### 3.2. Antioxidant Activities

Anwar and colleagues (2022) recently found a strong antioxidant activity in Ajwa date seed methanol extracts comparable to that of ascorbic acid [32]. In addition, research has also been conducted to explore the beneficial antioxidant effects of date seeds based on their rich source of phenolic compounds [34]. Salomón-Torres and colleagues (2019) found that the polyphenolic content in the seed of 'Medjool' dates is ten times higher than that in the pulp [35]. Presence of phenolic compounds such as tannins, flavonoids, saponins, anthraquinones, and coumarins agreed with the Shinoda assay contributed to the antioxidant properties of date seeds, resulting in a lower $IC_{50}$ of 0.0046 g/L by employing 2,2-diphenyl-1-picrylhydrazyl (DPPH) free radical method. Dietary polyphenolics from date seeds may supply substantial antioxidants, promoting health and preventing diseases.

Bentrad and colleagues (2020) found an antioxidant property of date seeds extracted from the Bent Kbala cultivar using the DPPH method [36]. DSEs demonstrated the highest free radical inhibition rate (89.89 $\pm$ 0.02%) and antioxidant activity ($IC_{50}$ = 0.031 μg/mL) compared to other natural antioxidants. These antioxidant properties are associated with phenolic compounds, such as phenolic acids, catechin tannins, and flavonoids contained in the date seeds, and their ability to donate hydrogen to DPPH radicals effectively.

Hilary and colleagues (2021) investigated the total polyphenol content of DSE, DSP, and date seed Arabic pita bread (DSB). It was found that the polyphenol content increased following digestion of DSB [9]. Polyphenols protect cell constituents such as DNA, protein, and lipid against oxidative damage by eliminating free radicals. The DPPH assay and Trolox Equivalent Antioxidant Capacity (TEAC) assay reported an enhanced antioxidant power in DSB after gastric digestion. However, the Ferric Reducing Antioxidant Power (FRAP) assay and the TEAC assay recorded a reduction in the antioxidant activity of DSE at the end of intestinal digestion. The DPPH assay remained unchanged despite the decrease in total polyphenol content. In the case of DSP, only the FRAP assay revealed a reduction in the antioxidant power as the digestion progressed from the gastric to intestinal phase.

In contrast, Platat and colleagues (2019) reported a significant increase in antioxidant effect in human subjects with a single dose of DSE, DSP, and DSB [37]. After 8 and 24 h of ingestion of date seed products, protection against protein and lipid peroxidation was increased. Presence of metabolites such as hydroxybenzoic acids, vanillic acids, protocatechuic acids, and vanillic acid sulphate in the excreted urine validates the role of date seeds in antioxidant activity.

### 3.3. Anticancer Activities

Phytotherapeutic agents such as plant-based cancer treatment are the preferable option, since they are natural, readily available, easily assimilated into the body, and have fewer side effects and lower toxicity [38]. As demonstrated in experimental studies, date seeds can be used as an anticancer treatment since they can successfully restrain malignancies. Hilary and colleagues (2021) revealed the potential of DSEs in reducing the viability of cancer cell lines at high concentrations (1000 μg/mL or above) after long hours of exposure [9]. After 48 h of treatment, the anticancer effect of reduced viability was found in the triple-negative breast cancer cell line (MDA-MB-231), breast cancer cell line (MCF-7), and colon adenocarcinoma cell line (Caco-2). For liver cancer cell line (HepG2) and prostate adenocarcinoma cell line (PC-3), a significant reduction in cell viability was observed with ≥1000 μg/mL of DSE treatment after 24 h of treatment.

Khan and colleagues (2022) reported that *P. dactylifera* seed extracts played a role in inducing cellular apoptosis through intrinsic pathways in the MDA-MB-231 cell line, MCF-7 cell line, and HepG2 cell line [38]. After 24 h of treatment, the cancer cells displayed morphological changes such as non-adherent, detached, and rounded shapes. Apoptotic features were observed after 48 h of exposure. The time- and dose-dependent manner of DSEs were shown by cell viability tests with an observation of a significant reduction in cancer cell growth. This result is suggested to link with the synergistic effect of bioactive agents—rutin and quercetin—present in date seeds. The morphology variation and cell viability remained unchanged in the normal Vero cell line following 24 h and 48 h of exposure. To the best of our knowledge, there is scarce literature available concerning the in vivo experiments for the anticancer properties of date seeds. Hence, it is recommended for future researchers to investigate this issue.

### 3.4. Antidiabetic Activities

Date seed active compounds, such as polyphenols and phenolics, aid in antidiabetic activity by scavenging free radicals, repairing pancreatic cells, reducing lipid peroxidation, and inhibiting enzymes on intestinal glucose absorption [39]. Shakoor and colleagues (2020) found that DSEs regulate glucose homeostasis with its high phenolics and flavanols content [40]. DSEs inhibited two digestive enzymes, pancreatic α-amylase and intestinal α-glucosidase, which are responsible for starch digestion and modulation of the glucose uptake by HepG2. At 400 μg/mL and 900 μg/mL of DSE, maximum enzyme inhibition was observed in α-amylase and α-glucosidase. Increased glucose uptake by HepG2 cells was observed at 40 μg/mL and 100 μg/mL of DSE compared to the control without insulin.

Hilary and colleagues (2021) investigated the metabolic changes in the preadipocyte (3T3-L1) and myoblast cells (C2C12) with DSE treatment and found enhanced glucose uptake in 3T3-L1 cells [9]. 3T3-L1 cells reported increased glucose uptake with 200 μg/mL of DSE treatment combined with insulin. The presence of polyphenols in date seeds plays a crucial role in enhancing the insulin signaling pathway. C2C12 muscle cells showed an improved glucose uptake with DSE treatment, but the results were inconsistent with different replicates.

Abdelaziz and colleagues (2015) revealed the potential of aqueous suspension of *P. dactylifera* seeds (aqPDS) in protecting streptozotocin (STZ)-induced diabetic rats from early diabetic complications in both liver and kidney [41]. Significant reduction of blood glucose level by 51% was observed in groups subjected to aqPDS treatment compared to the untreated diabetic group, but the normal blood glucose values were not restored. Four weeks of continuous aqPDS administration in STZ-induced diabetic rats not only decreased the levels of total cholesterol, triglycerides, serum urea, and creatinine, but also restored AST and ALT levels to normal. This indicated that intake of *P. dactylifera* seeds could lessen the renal damage induced by diabetes.

### 3.5. Anti-Inflammatory Activities

Oxidative stress such as protein peroxidation, lipid peroxidation, and DNA mutagenesis induces overproduction of reactive oxygen species and activates inflammatory mediators involved in the development and progression of chronic diseases, including cardiovascular diseases [42,43]. Saryono and colleagues (2020) carried out a study to investigate the anti-inflammatory mechanism of date seeds in middle-aged women. They found a significant reduction in pro-inflammatory cytokines and cyclooxygenase (COX) expression [8]. The level of interleukin-1β, transforming growth factor-β, COX-1, and COX-2 decreased upon uptake of date seeds treatment in middle-aged female subjects.

Bouhlali and colleagues (2020) revealed the anti-inflammatory properties of DSE by inhibiting protein denaturation, stabilising the lysosomal membranes, scavenging nitric oxide free radicals, and inhibiting C-reactive protein and fibrinogen production [44]. It is suggested that these mechanisms may be linked to the presence of phenolic compounds such as quercetin, rutin, caffeic acids, p-coumaric, and gallic acids in the date seeds, as observed through HPLC-DAD analysis. Among the four varieties of date seeds (Boufgous, Bousthammi, Jihl, and Majhoul), Jihl cultivar represents the most effective date seed by showing the strongest anti-inflammatory ability. Table 2 summarizes the biological activities of date seeds, while Figure 1 shows the schematic representation of biological activities of date seeds appraised in this review.

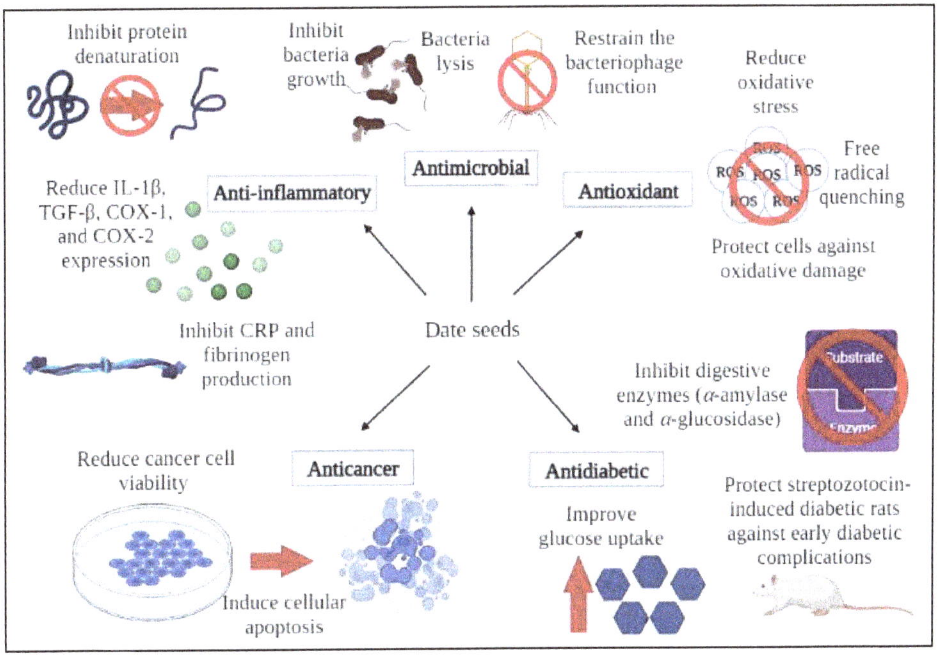


**Figure 1.** A schematic representation of the biological activities of date seeds. The highly valuable constituents of date seeds contribute to its antimicrobial, antioxidant, anticancer, antidiabetic, and anti-inflammatory properties.

**Table 2.** A summary of biological activities of date seeds.

| Biological Activities | Date Seed Cultivar | Type and List of Samples | Findings | References |
|---|---|---|---|---|
| Antimicrobial | Not stated | Bacterial strains:<br>• *P. aeruginosa*<br>• *E. coli*<br>• *K. pneumonia*<br>• *P. vulgaris*<br>• *P. mirabilis*<br>• MRSA<br>• *E. faecalis* | • The highest sensitivity was observed in MRSA, with an inhibition zone of 20 mm diameter and was more effective than the common antibiotic gentamycin, which showed only a diameter of 5 mm inhibition.<br>• Among the tested gram-negative bacteria, only *P. mirabilis* showed significant inhibition towards DSE. | [31] |
| | Ajwa | Bacterial strains:<br>• *S. aureus* ATCC 29213<br>• *E. coli* ATCC 25922<br>• *K. pneumonia* ATCC 700603<br>• *P. aeruginosa* ATCC 27853<br>• *E. faecalis* ATCC 29212 | • Notable inhibition zone by 100 mg/mL of Ajwa date seed against *S. aureus*, *E. coli*, *K. pneumonia*, *P. aeruginosa*, and *E. faecalis* was observed at 19 mm, 15 mm, 16 mm, 14 mm, and 16 mm, respectively. | [32] |
| | Not stated | Bacteria strain:<br>• *P. aeruginosa* ATCC 25668<br>Bacteriophage:<br>• Lytic *Pseudomonas* phage ATCC 14209-B1 | • DSEs significantly inhibited infectivity of *Pseudomonas* phage ATCC 14209-B1 to *P. aeruginosa*, reduced phage function, and completely disrupted the bacterial lysis. | [33] |

**Table 2.** *Cont.*

| Biological Activities | Date Seed Cultivar | Type and List of Samples | Findings | References |
|---|---|---|---|---|
| Antioxidant | Ajwa | Not stated | • Strong antioxidant nature of Ajwa date seed methanol extracts was comparable to that of ascorbic acid. | [32] |
| | Medjool | Not stated | • Approximately ten times the total polyphenolic content in the seed of 'Medjool' dates grown in Mexico compared to the quantity in the pulp.<br>• Shinoda assay confirmed presence of phenolic compounds such as tannins, flavonoids, saponins, anthraquinones, and coumarins.<br>• By employing DPPH free radical method, date seeds showed a much lower $IC_{50}$ result of 0.0046 g/L. | [35] |
| | Bent Kbala | Not stated | • DSE demonstrated the highest free radical inhibition rate (89.89 ± 0.02%) and the highest antioxidant activity ($IC_{50}$ = 0.031 μg/mL) compared to other natural antioxidants.<br>• These antioxidant properties are related to phenolic compounds, such as phenolic acids, catechin tannins, and flavonoids present in the date seeds, and their ability to donate hydrogen to DPPH radicals. | [36] |
| | Khalas | Not stated | • Increased polyphenol content following digestion of DSB.<br>• Both the DPPH and TEAC assays demonstrated a significant improvement in antioxidant power in DSB after gastric digestion.<br>• FRAP assay and TEAC assay recorded a significant decrease in the antioxidant activity of DSE at the end of intestinal digestion. The DPPH assay showed unchanged results in the antioxidant activity despite the decrease in total polyphenol content.<br>• In the case of the DSP, only the FRAP assay revealed a reduction in the antioxidant power as the digestion progressed from the gastric to intestinal phase. | [9] |

**Table 2.** *Cont.*

| Biological Activities | Date Seed Cultivar | Type and List of Samples | Findings | References |
| --- | --- | --- | --- | --- |
| | Khalas | 16 healthy adults (8 women and 8 men)<br>• Human serum and plasma<br>• Human urine | • Significant increase in antioxidant effect with a single dose of DSE, DSP, and DSB.<br>• Increased protection against protein and lipid peroxidation was observed up to 8 and 24 h after ingestion of date seed products.<br>• Presence of metabolites such as hydroxybenzoic acids, vanillic acids, protocatechuic acids, and vanillic acid sulphate in the excreted urine supports the involvement of date seeds in antioxidant activity. | [37] |
| Anticancer | Khalas | Cancer cell lines:<br>• human breast adenocarcinoma cell lines MCF-7 and MDA-MB-231<br>• human colon adenocarcinoma cell line Caco-2<br>• human hepatocyte carcinoma cell line HepG2<br>• human prostate adenocarcinoma cell line PC-3 | • DSEs reduced viability of cancer cell lines at high concentrations (1000 µg/mL or above) following long hours of exposure.<br>• The anticancer effect of reduced viability was observed in the MDA-MB-231, MCF-7, and Caco-2 after 48 h of treatment.<br>• HepG2 and PC-3 showed a significant cell viability reduction with ≥1000 µg/mL of DSE treatment after 24 h. | [9] |
| | Ajwa | Cancer cell lines:<br>• human triple-negative breast cancer cell line MDA-MB-231<br>• human estrogen receptor and progesterone receptor-positive breast cancer cell line MCF-7<br>• human liver carcinoma cell line HepG2<br>• normal kidney epithelial Vero cell lines | • *P. dactylifera* seed extracts induced cellular apoptosis through intrinsic pathways in the MDA-MB-231 cell line, MCF-7 cell line, and HepG2 cell line.<br>• Treated cancer cells displayed morphological changes such as non-adherent, detached, and rounded shapes after 24 h of exposure.<br>• Apoptotic features were observed after 48 h of exposure.<br>• The time- and dose-dependent manner of DSEs were shown by cell viability tests with an observation of a significant reduction in cancer cell growth.<br>• This cytotoxic activity was attributed to the synergistic effect of bioactive agents—rutin and quercetin—present in date seeds.<br>• No changes in morphology variation and cell viability were observed in the normal Vero cell line following 24 h and 48 h of exposure. | [38] |

**Table 2.** *Cont.*

| Biological Activities | Date Seed Cultivar | Type and List of Samples | Findings | References |
|---|---|---|---|---|
| Antidiabetic | Khalas | HepG2 liver cells culture | • DSEs inhibited two digestive enzymes, pancreatic α-amylase and intestinal α-glucosidase, which are responsible for starch digestion and modulation of the glucose uptake by HepG2.<br>• Maximum enzyme inhibition was observed in α-amylase and α-glucosidase at 400 µg/mL and 900 µg/mL of DSE, respectively.<br>• Increased glucose uptake by HepG2 cells was observed with doses of DSEs at 40 µg/mL and 100 µg/mL compared to the control without insulin. | [40] |
| | Khalas | C2C12 myoblast cell lines, 3T3-L1 preadipocytes | • The increase in glucose uptake was observed in 3T3-L1 cells with 200 µg/mL of DSE treatment in combination with insulin.<br>• Presence of polyphenols in date seeds is crucial for enhancing the insulin signaling pathway.<br>• An improved glucose uptake with DSE treatment was observed in C2C12 muscle cells, but the results were insignificant due to the high variability in the replicates. | [9] |
| | Hayani | 21 Adult male Wistar rats:<br>Group I (Control group): 7 rats received a single intraperitoneal injection of citrate buffer vehicle.<br>Group II (Untreated diabetic group): 7 rats received a single intraperitoneal injection of STZ (50 mg/kg).<br>Group III (aqPDS-treated group): 7 rats received a single intraperitoneal injection of STZ (50 mg/kg) and daily administration of aqPDS (1 g/kg/d) by oral gavage for 4 weeks. | • Potential of aqPDS in protecting STZ-induced diabetic rats from early diabetic complications in both liver and kidney.<br>• Significant reduction of blood glucose level by 51% was observed in groups subjected to aqPDS treatment compared to the untreated diabetic group, but the normal blood glucose values were not restored.<br>• Four weeks of continuous aqPDS administration in STZ-induced diabetic rats not only decreased the levels of total cholesterol, triglycerides, serum urea, and creatinine, but also restored AST and ALT levels to normal.<br>• This indicated that intake of *P. dactylifera* seeds could lessen the renal damage induced by diabetes. | [41] |

**Table 2.** *Cont.*

| Biological Activities | Date Seed Cultivar | Type and List of Samples | Findings | References |
| --- | --- | --- | --- | --- |
| Anti-inflammatory | Deglet Nour | Blood samples of 30 middle-aged women | • Significant reductions of pro-inflammatory cytokines and cyclooxygenase (COX) expression were observed.<br>• The level of interleukin-1β, transforming growth factor-β, COX-1, and COX-2 decreased following uptake of date seeds treatment in middle-aged female subjects. | [8] |
| | Boufgous, Bousthammi, Jihl, and Majhoul | 36 Wistar strain rats and 30 Swiss abino mice | • DSE inhibited protein denaturation, stabilised the lysosomal membranes, showed nitric oxide free radicals scavenging ability, and inhibited C-reactive protein and fibrinogen production.<br>• It is suggested that these mechanisms may be linked to the presence of phenolic compounds, such as quercetin, rutin, caffeic acids, p-coumaric, and gallic acids, in the date seeds, as observed through HPLC-DAD analysis.<br>• Among the four varieties of date seeds (Boufgous, Bousthammi, Jihl, and Majhoul), Jihl cultivar represents the most effective date seed by showing the strongest anti-inflammatory ability. | [44] |

## 4. Industrial Value of Date Seeds in a Commercial Application

DSP has been used in traditional remedies and for synthesizing citric acid and protein by *Candida lipolytica*, *Aspergillus oryzae*, and *Candida utilis* [19]. The date seed oil contains saturated and unsaturated fatty acids such as lauric and oleic acid [4]. Mrabet and colleagues (2020) reported the benefits of date seed oils for cooking, frying, or seasoning as a result of its excellent stability against oxidative rancidity and high temperature. Date seed oils were also used to replace conventional corn oil for mayonnaise production, and the mayonnaise outcome has better sensory characteristics [45].

Owing to its versatile antioxidant activities, date seeds can protect human skin against UV-B and UV-A radiation which can damage the skin [4]. Bioactive constituents such as flavonoids, phenolics, and phenolic acids inhibit free radicals production that causes cellular damage to the skin [17]. Lecheb and colleagues (2015) proposed a cosmetic cream using aqueous extract and oil from date seeds [46]. The spreadability, viscosity, and rheological behavior of the formulated cream were similar to those of the commercial cosmetic cream. Phytohormones in date seed oil possess anti-aging properties, reducing wrinkles and improving skin elasticity [47]. In recent years, date seed oil has been used to formulate body creams, soaps, hair products, and sunscreens for daily usage. The date seed oil contains vitamins, carotenes, and other nutrients that help maintain a healthy scalp, promote normal hair growth, and support the nutritional functions of hair follicles [20].

Moreover, date seed oil can be used as biodiesel owing to its low content of free fatty acids [4]. Using date seeds as a feedstock would bring enormous economic benefits and potential for future industrial operations [48].

## 5. Future Prospect

Long-term exposure to high oxidative stress plays a crucial role in the development of various chronic non-communicable diseases [49]. The ability of date seeds to scavenge free radicals and protect cells from oxidative damage might be the key mechanism of remedial action toward chronic diseases. In addition, the oil extracted from date seeds contains monounsaturated fatty acids and lipid-soluble antioxidants such as phenols and phytosterols that can reduce the risk of multiple diseases. With an abundance of date seeds, the oil extracted from the seeds might be useful to provide a chance for future enterprises in areas such as human foods application, medicine, biodiesel, and plastic manufacturing. Date seeds would evolve from waste to feedstock to promote economic and social growth in the date fruit production and manufacturing fields [48]. Innovative and optimized technology that improves the recovery and quality of DSEs will make the date seed a valuable source of antioxidants for nutraceuticals and functional food applications [3]. Although date seeds have been consumed over the centuries by humans and animals, controlled toxicity and bioavailability studies are still necessary to gauge the effectiveness of the medications and their potential to elicit a toxic response as the standard for every clinical trial.

## 6. Conclusions

Date seeds contain rich sources of nutritional functional components that may help to combat deficiency-related diseases, infections, and disorders related to oxidative stress. Current pharmacological studies have found that DSEs can regulate human health through several biological activities. However, an in-depth investigation is required to characterize polyphenols and other unique bioactive compounds, including their mechanism, as well as to establish ideal doses without causing harm. The potential of date seeds as a treatment against chronic diseases brings enormous promises and possibilities for future research.

**Author Contributions:** Conceptualization, A.A. and K.-S.L. writing—original draft preparation, A.S.-Y.K., M.A.A. and M.N.A. writing—review and editing, A.S.-Y.K., K.-S.L. and S.-H.E.L., supervision, C.-M.C., K.-S.L. and S.-H.E.L. funding acquisition, A.A. and W.-H.C. All authors have read and agreed to the published version of the manuscript.

**Funding:** This research was funded by ASPIRE, the technology program management pillar of Abu Dhabi's Advanced Technology Research Council (ATRC), via the ASPIRE "AARE (ASPIRE Awards for Research Excellence)" (Ref no. AARE20-185).

**Conflicts of Interest:** The authors declare no conflict of interest.

## References

1. Qadir, A.; Aqil, M.; Ahmad, U.; Khan, N.; Warsi, M.H.; Akhtar, J.; Arif, M.; Ali, A.; Singh, S.P. Date seed extract-loaded oil-in-water nanoemulsion: Development, characterization, and antioxidant activity as a delivery model for rheumatoid arthritis. *J. Pharm. Bioallied Sci.* **2020**, *12*, 308–316. [PubMed]
2. FAOSTAT. Available online: https://www.fao.org/faostat/en/#data/QV (accessed on 13 July 2022).
3. Ghafoor, K.; Sarker, M.Z.I.; Al-Juhaimi, F.Y.; Babiker, E.E.; Alkaltham, M.S.; Almubarak, A.K. Extraction and Evaluation of Bioactive Compounds from Date (Phoenix dactylifera) Seed Using Supercritical and Subcritical $CO_2$ Techniques. *Foods* **2022**, *11*, 1806. [CrossRef] [PubMed]
4. Mrabet, A.; Jiménez-Araujo, A.; Guillén-Bejarano, R.; Rodríguez-Arcos, R.; Sindic, M. Date Seeds: A Promising Source of Oil with Functional Properties. *Foods* **2020**, *9*, 787. [CrossRef] [PubMed]
5. Al-Dashti, Y.A.; Holt, R.R.; Keen, C.L.; Hackman, R.M. Date Palm Fruit (*Phoenix dactylifera*): Effects on Vascular Health and Future Research Directions. *Int. J. Mol. Sci.* **2021**, *22*, 4665. [CrossRef]
6. Ataei, D.; Hamidi-Esfahani, Z.; Ahmadi-Gavlighi, H. Enzymatic production of xylooligosaccharide from date (*Phoenix dactylifera* L.) seed. *Food Sci. Nutr.* **2020**, *8*, 6699–6707. [CrossRef]
7. Moslemi, E.; Dehghan, P.; Khani, M. The effect of date seed (*Phoenix dactylifera*) supplementation on inflammation, oxidative stress biomarkers, and performance in active people: A blinded randomized controlled trial protocol. *Contemp. Clin. Trials Commun.* **2022**, *28*, 100951. [CrossRef]
8. Saryono; Warsinah; Isworo, A.; Sarmoko. Anti-inflammatory activity of date palm seed by downregulating interleukin-1β, TGF-β, cyclooxygenase-1 and -2: A study among middle age women. *Saudi Pharm. J.* **2020**, *28*, 1014–1018. [CrossRef]
9. Hilary, S.; Kizhakkayil, J.; Souka, U.; Al-Meqbaali, F.; Ibrahim, W.; Platat, C. In-vitro Investigation of Polyphenol-Rich Date (*Phoenix dactylifera* L.) Seed Extract Bioactivity. *Front. Nutr.* **2021**, *8*, 667514. [CrossRef]
10. Niazi, S.; Khan, I.; Pasha, I.; Rasheed, S.; Ahmad, S.; Shoaib, M. Date Palm: Composition, Health Claim and Food Applications. *Int. J. Pub. Health Health Sys.* **2017**, *2*, 9–17.
11. Shina, S.; Izuagie, T.; Shuaibu, M.; Dogoyaro, A.I.; Garba, A.; Abubakar, S. The Nutritional Evaluation and Medicinal Value of Date Palm (*Phoenix dactylifera*). *Int. J. Mod. Chem.* **2013**, *4*, 147–154.
12. Ali, H.S.; Alhaj, O.A.; Al-Khalifa, A.S.; Brückner, H. Determination and stereochemistry of proteinogenic and non-proteinogenic amino acids in Saudi Arabian date fruits. *Amino Acids* **2014**, *46*, 2241–2257. [CrossRef] [PubMed]
13. Khalid, S.; Khalid, N.; Khan, R.S.; Ahmed, H.; Ahmad, A. A review on chemistry and pharmacology of Ajwa date fruit and pit. *Trends Food Sci. Technol.* **2017**, *63*, 60–69. [CrossRef]
14. Brouk, M.; Fishman, A. Antioxidant Properties and Health Benefits of Date Seeds. In *Functional Properties of Traditional Foods*; Kristbergsson, K., Ötles, S., Eds.; Springer: Boston, MA, USA, 2016; pp. 233–240.
15. Attia, A.I.; Reda, F.M.; Patra, A.K.; Elnesr, S.S.; Attia, Y.A.; Alagawany, M. Date (*Phoenix dactylifera* L.) by-Products: Chemical Composition, Nutritive Value and Applications in Poultry Nutrition, an Updating Review. *Animals* **2021**, *11*, 1133. [CrossRef]
16. Nehdi, I.A.; Sbihi, H.M.; Tan, C.P.; Rashid, U.; Al-Resayes, S.I. Chemical Composition of Date Palm (*Phoenix dactylifera* L.) Seed Oil from Six Saudi Arabian Cultivars. *J. Food Sci.* **2018**, *83*, 624–630. [CrossRef] [PubMed]
17. Alharbi, K.L.; Raman, J.; Shin, H.-J. Date Fruit and Seed in Nutricosmetics. *Cosmetics* **2021**, *8*, 59. [CrossRef]
18. Hinkaew, J.; Aursalung, A.; Sahasakul, Y.; Tangsuphoom, N.; Suttisansanee, U. A Comparison of the Nutritional and Biochemical Quality of Date Palm Fruits Obtained Using Different Planting Techniques. *Molecules* **2021**, *26*, 2245. [CrossRef]
19. Tafti, A.G. Solaimani Dahdivan, N. Yasini Ardakani, S.A. Physicochemical properties and applications of date seed and its oil. *Int. Food Res. J.* **2017**, *24*, 1399–1406.
20. Walke, D.; Daud, F. Date palm fruit (*Phoenix dactylifera* L.) as a cosmetic ingredient. *JETIR* **2018**, *5*, 755–762.
21. Al-Farsi, M. Date seeds: Usage of date seed extract in health. In *Nuts and Seeds in Health and Disease*; Elsevier: Amsterdam, The Netherlands, 2011.
22. Habib, H.M.; Kamal, H.; Ibrahim, W.H.; Dhaheri, A.S.A. Carotenoids, fat soluble vitamins and fatty acid profiles of 18 varieties of date seed oil. *Ind. Crops Prod.* **2013**, *42*, 567–572. [CrossRef]
23. Juhaimi, F.A.; Ghafoor, K.; Özcan, M.M. Physical and chemical properties, antioxidant activity, total phenol and mineral profile of seeds of seven different date fruit (*Phoenix dactylifera* L.) varieties. *Int. J. Food Sci. Nutr.* **2012**, *63*, 84–89. [CrossRef]
24. Anbarshahi, R. The Amazing Health Benefit of Date Seed Powder. Available online: https://date-seed.com/the-amazing-health-benefit-of-date-seed-powder/ (accessed on 11 July 2022).
25. El Fouhil, A.F.; Ahmed, A.M.; Atteya, M.; Mohamed, R.A.; Moustafa, A.S.; Al-Roalle, A.H.; Darwish, H.H. Hypoglycemic effects of date seed extract. Possible mechanism of action, and potential therapeutic implications. *Saudi Med. J.* **2013**, *34*, 1125–1132. [PubMed]

26. Biglar, M.; Khanavi, M.; Hajimahmoodi, M.; Hassani, S.; Moghaddam, G.; Sadeghi, N.; Oveisi, M.R. Tocopherol content and Fatty Acid profile of different Iranian date seed oils. *Iran. J. Pharm. Res.* **2012**, *11*, 873–878. [PubMed]
27. Habib, H.M.; Platat, C.; Meudec, E.; Cheynier, V.; Ibrahim, W.H. Polyphenolic compounds in date fruit seed (*Phoenix dactylifera*): Characterisation and quantification by using UPLC-DAD-ESI-MS. *J. Sci. Food Agric.* **2014**, *94*, 1084–1089. [CrossRef] [PubMed]
28. Bouhlali, E.d.T.; Hmidani, A.; Bourkhis, B.; Khouya, T.; Harnafi, H.; Filali-Zegzouti, Y.; Alem, C. Effect of *Phoenix dactylifera* seeds (dates) extract in triton WR-1339 and high fat diet induced hyperlipidaemia in rats: A comparison with simvastatin. *J. Ethnopharmacol.* **2020**, *259*, 112961. [CrossRef]
29. Rashmika, P.C.K.J. Swetha Sundararajan, Formulation and analysis of spread using dates seed. *Int. J. Food Sci. Nutr.* **2018**, *3*, 16–18.
30. Habib, H.M.; Ibrahim, W.H. Nutritional quality evaluation of eighteen date pit varieties. *Int. J. Food Sci. Nutr.* **2009**, *60* (Suppl. 1), 99–111. [CrossRef]
31. Alrajhi, M.; Al-Rasheedi, M.; Eltom, S.E.M.; Alhazmi, Y.; Mustafa, M.M.; Ali, A.M. Antibacterial activity of date palm cake extracts (*Phoenix dactylifera*). *Cogent Food Agric.* **2019**, *5*, 1625479. [CrossRef]
32. Anwar, S.; Raut, R.; Alsahli, M.A.; Almatroudi, A.; Alfheeaid, H.; Alzahrani, F.M.; Khan, A.A.; Allemailem, K.S.; Almatroodi, S.A.; Rahmani, A.H. Role of Ajwa Date Fruit Pulp and Seed in the Management of Diseases through In Vitro and In Silico Analysis. *Biology* **2022**, *11*, 78. [CrossRef]
33. Jassim, S.A.; Naji, M.A. In vitro Evaluation of the Antiviral Activity of an Extract of Date Palm (*Phoenix dactylifera* L.) Pits on a Pseudomonas Phage. *Evid. Based Complement. Altern. Med.* **2010**, *7*, 57–62. [CrossRef]
34. Hussain, M.I.; Farooq, M.; Syed, Q.A. Nutritional and biological characteristics of the date palm fruit (*Phoenix dactylifera* L.)—A review. *Food Biosci.* **2020**, *34*, 100509. [CrossRef]
35. Salomón-Torres, R.; Ortiz-Uribe, N.; Valdez-Salas, B.; Rosas-González, N.; García-González, C.; Chávez, D.; Córdova-Guerrero, I.; Díaz-Rubio, L.; Haro-Vázquez, M.D.P.; Mijangos-Montiel, J.L.; et al. Nutritional assessment, phytochemical composition and antioxidant analysis of the pulp and seed of medjool date grown in Mexico. *PeerJ* **2019**, *7*, e6821. [CrossRef] [PubMed]
36. Bentrad, N.; Gaceb-Terrak, R. Evaluation of the level of biomolecules isolated from date palm seeds (*Phoenix dactylifera*) and in vitro Antioxidant property. *BioMedicine* **2020**, *10*, 23–29. [CrossRef] [PubMed]
37. Platat, C.; Hillary, S.; Tomas-Barberan, F.A.; Martinez-Blazquez, J.A.; Al-Meqbali, F.; Souka, U.; Al-Hammadi, S.; Ibrahim, W. Urine Metabolites and Antioxidant Effect after Oral Intake of Date (*Phoenix dactylifera* L.) Seeds-Based Products (Powder, Bread and Extract) by Human. *Nutrients* **2019**, *11*, 2489. [CrossRef] [PubMed]
38. Khan, M.A.; Singh, R.; Siddiqui, S.; Ahmad, I.; Ahmad, R.; Upadhyay, S.; Barkat, M.A.; Ali, A.M.A.; Zia, Q.; Srivastava, A.; et al. Anticancer potential of *Phoenix dactylifera* L. seed extract in human cancer cells and pro-apoptotic effects mediated through caspase-3 dependent pathway in human breast cancer MDA-MB-231 cells: An in vitro and in silico investigation. *BMC Complementary Med. Ther.* **2022**, *22*, 68. [CrossRef] [PubMed]
39. Saryono, S. Date Seeds Drinking as Antidiabetic: A Systematic Review. *IOP Conf. Ser. Earth Environ. Sci.* **2019**, *255*, 012018. [CrossRef]
40. Shakoor, H.; Abdelfattah, F.; Albadi, K.; Adib, M.; Kizhakkayil, J.; Platat, C. Inhibition of Digestive Enzyme and Stimulation of Human Liver Cells (HepG2) Glucose Uptake by Date Seeds Extract. *Evid Based Complement. Altern. Med.* **2020**, *2020*, 4290702. [CrossRef]
41. Abdelaziz, D.H.; Ali, S.A.; Mostafa, M.M. *Phoenix dactylifera* seeds ameliorate early diabetic complications in streptozotocin-induced diabetic rats. *Pharm. Biol.* **2015**, *53*, 792–799. [CrossRef]
42. Hussain, T.; Tan, B.; Yin, Y.; Blachier, F.; Tossou, M.C.B.; Rahu, N. Oxidative Stress and Inflammation: What Polyphenols Can Do for Us? *Oxidative Med. Cell. Longev.* **2016**, *2016*, 7432797. [CrossRef]
43. Kong, A.S.-Y.; Lai, K.S.; Hee, C.-W.; Loh, J.Y.; Lim, S.H.E.; Sathiya, M. Oxidative Stress Parameters as Biomarkers of Cardiovascular Disease towards the Development and Progression. *Antioxidants* **2022**, *11*, 1175. [CrossRef]
44. Bouhlali, E.D.T.; Hmidani, A.; Bourkhis, B.; Khouya, T.; Ramchoun, M.; Filali-Zegzouti, Y.; Alem, C. Phenolic profile and anti-inflammatory activity of four Moroccan date (*Phoenix dactylifera* L.) seed varieties. *Heliyon* **2020**, *6*, e03436. [CrossRef]
45. Basuny, A.M.M.; Al-Marzooq, M.A. Production of mayonnaise from date pit oil. *Food Nutr. Sci.* **2011**, *2*, 938–943. [CrossRef]
46. Lecheb, F.; Benamara, S. Feasibility study of a cosmetic cream added with aqueous extract and oil from date (*Phoenix dactylifera* L.) fruit seed using experimental design. *J. Cosmet. Sci.* **2015**, *66*, 1–12.
47. Younas, A.; Naqvi, S.A.; Khan, M.R.; Shabbir, M.A.; Jatoi, M.A.; Anwar, F.; Inam-Ur-Raheem, M.; Saari, N.; Aadil, R.M. Functional food and nutra-pharmaceutical perspectives of date (*Phoenix dactylifera* L.) fruit. *J. Food Biochem.* **2020**, *44*, e13332. [CrossRef] [PubMed]
48. Yousuf, R.G.; Winterburn, J.B. Waste date seed oil extract as an alternative feedstock for Poly(3-hydroxybutyrate) synthesis. *Biochem. Eng. J.* **2017**, *127*, 68–76. [CrossRef]
49. Seyedsadjadi, N.; Grant, R. The Potential Benefit of Monitoring Oxidative Stress and Inflammation in the Prevention of Non-Communicable Diseases (NCDs). *Antioxidants* **2020**, *10*, 15. [CrossRef]

biomolecules

Systematic Review

# The Efficacy of Gum Arabic in Managing Diseases: A Systematic Review of Evidence-Based Clinical Trials

**Yamamh Al-Jubori [1], Nazik Tayfour Babiker Ahmed [1], Rawan Albusaidi [1], James Madden [2], Srijit Das [3] and Srinivasa Rao Sirasanagandla [3,*]**

[1] College of Medicine and Health Sciences, Sultan Qaboos University, Muscat 123, Oman
[2] GKT School of Medicine, King's College London, Great Maze Pond, London SE1 1UL, UK
[3] Department of Human and Clinical Anatomy, College of Medicine and Health Sciences, Sultan Qaboos University, Muscat 123, Oman
* Correspondence: srinivasa@squ.edu.om; Tel.: +968-24141176

**Abstract:** Gum arabic (GA) is a natural product commonly used as a household remedy for treating various diseases in the Sub-Saharan Africa region. Despite its claimed benefits, there has been a lack of research on the findings of current clinical trials (CTs) that investigated its efficacy in the treatment of various medical diseases. The aim of this systematic review was to study CTs which focused on GA and its possible use in the management of various medical diseases. A search of the extant literature was performed in the PubMed, Scopus, and Cochrane databases to retrieve CTs focusing on evidence-based clinical indications. The databases were searched using the keywords ("Gum Arabic" OR "*Acacia senegal*" OR "*Acacia seyal*" OR "Gum Acacia" OR "Acacia Arabica") AND ("Clinical Trial" OR "Randomized Controlled Trial" OR "Randomized Clinical Trial"). While performing the systematic review, data were obtained on the following parameters: title, authors, date of publication, study design, study aim, sample size, type of intervention used, targeted medical diseases, and main findings. Twenty-nine papers were included in this systematic review. The results showed that ingestion of GA altered lipid profiles, renal profiles, plaque, gingival scores, biochemical parameters, blood pressure, inflammatory markers, and adiposity. GA exhibited anti-inflammatory, prebiotic, and antibacterial properties. GA has been successfully used to treat sickle cell anemia, rheumatoid arthritis, metabolic disorders, periodontitis, gastrointestinal conditions, and kidney diseases. Herein, we discuss GA with respect to the underlying mechanisms involved in each medical disease, thereby justifying GA's future role as a therapeutic agent.

**Keywords:** *Acacia senegal*; *Acacia seyal*; gum arabic; natural product; systematic review; treatment

## 1. Introduction

Gum arabic (GA) is an exudate with a gummy texture obtained from *Acacia seyal* and *Acacia senegal* umbrella-shaped branches. A cut is made on the branches by which the exudate is obtained from or naturally present in and is made to harden in the air. GA is mainly found in Sudan, Chad, and Nigeria [1,2] (Figure 1). Structurally, GA is an arabinogalactan–protein complex. This complex is composed of magnesium, calcium, and potassium salts of arabic acid. Arabic acid structure is made up of 1-3-linked β-D-galactopyranosyl units, along with branches that consist of two to five β-D-galactopyranosyl residues linked together through 1,3-ether linkages and connected to the fundamental β-D-galactopyranosyl chain by 1,6-linkages (Figure 2) [1].

GA is largely fermented in the large intestines into short-chain fatty acids by microorganisms [3]. Traditionally, GA has been used as an oral hygiene substance [4]. Health benefits were seen following GA treatment. Direct application of herbal formulation containing GA on teeth and gums significantly reduced gingival and plaque index scores [4]. GA contains high amounts of calcium and phosphate ions. In vitro studies demonstrated that it can prevent tooth enamel demineralization, in addition to enhancing its remineralization [5–7]. In mice, GA supplementation in drinking water or along with diet was observed to reduce obesity by altering the expression of lipid metabolic genes and age-dependent fat deposition in the visceral adipose tissue [8,9]. It lowers cholesterol levels, as it possesses a high amount of fiber. GA treatment, along with atorvastatin, reduced total cholesterol, LDL, and triglyceride levels in patients with hyperlipidemia [10]. These effects subsequently decrease the risk of heart disease [10].

**Figure 1.** Photographs showing (**A**) *Acacia seyal* and (**B**) Gum Arabic.

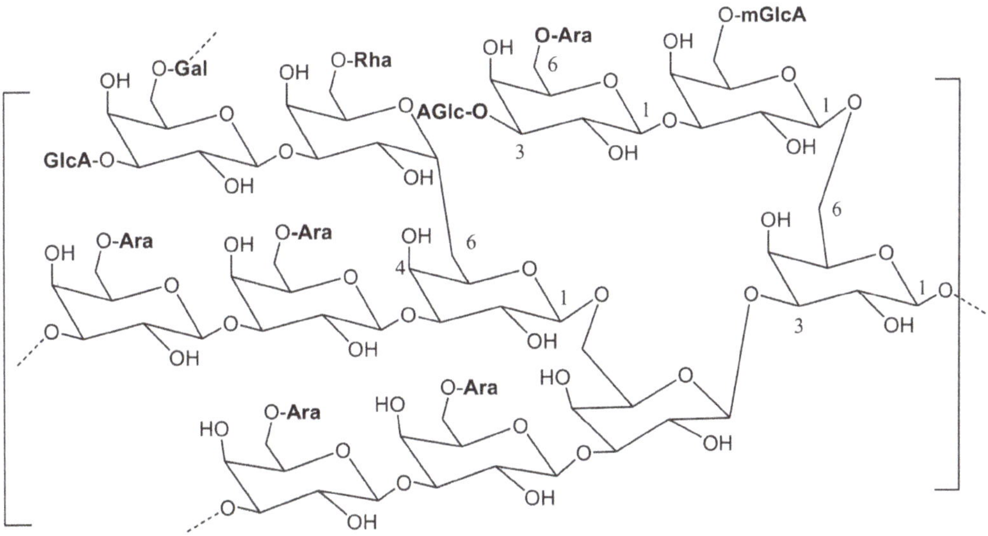


**Figure 2.** The basic chemical structure of arabic acid present in gum arabic [1].

GA also exhibited antioxidant properties by increasing superoxide dismutase, catalase, and glutathione peroxidase activity in the liver [3,11–13]. GA oral supplementation increased 24 h creatinine clearance and binds with free water, thereby reducing intestinal absorption and water content in urine [14,15]. Surprisingly, GA was ignored by the locals, and the tree branches were used for coal and fire with decreases in production [1]. Thus, the benefits were unknown until recently, when many experimental and CT studies revealed its benefits.

Several experimental studies have demonstrated the potential benefits of the use of GA in clinical practice [16–19]. In a recent study, GA treatment inhibited colorectal carcinogenesis in mice [16]. GA treatment reduced the formation of aberrant crypts foci in the colon, mainly by reducing local genotoxicity, as well as oxidative stress [16]. In the same study, reduced genotoxicity in the liver and bone marrow, as well as low oxidative stress in the liver and blood, were observed [16]. GA supplementation protected rat heart from ischemia/reperfusion injury by decreasing apoptotic enzyme levels, as well as from the formation of proinflammatory cytokines [20]. GA treatment alleviated $B_1$-induced hepatic injury through its antioxidant and anti-inflammatory properties [17].

Another research study showed that regular GA consumption stimulates innate immunity against various infections by inducing cathelicidin expression [18]. GA supplementation in type 2 diabetic rats prevented learning and memory loss. These effects were associated with increased expression of PGC-1a and ATP synthase β-subunit protein in the hippocampus [21]. GA administration in rats with dextran sodium sulfate-induced colitis resulted in a reduction in the severity of colitis, colonic fibrosis, and TGFβ1 expression [22]. GA pretreatment prevented butralin-exposure-induced renal damage by promoting antioxidants and increasing free radical scavenging activity [23]. Another experimental study on diabetic rats concluded that GA reduced the progression of chronic kidney disease [24].

Studies were published on GA and water pipe smoking (WPS) in mice. Researchers showed that in male mice exposed to hookah smoke for 30 minutes each day for 30 days with coadministration of GA, the negative effects of smoke exposure on the reproductive system were reduced [25]. Another study on mice exposed to WPS and GA concluded that GA reduced the harmful effects of WPS on thrombosis, cardiovascular toxicity, inflammation, and oxidative stress [26].

GA was found to be effective against diarrhea [27,28]. GA supplementation as an additive to oral rehydration solution significantly reduced the duration of diarrhea and frequency of defecation and improved the consistency of the stool [27]. Clinical trials (CTs) are being conducted in the human population to develop a new therapeutic drug in order to treat, prevent, or reduce the incidence of disease [29]. To the best of our knowledge, the present systematic review may be the first of its kind to discuss CTs conducted on human subjects exclusively focusing on GA and its beneficial effects in the management of various medical diseases.

## 2. Materials and Methods

### 2.1. Study Design

A systematic review of all human CTs was conducted to explore the current best evidence of the possible use of GA for various medical diseases.

### 2.2. Search Strategy

Relevant studies were identified through a thorough search of electronic databases such as PubMed, Scopus, and the Cochrane library. In addition, a snowballing method was employed whereby relevant articles were found by screening the reference list for any additional articles that met the eligibility criteria of the current study. The earlier published papers were screened during the period from April 2022 to August 2022. The databases were searched using the keywords: ("Gum Arabic" OR "*Acacia senegal*" OR "*Acacia seyal*" OR "Gum Acacia" OR "Acacia Arabica ") AND ("Clinical Trial" OR "Randomized Controlled Trial" OR "Randomized Clinical Trial"). Unpublished articles were excluded from the search strategy.

### 2.3. Inclusion Criteria

Published literature that fulfilled the following criteria was included: (i) all studies that were published in English language and reported CTs of GA treatment against targeted medical conditions in humans. All CTs of GA treatment, regardless of randomization, blinding, phase of trial, and statistical method, were used for assessment of outcome, irrespective of negative or positive results.

### 2.4. Exclusion Criteria

The exclusion criteria for the present systematic review were: "Studies that used other natural product/compound combined with GA for the treatment"; "any preclinical, unpublished, duplicated, and incomplete CTs"; "Studies that reported the GA CT without a specific targeted medical disease" and "Studies that were published in languages other than English".

### 2.5. Data Collection

Reviewers (Y.A.-J., N.T.B.A., R.A., J.M., and S.R.S.) first screened titles and abstracts of all retrieved papers for inclusion. The full texts of all screened papers were then studied independently in order to determine the final study selection. Duplicate information on the same studies was removed. Agreement on the inclusion and exclusion criteria was concordant, and discrepancies were resolved by consensus of all researchers. The following data were collected from included studies: authors and date of study, study design, targeted medical disease, sample size, and main results/findings (Table 1).

**Table 1.** Table showing the study characteristics of clinical trials conducted to show the beneficial effects of GA in various medical diseases.

| Sl No | Authors and Year | Treatment Duration | Dose and Mode of Treatment | Sample Size | Medical Disease | Clinical Tests | Main Findings |
|---|---|---|---|---|---|---|---|
| 1 | Larson et al., 2021 [30] | 6 weeks | 0, 20, and 40 g of GA in orange juice, along with a bagel and cream cheese, after a 12 h fast | 48 | Type 2 Diabetes | Computerized visual analog scales | After the individuals consumed a bagel and cream cheese, along with 40 g of GA, they reported feeling less hungry 15 min and 240 min later. In comparison to the control group, the post-acacia ingestion symptoms of bloating, gas, and GI rumbling were more severe. Although there was no significant difference in the area under the curve or changes in blood glucose response, blood glucose with 20 g of fiber at 30 min was considerably lower than the control. |
| 2 | Babiker et al., 2018 [31] | 3 months | 30 g per day of GA administered orally | 91 | Type 2 Diabetes | Anthropometric obesity markers, blood pressure, and lipid profile | The effects of GA reduced BMI by 2% and visceral body adiposity index by 23.7%. |
| 3 | Babiker et al., 2012 [32] | 6 weeks | 30 g per day of GA administered orally | 120 | Obesity | Weight and height using standardized scales, skin fold thickness using a Harpenden skin fold caliper, and fat percentage using the Jackson and Pollock 7 caliper method and Siri equation | Body fat percentage reduced by 2.18%. |

**Table 1.** *Cont.*

| Sl No | Authors and Year | Treatment Duration | Dose and Mode of Treatment | Sample Size | Medical Disease | Clinical Tests | Main Findings |
|---|---|---|---|---|---|---|---|
| 4 | Calame et al.,2011 [33] | 3 h post consumption | STUDY 1: 5 and 10 g of EmulGold (EG) and 10 and 40 g of PreVitae (PV) dissolved in 250 mL of water; STUDY 2: "glass of water containing 0, 10, 20 or 40 g of EG or PV" | 70 | Obesity | Caloric intake and visual analog scale | Three hours after ingesting GA, participants showed substantial reductions in calorie consumption and subjective feelings of fullness. |
| 5 | Jarrar et al., 2021 [34] | 12 weeks | 20 g daily oral dose of GA | 61 | Metabolic Syndrome | Anthropometric measurements, diet and physical assessment via The Food Processor® nutrition and fitness software, ESHA food-analysis program (version 10.4), the Kuwaiti Food Composition database, bowel movement and satiety questionnaire, fasting blood sample to measure lipid profile, HbA1c, and fasting blood glucose | Systolic and diastolic blood pressure, fat-free mass, hunger, and fasting plasma glucose were all lower in the participants. An increase in dietary fiber consumption was reported. Bowel motions and bloating were improved. |
| 6 | Babiker et al., 2017 [35] | 3 months | 30 g daily intake of GA administered orally | 91 | Type 2 Diabetes | BMI, lipid profile, fasting blood glucose, and HbA1c | BMI was significantly lower in the GA group by 2.1%, although there was no significant change in waist size. Regarding metabolic measures, the GA group presented with a substantial reduction in fasting plasma glucose of 6.24% and a reduction in glycosylated hemoglobin (HbA1c) of 8.8%. By reducing LDL cholesterol by 19.5%, total cholesterol by 8.28%, and triglycerides by 10.95% from baseline values, GA supplementation improved lipid profiles. Within the GA group, HDL cholesterol significantly increased by 19.89%. |

**Table 1.** *Cont.*

| Sl No | Authors and Year | Treatment Duration | Dose and Mode of Treatment | Sample Size | Medical Disease | Clinical Tests | Main Findings |
|---|---|---|---|---|---|---|---|
| 7 | Nasir et al., 2016 [36] | 16 weeks | 10 g daily intake of GA administered orally | 40 | Type 2 Diabetes | Anthropometric measurements, lipid profile, fasting blood glucose, HbA1c, urea nitrogen, and creatinine | Fasting blood sugar and HbA1c levels significantly decreased in all groups, which was followed by large drops in total protein and uric acid levels. In diabetics and those with diabetic nephropathy, there was a noticeable improvement in renal function following GA supplementation across all groups, with substantial reductions in blood urea nitrogen and creatinine levels. |
| 8 | Haskell et al., 1992 [37] | 12 weeks | 15 g of GA dissolved in water | 62 | Hyperlipidemia | Lipid profile | Water-soluble dietary fiber made up of psyllium husk, pectin, guar and locust bean gums was significantly reduced in cholesterol—comparable to changes observed with 10 g of water-soluble dietary fiber per day from high-viscosity guar gum. GA (low viscosity) did not produce a significant lipid-lowering effect compared to a placebo. |
| 9 | Mohamed et al., 2015 [10] | 4 weeks | 30 mg of GA taken orally | 110 | Hyperlipidemia | Lipid profile | Reduced levels of total cholesterol (25.9% to 7.8%), triglycerides (38.2% to 2.9%), and LDL (30.8% to 8.1%) were the significant findings. |
| 10 | Kaddam et al., 2019 [38] | 12 weeks | 30 g per day of GA administered orally | 47 | Sickle Cell Anemia | Lipid profile | GA significantly decreased levels of triglycerides (10%), LDL (8% reduction), and total cholesterol (7% reduction). |
| 11 | Kaddam et al., 2019 [39] | 12 weeks | 30 g per day of GA administered orally | 47 | Sickle Cell Anemia | Liver enzymes, total protein, albumin, electrolytes, urea, creatinine, and uric acid were determined in the serum | Direct bilirubin, serum alanine transaminase, and serum urea levels all decreased significantly after 4 weeks of treatment, and these results persisted through the 8th week. |
| 12 | Kaddam et al., 2017 [40] | 12 weeks | 30 g per day of GA administered orally | 47 | Sickle Cell Anemia | Total antioxidant capacity (TAC), malondialdehyde (MDA), and hydrogen peroxide (H2O2) levels were measured by spectrophotometric methods | Total antioxidant capacity was significantly increased by GA, and oxidative indicators of MDA and hydrogen peroxide were reduced. |

**Table 1.** *Cont.*

| Sl No | Authors and Year | Treatment Duration | Dose and Mode of Treatment | Sample Size | Medical Disease | Clinical Tests | Main Findings |
|---|---|---|---|---|---|---|---|
| 13 | Kamal et al., 2020 [41] | 1 year | 62.5 mg of GA dissolved in 1 ml distilled water | 63 | Dental Caries | Mitis Salivarius agar and Lactobacillus MRS agar were used, and microbial analysis was performed | Chlorhexidine, GA, and liquorice mouthwashes differed statistically significantly from each other in the decayed, missing, and filled index (DMF). However, there was no statistically significant difference between the two latter groups. Patients who took GA or liquorice mouthwash had considerably lower levels of Streptococcus mutans and Lactobacillus acidophilus. However, there have been reports of oral adverse effects with chlorhexidine mouthwash. |
| 14 | Gazi, 1991 [42] | 1 week | Chewing on GA for 10 min 5× a day | 20 | Plaque Formation | Gingival and plaque scores and daily photographic assessment of erythrocine-stained plaques | After GA compared to sugar-free gum, daily photographic examination of erythrocine-stained plaque revealed lower scores, as well as lower gingival and plaque scores. |
| 15 | Gafar et al., 2022 [43] | 2 months | 150 mg GA powder applied twice a day | 30 | Plaque-Induced Gingivitis | Plaque index, gingival index, and gingival crevicular fluid interleukin 1 beta were measured | It was discovered that GA worked well to reduce plaque and gingival irritation. The intervention group had considerably lower mean gingival index scores at 30 days, significantly lower mean plaque index scores at 30 and 60 days, and significantly lower gingival crevicular fluid interleukin 1 beta levels at 30 and 60 days. |
| 16 | Singhal et al., 2018 [44] | 90 days | Participants were asked to use GA on a set of 2–4 teeth (premolars and molars) in one specified quadrant | 59 | Chronic Periodontitis | Periodontal pocket depth, clinical attachment levels, and gingival indices (GI) and plaque indices (PI) | In the group that utilized GA gel, there was a greater reduction in periodontal pocket depth and an increase in clinical attachment level, as well as a lower index of plaque and gingival index. |
| 17 | Bielfeldt et al., 2021 [45] | 10 min | 10 g of pastille GA | 26 | Xerostomia | Gravimetric measurement of salivary flow rate and visual analog scale. Saliva surface tension was measured in pooled saliva samples (0–5 min of sampling); Raman spectroscopy | After chewing the pastille, the mean salivary flow rate considerably rose by 8.03 g/min compared to the mean changes after chewing the control product, which increased by 3.71 g/min. |

**Table 1.** *Cont.*

| Sl No | Authors and Year | Treatment Duration | Dose and Mode of Treatment | Sample Size | Medical Disease | Clinical Tests | Main Findings |
|---|---|---|---|---|---|---|---|
| 18 | Kamal et al., 2021 [46] | 12 weeks | 30 g per day of GA administered orally | 40 | Rheumatoid Arthritis | A Cobas C311 (Roche, Germany) automated chemistry analyzer was used to directly determine the values for total protein, albumin, alanine aminotransferase (ALT), aspartate aminotransferase (AST), alkaline phosphatase (ALP), and creatinine | The liver enzymes ALT and AST were dramatically reduced due to GA, whereas ALP did not change appreciably. The albumin level significantly increased, whereas the serum globulin level minimally changed. GA considerably lowered urea and salt levels, with little impact on creatinine and potassium levels in terms of electrolyte levels and renal function. |
| 19 | Kamal et al., 2018 [47] | 12 weeks | 30 g per day of GA administered orally | 40 | Rheumatoid Arthritis | Complete blood count and liver and kidney function test | The levels of tumor necrosis factor alpha, erythrocyte sedimentation, the number of swollen and painful joints, and the illness severity score (28 P.V:0.00) were all shown to be significantly reduced by GA. Blood index changes were minimal. |
| 20 | Eltayeb et al., 2004 [48] | One-time oral dose | N/A | 24 | Drug Efficacy | Blood samples | When comparing two groups (one that took GA 2 hours after ingesting amoxicillin and the other that took the medication at the same time as GA), it was discovered that the latter group's peak amoxicillin concentration was much lower. |
| 21 | Hosseinpour et al., 2012 [49] | 4 weeks | GA ointment | 60 | Peristomal Skin Inflammation | Skin inflammation was diagnosed clinically according to a study by Ruzicka et al.; severity was defined according to the Physician Global Assessment (PGA) | Comparing the GA ointment group to the control group using zinc sulfate ointment, the GA ointment group showed a substantially lower level of peristomal skin inflammation. |
| 22 | Ali et al., 2020 [13] | 12 weeks | 30 g per day of GA administered orally | 40 | Kidney Disease | Blood samples | GA decreased the levels of the oxidative markers of MDA and CRP and significantly increased overall antioxidant capacity. |
| 23 | Elamin et al., 2017 [50] | 4 weeks | 10, 20, or 40 grams per day of GA administered orally | 36 | Kidney Disease | Clinical interviews for symptom severity, fasting blood sample, and 24 h urine collections | Both sodium levels and C-reactive protein levels were decreased in individuals who took GA. Other electrolyte concentrations, urine volume, and indoxyl sulfate levels were not affected. |

**Table 1.** *Cont.*

| Sl No | Authors and Year | Treatment Duration | Dose and Mode of Treatment | Sample Size | Medical Disease | Clinical Tests | Main Findings |
|---|---|---|---|---|---|---|---|
| 24 | Ali et al., 2008 [51] | 3 months | 50 g daily oral dose of GA | 36 | Kidney Disease | Blood samples | In comparison to baseline, GA was observed to significantly lower blood concentrations of urea, creatinine, uric acid, and phosphorus. GA ingestion was also reported to increase serum calcium levels. |
| 25 | Bliss et al., 1996 [14] | 1 month | 50 g of GA | 20 | Chronic Renal Failure | Blood and stool samples | The amount of nitrogen and bacterial bulk in feces dramatically increased due to GA. Additionally, serum urea nitrogen was significantly decreased. |
| 26 | Bliss et al., 2014 [52] | 32 days | 16 g daily intake of GA administered orally | 189 | Fecal Inconti-nence | Fecal collection and fecal incontinence diary filled in by subjects | The frequency of feces incontinence was comparable between individuals who took GA and those who took a placebo. It has been claimed that psyllium supplements can lower the frequency of fecal incontinence. |
| 27 | Calame et al., 2008 [53] | 4 weeks | Daily doses (5, 10, 20, and 40 g) of GA (EmulGold®) dissolved in water | 54 | Prebiotic | Fecal samples and volunteer survey | In comparison to the negative control group, the counts of Bifidobacteria and Lactobacilli were significantly increased after GA administration—10 g in particular. |
| 28 | Salah et al., 2012 [54] | Until res-olution (range: 24 h–6 weeks; average: 5 days) | 5–10 mg GA solution administered daily | 180 | Acute Non-Bloody Diarrhea | Electrolytes, body weight, and clinical observation | 90% of the children who took GA had their diarrhea cease within 24 hours, and 80% had been released after one day. Within the first five days of their admission, each of them made progress and was released. None of them had shock or severe dehydration. Only three children experienced electrolyte imbalance. At the conclusion of the trial, the weight had increased by 47.8% and had dropped by just 5.5%. Only two (3.3%) of the 61 children who were still being monitored after 6 weeks had experienced diarrhea again. |
| 29 | Suresh et al., 2021 [55] | 1 month | 10 g of GA dissolved in 200 mL of water | 10 | Gastroparesis | Weight and height, glucose measurement, and ANMS GCSI-DD validated survey for symptom severity | In comparison to psyllium husk, blood sugar levels were controlled in patients receiving GA and partly hydrolyzed guar gum. For the test fibers, the mouth-to-cecum transit time was not significant. |

## 3. Results

### *3.1. Study Selection*

This review was performed in accordance to the PRISMA (Preferred Reporting Items for Systematic Reviews and Meta-Analyses) guidelines.

The literature search led to the identification of 631 studies (Figure 3). Following an initial screening for duplicates and after applying the inclusion criteria to the title, the abstracts of 38 articles were found to be suitable for full-text screening. During title screening, the agreement between the six reviewers was unanimous and conclusive. Twenty-nine articles were eligible, as they met the inclusion criteria and were therefor included in the systematic review.

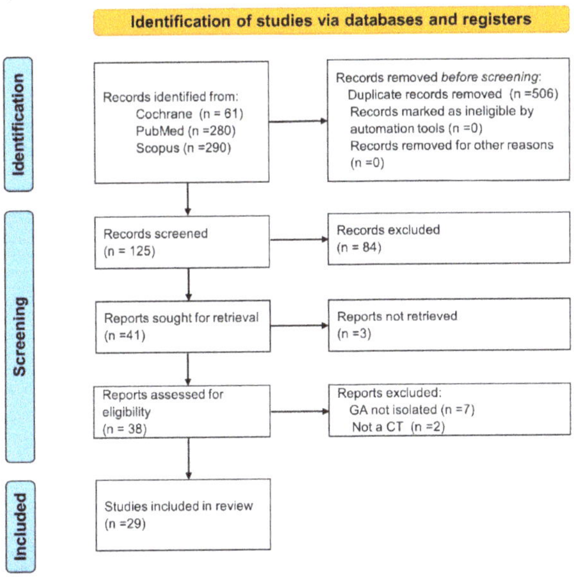


**Figure 3.** PRISMA flow diagram showing the selection, inclusion, and exclusion of clinical trial studies.

### *3.2. Study Characteristics*

The majority of the publications (79.3%) were published after 2010. The papers included information from nine different countries. The studies included in the systematic review were from worldwide populations in Asia, Africa, Europe, and the United States. All the studies included in this review were CTs. Of the 29 articles that were included, 31.0% were related to metabolic disorders (e.g., type 2 diabetes and hyperlipidemia) [10,30–37], 13.9% were related to kidney diseases [13,14,50,51], 17.2% were related to gastrointestinal health [49,52–55], 17.2% were related to oral health [41–45], 10.3% were related to sickle cell anemia [38–40], 6.9% were related to rheumatoid arthritis [46,47], and 3.4% were related to drug efficacy [48] (Table 1).

### *3.3. Gum Arabic and Other Diseases*

The findings from different published studies are discussed under different sections.

#### 3.3.1. Gum Arabic and Metabolic Disorders

GA was found to have a positive effect on satiety and appetite reduction [30]. Subjects reported a decrease in caloric intake and an increase in the their satiety following consumption of GA [33]. Furthermore, when studying the effects of ingesting GA on adults who were at high risk of developing metabolic syndrome, it was found that study subjects had reduced systolic and diastolic blood pressure, fat-free body mass, appetite, and fasting plasma glucose, along with an increased dietary fiber intake. Improvement in bloating and bowel movements was also reported [34].

Another study asserted similar findings; subjects who ingested GA exhibited a significant reduction in fasting plasma glucose and HbA1c [34–36]. Another trial, in which patients with hyperlipidemia were given GA alongside atorvastatin medications, reported that the reduction in their lipid profile was significantly improved [10]. A significant increase in the HDL cholesterol level was also noted [35].

These findings were contradicted by one study that investigated the effects of the viscosity of fiber supplements on the lipid profile. The authors discovered that GA, owing to its low viscosity, did not significantly lower cholesterol levels compared to other fiber supplements containing medium- to high-viscosity water-soluble dietary fiber [37]. Moreover, GA was deemed useful for the reduction in weight gain in patients with type 2 diabetes, as it helped to decrease the body adiposity index by 23.7% and BMI by 2% [31]. Another study yielded similar findings; a reduction of 2.18% of body fat was reported in patients who ingested GA [32]. It was further observed that consumption of GA significantly decreases fasting blood glucose, HbA1c, total protein, and uric acid concentration [36]. It was also observed that GA is a helpful supplement for diabetic patients, specifically those who have diabetic nephropathy, as it decreases blood urea nitrogen and creatinine concentrations [36].

#### 3.3.2. Gum Arabic and Sickle Cell Anemia

In sickle cell anemia, dyslipidemia is a common occurrence as a result of oxidative stress reactions [38]. It was revealed that GA significantly reduced total cholesterol, LDL, and triglyceride levels [38]. A team of researchers who conducted a CT study found that sickle cell patients who consumed GA experienced a significant reduction in their direct bilirubin, serum alanine transaminase, and serum urea levels [39]. Another beneficial effect of GA was an increase in total antioxidant capacity and a reduction in MDA and $H_2O_2$ oxidative markers in sickle cell anemia patients [40].

#### 3.3.3. Gum Arabic and Oral Health

CTs were conducted to test the efficacy of GA as an antibacterial in comparison to liquorice and chlorhexidine mouthwashes. Results revealed a statistically significant decrease in the counts of *Streptococcus mutans* and *Lactobacillus acidophilus* for both GA and liquorice mouthwash without any oral side effects. Moreover, resistance was observed in subjects who used chlorhexidine mouthwash. No significant difference was found between GA and liquorice mouthwash, implying that they can both be used to effectively prevent dental caries [41].

Furthermore, another research finding in the field of oral health showed the effective use of GA in the prevention of dental plaques [42]. In comparison to sugar-free gum, daily photographic assessment of erythrosine-stained plaque showed lower scores following consumption of GA [42]. A group of researchers observed similar findings with respect to reduction in plaque and gingival inflammation [43]. Subjects who applied GA powder had a significantly lower mean gingival index score, mean plaque index score, and gingival crevicular fluid interleukin-1β [43].

A CT study reported that GA use has been of immense benefit for patients with xerostomia. Subjects who were in the GA group had a significantly higher salivary flow rates by 8.03 g within 10 min compared to the control group [45]. Another CT study observed a significant reduction in PPD and a gain in CAL in subjects who used GA gel. Additionally, improved plaque and gingival index scores were also noted [44].

#### 3.3.4. Gum Arabic and Rheumatoid Arthritis

It was found that GA had a positive effect on restoring the baseline liver and kidney profiles in patients with rheumatoid arthritis [46]. GA significantly decreased liver enzymes, with the exception of alkaline phosphatase, urea, and sodium levels, and significantly increased albumin levels, with a minor impact on the serum globulin level [46]. Another research study showed that GA significantly decreased TNF-alpha, the erythrocyte sedimentation rate, and the number of swollen and tender joints, as well as the disease severity in rheumatoid arthritis patients [47].

#### 3.3.5. Gum Arabic and Drug Interactions

GA was also found to enhance drug efficacy, specifically the absorption of amoxicillin [48]. When measuring the peak amoxicillin concentration between two groups (one that took GA 2 h post amoxicillin ingestion and the other that took the drug simultaneously with GA), it was found to be significantly lower in the group that took GA simultaneously [48]. In an experimental study, the effect of GA on gastric ulcers and its interaction with the antiulcer effect of ranitidine was studied in rats. GA significantly potentiated the antiulcer effect of ranitidine [56]. In another study, oral administration of GA was shown to accelerate the absorption of certain solutes [57]. Meloxicam was used as an anti-Cox-1 and Cox-2 non-steroidal anti-inflammatory drug [58]. In rats, GA supplementation showed protective effects against meloxicam-induced gastrointestinal insult. In this study, there were no pharmacological interactions with meloxicam [58].

#### 3.3.6. Gum Arabic and Gastrointestinal Conditions

Researchers observed that administration of GA positively improved acute non-bloody diarrhea in children in terms of symptoms, weight improvements, and the prevention of marked severe dehydration [54]. Treatment with GA was found to be ineffective in patients with fecal incontinence. Patients who received GA supplementation had a fecal incontinence frequency that was not statistically different from the group that received a placebo. Instead, psyllium as a supplement was found to be of beneficial in reducing the frequency of fecal incontinence [52]. Another CT study revealed a beneficial effect of GA in children who had colostomies. It was revealed that the group with GA ointment experienced a significant reduction in peristomal skin inflammation in comparison to the control group [49]. GA was also shown to have potential prebiotic benefits. Researchers found that following ingestion of GA, the count of Bifidobacterium and Lactobacilli increased significantly [53]. In a CT, patients with gastroparesis were shown to benefit from GA administration, as it played a role in the regulation of their blood glucose levels. However, no significant findings were noted with respect to the mouth-to-cecum transit delay [55].

#### 3.3.7. Gum Arabic and Chronic Kidney Diseases

GA was also found to alleviate the adverse effects of chronic renal failure [13,14,50,51]. Patients with chronic renal failure who received GA showed significant decreases in serum urea levels compared to the baseline and the control group [51]. Serum creatinine levels also significantly decreased in the groups of gum users compared to the control group. There was a significant decrease in the serum uric acid level compared to baseline. Serum calcium levels increased, and this increase was significantly different from the baseline and control groups. Serum phosphorus levels decreased significantly compared to baseline [51]. A reduction in serum urea level was also reported in another CT, alongside an increase in fecal bacterial mass and nitrogen content in patients following GA consumption [14]. A study revealed that GA significantly increased total antioxidant capacity levels and reduced oxidative markers MDA and C-reactive protein in patients undergoing hemodialysis, serving as evidence of potent anti-inflammatory properties of GA [13]. These results are similar to those reported in another study showing that consumption of GA significantly decreased C-reactive protein and sodium levels without affecting the levels of other electrolytes, urine volume, or indoxyl sulfate [50]. A schematic diagram depicting the beneficial effects of GA on various medical diseases is shown in Figure 4.

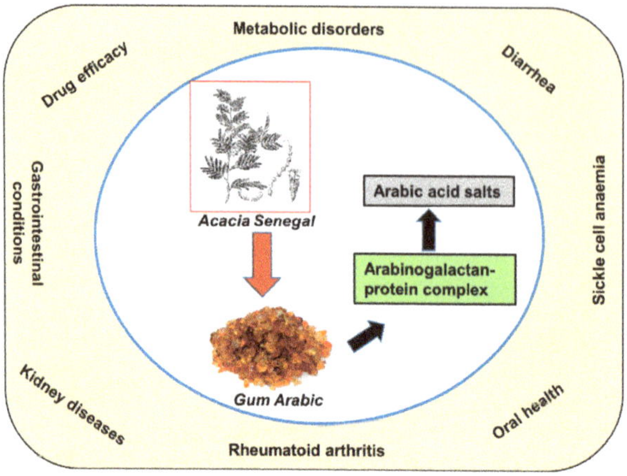


**Figure 4.** A schematic diagram depicting the beneficial effects of GA on various medical diseases.

## 4. Discussion

Based on the existing findings, we highlight the importance of GA in various diseases. GA supplementation was shown to be effective against metabolic disorders [10,30–37], sickle cell anemia [38–40], kidney disease [13,14,50,51], and rheumatoid arthritis [46,47], as well as for the management of oral health [41–45] and gastrointestinal conditions [49,52–55]. GA treatment did not show any potential drug interactions [48]. Additionally, research studies revealed positive effects on appetite and a subsequent significant effect on lowering patients' caloric intake following GA consumption [30,33,36]. Other studies demonstrated similar effects for other gum products [59,60]. Hence, further research is warranted to explore the exact mechanism of action of the active components present in GA. Contradictory results related to GA's effect on blood glucose levels have been reported [30,34,35]. Hence, addition CTs in a larger population need to be conducted in order to determine the exact role of GA in blood glucose levels.

GA is a soluble fiber; dietary fiber increases fecal bulk [61] and reduces the transit time [62]. Dietary fiber is important for combating obesity, and increased intake of dietary fiber has been associated with a reduction in BMI [63]. Dietary fibers have effects on satiety and blood glucose levels. An previous study showed that GA exhibited significant positive effects on satiety measures 15, 30, and 240 min following consumption [30]. The type and viscosity of fiber can have an impact on blood glucose levels after consumption, and variation in the amount of fiber consumed cannot consistently predict the resulting reduction in postprandial glycemic response [30].

In in animals [9,64] and humans, GA treatment has been shown to be effective against body weight and adiposity. Researchers found that GA supplementation significantly reduced BMI, body fat percentage, hip circumference, lipid accumulation product, and visceral adiposity index (VAI) [31,32]. Furthermore, reduced VAI was associated with impaired glucose and lipid metabolism, insulin resistance, and hypertension [31].Thus, it was concluded that GA's positive effects in combating obesity may be related to its positive effect on satiety [30].

Studies confirmed GA effects on chronic conditions, demonstrating their repeatability [31,33]. These effects include a reduction in weight gain, blood pressure, and BMI, which are all positive indicators, strongly suggesting the use of GA as a supplement [31,33]. In addition, it has been demonstrated that GA ingestion leads to a decrease in total cholesterol, LDL, and triglycerides in patients with sickle cell anemia [38]. This lipid-lowering effect is useful, as dyslipidemia is common in patients with sickle cell anemia resulting from oxidative stress [38].

It has been suggested that GA lowers lipid levels. The mechanism behind this phenomenon is that GA binds to bile acids and reduces their absorption from the terminal ileum [65]. The fermentation process in the large intestine then makes the bile acids insoluble, thus promoting their excretion in stool [65]. De novo production of bile acids by the liver requires serum cholesterol. Thus, prolonged ingestion of GA may lead to a reduction in the cholesterol level in plasma [65]. Whereas this information supports the former hypothesis, the findings from the published literature retrieved for this review do not necessarily confirm or deny this phenomenon. Further studies with sufficient clinical data need to be considered when prescribing as a supplement for the treatment of medical diseases.

A study was conducted on 47 patients carrying hemoglobin SS (HbS); GA was administered in a dose of 30 g/day for a period of 12 weeks [39]. In patients with sickle cell anemia, GA administration decreased bilirubin, AST, and serum urea [39]. The results show that GA increased the level of HbF and significantly decreased the level of HbS [40]. The positive effect of GA was explained by the fact that GA degradation resulted in short-chain fatty acids, which, in turn, stimulated HbF expression in the red blood cells [40].

It has not yet been confirmed whether GA directly causes these effects or whether they are byproducts of the repeatedly demonstrated effects on dietary content and overall caloric intake. However, this may be less likely, given that the effects lasted for eight weeks after patients discontinued treatment. Additionally, GA was found to significantly increase total antioxidant capacity and decrease MDA and $H_2O_2$ levels in patients with sickle cell anemia [40]. A similar effect was also found in patients undergoing dialysis, whereby treatment with GA led to decreased CRP and increased antioxidant capacity and MDA levels [51]. GA also resulted in decreased post-colostomy peristomal skin inflammation in pediatric patients treated with acacia ointment compared to those treated with zinc sulfate ointment [49]. Although this does not directly apply to the other studies because of the different mode of delivery, it does potentially provide insight into the mechanism of action of the active compounds in GA, which need to be studied in detail [49].

In addition to specific mechanisms, GA's general anti-inflammatory [66] and antioxidant effects [11,13] are important in clinical conditions, especially following surgery. GA was also demonstrated to decrease serum urea, creatinine, uric acid, and phosphorus levels, and to increase serum calcium [51], in accordance with its positive effects on kidney product profiles reported in other studies involving various chronic kidney diseases [50].

It is pertinent to mention that DNA damage in kidney disease was first detected in the deoxycorticosterone acetate (DOCA)/salt model, and researchers found DNA single- and double-strand breaks [67]. Oxidative stress leads to damage of the kidneys [68]. The antioxidative properties of GA can verify the complete formation of superoxide and oxidative-stress-induced DNA double-strand breaks [11]. The potent antioxidant properties of GA can be used in vulnerable patient populations with various clinical conditions characterized by increased lipid peroxidation and tissue injuries [37]. However, potential drug interactions need to be better characterized, findings need to be explored in CTs with larger sample sizes.

GA was found to have a potential prebiotic effect, as researchers have demonstrated increased counts of Bifidobacteria and Lactobacilli in patients taking GA [53]. This can be explained by the fact that GA is only degraded in the cecum, where it undergoes complete fermentation and therefore promotes bacterial proliferation [53]. Fecal incontinence (FI) is loss of control of bowel contents, leading to discharge of fecal matter. Dietary fiber can lessen FI through its withstanding capacity to fermentation by colonic bacteria, as well as its solubility and degradation [52,69]. In a single-blind RCT, GA supplementation did not significantly reduce FI frequency compared to *psyllium* supplementation [52]. Researchers hypothesized that the high degradation of GA by colonic bacteria [69] and its subsequently reduced content in feces could be the reason for its lack of clinical effect on FI [52].

Inflammatory state affects the synovial joints [70]. In patients with rheumatoid arthritis, GA treatment improved liver and kidney enzyme profiles, with positive subsequent improvement in their condition [46]. Other findings reported decreased TNF-$\alpha$ and ESR; patients also experienced fewer swollen and tender joints and lower disease severity scores [47]. GA may act as a positive immunomodulator. Butyrate is an end product of dietary fiber and starch after their aerobic fermentation by colonic bacteria [71]. Butyrate is a well-known potent anti-inflammatory agent. It suppresses the expression of proinflammatory cytokines by inhibiting NF$\kappa$B activation [71].

The anti-inflammatory property of GA manifested through its derivative, butyrate. GA can be used as a natural means of increasing the level of short-chain fatty acids, which have an immunomodulatory effect that is helpful in reducing inflammation and improving patients' quality of life [47].

GA was demonstrated to improve amoxicillin absorption [48]. Although no specific mechanism was suggested in the study, this result potentially indicates that GA affects the absorption of biomolecules in the gut, which may explain some of the effects observed in other studies [48]. This may amplify allergies to amoxicillin by increasing its serum concentrations with the same dose, which still needs to be considered if GA becomes more widely used. This effect was further substantiated by a study that found that the coexistence of GA and amoxicillin in the upper gastrointestinal tract significantly decreased the absorption of amoxicillin [48]. This might lead to therapeutic failure and the development of drug resistance [48].

Numerous research papers have found a possible relationship between GA and oral health [41–45]. A recent CT on GA mouthwash showed promising caries-preventive and antibacterial effects with no oral side effects. Furthermore, a lack of significant difference between oral *Streptococcus mutans* and *Lactobacillus acidophilus* counts, as well as DMF index scores, in patients treated with either a licorice mouthwashes or GA mouthwash indicated equivalent capabilities of the two interventions in preventing caries [41]. The same study demonstrated bacterial resistance and oral side effects to a chemical agent, chlorhexidine mouthwash, after 9 and 12 months of use. This demonstrates that GA may be used as a natural mouth wash to prevent caries and may have benefits associated with improved adherence. However, in our opinion, future studies are needed to ensure that other bacteria-causing caries are also prevented by GA. In addition, GA supplementation was found to be associated with increased enamel hardness, which may be explained by the presence of polysaccharides and the high concentration of minerals (calcium, magnesium, and sodium) in GA [5–7]. Other findings are associated with oral health, including a decrease in plaque formation compared to intervention with sugar-free gum [42–44]. This suggests that GA is also a suitable supplement that can be used to effectively prevent oral infections, in particular in patients with difficulty in brushing teeth (e.g., Parkinson's patients living alone) or in areas where access to running water or dental hygiene products is limited [72].

The present study is subject to some limitations. We did not assess the reported CTs in terms of randomization, blinding, or phase of trial (0-V). Furthermore, the CTs were not scored in terms of quality. We did not limit the literature search to any particular time period but included as many studies as possible. Moreover, we did not assess the included published papers for publication bias.

## 5. Recommendations

The majority of the reported CTs on GA are limited by the following principal factors: a lack of a specifically demonstrated mechanism of action; a lack of repeatable findings; and studies involving special populations, for example, pregnant and breast-feeding women, children, and elderly patients. Addressing these issues will provide robust evidence for the use of GA in therapeutic applications. GA has been reported to possess various compounds. The isolation of active compounds from GA is highly recommended because the efficacy of active compounds can be easily studied at the molecular or genetic level. Furthermore, this can ease the marketing of the drug as an effective candidate for therapeutic use. The reported CTs used different doses of GA. Further research is needed on the selection and preparation of the final dosage of GA specific to each disease. For effective results, additional focus should be placed on nanoformulation-based drug delivery of GA. An example is resveratrol nanoformulations, which have been reported to lead to remarkable results [73]. The CTs include in this review did not report any toxicities or side effects associated with GA treatment. The majority of included CTs are of short duration and therefore may not have revealed toxicities or side effects. Hence, future multicenter studies with longer durations of treatment and larger sample sizes are warranted.

## 6. Conclusions

This systematic review represents a humble attempt to justify the role of GA in complementary medicine with evidence from published CTs. The results of the included CTs reported on the efficacy of GA against various diseases, such as sickle cell anemia, rheumatoid arthritis, periodontitis, metabolic disorders, kidney disease, oral health problems, gastrointestinal conditions, and peristomal skin inflammation. These findings indicate that GA exerted quantifiable benefits in a number of inpatient and outpatient cohorts. Admittedly, CTs did not provide sufficient data on the adverse effects of GA. Future studies need to explore each active compound present in GA, which has various protective functions in different medical diseases. Easy availability, compliance, and cost-effectiveness could encourage the use of GA with evidence from larger studies from different parts of the world. Additionally, toxicity changes in the liver and kidney need to be explored in detail.

**Author Contributions:** Conceptualization, S.R.S., S.D. and Y.A.-J.; methodology, S.R.S. and S.D.; software, Y.A.-J., N.T.B.A., R.A., J.M. and S.R.S.; data collection, Y.A.-J., N.T.B.A., R.A., J.M. and S.R.S.; writing, Y.A.-J., N.T.B.A., R.A., J.M. and S.R.S.; editing, S.D. and S.R.S. All authors have read and agreed to the published version of the manuscript.

**Funding:** The present study received no external funding.

**Institutional Review Board Statement:** Not applicable.

**Informed Consent Statement:** Not applicable.

**Data Availability Statement:** Not applicable.

**Acknowledgments:** The authors acknowledge the kind help received from Hassan Al-Lawati for the photographs taken from a Shutterstock subscription.

**Conflicts of Interest:** The authors declare no conflict of interest.

## Abbreviations

| | |
|---|---|
| ATP | Adenosine triphosphate |
| BMI | Body mass index |
| CTs | Clinical trials |
| CRP | C-reactive protein |
| DMF | Decay–missing–filled |
| ESR | Erythrocyte sedimental rate |
| GA | Gum Arabic |
| HbA1c | Hemoglobin A1C |
| HDL | High-density lipoprotein |
| LDL | Low-density lipoprotein |
| MDA | Malondialdehyde |
| NPs | Natural products |
| NF-κB | Nuclear factor-κB |
| PGC-1alpha | Peroxisome proliferator-activated receptor-gamma coactivator |
| TGFβ1 | Transforming growth factor beta 1 |
| TNF | Tumor necrosis factor |
| WPS | Water pipe smoking |

## References


1. Ashour, M.A.; Fatima, W.; Imran, M.; Ghoneim, M.M.; Alshehri, S.; Shakeel, F. A Review on the Main Phytoconstituents, Traditional Uses, Inventions, and Patent Literature of Gum Arabic Emphasizing *Acacia seyal*. *Molecules* **2022**, *27*, 1171. [CrossRef] [PubMed]
2. Nasir, O. Renal and extrarenal effects of gum arabic (*Acacia senegal*)-what can be learned from animal experiments? *Kidney Blood Press Res.* **2013**, *37*, 269–279. [CrossRef] [PubMed]
3. Ali, B.H.; Ziada, A.; Blunden, G. Biological Effects of Gum Arabic: A Review of Some Recent Research. *Food Chem. Toxicol.* **2009**, *47*, 1–8. [CrossRef]
4. Pradeep, A.R.; Agarwal, E.; Bajaj, P.; Naik, S.B.; Shanbhag, N.; Uma, S.R. Clinical and microbiologic effects of commercially available gel and powder containing Acacia arabica on gingivitis. *Aust. Dent. J.* **2012**, *57*, 312–318. [CrossRef]
5. Beyer, M.; Reichert, J.; Heurich, E.; Jandt, K.D.; Sigusch, B.W. Pectin, Alginate and Gum Arabic Polymers Reduce Citric Acid Erosion Effects on Human Enamel. *Dent. Mater.* **2010**, *26*, 831–839. [CrossRef]
6. Mount, G.J.; Hume, W.R. *Preservation and Restoration of Tooth Structure*; Knowledge Books and Software: Queensland, Australia, 2005.
7. Onishi, T.; Umemura, S.; Yanagawa, M.; Matsumura, M.; Sasaki, Y.; Ogasawara, T.; Ooshima, T. Remineralization Effects of Gum Arabic on Caries-like Enamel Lesions. *Arch. Oral Biol.* **2008**, *53*, 257–260. [CrossRef]
8. Ahmed, A.A.; Musa, H.H.; Fedail, J.S.; Sifaldin, A.Z.; Musa, T.H. Gum arabic suppressed diet-induced obesity by alteration the expression of mRNA levels of genes involved in lipid metabolism in mouse liver. *Bioact. Carbohydr. Diet Fibre* **2016**, *7*, 15–20. [CrossRef]
9. Ushida, K.; Hatanaka, H.; Inoue, R.; Tsukahara, T.; Phillips, G.O. Effect of long term ingestion of gum arabic on the adipose tissues of female mice. *Food Hydrocoll.* **2011**, *25*, 1344–1349. [CrossRef]
10. Mohamed, R.E.; Gadour, M.O.; Adam, I. The lowering effect of Gum Arabic on hyperlipidemia in Sudanese patients. *Front. Physiol.* **2015**, *6*, 160. [CrossRef]
11. Ali, B.H.; Al-Husseni, I.; Beegam, S.; Al-Shukaili, A.; Nemmar, A.; Schierling, S.; Queisser, N.; Schupp, N. Effect of gum Arabic on oxidative stress and inflammation in adenine–induced chronic renal failure in rats. *PLoS ONE* **2013**, *8*, e55242. [CrossRef]
12. Kong, H.; Yang, J.; Zhang, Y.; Fang, Y.; Nishinari, K.; Phillips, G.O. Synthesis and Antioxidant Properties of Gum Arabic-Stabilized Selenium Nanoparticles. *Int. J. Biol. Macromol.* **2014**, *65*, 155–162. [CrossRef] [PubMed]
13. Ali, N.E.; Kaddam, L.A.; Alkarib, S.Y.; Kaballo, B.G.; Khalid, S.A.; Higawee, A.; Abdelhabib, A.; Alaaaldeen, A.; Phillips, A.O.; Saeed, A.M. Gum Arabic (*Acacia senegal*) Augmented Total Antioxidant Capacity and Reduced C-Reactive Protein among Haemodialysis Patients in Phase II Trial. *Int. J. Nephrol.* **2020**, *2020*, 7214673. [CrossRef] [PubMed]
14. Bliss, D.Z.; Stein, T.P.; Schleifer, C.R.; Settle, R.G. Supplementation with Gum Arabic Fiber Increases Fecal Nitrogen Excretion and Lowers Serum Urea Nitrogen Concentration in Chronic Renal Failure Patients Consuming a Low-Protein Diet. *Am. J. Clin. Nutr.* **1996**, *63*, 392–398. [CrossRef] [PubMed]
15. Nasir, O.; Artunc, F.; Saeed, A.; Kambal, M.A.; Kalbacher, H.; Sandulache, D.; Boini, K.M.; Jahovic, N.; Lang, F. Effects of gum arabic (*Acacia senegal*) on water and electrolyte balance in healthy mice. *J. Ren. Nutr.* **2008**, *18*, 230–238. [CrossRef] [PubMed]
16. Avelino, A.L.N.; Silva, N.V.R.E.; Oliveira, G.B.; Silva, A.A.S.; Cavalcanti, B.C.; Jamacaru, F.V.F.; Dornelas, C.A. Antioxidant and Antigenotoxic Actions of Gum Arabic on the Intestinal Mucosa, Liver and Bone Marrow of Swiss Mice Submitted to Colorectal Carcinogenesis. *Nutr. Cancer* **2022**, *74*, 956–964. [CrossRef] [PubMed]
17. Ahmed, N.; El-Rayes, S.M.; Khalil, W.F.; Abdeen, A.; Abdelkader, A.; Youssef, M.; Maher, Z.M.; Ibrahim, A.N.; Abdelrahman, S.M.; Ibrahim, S.F.; et al. Arabic Gum Could Alleviate the Aflatoxin B1-provoked Hepatic Injury in Rat: The Involvement of Oxidative Stress, Inflammatory, and Apoptotic Pathways. *Toxins* **2022**, *14*, 605. [CrossRef]
18. Siednamohammeddeen, N.; Badi, R.; Mohammeddeen, T.; Enan, K.; Saeed, A. The effect of gum Arabic supplementation on cathelicidin expression in monocyte derived macrophages in mice. *BMC Complement. Med. Ther.* **2022**, *22*, 149. [CrossRef]
19. Fareed, S.A.; Almilaibary, A.A.; Nooh, H.Z.; Hassan, S.M. Ameliorating effect of gum arabic on the liver tissues of the uremic rats; A biochemical and histological study. *Tissue Cell* **2022**, *76*, 101799. [CrossRef]
20. Gouda, E.; Babiker, F. Gum Arabic protects the rat heart from ischemia/reperfusion injury through anti-inflammatory and antioxidant pathways. *Sci. Rep.* **2022**, *12*, 17235. [CrossRef]
21. Rajab, E.; Keogh, M.B.; Laiva, A.L.; Al Hannan, F.; Fateel, M.; Abdulwahab, F.; Al Madhoob, F.; AlHamer, B.; Ghazzal, S.Y.; Dawaishan, A.; et al. Gum Arabic supplementation prevents loss of learning and memory through stimulation of mitochondrial function in the hippocampus of type 2 diabetic rats. *J. Funct. Foods* **2021**, *87*, 104757. [CrossRef]
22. Al-Araimi, A.; Al Kindi, I.A.; Bani Oraba, A.; AlKharusi, A.; Ali, B.H.; Zadjali, R.; Al Sinawi, S.; Al-Haddabi, I.; Zadjali, F. Gum Arabic Supplementation Suppresses Colonic Fibrosis After Acute Colitis by Reducing Transforming Growth Factor B1 Expression. *J. Med. Food* **2021**, *24*, 1255–1263. [CrossRef] [PubMed]
23. Refaie, A.A.; Shalby, A.B.; Kassem, S.M.; Khalil, W.K.B. DNA Damage and Expression Profile of Genes Associated with Nephrotoxicity Induced by Butralin and Ameliorating Effect of Arabic Gum in Female Rats. *Appl. Biochem. Biotechnol.* **2021**, *193*, 3454–3468. [CrossRef] [PubMed]
24. Ali Alshehry, G. The Role of Gum Arabic for a Protective Kidney Dysfunction Induced Gentamicin on Diabetes Rats. *Adv. Mater. Sci. Eng.* **2022**, *2022*, 8617445. [CrossRef]

25. Ali, B.H.; Al-Salam, S.; Balushi, K.A.A.; Za'abi, M.A.; Adham, S.A.; Beegam, S.; Yuvaraju, P.; Manoj, P.; Nemmar, A. Ameliorative Effect of Gum Acacia on Hookah Smoke-Induced Testicular Impairment in Mice. *Biomolecules* **2020**, *10*, 762. [CrossRef]
26. Nemmar, A.; Al-Salam, S.; Beegam, S.; Yuvaraju, P.; Ali, B.H. Gum Arabic Ameliorates Impaired Coagulation and Cardiotoxicity Induced by Water-Pipe Smoke Exposure in Mice. *Front. Physiol.* **2019**, *10*, 53. [CrossRef] [PubMed]
27. Salwan, H.; Isnada, I.; Bakri, A.; Ismail, R.; Bahar, E. Benefits of gum arabic supplementation to oral rehydration solution in managing acute diarrhea. *Paediatr. Indones.* **2007**, *47*, 265–269. [CrossRef]
28. Teichberg, S.; Wingertzahn, M.A.; Moyse, J.; Wapnir, R.A. Effect of gum arabic in an oral rehydration solution on recovery from diarrhea in rats. *J. Pediatr. Gastroenterol. Nutr.* **1999**, *29*, 411–417. [CrossRef] [PubMed]
29. Mullins, C.D.; Vandigo, J.; Zheng, Z.; Wicks, P. Patient-centeredness in the design of clinical trials. *Value Health* **2014**, *17*, 471–475. [CrossRef]
30. Larson, R.; Nelson, C.; Korczak, R.; Willis, H.; Erickson, J., Wang, Q.; Slavin, J. Acacia gum is well tolerated while increasing satiety and lowering peak blood glucose response in healthy human subjects. *Nutrients* **2021**, *13*, 618. [CrossRef]
31. Babiker, R.; Elmusharaf, K.; Keogh, M.B.; Saeed, A.M. Effect of Gum Arabic (*Acacia senegal*) supplementation on visceral adiposity index (VAI) and blood pressure in patients with type 2 diabetes mellitus as indicators of cardiovascular disease (CVD): A randomized and placebo-controlled clinical trial. *Lipids Health Dis.* **2018**, *17*, 56. [CrossRef]
32. Babiker, R.; Merghani, T.H.; Elmusharaf, K.; Badi, R.M.; Lang, F.; Saeed, A.M. Effects of gum Arabic ingestion on body mass index and body fat percentage in healthy adult females: Two-arm randomized, placebo controlled, double-blind trial. *Nutr. J.* **2012**, *11*, 111. [CrossRef] [PubMed]
33. Calame, W.; Thomassen, F.; Hull, S.; Viebke, C.; Siemensma, A.D. Evaluation of Satiety Enhancement, Including Compensation, by Blends of Gum Arabic. A Methodological Approach. *Appetite* **2011**, *57*, 358–364. [CrossRef] [PubMed]
34. Jarrar, A.H.; Stojanovska, L.; Apostolopoulos, V.; Feehan, J.; Bataineh, M.F.; Ismail, L.C.; Al Dhaheri, A.S. The Effect of Gum Arabic (*Acacia senegal*) on Cardiovascular Risk Factors and Gastrointestinal Symptoms in Adults at Risk of Metabolic Syndrome: A Randomized Clinical Trial. *Nutrients* **2021**, *13*, 194. [CrossRef] [PubMed]
35. Babiker, R.; Elmusharaf, K.; Keogh, M.B.; Banaga, A.S.I.; Saeed, A.M. Metabolic effects of Gum Arabic (*Acacia senegal*) in patients with Type 2 Diabetes Mellitus (T2DM): Randomized, placebo controlled double blind trial. *Funct. Foods Health Dis.* **2017**, *7*, 219–231. [CrossRef]
36. Nasir, O.; Babiker, S.; Salim, A.M.M. Protective Effect of Gum Arabic Supplementation for Type-2 Diabetes Mellitus and Its Complications. *Int. J. Multidiscip. Curr. Res.* **2016**, *4*, 2321–3121.
37. Haskell, W.L.; Spiller, G.A.; Jensen, C.D.; Ellis, B.K.; Gates, J.E. Role of Water-Soluble Dietary Fiber in the Management of Elevated Plasma Cholesterol in Healthy Subjects. *Am. J. Cardiol.* **1992**, *69*, 433–439. [CrossRef]
38. Kaddam, L.; Fadl-Elmula, I.; Eisawi, O.A.; Abdelrazig, H.A.; Saeed, A.M. *Acacia senegal* (Gum Arabic) Supplementation Modulate Lipid Profile and Ameliorated Dyslipidemia among Sickle Cell Anemia Patients. *J. Lipids* **2019**, *2019*, 3129461. [CrossRef]
39. Kaddam, L.A.; Fdl-Elmula, I.; Eisawi, O.A.; Abdelrazig, H.A.; Elnimeiri, M.K.; Saeed, A.M. Biochemical effects and safety of Gum Arabic (*Acacia senegal*) supplementation in patients with sickle cell anemia. *Blood Res.* **2019**, *54*, 31–37. [CrossRef]
40. Kaddam, L.; Fadl-Elmula, I.; Eisawi, O.A.; Abdelrazig, H.A.; Salih, M.A.; Lang, F.; Saeed, A.M. Gum Arabic as novel anti-oxidant agent in sickle cell anemia, phase II trial. *BMC Hematol.* **2017**, *17*, 4. [CrossRef]
41. Kamal, D.; Hassanein, H.; Akah, M.; Abdelkawy, M.A.; Hamza, H. Caries Preventive and Antibacterial Effects of Two Natural Mouthwashes vs Chlorhexidine in High Caries-risk Patients: A Randomized Clinical Trial. *J. Contemp. Dent. Pract.* **2020**, *21*, 1316–1324. [CrossRef]
42. Gazi, M.I. The finding of antiplaque features in Acacia Arabica type of chewing gum. *J. Clin. Periodontol.* **1991**, *18*, 75–77. [CrossRef] [PubMed]
43. Gafar, A.M.; Ramadan, A.M.; ElSaid, N.A.; Nurelhuda, N.M. Effect of Gum Arabic on Plaque-Induced Gingivitis: A Randomised Controlled Trial. *Saudi Dent. J.* **2022**, *34*, 494–502. [CrossRef] [PubMed]
44. Singhal, R.; Agarwal, V.; Rastogi, P.; Khanna, R.; Tripathi, S. Efficacy of Acacia arabica gum as an adjunct to scaling and root planing in the treatment of chronic periodontitis: A randomized controlled clinical trial. *Saudi Dent. J.* **2018**, *30*, 53–62. [CrossRef] [PubMed]
45. Bielfeldt, S.; Wilhelm, D.; Neumeister, C.; Schwantes, U.; Wilhelm, K.P. Effect of a newly developed pastille on the salivary flow rate in subjects with dry mouth symptoms: A randomized, controlled, monocentric clinical study. *BMC Oral Health* **2021**, *21*, 117. [CrossRef] [PubMed]
46. Kamal, E.; Kaddam, L.A.G.; Alagib, A.; Saeed, A. Dietary Fibers (Gum Arabic) Supplementation Modulates Hepatic and Renal Profile Among Rheumatoid Arthritis Patients, Phase II Trial. *Front. Nutr.* **2021**, *8*, 552049. [CrossRef]
47. Kamal, E.; Kaddam, L.A.; Dahawi, M.; Osman, M.; Salih, M.A.; Alagib, A.; Saeed, A. Gum Arabic Fibers Decreased Inflammatory Markers and Disease Severity Score among Rheumatoid Arthritis Patients, Phase II Trial. *Int. J. Rheumatol.* **2018**, *2018*, 4197537. [CrossRef]
48. Eltayeb, I.B.; Awad, A.I.; Elderbi, M.A.; Shadad, S.A. Effect of gum arabic on the absorption of a single oral dose of amoxicillin in healthy Sudanese volunteers. *J. Antimicrob. Chemother.* **2004**, *54*, 577–578. [CrossRef]
49. Hosseinpour, M.; Fazeli, A.; Agabeigi, M. Efficacy of *Acacia senegal* for stoma care in children with colostomy. *Eur. J. Pediatr. Surg.* **2012**, *22*, 234–237. [CrossRef]

50. Elamin, S.; Alkhawaja, M.J.; Bukhamsin, A.Y.; Idris, M.A.; Abdelrahman, M.M.; Abutaleb, N.K.; Housawi, A. A.Gum Arabic Reduces C-Reactive Protein in Chronic Kidney Disease Patients without Affecting Urea or Indoxyl Sulfate Levels. *Int. J. Nephrol.* **2017**, *2017*, 9501470. [CrossRef]
51. Ali, A.A.; Ali, K.E.; Fadlalla, A.E.; Khalid, K.E. The effects of gum arabic oral treatment on the metabolic profile of chronic renal failure patients under regular haemodialysis in Central Sudan. *Nat. Prod. Res.* **2008**, *22*, 12–21. [CrossRef]
52. Bliss, D.Z.; Savik, K.; Jung, H.J.G.; Whitebird, R.; Lowry, A.; Sheng, X. Dietary fiber supplementation for fecal incontinence: A randomized clinical trial. *Res. Nurs. Health* **2014**, *37*, 367–378. [CrossRef] [PubMed]
53. Calame, W.; Weseler, A.R.; Viebke, C.; Flynn, C.; Siemensma, A.D. Gum arabic establishes prebiotic functionality in healthy human volunteers in a dose-dependent manner. *Br. J. Nutr.* **2008**, *100*, 1269–1275. [CrossRef] [PubMed]
54. Salah, S.; Salih, A.; Sabir, O.M.; Mshelbwala, M.; Gadour, M.O.E.H. Gum Arabic a superb anti-diarrheal agent. *Sudan J. Med. Sci.* **2012**, *7*, 83–88.
55. Suresh, H.; Zhou, J.; Ho, V. The short-term effects and tolerability of low-viscosity soluble fibre on gastroparesis patients: A pilot clinical intervention study. *Nutrients* **2021**, *13*, 4298. [CrossRef] [PubMed]
56. Al-Yahya, A.A.; Asad, M. Antiulcer activity of gum Arabic and its interaction with antiulcer effect of ranitidine in rats. *Biomed. Res.* **2016**, *27*, 1102–1106.
57. Codipilly, C.N.; Wapnir, R.A. Proabsorptive action of gum arabic in isotonic solutions orally administered to rats. II. Effects on solutes under normal and secretory conditions. *Dig. Dis. Sci.* **2004**, *49*, 1473–1478. [CrossRef]
58. Abd El-Mawla, A.M.; Osman, H.E. Effects of Gum acacia aqueous extract on the histology of the intestine and enzymes of both the intestine and the pancreas of albino rats treated with Meloxicam. *Pharmacogn. Res.* **2011**, *3*, 114–121. [CrossRef]
59. Hetherington, M.M.; Boyland, E. Short-term effects of chewing gum on snack intake and appetite. *Appetite* **2007**, *48*, 397–401. [CrossRef]
60. Hetherington, M.M.; Regan, M.F. Effects of chewing gum on short-term appetite regulation in moderately restrained eaters. *Appetite* **2011**, *57*, 475–482. [CrossRef]
61. Eggum, B.O. The influence of dietary fibre on protein digestion and utilization in monogastrics. *Arch. Tierernahr.* **1995**, *48*, 89–95. [CrossRef]
62. Roberfroid, M. Dietary fiber, inulin, and oligofructose: A review comparing their physiological effects. *Crit. Rev. Food Sci. Nutr.* **1993**, *33*, 103–148. [CrossRef] [PubMed]
63. Menni, C.; Jackson, M.A.; Pallister, T.; Steves, C.J.; Spector, T.D.; Valdes, A.M. Gut microbiome diversity and high-fibre intake are related to lower long-term weight gain. *Int. J. Obes.* **2017**, *41*, 1099–1105. [CrossRef]
64. Ahmed, A.A.; Musa, H.H.; Fedail, J.S.; Sifaldin, A.Z.; Musa, T.H. Gum arabic decreased visceral adipose tissue associated with downregulation of 11β-hydroxysteroid dehydrogenase type I in liver and muscle of mice. *Bioact. Carbohydr. Diet Fibre* **2015**, *6*, 31–36. [CrossRef]
65. Moundras, C.; Behr, S.R.; Demigné, C.; Mazur, A.; Rémésy, C. Fermentable polysaccharides that enhance fecal bile acid excretion lower plasma cholesterol and apolipoprotein E-rich HDL in rats. *J. Nutr.* **1994**, *124*, 2179–2188. [CrossRef] [PubMed]
66. Jhundoo, H.D.; Siefen, T.; Liang, A.; Schmidt, C.; Lokhnauth, J.; Moulari, B.; Béduneau, A.; Pellequer, Y.; Larsen, C.C.; Lamprecht, A. Anti-inflammatory effects of acacia and guar gum in 5-amino salicylic acid formulations in experimental colitis. *Int. J. Pharm. X* **2021**, *3*, 100080. [CrossRef] [PubMed]
67. Schupp, N.; Kolkhof, P.; Queisser, N.; Gärtner, S.; Schmid, U.; Kretschmer, A.; Hartmann, E.; Oli, R.G.; Schäfer, S.; Stopper, H. Mineralocorticoid receptor-mediated DNA damage in kidneys of DOCA-salt hypertensive rats. *FASEB J.* **2011**, *25*, 968–978. [CrossRef]
68. Cachofeiro, V.; Goicochea, M.; De Vinuesa, S.G.; Oubiña, P.; Lahera, V.; Luno, J. Oxidative stress and inflammation, a link between chronic kidney disease and cardiovascular disease: New strategies to prevent cardiovascular risk in chronic kidney disease. *Kidney Int.* **2008**, *74*, S4–S9. [CrossRef]
69. Bliss, D.Z.; Jung, H.-J.; Savik, K.; Lowry, A.; LeMoine, M.; Jensen, L.; Werner, C.; Schaffer, K. Supplementation With Dietary Fiber Improves Fecal Incontinence. *Nurs. Res.* **2001**, *50*, 203–213. [CrossRef]
70. June, R.K.; Liu-Bryan, R.; Long, F.; Griffin, T.M. Emerging role of metabolic signaling in synovial joint remodeling and osteoarthritis. *J. Orthop. Res.* **2016**, *34*, 2048–2058. [CrossRef]
71. Lührs, H.; Gerke, T.; Schauber, J.; Dusel, G.; Melcher, R.; Scheppach, W.; Menzel, T. Cytokine-activated degradation of inhibitory κB protein α is inhibited by the short-chain fatty acid butyrate. *Int. J. Colorectal. Dis.* **2001**, *16*, 195–201. [CrossRef]
72. Coleman, P. Improving oral health care for the frail elderly: A review of widespread problems and best practices. *Geriatr. Nurs.* **2002**, *23*, 189–198. [CrossRef] [PubMed]
73. Sharifi-Rad, J.; Quispe, C.; Mukazhanova, Z.; Knut, E.; Turgumbayeva, A.; Kipchakbayeva, A.; Seitimova, G.; Mahomoodally, M.F.; Lobine, D.; Koay, A.; et al. Resveratrol-Based Nanoformulations as an Emerging Therapeutic Strategy for Cancer. *Front. Mol. Biosci.* **2021**, *8*, 649395. [CrossRef] [PubMed]

MDPI AG
Grosspeteranlage 5
4052 Basel
Switzerland
Tel.: +41 61 683 77 34

*Biomolecules* Editorial Office
E-mail: biomolecules@mdpi.com
www.mdpi.com/journal/biomolecules

www.ingramcontent.com/pod-product-compliance
Lightning Source LLC
Chambersburg PA
CBHW040820250926
51850CB00007BA/1218
*9783725851959*